Equation of time and analemmatic curve

The two-thousand-year history and theory of the equation of time

Part 1

Equation of time and analemmatic curve

The two-thousand-year history and theory of the equation of time

Third edition

Author Denis Lenardič, contact@pvresources.com

Design and cover page Atelje Globočnik Andreja Globočnik s.p.

Translations Alkemist d.o.o., Ljubljana

Proofreading Samantha Lyle, www.englishproofreading.co.uk

Lenardič, Denis
Equation of time and analemmatic curve
The two-thousand-year history and theory of the equation of time
Band 1

3. Auflage 2025

in der Mediengruppe Westarp
Kirchstr. 5 - 39326 Hohenwarsleben
www.westarp.de, www.westarp-bs.de, www.book-on-demand.de
produkthaftung@westarp.de

ISBN: 978-3-96004-176-4

Printed in Germany.

This work did not receive any grants from funding agencies in the public, commercial, or not-for-profit sectors.

Denis Lenardič

Equation of time and analemmatic curve

The two-thousand-year history and theory of the equation of time

Third edition ▪ ***Part 1***

Lucius, the hero of Apuleius' novel *Metamorphoses*, was a rather tragic character. He wished to become an owl but the magic went awry and turned him into an ass instead. He was completely ruined; his untimely transformation had rendered him incapable of speech, and he was only able to redeem himself after a series of traumatic adventures, after eating roses ...

Apuleius was from North Africa, specifically from today's Algeria, and wrote *Metamorphoses* or *The Golden Ass* in the second century, during the reign of Antoninus Pius or Marcus Aurelius. In the middle of the second century, mathematics and astronomy in Alexandria had peaked in Claudius Ptolemy and his writing of the *Almagest*, one of the most important preserved antique works and which forms the basis of the first few chapters of this book.

A special thank you

Special thanks to the guardians of the magnificent Library of Alexandria, who watched over ancient treasures for centuries, compiling sources and preserving it for future generations. Without them, our knowledge of the early history of the equation of time would be severely limited if not entirely lost.

Great credit in enabling the creation of this book also goes to the following two libraries:

- **Deutsches Museum Library, Munich** – without this institution, this book would not exist. Special thanks go to the library, because its doors are open throughout the year for anyone with an interest in the history of exact sciences or engineering. During my years writing this book, I spent many weekends in the library reading room, gaining unforgettable memories.
- **The Central Technological Library of the University of Ljubljana** – a particularly strong mark on the book was also left by the staff at the CTK Interlibrary Loan Department of the Central Technological Library of the University of Ljubljana. They always successfully fulfilled even the most unusual enquiries from this rather demanding author. Without them, this book would not be what it is today.

Credit for searching the literature also goes to Antiquariat Renner, Albstadt, Germany.

Special thanks go to Miran Kuhar from the Faculty of Civil and Geodetic Engineering, University of Ljubljana, who reviewed Appendix D of the third edition.

Special thanks for contributions to the first edition go to two anonymous experts from the Bavarian Academy of Sciences and Humanities and from the Paris Observatory, for their interpretation of corrections in the third chapter and additional explanations in the eighteenth chapter of the book.

Thanks also to the anonymous reader who pointed out the errors in Chapter 13 of the first edition.

For the transmission of reproductions, mainly of image material, which provided the basis for some of the images in this book, credit also goes to the following institutions or web services, listed in alphabetical order:

- **Bayerische Staatsbibliothek München**
- **Bodleian Library Oxford**
- **Niedersächsische Staats- und Universitätsbibliothek Göttingen**
- **Staats- und Stadtbibliothek Augsburg**
- **Stadtbibliothek Nürnberg**
- **Universitätsbibliothek Erfurt**

For providing information:

- **Staatsarchiv Sachsen, Staatsarchiv Leipzig**
- **Staatsbibliothek Bamberg**

Credit to all libraries that strive for high-quality and professional digitisation of old works. These publicly available book services also greatly facilitated the creation of this book. Particular special thanks for the availability of digitised publications go to the following libraries or services:

- **Bayerische Akademie der Wissenschaften München**
- **Bayerische Staatsbibliothek München**
- **Berlin-Brandenburgische Akademie der Wissenschaften, Deutsches Textarchiv**
- **Bibliothèque nationale de France, project Gallica**
- **Deutsche Digitale Bibliothek, Kultur und Wissen online**
- **Deutsches Museum München**
- **Digitala Vetenskapliga Arkivet, DiVA Portal**
- **e manuscripta, the digital platform for manuscript material from Swiss libraries and archives**
- **e-rara.ch, the platform for digitised rare books from Swiss libraries**
- **ECHO – Cultural Heritage Online**
- **Goethe Universität, Frankfurt am Main, Universitätsbibliothek**
- **Google Books**
- **Internet Library, Archive.org**
- **Martin-Luther-Universität Halle-Wittenberg**
- **National Library of the Czech Republic, Manuscriptorium**
- **Niedersächsische Staats- und Universitätsbibliothek Göttingen**
- **Sächsische Landesbibliothek - Staats- und Universitätsbibliothek Dresden**
- **Thüringer Universitäts- und Landesbibliothek Jena**
- **Universitäts- und Landesbibliothek Düsseldorf**
- **Universitäts- und Landesbibliothek Münster**
- **Universitäts- und Landesbibliothek Sachsen-Anhalt**
- **Universitätsbibliothek Leipzig**
- **Vatican Apostolic Library**

Preface

May the good people be pleased with the particulars of my contribution! May the ill-minded people also derive pleasure out of ridiculing me with ignorance unable to understand my contribution.

Bhāskara II, Siddhāntaśiromaṇi, Verse 5
(Siddhāntaśiromaṇi, Arkasomayaji, trans., Rashtriya Sanskrit Vidyapeetha, 2000, p.2)

Preface to the third edition

The first edition of the book, *Equation of Time and Analemmic Curve*, which was still imperfect in certain respects, was promptly superseded by a second, revised, edition. This has now been further revised and expanded into the third edition which is now before you. Much more comprehensive than the second edition, this third edition has seven additional chapters. In addition to these, errors and historical inaccuracies that had crept into the second edition have been corrected. I am indebted to the technical editor Alistair Kwan for his invaluable contributions, which have enhanced the reader-friendliness of the text. Some of his comments have also been incorporated into the introduction.

The third edition also aims to provide readers with a deeper understanding of the broader cultural history and context, as well as the significant achievements that have led to discoveries in the fields of mathematics and astronomy, and that have in one way or another influenced our understanding of the equation of time. It also dedicates a significant portion of the content to the development of trigonometry, in ancient times and in subsequent eras. In order to facilitate comprehension, modern mathematical notation is frequently employed in the description of ancient trigonometric procedures. These instances will be readily identifiable for the reader who is attentive to detail. Additionally, this edition presents illustrative examples of the utilisation of the equation of time, including its application in astronomical observations during the 15th century, RØMER's determination of the speed of light, and the determination of longitude.

The introduction to this third edition guides the reader through the origins of trigonometry to the ancient mathematical procedures for calculating the length of the day, which are given in modern mathematical notation as they would be expressed today. The revised introduction now includes a concise overview of the Aramaic Book of *Enoch* and its astronomical implications, as well as a novel account or illustration of HIPPARCHUS' method for determining the longitude of the apogee. Furthermore, the introduction presents a more comprehensive reconstruction of PTOLEMY's method for calculating the chord function. Additionally, the supplement includes a complete exposition of MENELAUS' theorem, which now contains the full range of mathematical expressions, as well as an explanation of several other theorems from MENELAUS' *Sphaerics*. The introduction also presents new explanations of PTOLEMY's climata and ortive amplitude, with a more detailed analysis of oblique ascension. These concepts were instrumental in the calculation of the length of the day in antiquity. The initial chapter provides further clarification of the individual computational examples, which should facilitate comprehension of the subject matter, specifically the initial or fundamental stages. The second chapter provides a detailed account of ĀRYABHAṬA's trigonometric procedures for determining the value of the function *jya*, which is also known as the sine. Additionally, inconsistencies in the vocabulary of the second edition, derived from various sources, have been rectified. The third chapter introduces new elements pertaining to the early era of Arab astronomy and the contributions of AL-HĀSHIMĪ and ABŪ MA'SHĀR. In order to gain a deeper insight into the development of spherical trigonometry during this period, it is also necessary to consider the contributions of ABŪ AL-WAFĀ'. It is notable that the radius of a circle of unit length ($R = 1$) first appears in written sources during this period. Additionally, the significant contribution of AL-JAYYĀNĪ to the advancement of spherical trigonometry is duly acknowledged, and a detailed account of AL-KĀSHĪ's intriguing trigonometric methodologies is provided. The fourth new chapter provides a concise overview of ancient Chinese astronomy, with particular emphasis on the origins of calendarisation, or the determination of the length of days and the equation of the centre. Alistair Kwan is to be commended for ensuring that the notation of Chinese personal names and the names of ancient Chinese works in the fourth chapter are consistent with that of Hanyu Pinyin. The fourth chapter in the second edition has been expanded with extensive additions, forming new fifth, sixth and seventh chapters. The fifth chapter on the *Toledan* and *Alfonsine Tables* has been significantly expanded and also includes the achievements of the Carolingian period and later astronomers. In the sixth chapter, the section on the Hebrew astronomers is expanded and a wholly new chapter is devoted to the achievements of astronomers in Europe, including their knowledge of the equation of time. Among these astronomers, JOHN OF GMUNDEN and BERNHARD WALTHER are particularly noteworthy. This chapter also provides a description of BIANCHINI's tables, which were a valuable tool for solving astronomical mathematical calculations in practice. The contributions of significant ephemeris makers of the period are also listed. The seventh chapter features comprehensive expansions, while a concise subchapter on the mathematical contributions of JOST BÜRGI and his clocks and globes represents a novel contribution. The eighth chapter, which describes the period in which HUYGENS and FLAMSTEED worked, includes new sections on the work of WILHELM SCHICKARD, JEREMIAH HORROCKS, ALBERT CURTZ, ISMAËL BOULLIAU and JEREMY SHAKERLEY. This chapter also describes RØMER's method for determining the speed of light from the position of Jupiter's moons, for which he needed CASSINI's tables and values of the equation of time. At the end of the chapter, the contributions or written works of some other lesser-known authors of the period are briefly mentioned. The ninth chapter gives a fairly detailed description of EULER's procedures for calculating the equation of noon – the procedures for calculating the equation of noon had already been given by DE LA HIRE before EULER – and the equation of the centre. The use of EULER's formula, which is also used in some of the derivations in this book, is also briefly explained. This chapter shifts the focus somewhat from astronomy to introduce some of EULER's mathematical procedures, which will be employed in subsequent chapters. The ninth chapter concludes with a description of the work of TOBIAS MAYER and a new short section on the contribution of ROGER BOSCOVICH to the derivation of differential expressions of spherical trigonometry. The tenth chapter introduces a new section on the work of astronomers at the University of Uppsala, which resulted in the development of differential expressions in spherical trigonometry. Among the most significant astronomers who studied at Uppsala are SAMUEL KROOK, JOHAN THUN, ANDREAS BOSTRÖM and, most notably, FREDERICK GABRIEL CRAELIUS. This chapter also provides a concise overview of the origins of the earliest periodical in astronomy, the French almanac *Connoissance des temps*, the inaugural edition of which was released in March 1679. The eleventh chapter, composed in a wholly original style, elucidates the methodology for calculating the altitude of the Sun, or the Sun-altitude table, to which LAMBERT and BODE made significant contributions. The tables facilitated the determination of the solar noon and, consequently, the regulation or setting of hours through the measurement of the Sun's altitude. The twelfth chapter provides further insight into LAGRANGE's methodologies and offers a detailed account of JOHANN SCHULZE's intricate series expansion process, which is not particularly accessible from a mathematical standpoint. To fully comprehend SCHULZE's procedures, it is essential to possess a comprehensive understanding of trigonometry. It was SCHULZE's pioneering work in series expansion that constituted an important milestone, shortly followed by the advent of the modern form of the equation of time. This development can be attributed, in part, to LAGRANGE, who established its mathematical foundations; and to DELAMBRE and LITTROW, who "refined" it into its current form. The thirteenth chapter presents novel descriptions of the work of astronomers associated with the Vienna School, most notably MAXIMILIAN HELL and TOBIAS BÜRG. The history of the *Vienna Ephemerides*, which also includes the altitude of the Sun and the time of solar noon, is also described. These tables were also of considerable use in calibrating clocks. The fourteenth chapter presents a number of new entries in the historical tables, giving the individual authors of the equation of time. Additionally, commentary has been included on the equation of time in a circular eccentric orbit. The second, modern mathematical section of the equation of time treatment commences in the fifteenth chapter, which is further augmented by a more comprehensive delineation of BESSEL function utilisation in elliptical orbit parameter calculation. Furthermore, illustrative examples of approximate equation of time expressions and diurnal length variation expressions are provided. The content of the seventeenth chapter, which addresses the rotation of the Earth, has also been significantly revised. The concepts of precession, nutation and the motion of the Earth's poles are elucidated in a succinct manner. This chapter presents expressions for calculating the deceleration of the Earth's rotation and a separate subsection outlines the procedures by which the expressions for nutation are derived, at least tentatively. Nevertheless, readers seeking to comprehend this chapter or the derivation of the nutation expressions may require supplementary literature; suggestions for further reading are given at the end of the book. The precise calculation of the equation of time is explained in a separate eighteenth chapter in this edition. Appendix D has also been significantly updated and now contains all the principal expressions of spherical trigonometry and spherical triangles. Appendix D now also includes modern derivations of ancient problems of spherical astronomy, such as the derivations of oblique ascension and the ortive amplitude using spherical trigonometry.

The third edition also concludes with an expanded historical timeline in comparison to the previous edition, an index of persons and terms, and the most important corrections of errors in the second edition. To facilitate comprehension, this book incorporates a number of computational examples that elucidate the astronomers' methodologies over time. All the examples have been calculated using standard spreadsheet software. While programming expertise or the utilisation of dedicated mathematical software is not a prerequisite for calculating the examples, it is highly beneficial in certain instances, particularly for numerical calculations and the precise determination of the equation of time.

This third edition is a comprehensive resource that goes beyond mere reading; it is designed to facilitate in-depth study of the subject matter. This edition features a more rigorous mathematical treatment in several chapters and introduces new mathematical derivations and procedures. Therefore, it is essential for readers to carefully examine this edition to gain a comprehensive understanding of the subject.

The majority of the errors identified in the previous edition have been rectified. However, it is possible that some errors may have been introduced in the third edition, along with the new content. I extend my apologies to readers for any inconvenience caused by these errors and ask for their understanding.

This edition also includes some quotations from literature in other languages for illustrative purposes. As in the second edition, I have not translated the source text, as a poor translation could potentially have a detrimental impact on the overall comprehension of the technical content of the book. A reader who does not understand the quotations in the foreign language can omit them without it affecting their understanding of the book's technical content.

It is my hope that this third edition will prove a valuable resource for readers.

In the third edition, which, like the previous two, is accompanied by Lucius, readers are invited to contribute comments, even if they pertain to roses.

Jesenice, 24 August 2024,
$JD \approx 2460546.9402778$

Preface to the second edition

Making up for past mistakes is not a pleasant task, but it is necessary. I would like to thank the readers who pointed out the errors in the first edition. Writing a book like this is not simple, otherwise someone would probably have written a book on the equation of time by now. Good articles on the equation of time are rare, but when it comes to books, this one is the first attempt to present the historical background and theoretical treatment of the equation of time in a comprehensive way. The second edition, like the first, is written more like a lecture than a textbook, with more mathematical derivations, which will appeal to some readers perhaps more than others. In addition to corrections to the first edition, the second edition also contains some new sections, the text has been slightly rearranged for more fluid reading, the biographical information about the personalities is part of the main text in this edition and the timeline is fully compiled in an appendix. The introduction gives a brief new guide to some of the procedures of sexagesimal arithmetic, and a brief history of the Library of Alexandria is also a fresh addition. In the first chapter, the reader will find PTOLEMY's procedures for spherical astronomy, or the determination of the true longitude and right ascension. The second chapter includes a brief history of trigonometry in India and ĀRYABHAṬA's procedures for calculating the sine function. Also new is a short section in Chapter 3 on spherical trigonometry in the Arab world, where the contributions of AL-JAYYĀNĪ and AL-ṬŪSĪ are particularly important. Particularly through the works of IBN AFLAH or GEBER, much of this knowledge reached medieval Europe, where REGIOMONTANUS is greatly credited with the further development of trigonometry and the dissemination of knowledge about it. Chapter 4 contains more information on the *Toledan* and *Alphonsine Tables*, together with simpler examples, and the section on Kepler's *Rudolphine Tables* and his equation of time has been supplemented. Chapter 5 is enriched by a contribution on the work of GABRIEL MOUTON and also contains some brief references to clocks showing solar time and the equation of time that were produced in the seventeenth century. In describing the work of SAMUEL VINCE, Chapter 6 adds some brief comments on spherical trigonometry, which will help the reader understand the mathematical procedures. The bulk of the additions are in Chapter 7. In the work of LAGRANGE, the procedure, or rather the derivation of the Lagrange reversion, is described in more detail. The procedure for deriving JOHANN SCHULZE's equation of time, which is mathematically quite demanding and was sparsely treated in the first edition, is now described in much greater detail. A short section on the work of GEORG VEGA has been added, along with VEGA's peculiar example of the calculation of the equation of time. There are also some corrections in the section on the work of JOHANN LITTROW. In his equations, LITTROW used the symbol ε for eccentricity and e for the obliquity of the ecliptic, which is misleading. In this edition, the nomenclature in LITTROW's work is consistent throughout the book, rather than following LITTROW's notation, and is in line with modern mathematical notation (e eccentricity, ε obliquity of the ecliptic). Chapter 8 contains some additions to the tables, and of particular note are the additions to the derivation of the equation of time for an eccentric orbit. The last part of Chapter 9 has been partly rearranged for better understanding, and expressions for the expansion of Bessel functions into series have been added for the calculation of the

Kepler equation. Chapter 9 includes Euler's kinematic equations, albeit without derivation, from which expressions for nutation are derived, and polynomial expressions for the deceleration of the rotation of the earth, which were not present in the first edition, have also been added. In the thirteenth chapter, several errors in the equations of the first edition are corrected, a few minor additions are made and a timeline is added at the end. Finally, publications on the position of the Sun are collected, and new web links and standards are added. The second edition is also rounded off by indexes of persons and terms. The format of the book has also been slightly modified or adapted, resulting in a more favourable price for the reader of the second edition.

Special thanks go to anonymous reviewer who found many small mistakes in the manuscript of the second edition. All mistakes left in the text are my fault.

I hope that the second edition will serve readers well. I will be happy to receive comments from readers on this edition as well, even if they may be about roses.

In Jesenice, November 2021,
$JD \approx 2459534.0277431$

Table of contents

Symbols, abbreviations and acronyms

SIGNS OF THE ZODIAC

Aries	0^S	♈	the Ram
Taurus	1^S	♉	the Bull
Gemini	2^S	♊	the Twins
Cancer	3^S	♋	the Crab
Leo	4^S	♌	the Lion
Virgo	5^S	♍	the Virgin
Libra	6^S	♎	the Scales
Scorpio	7^S	♏	the Scorpion
Sagittarius	8^S	♐	the Archer
Capricorn	9^S	♑	the Sea goat
Aquarius	10^S	♒	the Water bearer
Pisces	11^S	♓	the Fish

ASTRONOMICAL CONSTELLATIONS

Aries	♈	19 April – 13 May
Taurus	♉	14 May – 19 June
Gemini	♊	20 June – 20 July
Cancer	♋	21 July – 9 August
Leo	♌	10 August – 15 September
Virgo	♍	16 September – 30 October
Libra	♎	31 October – 22 November
Scorpio	♏	23 November – 29 November
Ophiuchus	⛎	30 November – 17 December
Sagittarius	♐	18 December – 18 January
Capricorn	♑	19 January – 15 February
Aquarius	♒	16 February – 10 March
Pisces	♓	11 March – 17 April

ASTRONOMICAL SYMBOLS

Mercury	☿
Venus	♀
Earth	⊕
Moon	☽
Mars	♂
Jupiter	♃
Saturn	♄
Uranus	♅
Neptune	♆
Me☿	heliocentric longitude of Mercury
V♀	heliocentric longitude of Venus
E⊕	heliocentric longitude of the Earth
M♂	heliocentric longitude of Mars
J♃	heliocentric longitude of Jupiter
S♄	heliocentric longitude of Saturn
U♅	heliocentric longitude of Uranus
N♆	heliocentric longitude of Neptune
♈	vernal equinox
☊	nutation, ascending node

MATHEMATICAL SYMBOLS, TRIGONOMETRY

∢, ⌒	spherical angle, arc
∠, ⌒	angle
△	triangle
⊿	right triangle
°	degrees
′	arc minutes
″, as	arc seconds
mas	milliarc seconds
µas	microarc seconds

MATHEMATICAL SYMBOLS, HISTORICAL PART

b	number of additional lunar months, adhikamāsās in a kalpa
c	the equation of the centre or solar equation
c_{max}	the maximum of the solar equation
c_s	shifted equation of centre
c_1, c_2	equation of centre for different temporal moments
d	number of solar days from the beginning of a kalpa to the beginning of a kaliyuga

d_k	number of solar days in a kalpa
d_t	number of lunar days since the beginning of a kalpa
$d_\odot$	diurnal motion of the Sun
e	eccentricity
E	equation of time
E_s	equation of time as a function of shifted longitude
E_0	equation of time for epoch or reference moment
k	epoch constant
k_d	difference between solar and lunar days in a kalpa, kshayās
m_l	number of additional lunar months or adhikamāsās
n	daily motion of the mean Sun
p	precession
r	radius, circumference of the epicycle
R	radius, circumference of the orbit
t	number of lunar days in a kalpa, tithis
t	number of years
α	right ascension
$\alpha_\odot$	right ascension of the apparent Sun
$\alpha_{\odot 1}$, $\alpha_{\odot 2}$	right ascension for different temporal moments
α_0	right ascension, for epoch or reference moment
α_M	right ascension of the mean Sun
δ	declination
Δ	longitude shift
Δ_1	contribution to the equation of time due to non-uniform motion
Δ_2	contribution to the equation of time due to obliquity of the ecliptic
Δn	correction of daily motion of the Sun
$\Delta t_\odot$	difference between the passages of the true Sun across the meridian at moments *A* and *B* at midday
Δt_M	difference between the passages of the mean Sun across the meridian at moments *A* and *B* at midday
$\Delta\psi$	lunisolar precession
$\Delta\kappa_M/\Delta t$	daily change of the mean anomaly
ε	obliquity of the ecliptic
κ	true anomaly
κ_M	mean anomaly
λ_0	true longitude, for epoch or reference moment
λ_M	mean longitude of ecliptical mean Sun
λ_{Ms}	mean longitude shifted
λ_{M0}	mean longitude, for epoch or reference moment
λ_Γ	longitude of apogee
λ_Π	longitude of perigee
$\lambda_\odot$	true longitude
$\lambda_{\odot s}$	shifted true longitude
$\lambda_{\odot 1}$, $\lambda_{\odot 2}$	true longitude for different temporal moments
$\bar{\lambda}_1$, $\bar{\lambda}_2$	mean longitude for different temporal moments
Γ	apogee
Π	perigee

MATHEMATICAL SYMBOLS, MODERN PART

a	major semi-axis of the ellipse
A	amplitude, coefficients given in VSOP87
A_i	arguments, functions of the fundamental arguments
b	minor semi-axis of the ellipse
B	phase, coefficients given in VSOP87
C	frequency, coefficients given in VSOP87
C	equation of the centre
C'_k	coefficients describing the coupled effects of precession and nutation
d	number of days from perihelion
d_n	consecutive day of the year
d_p	day of the perihelion
d_w	day of the winter solstice
D	mean elongation of the Moon, angle between the Moon and the Sun
D_{UT}	interval in days from 1 January 12:00 $UT1$
e	Euler's number
e	eccentricity of Earth's orbit
E	equation of time
E	eccentric anomaly
E_0	equation of the origins
E_e	equation of the equinoxes
E_{prec}	precession
F	argument of the longitude of the Moon
h	hour angle or the path that the Sun travels in one hour
$h_{\odot G}$	Greenwich hour angle, GHA
$\bar{h}_G$	Greenwich mean Sun hour angle
i	inclination of orbit in relation to the reference plane
K	amplitude, coefficients given in VSOP87
l	local longitude
l_{TZ}	standard latitude for a given time zone
$L_☽$	mean longitude of the Moon
M	mean anomaly of the Sun
$M_☽$	mean anomaly of the Moon
n	mean angular velocity of radius vector r
n	powers of the time variable τ, VSOP87
N_i	frequencies, coefficients given in VSOP87
p	ordinate in the ellipse focus
p_a	approximation of the general precession of longitude
P	perturbations due to the planets
r	radius of the ellipse
R	radius
R	reduction of the ecliptic to the equator
s	arc length in radians
S	the Sun

S	amplitude, coefficients given in VSOP87
S'_k	coefficients describing the coupled effects of precession and nutation
t	time in Julian decamillennia
t	position of the Sun (time)
t_{UT}	universal time
T	time when the planet is in perihelion
T	time in Julian centuries
T_c	time in Julian centuries, corrected due to rotation of the Earth
T_{UT}	interval in Julian centuries from 1 January 12:00 $UT1$
Z	Earth
α_i	intermediate right ascension
$\alpha_\odot$	right ascension
$\bar{\alpha}$	right ascension of the (equatorial) mean Sun
δ	declination
ΔCT	time elapsed since the winter solstice
ΔD	daily change in day length
ΔDST	difference between local standard time LST and daylight savings time DST (if applicable)
ΔT	difference between dynamic and universal time
$\Delta\varepsilon$	nutation in obliquity
$\Delta\psi$	nutation in longitude
ε	obliquity of the ecliptic
ε_A	mean obliquity of the ecliptic for a given moment
θ	Earth rotation angle
θ'	angle between winter solstice and perihelion
κ	constant of aberration
κ_c	correction due to aberration
$\lambda_\odot$	longitude of the Sun
$\bar{\lambda}$	longitude of the mean Sun
λ_i^0	phases, coefficients given in VSOP87
λ_M	longitude of the first mean Sun circling the ecliptic
ν	true anomaly of the Sun
♈	vernal equinox
Π	perihelion
τ	time measured in Julian millennia
φ	VSOP87 argument, angle
φ	latitude
ϕ	complement of latitude
ψ	precession of the equator (lunisolar precession)
ω	argument of perihelion
Ω	mean longitude of the ascending node of the Moon
Ω	nutation
Ω_R	angular velocity due to rotation
ϖ	longitude of perihelion

ABBREVIATIONS AND ACRONYMS

as	arc second
BCE	before common era
CE	common era
CIO	celestial intermediate origin
DST	daylight savings time
FK5	Fifth Fundamental Catalogue
GAST	Greenwich apparent sidereal time
GHA	Greenwich hour angle
GMST	Greenwich mean sidereal time
GMT	Greenwich mean time
GST	Greenwich sidereal time
IAU	International Astronomical Union
IERS	International Earth Rotation and Reference Systems Service
JD	Julian date
JDN	Julian day number
JDT	Julian date, based on terrestrial dynamical time (*TD*)
LST	local standard time (local time)
LTST	local true solar time (local solar time)
mas	milliarc second
MJD	modified Julian date
rad	radian
TAI	Time, atomic international
TD	Terrestrial dynamical time
TT	Terrestrial time
$UT, UT1$	Universal time
VLBI	very long baseline interferometry
VSOP	Variations Séculaires des Orbites Planétaires

The path to spherical trigonometry

The sun does never set nor rise. When people think the sun is setting (it is not so). For having arrived at the end of the day, it makes itself produce two opposite effects, making night for what is below and day for what is on the other side.

Aitareya Brāhmaṇa 3.44
(Haug, 1863, Aitareya Brahmanam of the Rigveda, p.242)

Gnomon

The invention of the *gnomon*, a shadow casting rod, is attributed a similar role for the development of natural sciences as the invention of the wheel on an axle for the development of mechanics in general.[1] It is not known where the gnomon, one of the oldest astronomical devices or tools, was first used for astronomical purposes, but we know that it was used very early on in Egypt, Mesopotamia, China and India. In about the 9th or 10th millennium BCE, the summer monsoon from central Africa moved north and provided a modest but sufficiently large amount of rainfall in the southwestern Egyptian desert, which allowed for the survival of nomadic tribes and their livestock. In the late Neolithic period, in Nabta, the landscape was covered with monsoonal lakes, about a hundred kilometres west of Abu Simbel, a sort of ritual centre, and the nomadic tribes used a stone megalith, perhaps as a gnomon, to mark direction, which assisted them in migrations.[2] Nabta was inhabited until the 4th millennium BCE, when it started to transform into a desert again as a result of climate change. One of the oldest known Egyptian records on the measurement of the length of a shadow originates from the period of the Middle Egyptian kingdom and speaks of the inability to measure the length of a shadow in the case of a solar eclipse.[3]

In the Indus Valley civilization of Ancient India, directions in the sky may have been determined with the naked eye by observing the sunrise and the setting of the stars or constellations. From the oldest surviving written sources of *Vedas* and *Brahmanas*, the use of astronomical instruments for observations is not apparent.[4] The oldest known Indian record of the use of the gnomon, named *śaṅku* or *śaṅku-yantra* or *nara-yantra* in Sanskrit, is a description of determining directions in the sky in the *Kātyāyana Śulba Sūtra*, written around 200 BCE.[5] In view of the numerous geometric patterns on older archaeological finds, the gnomon was probably known in India well before then. From the *Vedāṅga* texts, which are an addition to the *Vedas* and describe mathematical and astronomical concepts, the use of water clocks in astronomical observations is also

1 Couprie, 2011, Heaven and Earth in Ancient Greek Cosmology: From Thales to Heraclides Ponticus, p.41.

2 Wendorf, Schild, 2001, Holocene Settlement of the Egyptian Sahara: Volume 1: The Archaeology of Nabta Playa, pp.489–502. Malville, 2008, The Astronomical Gnomon. African Cultural Astronomy, Holbrook et al. (eds.), pp.39–51.

3 Daressy, 1901, Catalogue général des antiquités égyptiennes du Musée du Caire 25001–25385, Ostraca. See Borchard, 1920, Altägyptische Zeitmessung, pp.27 and 64–65, text and footnote 1. The Middle Kingdom dates from the 11th dynasty when NEBHEPETRE MENTUHOTEP II brought together Lower and Upper Egypt, up to the 13th dynasty, about the middle of the 21st century BCE to the middle of the 17th century BCE.

4 *Ṛgveda*, *Sāmaveda*, *Yajurveda*, *Atharvaveda* and *Brahmanas*, the prose comments on the *Vedas* do not make notes on astronomical instruments. Ôhashi, 1993, Development of Astronomical Observation in Vedic and Post-Vedic India, pp.193–197. The text also mentions some false interpretations of references to gnomons from older texts. For an additional explanation, see also Holbrook, Baleisis, 2008, Naked-eye Astronomy for Cultural Astronomers. African Cultural Astronomy, pp.53–75.

5 Ôhashi, 1993, pp.206–207. The text of *Kātyāyana Śulba Sūtra* is supposed to have been written by KĀTYĀYANA, about whom we know practically nothing, but most likely he was not the (only) author. *Kātyāyana Śulba Sūtra* was written much later than the similar, but mathematically more significant texts of the *Āpastamba Śulba Sūtra* (6th century BCE) and *Baudhāyana Śulba Sūtra* (8th century BCE), which is also the oldest well-known mathematical text from India. In literature, references from *Aitareya Brāhmaṇa*, XVIII.4 are mentioned as the oldest example of the gnomon in India, which however cannot be confirmed, Ôhashi, 1993, pp.204–205, also note 12, p.240. For more information on the gnomon in India, see Subbarayappa, Sarma, 1985, Indian Astronomy - A Source Book, Chapter 15, pp.182–195, Sen, Shukla (eds.), 1985, History of Astronomy in India, pp.346–351, and Ôhashi, 1994, Astronomical Instruments and Classical Siddhāntas, pp.168–196. For water clocks, see Mishra, 2005, Vedanga Jyotisham, pp.104–107 and 170–173.

evident.[6] Astronomical concepts are addressed by *Jyotiṣa Vedāṅga*, and the shadow of the gnomon (*pauruṣī*) is also mentioned in other old Indian texts such as *Candra-prajnapti* and *Uttarādhyayana*.[7]

The best sources of Mesopotamian astronomical heritage are records of various astronomical texts, such as *Enūma Anu Enlil*[8] from the period around 1200 BCE and the MUL.APIN collection of clay tablets from the 7th century BCE.[9] In the MUL.APIN, notes on the length of the shadow of a gnomon with the length of one cubit are also preserved. Despite some uncertainties, there is no doubt that the values are valid for the vertical gnomon with a length of one cubit (about 0.5 m), throwing a shadow on a horizontal surface.[10] The gnomon shadow length chart applies fairly well for locations between 35° and 36° N, which corresponds to the position of the Aššur site, from where the results of the astronomical observations originally written by Assyrian astronomers and recorded on later tablets, are supposed to have originated.[11] According to the sources, the gnomon from Mesopotamia, where it was certainly used by the 15th century BCE, is supposed to have been brought to Greece, along with the sundial, by ANAXIMANDER. HERODOTUS states that the Greeks were acquainted from the Babylonians with the polos (Gr. *πόλος*), gnomon (Gr. *γνώμων*) and twelve parts of the day,[12] and SUIDAS[13] states that ANAXIMANDER is thought

6 The *Vedāṅga* texts describe six branches of knowledge: phonetics, *Śikṣā*, grammar, *Vyākoraṅa*, etymology, *Nirukta*, metrics, *Chandas*, astronomy, *Jyotiṣa*, and rules of rituals and celebrations, *Kalpa*, divided into four parts - public celebrations, *Śrauta Sūtras*, celebrations, *Gṛyhya Sūtras*, law, *Dharma Sūtras*, and instructions for making altars, *Śulba Sūtras*. *Jyotiṣa Vedāṅga* is the main source of astronomical data from the Vedic period, while the *Śulba Sūtras* describe Vedic geometry. *Jyotiṣa Vedāṅga* may have been created around the year 1400 BCE. The exact time of its composition is unknown and the origins are probably somewhere in the north of India, between 27° and 29° N; see, Ôhashi, 1993, p.202.

7 *Candra-prajnapti* probably originates from the period between the 2nd and 4th centuries BCE. Also, *Uttarādhyayana*, the first of the texts of *Mūla Sūtras*, XXVI, 13–14, Ôhashi, 1993, p.202.

8 *Enūma Anu Enlil* is a collection of 70 (or 68, depending on the interpretation) tables with numerous astronomical notes. The source of the oldest records dates back to the beginning of the first ancient Babylonian dynasty (beginning of the 2nd millennium BCE). The translation of the title *Enūma Anu Enlil* reads: *"In the time of Anu and Enlil ..."* The Sumerian god AN, Akadic ANU, was the father of the gods and the god of heaven, the heavens. ENLIL also belongs at the very top of the Mesopotamian deities; he was the god of the atmosphere and the winds. For a more detailed explanation of Table XIV, see Al Rawi, George, 1991/1992, Enūma Anu Enlil XIV and Other Early Astronomical Tables, Archiv für Orientforschung, Bd. 38/39, pp.52–73.

9 The name MUL.APIN is the incipit - the opening words - on the tables. It means the constellation of the Triangle and the star δ Andromedae or Delta Andromedae. Weidner, 1924, Ein Babylonisches Compendium Der Himmelskunde, The American Journal of Semitic Languages and Literatures, vol. 40, no. 3, pp.186–208. For a detailed explanation, see Hunger, Pingree, 1989, MUL.APIN. An Astronomical Compendium in Cuneiform. Archiv für Orientforschung Beiheft 24 and Watson, Horrowitz, 2011, Writing Science before the Greeks: A Naturalistic Analysis of the Babylonian Astronomical Treatise MUL.APIN. The data, which was probably written on tablets around 700 BCE, is an older source.

10 The cubit, Lat. *cubitum*, an ancient measure of length, corresponds to approximately 0.5 m. The lengths of the ancient cubits differ from one another. For the Mesopotamian cubit, see Reallexikon der Vorgeschichte/8, 1927, p.60, or Unger, 1916, Zwei Babylonische Antiken aus Nippur, Publikationen der Kaiserlich Osmanischen Museen 1. For the lengths of the shadow, see Van der Waerden, 1968, Anfänge der Astronomie, p.80.

11 Van der Waerden, 1949, Babylonian Astronomy II, p.15, indicates a period between 1400 and 900 BCE. Schaefer, 2007, The Latitude and Epoch for the Origin of the Astronomical Lore in MUL.APIN, the period around the year 1370 BCE.

12 The oldest known source from 430 BCE is HERODOTUS, who cites this in Chapter 109 of book II (2.109.3). See also EUCLID, *Elements*, book II, pp.370–371. For more historical sources, see also Gibbs, 1973, Greek and Roman Sundials, pp.5–11. The oldest partially preserved Greek sundial is planar and originates from the 4th century BCE, probably from Athens or the surrounding area. It contains an indication of the name THEOPHILUS. THEOPHILUS ruled Athens in 348/347 BCE. Schaldach, 2004, The Arachne of the Amphiareion and the Origin of Gnomonics in Greece, particularly p.247. The word *polos* has several meanings, among others also a (celestial) hemisphere, see Szabó, 1992, Das geozentrische Weltbild, p.139.

13 SUIDAS (SUDA), Gr. *Σουΐδας*, a Byzantine encyclopaedia from the 10th century, which contains a lot of information about the later lost ancient works. The authors of the encyclopaedia are not known, nor is it clear whether SUIDAS (SUDA) was an actual person or just the title of the encyclopaedia.

to have brought this knowledge to Greece. From DIOGENES LAERTIUS' work, which mentions FAVORINUS,[14] it is evident that ANAXIMANDER used a shadow catcher (Gr. *σκιοθήρων*) in Sparta, which is understood to be a synonym for the gnomon.[15] ANAXIMANDER is supposed to have used the gnomon primarily to observe equinoxes and solstices. The gnomon was also certainly known to THALES, as he was supposed to have measured the height of the pyramid in Egypt, which he could only have done with a gnomon, though it is not clear whether THALES became acquainted with the gnomon in Egypt or elsewhere.

The sexagesimal system

The first written evidence of a measurement system in Mesopotamia, originally intended to measure distances and afterwards to measure angles or arcs, is Table XIV of the *Enūma Anu Enlil* collection. We do not know the Sumerian measurement system in its entirety; however, the significance of their measurement quantities and units is fairly well understood from archaeological fragments.[16] Probably by the early 3rd millennium BCE, the Sumerians were using different basic measurement units, including the *danna*, which measured distance indirectly by the time needed to travel across it.[17] The *danna* corresponds to approximately 11 kilometres or seven miles, with 12 such units required for a day-long walk or for one turn of the celestial arch.[18] The day was divided into twelve double hours, which were also called *danna*. At least from the 7th century BCE onwards, these units described the time elapsed in relation to the time of the sunrise or sunset, which is evident from royal records and letters, as well as from ritual, astrological and other records.[19]

These units were then subdivided into 30 smaller units, called UŠ,[20] so that one turn of the celestial arch covered 360 UŠ, where 1 UŠ ≡ 1° or 4 minutes, corresponding to the division of the circle into 360° as we know it today.[21] The measurement system in Mesopotamia, which was initially

14 Diogenes Laertius II.1. Lat. FAVORINUS, a Roman philosophical sceptic and rhetorician, worked in the 2nd century.

15 Szabó, 1992, pp.71–72. Vitruvius, *De Architectura*, 1.6, p.26. DIOGENES otherwise said that ANAXIMANDER invented the gnomon, which is not true since it was known and used in Egypt and Mesopotamia, as well as in China and in India.

16 Powell, 1990, Masse und Gewichte, Reallexikon der Assyriologie und vorderasiatischen Archäologie, pp.457–468, Powell, 1976, Sumerian area measures and the alleged decimal substratum, pp.165–221, Friberg et al., 1990, Seed and Reeds – A Metro-Mathematical Topic Text from Late Babylonian Uruk, pp.482–557.

17 For the preserved table, see Thureau-Dangin, 1910, Inventaire des Tablets de Tello conservées au Musée Impérial Ottoman, p.11, 1175.

18 Also Akadic bēru ≈ 10.8 km or ≈ 60 Greek stadia. Neugebauer, 1983, Astronomy and History, selected essays, pp.16–17.

19 The lengths of daylight are given on cuneiform tablet *Enūma Anu Enlil* 14 (EAE 14). Steele, Brack-Bernsen, 2008, p.261.

20 UŠ means length. Neugebauer, 1938, Untersuchungen zur Geschichte der Antiken Astronomie III, or Neugebauer, 1941, Some Fundamental Concepts in Ancient Astronomy, p.16, reprinted in Neugebauer, 1983, p.8. Sachs, Hunger, 1988, Astronomical Diaries and Related Texts. Vol I, Diaries from 652 B. C. to 262 B. C. Hunger, Pingree, 1999, Astral Sciences in Mesopotamia, pp.44–50.

21 360° = 24 hours, see also Equation 0.16 and explanation or note 112. For the development of the sexagesimal system, see Neugebauer, 1927, Zur Entstehung des Sexagesimalsystems and Thureau-Dangin, 1939, Sketch of a history of the sexagesimal system.

used for counting and measuring, evolved for centuries, and the beginnings of the sexagesimal system, which appear to be based on the Sumerian measurement system, are still poorly understood. Among the oldest preserved records written in base 60 are counts of people and sheep. In about 2500 BCE, the ratio between the units for weight, the mina and the shekel, was $2 \times \text{mina} = 60 \times \text{shekel}$. Similarly, the measurement system for measuring length also developed over the centuries. The Sumerian measurement system underwent several changes and supplements leading to the sexagesimal system probably around 2050 BCE, which is also useful for the purposes of calculation.[22]

Stars and the zodiac

We conclude from cave paintings that our ancestors were following the stars in the Stone Age, and it is certain that calendar making or the monitoring of the seasons was already used in the era of foraging. For the needs of agriculture, ancient people used simple calendars and times, useful for agricultural purposes, based on the (heliacal) rising of stars or constellations. Similar to the measurement system, agronomy also developed over the centuries, and along with knowledge and understanding of the ecliptic, conditions were also given for the division of the ecliptic, first into parts and then into 30° portions. The history of the division of the ecliptic into parts dates back to the 2nd millennium BCE and it is probably also related to the division of the day into time units.[23]

In Ancient India, the stars were well-known at least by 3000 BCE to the astronomers of the Harappan culture.[24] As knowledge increased over the centuries, the ecliptic came to be divided into even parts. In Indian sources, the division is described by the concept of *nakṣatras*.[25] Even in the oldest Vedic sources, the term *nakṣatra* is also used for stars or for 27 (or sometimes 28) groups of stars with which the Moon is in conjunction on a given day.[26] We know almost nothing about the calendars of the oldest cultures, such as the Indus Valley cultures, but there are preserved Indian hymns from the Vedic period proving that the Indians divided the year

22 Robson, 2008, Mathematics in Ancient Iraq, pp.75–84. See also Friberg, 2019, Three thousand years of sexagesimal numbers in Mesopotamian mathematical texts.

23 Recorded in the table VAT 4924. Van der Waerden, 1952–1953, History of the Zodiac and Van der Waerden, 1951a, Babylonian Astronomy III. From the MUL.APIN tables, it follows that the Babylonians knew the trajectory of the Moon through 17 constellations, which were replaced by 12 constellations by the 5th century BCE. Brack-Bernsen, Hunger, 1999, p.280.

24 Parpola, 2013, Beginnings of Indian Astronomy, with Reference to a Parallel Development in China.

25 *Nakṣatras* are already mentioned by name in *Ṛgveda* X.85.13, in *Atharvaveda* XIX.6.2, and in full in *Atharvaveda* XIX.7 and in *Taittirīya Saṁhitā* IV.4.10. Yampolsky, 1950, The Origin of the Twenty-Eight Lunar Mansions. Weinstock, 1949, Lunar Mansions and Early Calendars.

26 *Nakṣatras* refer to *Jyotiṣa Vedāṅga* as parts of the ecliptic. Gondhalekar, 2013, The Time Keepers of the Vedas, pp.141–143 and 155–162. Mishra, 2005, pp.187–190. *Nakṣatras* probably originate from the time before the Vedic period.

into 12 and 360 parts very early on.[27] From the sources from the Vedic period, the division of the day into 30 parts is also derived, which is unique to Indian culture.[28]

The earliest recorded signs of the zodiac originate from Mesopotamia, and date from 419 BCE.[29] Unlike the Mesopotamians, whose zodiac evolved for centuries, the Greeks summarised the basics of the zodiac with some changes taken from the Babylonians, some of which are most likely to have originated in Egypt though some are Greek. The notion of the Greek zodiac is found in *Parapēgmas*, the precursors of the modern almanacs, of EUCTEMON and EUDOXUS, which determine the seasonal calendar on the basis of the ascensions of the constellations.[30] The first person whom we know to have systematically used the signs of the zodiac in the *Parapēgma* was CALLIPPUS, in about 330 BCE. A very good description of the meaning of the signs of the zodiac was presented by GEMINUS in chapter 1 of his *Introduction to the Phenomena*.[31]

The Book of Enoch

The Book of Enoch is a group of manuscripts probably written between 300 and 100 BCE. Early but incomplete version of the text survive in the Ge'ez language, which is nowadays used only for liturgical purposes by the Ethiopian Orthodox Church.[32] The original writings were probably written in a Semitic language, presumably Aramaic, and later translated into Hebrew and Greek. *The Book of Enoch* is a religious text, but it also contains chapters on astronomical concepts. The Ethiopian text shows knowledge of the path of the Sun, the division of the ecliptic into parts,[33] the changing length of the days of the year, the equinoxes and solstices and the associated seasonal or equinoctial hours.[34] The Ethiopian text describes the variation of the length of the

27 Division of the year into four parts in *Ṛgveda* I.155.6, division into 360 parts I.164.48, also in *Atharvaveda* X.8.4. Division into 360 and 720 parts also in *Śatapatha Brāhmaṇa* X.5.4-4. The emergence of the oldest of the Indian hymns of the *Ṛgveda* probably dates back to the 2nd millennium BCE. Ôhashi, 1993, Development of Astronomical Observation in Vedic and Post-Vedic India. From the verses of the *Ṛgveda* comes the knowledge of the calendar, with the months and seasons, apparently based on observation of the Moon. *Ṛgveda* VII.103.7-9, I.25.8, 10.85.5. For the seasons, see *Ṛgveda* X.90.6, *Śatapatha Brāhmaṇa* I.6.1.2-3 and *Taittirīya Saṁhitā* IV.4.11 (*Yajurveda*).

28 $1/30$ of the day or 48 minutes is a *muhūrta*. Gondhalekar, 2013, pp.92–97. *Śatapatha Brāhmaṇa* X.4.2.18, for an explanation see Eggeling, 1897, The Śataphata Brāhmana According to the Text of the Mādhyamdina School, Part IV, pp.351–352. Also *Ṛgveda* X.189.3. Michel (ed.), 2008, Rig-Veda, das Heilige Wissen Indiens, 2, p.403.

29 For an explanation of the possible origin of the symbols of the signs of the zodiac, see Dewdney, 1931, The Zodiac and Early Astronomy, and Webster, 1940, The Origin of the Signs of the Zodiac: An Interpretation from the Psychological Viewpoint.

30 EUCTEMON's *Parapēgma* is dated about 430 BCE. Van der Waerden, 1984, Greek Astronomical Calendars, I, p.103. EUDOXUS' work is said to be based on data from the 11th century BCE, i.e. from the time of Assyrian rule. Schaefer, 2004, The latitude and epoch for the origin of the astronomical lore of Eudoxus. For more information see Rehm, 1913, Griechische Kalender III. Das Parapegma des Euktemon.

31 Evans, Berggren, 2006, Geminos's Introduction to Phaenomena, pp.113–136.

32 For an English translation, see Schodde (trans.), 1882, The Book of Enoch.

33 The word *'gate'* is used in the text for each part, something like the division into the signs of the zodiac.

34 Schodde (trans.), 1882, Section XIII.

day in an approximate and apparently linear manner, reminiscent of the astronomical records of Mesopotamia. Some scrolls, only recently discovered in the Qumran Caves, contain much more astronomical information.[35] The inscriptions on these scrolls are in Aramaic. From an astronomical point of view, the four scrolls 4Q208 to 4Q211 are important for describing the motions of the Moon and stars. The scrolls suggest a link with Babylonian astronomy, and knowledge of the Moon's movements was particularly important for the construction of the lunar or lunisolar calendar.[36]

And on that day the night becomes longer and amounts to the double of the day: and the night amounts exactly to twelve parts and the day to six.

The Book of Enoch, Book 4, The Luminaries, 26 (Winter, trans., 2015, The Complete Book of Enoch)

The birth of Greek astronomy

From the PYTHAGOREANS[37] onwards, the model of circular motion, perhaps even epicycles, emerged, according to which the celestial bodies move on smaller circles, i.e. epicycles, in addition to moving around the Earth, which made it possible to explain the uneven motion of the planets, which we understand today as being due to moving around the Sun at variable speeds on elliptical orbits. Historical sources reveal a simple system of epicycles, according to which Mars, Jupiter and Saturn moved clockwise,[38] which was in the opposite direction to later, when, towards the end of the 3rd century BCE, APOLLONIUS designed a system of epicycles in which all the planets moved counterclockwise. Uneven movement, perhaps even an epicycle system, was already known by PLATO, which we conclude from the dialogues *The Republic* and *Timaeus*.[39] It is clear from the statements of SIMPLICIUS that PLATO's way of thinking played an important role in the further development of models of planetary motion. PYTHAGOREAN mathematics or geometry was at a sufficiently high level that mathematical support or geometric constructions are not excluded in this respect. For example, EUCLID summarised part of his *Elements*, in particular the second and fourth books, from sources from the PYTHAGOREAN period.[40]

35 Drawnel, 2011, The Aramaic Astronomical Book (4Q208-4Q211) from Qumran: Text, Translation, and Commentary. Records discovered in the Qumran Caves, a rich archaeological site near the Dead Sea coast.

36 For a possible interpretation of the Aramaic calendar, see Jacobus, 2022, How 4QAstronomical Enoch[a–b] (4Q208–209) Transformed Elements of Late Babylonian Magical Hemerological Texts into a Synchronistic Calendar.

37 The PYTHAGOREANS were philosophers who followed the teachings of PYTHAGORAS; they worked in the 5th century and the first half of the 4th century BCE. More than two hundred PYTHAGOREAN philosophers are known, but many more have probably been forgotten, and some of the names of the respectable wives of PYTHAGOREANS have also been preserved. Representatives of PYTHAGOREANS are, for example, EMPEDOCLES, Gr. *Ἐμπεδοκλῆς*, ECPHANTUS, Gr. *Ἔκφαντος*, PHILOLAOS, Gr. *Φιλόλαος*, DEMOSTHENES, Gr. *Δημοσθένης*, and others.

38 See Neugebauer, 1972, Planetary Motion in P. Mich. 149, p.22, Van der Waerden, 1951b, Die Astronomie der Pythagoreer and Van der Waerden, 1975, The Earliest Form of the Epicycle Theory.

39 Knorr, 1990, Plato and Eudoxus on the Planetary Motions. Journal for the History of Astronomy, vol. 21, no. 4, pp.313–329.

40 Neuenschwander, 1973, Die ersten vier Bücher der Elemente Euklids. Archive for History of Exact Sciences, vol. 19, pp.325–380.

In the 4th century BCE, HERACLIDES PONTICUS advocated that the Earth retated around its axis. Similarly to HERACLIDES, the Earth's rotation was also supposed to have been argued by the philosophers HICETAS and ECPHANTUS. The movement of the Earth around the Sun was advocated by ARISTARCHUS OF SAMOS. Unfortunately, his work describing the heliocentric model is lost, and we know about him only indirectly, especially from the works of ARCHIMEDES and PLUTARCH.[41] ARISTARCHUS is also very likely to have followed HERACLIDES' idea of the Earth rotating around its own axis.[42] As far as we know, in addition to ARISTARCHUS, the Sun being at the centre of the universe was only advocated by the astronomer SELEUCUS OF SELEUCIA.[43]

Trigonometry before Hipparchus

The beginnings of spherical trigonometry are lost in time; credit for its beginnings probably goes to EUDOXUS or the mathematical school of Cyzicus. Among the mathematicians whom we assume inherited the knowledge of their predecessors, we certainly categorize EUCLID, with his work *Phaenomena*, and AUTOLYCUS. We do not know anything reliable about the life of AUTOLYCUS OF PITANE (Gr. *Αυτόλυκος ὁ Πιταναίος*) beyond that he was a teacher to ARCESILAUS. We suppose that he was slightly older than EUCLID, and that he worked at the transition from the 4th to the 3rd century BCE, though his dates are not well known. The year of ARCESILAUS' birth, 315 BCE, is a helpful piece of information. His works *On the Moving Sphere* and *On the Rising and Setting of Stars* are of utmost importance, since they are the oldest fully preserved Greek mathematical works. With similar themes to AUTOLYCUS' book *On the Moving Sphere*, THEODOSIUS' *Sphaerics*[44] also deals with the moving sphere; we assume that both of these works are based on an older, forgotten work, possibly by EUDOXUS.[45] According to sources, THEODOSIUS probably lived around 100 BCE or about the same time as HIPPARCHUS. He was from Bithynia, a province in present-day Turkey, and some sources, due to an obvious error in the *Suda encyclopaedia*, also referred to him as THEODOSIUS OF TRIPOLIS.[46]

41 ARISTARCHUS' work *On the Sizes and Distances of the Sun and the Moon* is preserved. Heath, 2007, Aristarchus of Samos, the Ancient Copernicus. Heath, 1897, The Works of Archimedes, pp.221–232. For more information on ARISTARCHUS, see also Heath, 2006, A History of Greek Mathematics, Volume II, pp.1–15.

42 Heath, 2007, p.304.

43 SELEUCUS worked around 150 BCE. He was also the first we know to have discovered the connection between the tides and the Moon. See Strabo, Geographica, section 174, Forbiger (transl.), 2007, p.238.

44 Kunitzsch, Lorch, 2010, Theodosius Sphaerica.

45 Tropfke, 1923, Geschichte der Elementar-Mathematik, Band 5, p.102 or Heath, 2006, I, pp.348–353. See also note 87.

46 Bulmer-Thomas, 1976, Theodosius of Bithynia, Dictionary of Scientific Biography.

Apollonius and Hipparchus

APOLLONIUS, better known for his work on geometry and conic sections, undoubtedly knew ARISTARCHUS' work. He perfected the system of epicycles, which explained the motion of the planets and the Sun in a much more precise manner than the system of concentric spheres, and assumed that all the planets in the epicycles moved counterclockwise, which was much more in line with the results of observations than EUDOXUS' spheres.[47] He also confirmed that the model of epicycles is mathematically equivalent to an eccentric circle (Figure 0.1).[48]

Epicyclic and eccentric motion models (Figure 0.1) are equivalent if the radius ES of the epicycle is equal to the eccentricity e or OC, and if the movement is such that the angles α (mean anomaly, see Chapter 1) and β are always the same.[49] This is the case if the epicycle radius ES is always parallel to the connecting line CA between the centre and apogee, or if the condition (Figure 0.1) is satisfied:[50]

$$\frac{ES}{OE} = \frac{e}{CA} \qquad \text{or} \qquad \frac{r}{R} = \frac{e}{R'}$$

The first person who systematically used mathematical computational models in astronomical problems was probably HIPPARCHUS, who is thought of as the father of trigonometry. We know that HIPPARCHUS, considered one of the greatest names of ancient astronomy, possessed a wealth of Babylonian knowledge.[51] Most likely, he personally came into contact with Babylonian astronomers, who recorded extensive data and predicted phenomena but generally did not look for their causes. That the Babylonian achievements were known in Greek Egypt is evidenced by papyruses containing the ephemerides, which were compiled in Babylon.[52] HIPPARCHUS, who was very precise, made his most important observations in Rhodes, although he is also credited with observing the vernal equinox in Alexandria in March 146 BCE.[53] He introduced into Greek mathematics the division of the circle into 360° and also drew up the first trigonometric tables of the chord function. He wrote several works, but unfortunately all of them, with the exception of one minor treatise, have been lost, and even this has probably survived only because of ARATUS' poetic abilities, since ARATUS' extant treatise refers to the work of HIPPARCHUS. Most of our knowledge of HIPPARCHUS' work comes from the *Almagest*. His greatest contribution to the development of Greek mathematics may be that he took mathematics from an idealistic and theoretical concept to the path of rigorous

47 Neugebauer, 1959a, The Equivalence of Eccentric and Epicyclic Motion According to Apollonius, p.7.

48 Neugebauer, 1959a, p.8, note 15.

49 An object on the deferent or centre of the epicycle rotates counterclockwise. An object on the epicycle can rotate clockwise or counterclockwise. For some comments or explanations see Crowe, 2001, Theories of the World from Antiquity to the Copernican Revolution, pp. 32–34 or Pedersen, 2011, pp.136–137.

50 Neugebauer, 1959a, p.8. See also Toomer, 1998, Ptolemy's Almagest, pp.147–150 or Evans, 1998, The History and Practice of Ancient Astronomy, pp.212–213.

51 Heath, 2006, II, p.255. Aaboe, 1955, On the Babylonian Origin of Some Hipparchian Parameters.

52 Neugebauer, 1969, The Exact Sciences in Antiquity, pp.161–164.

53 *Almagest* III.1, Toomer, 1998, p.133.

empirical practical computation. He compiled a catalogue of hundreds of stars,[54] made accurate estimates of the length of the lunar month and introduced the 304-year HIPPARCHIC cycle. He also improved ARISTARCHUS' estimate of the distances between the Earth and the Sun and tried to improve the system of epicycles.[55] He defended the use of astronomy in geography and criticised ERATOSTHENES, partly unjustly. His greatest discovery, however, was the discovery of precession.

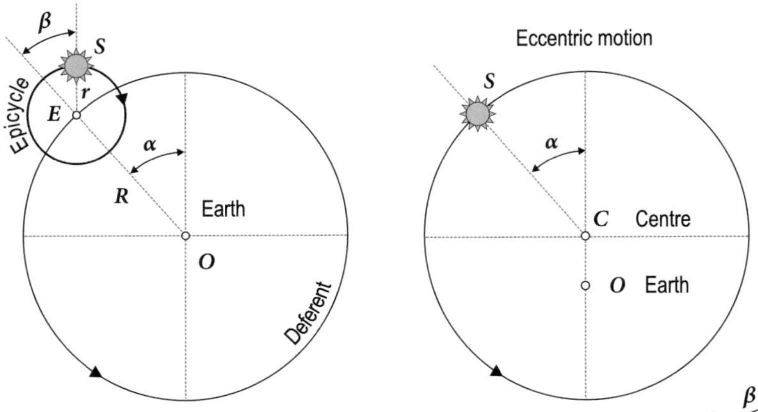

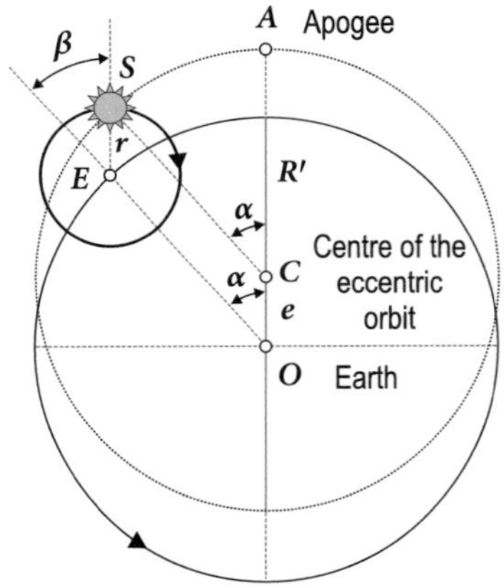

Figure 0.1 The model of epicycles (image on the left), in which the planet or the Sun follows a large orbit, the deferent, and a smaller one, the epicycle. After Neugebauer, 1969, p.155, Fig. 25 and Neugebauer, 1975, III, p.1220, Fig. 50 and 51.

How HIPPARCHUS came to achieve his highly accurate results is not known; perhaps instead of using spherical trigonometry, he used the geometric construction of an *analemma*[56] or stereographic projection.[57] From the text in VITRUVIUS' *De Architectura*, where VITRUVIUS describes an astronomical instrument named the *arachne*,[58] it follows that the basics of

54 Gysembergh, Williams and Zingg, 2022, New evidence for Hipparchus' Star Catalogue revealed by multispectral imaging. See also Nadal, Brunet, 1989, Le "Commentaire" d'Hipparque II. Position de 78 étoiles.

55 See also Nadal, Brunet, 1984, Le "Commentaire" d'Hipparque I. La sphère mobile.

56 An *analemma* is an orthographic or stereographic projection of a sphere or, more generally, the celestial arch onto one of the rectangular planes, either the horizontal plane, the meridian plane or the vertical plane. In antiquity, various analemmas were used to solve problems in astronomy and for the planning of sundials. Under the term *analemma* or *analemmic curve*, we also understand the position of the Sun at the same times during the course of the year, see Chapter 20.

57 For a possible explanation of HIPPARCHUS' calculation by using spherical trigonometry see Sidoli, 2004, Hipparchus and the Ancient Metrical Methods on the Sphere.

58 *Arachne*, Gr. ἀράχνη = spider. A Greek mythological personality, a weaver.

stereographic projection were known as early as the 1st century BCE.[59] A quotation from SYNESIUS' letter indicates that HIPPARCHUS also knew about projection onto a plane.[60]

From the measured time intervals between equinoxes and solstices it was already clear in antiquity that the uniform motion of the Sun along a circular orbit with the Earth at the centre does not correspond to the actual state, since the intervals between equinoxes and solstices are not equally long. HIPPARCHUS circumvented this problem either by epicycles or by introducing an eccentric circular orbit, where the Earth was no longer at the centre, and the Sun was still moving along a circle, but with apparently varying speed due to eccentricity.[61]

We know from PTOLEMY's references in the *Almagest* III.4 that HIPPARCHUS knew the lengths of the seasons, specifically spring and summer.[62] For the period between the vernal equinox and the summer solstice, he recorded a length of 94½ days, and for the period between the summer solstice and the autumnal equinox, 92½ days (Figure 0.2). The length of the year, or the number of days between the equinoxes and the solstices, can be measured by measuring the length of the shadow with a gnomon on an equinox or solstice, a process that was well-known by the time of HIPPARCHUS. From this data and the length of the tropical year, HIPPARCHUS concluded a period of 365.2467 days, and the longitude of the apogee and the ratio between eccentricity and radius e/R can be calculated.[63] HIPPARCHUS did not necessarily determine the length of the year through observation; his estimate of the tropical year length is an average over a period of 304 years, or 4 × 76 years = 4 × the CALIPPIC cycle:

$$l = 365\tfrac{1}{4} - \frac{1}{300}\text{days} = 365.2467$$

or:[64]

$$\begin{aligned}
19 \text{ years} &= 235 \text{ months} \\
1 \text{ month} &= 29;31,50,8,20 \text{ days} \\
1 \text{ year} &= \frac{235}{19} \cdot 29;31,50,8,20 \text{ days} \\
&= 365;14,48,33,35,47\ldots \text{ days} \approx 365\tfrac{1}{4} - \frac{1}{314}\text{ days} \\
&\approx 365;14,48 \text{ days} = 365\tfrac{1}{4} - \frac{1}{300}\text{ days}
\end{aligned}$$

59 VITRUVIUS, *De Architectura* IX.8. Morgan, 1960, Vitruvius the Ten Books on Architecture, p.273.

60 SYNESIUS, in a letter to PAEONIUS, see Fitzgerald, 1926, Letters of Synesius, p.263.

61 Jones, 1991a, Hipparchus's Computation of Solar Longitudes, p.102.

62 Toomer, 1998, p.153. For information on PTOLEMY and the *Almagest*, see also Chapter 1.

63 A tropical year is the time period between two consecutive passages of the Sun through the same point on the ecliptic. The mean longitude of the Sun increases by 360° during this period. The name of the tropical year originates from the Greek word *tropikós* (Gr. *τροπικός*, to turn) and means the return to the same tropic (Tropic of Cancer, Tropic of Capricorn). Petersen, Schmidt, 1968, The Determination of the Longitude of the Apogee of the Orbit of the Sun according to Hipparchus and Ptolemy, p.78, Table 1. For further comments, see Neugebauer, 1959b, Ptolemy's Geography, Book VII, Chapters 6 and 7.

64 Swerdlow, 1980, Hipparchus's Determination of the Length of the Tropical Year and the Rate of Precession, p.299. For CENSORINUS' comments on the length of HIPPARCHUS' year, see *De di Natali* 18,9 or Censorinus, 1887, Hultsch (ed.), p.38. Also Censorinus, 2011, Brodersen (trans.), p.84. Swerdlow, 1980, p.294 note 4.

HIPPARCHUS is regarded as one of the best observers among ancient astronomers and his work was surpassed only by CLAUDIUS PTOLEMY three hundred years later, but apparently even PTOLEMY used the work of Babylonian astronomers or HIPPARCHUS' observations in his work.

Example 0.1 Using the length of the seasons, the length of the year and the geometric construction in Figure 0.2, calculate HIPPARCHUS' value of the longitude of the apogee λ_Γ and the e/R ratio.[65]

Using the triangles $\triangle OC\Upsilon$ and $\triangle OCT$ (Figure 0.2), we can write the following equations:

$$h = e \sin \lambda_\Gamma = R \sin x \qquad \text{or} \qquad e \sin(90° - \lambda_\Gamma) = R \sin y$$

therefore:

$$\frac{e}{\sin x} = \frac{R}{\sin \lambda_\Gamma} \qquad \text{or} \qquad \frac{e}{\sin y} = \frac{R}{\cos \lambda_\Gamma}$$

extract ***e*** from the above equations, equate the two equations and get:[66]

$$\tan \lambda_\Gamma = \frac{\sin x}{\sin y} \qquad \text{and} \qquad \frac{e}{R} = \frac{\sin x}{\sin \lambda_\Gamma}$$

The angles x and y can be calculated by taking into account the daily movement of the Sun, which is

$$v = \frac{360°}{365.2467 \text{ days}} \approx 59°8'/\text{day}$$

and by the length of the spring and summer, which are known. The following two equations follow from Figure 0.2 (bottom):

$$x + 180° + x = (94.5 \text{ days} + 92.5 \text{ days}) \cdot v \qquad \text{and} \qquad x + 90° + y = 94.5 \text{ days} \cdot v$$

For the angles x and y, we obtain the values $\approx 2°9'$ and $\approx 59'.133$ respectively. Using these two values, we calculate the longitude of the apogee, which is $65°22'$, and for the ratio e/R, we obtain a value $\approx 1/24$.

65 The example is summarised by Pedersen, Pihl, 1974, p.86. See also Neugebauer, 1975, p.58 and p.1221 Fig. 53.
66 HIPPARCHUS had to solve the problem using the chord function, since the tangent function was not known in his time.

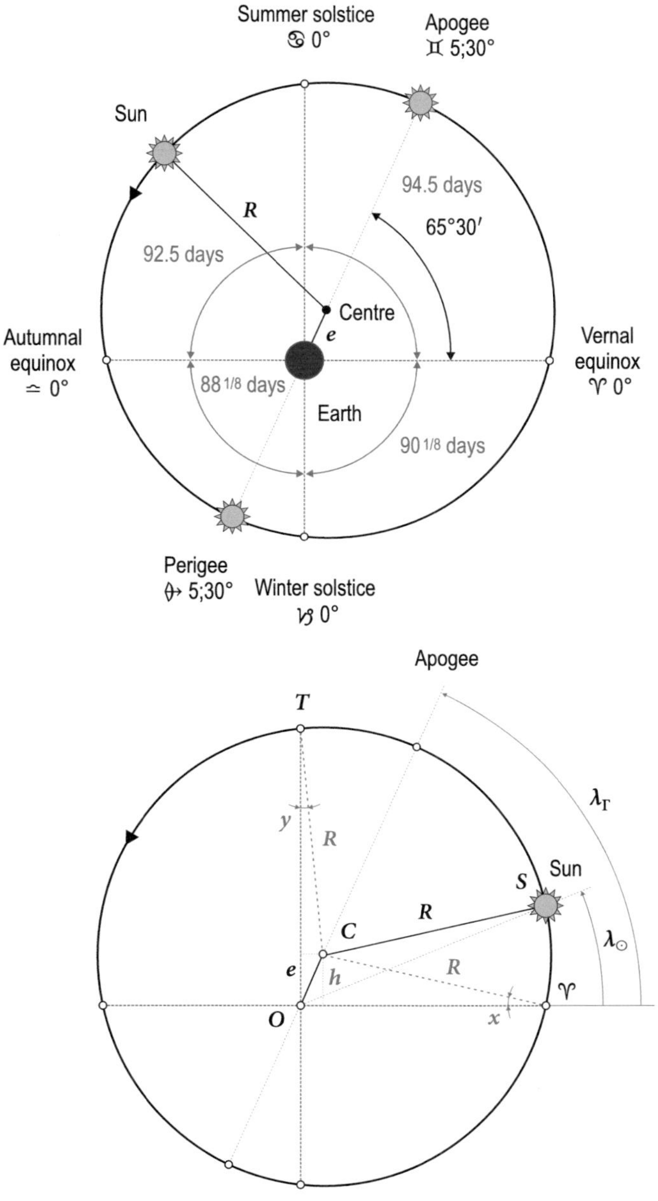

♈	vernal equinox
$\lambda_{\odot}$	true longitude
λ_{Γ}	the longitude of the apogee, HIPPARCHUS' value = ♊ 5;30° or 65;30°
R	radius
e	eccentricity = OC
O, C	the observer (Earth) and the centre of the eccentric circle

Figure 0.2 HIPPARCHUS' model of the eccentric movement of the Sun. Ratio $e : R = 1 : 24$. Summarised from Jones, 1991a, p.102, Fig. 1, and Pedersen, Pihl, 1974, p.87, Fig. 7.12. See also Neugebauer, 1975, pp.57–58, 1221, Fig. 53 and Toomer 1998, p.154, Fig. 3.9. Current value between the winter solstice and spring equinox is 88.99 days, between the spring equinox and summer solstice 92.74 days, between the summer solstice and autumn equinox 93.65 days, and between the autumn equinox and winter solstice 89.86 days (Astronomical Phenomena for the Year 2025, A1).

Sexagesimal Arithmetic

The procedures for multiplying, dividing and extracting square roots from sexagesimal numbers were known to Babylonian mathematicians and astronomers. The Babylonians were also familiar with quadratic and cubic equations, and with solving various linear and non-linear equations with one or more variables.[67] Much of this wealth of knowledge was later taken up by the Greeks.[68] The mathematical procedures of multiplication, division and root extraction, as used in the time of CLAUDIUS PTOLEMY in Alexandria, are described by THEON in his commentary on the first and second books of the *Almagest*.[69] The arithmetical procedures in Greek mathematics were not the simplest, since the Greeks used the letters of the Greek alphabet as signs for digits or numbers. The following procedures are illustrated using modern mathematical notation and correspond essentially to the procedures used by mathematicians and astronomers in ancient Alexandria and Greece. The multiplication or division process is shown in the examples below, the calculation on the left is summarised from the literature, and intermediate results are added on the right for ease of understanding. The column markers °, ′, ″ ... correspond to the sexagesimal places e.g. the fractional degrees, minutes, seconds, etc.[70]

Example 0.2 An example of multiplying or squaring a sexagesimal number is to calculate the square of 37;4,55, which is the value of Crd 36°. Summarised from the literature, see Ideler, 1812, p.13, and Halma, 1821, pp.117–118. On the left is the procedure summarised from the literature, and on the right are the results of intermediate calculations to aid understanding.

°	′	″		°	′	″
(37	4	55)	×	(37	4	55)
37	4	55				

°	′	″	‴	⁗			
1369	148	2035			37×37=1369	37×4=148	37×55=2035
	148	16	220		4×37=148	4×4=16	4×55=220
		2035	220	3025	55×37=2035	55×4=220	55×55=3025
1369	296	4086	440	3025			
1375	4	14					

67 For more information on Babylonian mathematics, see Neugebauer, 1934, Vorlesungen über Geschichte der antiken mathematischen Wissenschaften, pp.166–208 or Gheverghese, 2011, The Crest of the Peacock, pp.125–176. Some information is also given in Van der Waerden, 1983, Geometry and Algebra in Ancient Civilizations, pp.56–63.

68 For more information see Friberg, 2007, Amazing traces of a Babylonian origin in Greek mathematics.

69 Halma, 1821, pp.117–187.

70 Ideler, 1812, Über die Trigonometrie der Alten, pp.12–15. Heath, I, 1921, pp.57–63. For easier calculation in Mesopotamia different arithmetic lookup tables were used. For more information see Robson, 2008.

The result obtained by the above procedure is 1369;296,4086,440,3025. By taking multiples of 60, we convert it into sexagesimal form:

$$3025 = 50 \times 60 + 25$$

$$490 = 8 \times 60 + 10 \qquad (440 + 50 = 490)$$

$$4094 = 68 \times 60 + 14 \qquad (4086 + 8 = 4094)$$

$$364 = 6 \times 60 + 4 \qquad (296 + 68 = 364)$$

$$(1369 + 6 = 1375)$$

The final result is 1375;4,14,10,25 or, if the smaller terms are ignored, 1375;4,14.

Example 0.3 An example of dividing sexagesimal numbers 1515;20,15: 25;12,10. The example is taken from Ideler, 1812, p.13. It is also published in Halma, 1821, pp.118–119. See also Heath I, 1921, p.59.

°	′	″	‴		°	′	″		°	′	″	
(1515	20	15)		:	(25	12	10)	=	60	7	33	
1500				=	25			×	60			first term: 60°
	920											remainder = 15° = 900′ 900′ + 20′ = 920′
	720			=		12		×	60			
	200											remainder = 920′ − 720′ = 200′
	10			=			10	×	60			
	190			=								remainder = 200′ − 10′ = 190′
	175			=	25			×		7		second term: 7′
		900										remainder = 15′ = 900″
		915										
		84		=		12		×		7		
		831										remainder = 915″ − 84″ = 831″
		1	10	=			10	×		7		
		829	50									remainder = 831″ − 1″10‴ = 829″50‴
		825		=	25			×			33	third term: 33″ remainder = 4″50‴ = 290‴
		4	50									remainder = 4″50‴ = 290‴
			396			12		×			33	remainder = 106‴

The result of the division is thus an approximation of 60;7.33.

The procedure for extracting square roots, which is also of Babylonian origin, yields an approximation.[71] The process can be represented graphically by the area of the square whose square root is being sought. THEON gives the example of the square root of 4500, which

71 Flores, 2014, The Babylonian Method for Approximating Square Roots: Why Is It So Efficient? In: The Mathematics Teacher, Vol. 108, No. 3, pp.230–235. The procedure is iterative and yields a rapidly converging sequence of approximations that can be carried on to arbitrary precision.

corresponds to the area of the square □*ABCD* in Figure 0.3.[72] First we find the number or root whose square gives the integer solution closest to 4500; in this case it is 67, which is the length of the side of the square □*AEFG* with area 4489. The difference between the given area and the calculated approximation is 11(°) or 660(′) and represents the area of the gnomon ▟.[73] This figure contains two equal rectangles, namely ▭*HF* and ▯*FK*. The longer side of the two rectangles is 67, so the number 660 is divided by 67 × 2 or 134. The integer part of the quotient is 4, which represents the shorter sides *EH* and *GK* of the two squares, each of which has an area of 67 × 4 = 268. The smaller square □*FL* has an area of 16. The area of the square □*AHLK* is therefore equal to 4489;536,16 or, converted to sexagesimal notation, 4497;56,16. We then repeat the process with the remainder, which is 2;3,44 or 7424(‴) and is the area of the gnomon ▟*BLD*. Again we have two identical rectangles, namely ▭*BL* and ▯*LD* with longer side 67;4. We divide 7424 by twice the value of the longer side, which is now 134;8. The integer part of the quotient is 55, which for □*ABCD* with area 4500 therefore gives an approximation of the square root of 67;4,55.

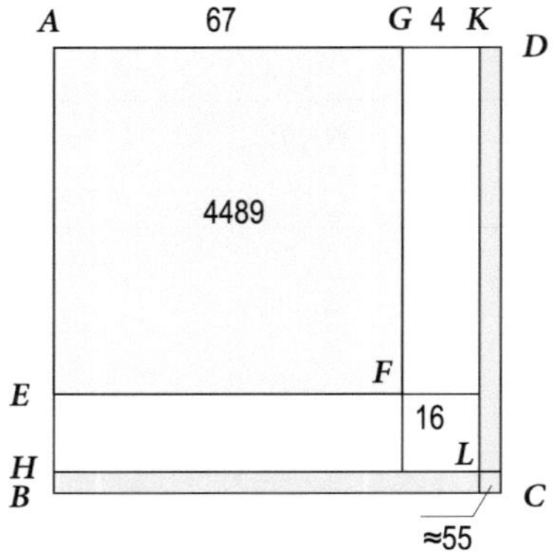

Figure 0.3 Graphical representation of the roots of the sexagesimal 4500. Taken from Ideler, 1812, p.13. Also published in Halma, 1821, pp.118–119.

72 Ideler, 1812, pp.14–15. The value 4500 was not chosen by THEON at random; it is the square of the distance *EB*, see Figure 0.5, and it is necessary in calculation of the chord of an arc of 36°, which is one of the basic values from which the other values of the chord function are derived. See Toomer, 1998, p.49 or *Almagest* 1.10.

73 The word *gnomon* has multiple meanings. In this case geometrical structure is meant. See also Figure 0.3.

The trigonometric chord function

HIPPARCHUS knew about the chord function; according to sources, he also wrote a book on it which, however, has been lost.[74] In modern mathematical terms, the chord function Crd[75] is related to the sine as (Figure 0.4):

$$\text{Crd}\,\alpha = 2R\sin\frac{\alpha}{2} \qquad [0.1]$$

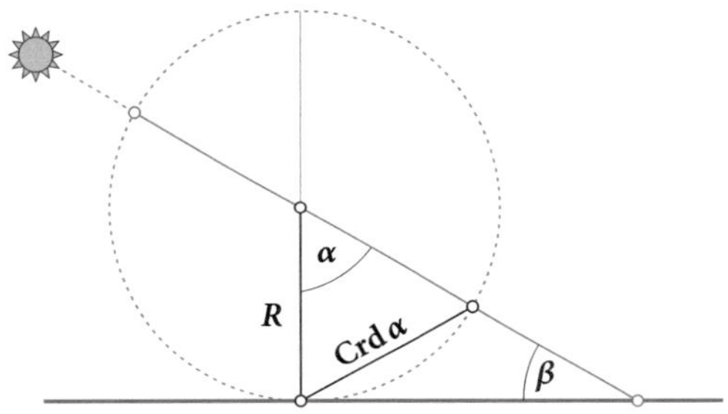

Figure 0.4 The chord function. Summarised from Neugebauer, 1975, III, Figures 1 and 10.

The calculation of the chord function and a table of values for a circle with radius $R = 60$ was later given by CLAUDIUS PTOLEMY in Alexandria around 150 CE, in chapters 10 and 11 of the first book of the *Almagest*.[76] *The Almagest* also shows a practical use of calculating shadow length.

The calculation of PTOLEMY's values of the chord function is based on the values of the function for angles 36°, 72°, 60°, 90° and 120° calculated using the laws of plane geometry for a circle of radius $R = 60$. PTOLEMY derived the values of the chord function for angles 36° and 72° using the rules of EUCLID's *Elements* XIII.9 and XIII.10, starting from the geometrical construction in Figure 0.5.[77] The radius DG of the circle is divided in half at point E, and the distance EB is equal to the distance EZ. The resulting figure has the following properties:

74 Toomer, 1974, The Chord Table of Hipparchus and the Early History of Greek Trigonometry; Centaurus, vol. 18, no. 1, pp.6–28.

75 Trigonometric functions where the radius of the circle is not equal to 1 - generally the radius in antiquity of 60 units - is indicated by a capital letter, for example, **Crd** for the chord function, **Sin** for the sine function, and so on.

76 Toomer, 1998, pp.57–60. PTOLEMY gave the values for the chord function for angles between $^1/_2$° and 180° in steps of $^1/_2$°.

77 Bulmer-Thomas, 1939, pp.415–423 or Pedersen, 2011, pp.56–57. See also Heath, 1956, The Thirteen Books of The Elements, vol. 3, reprint, XIII.9 and XIII.10, pp.455–461. For a description of PTOLEMY's calculation procedures and for the newly calculated values of the chord functions, see Glowatzki, Göttsche, 1976, Die Sehnentafel des Klaudios Ptolemaios, nach den historischen Formelplänen neu berechnet, pp.21–29 and 60–69. A very good summary of PTOLEMY's procedures is also given in Chatterjeee, 1970, I, Appendix VI, pp.176–188. For tabulated sexagesimal values, see Manitius, Des Claudius Ptolemäus Handbuch der Astronomie, 1. und 2. Band. Leipzig, 1963, reprint or Halma, 1813, pp.38–45 or Toomer, 1998, pp.56–61. For recalculated values see Glowatzki, Göttsche, 1976, pp.35–42.

- Distance BZ is the length of the side of the regular circle-inscribed pentagon or Crd 72°.
- Distance DZ is the length of the side of the regular circle-inscribed decagon or Crd 36°.
- Distance BD is the radius ($R = 60$) equal to the side of the regular circle-inscribed hexagon.
- Distance AB is the side of the circle-inscribed square or Crd 90°.
- The side of the circle-inscribed equilateral triangle corresponds to Crd 120°.
- The diameter of the circle or $2 \times R$ is Crd 180°.

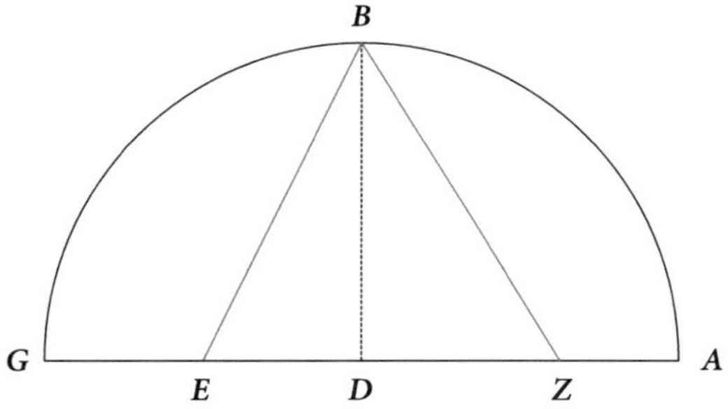

Figure 0.5 Geometric construction from which PTOLEMY derived values of the chord functions of angles 36° and 72°. Source: Halma, 1821, pp.120–186, Figure 44.

PTOLEMY thereby calculated the following chords for the above angles:[78]

$$\text{Crd } 36° = 37; 4{,}55$$

$$\text{Crd } 72° = 70; 32{,}3$$

$$\text{Crd } 60° = 60$$

$$\text{Crd } 90° = 84; 51{,}10$$

$$\text{Crd } 120° = 103; 55{,}23$$

From these values, he then derived all the others. For example, for the sides of the circle-inscribed square or triangle corresponding to Crd 90° and Crd 120°, the following relationships apply (see also Figure 0.6):

[0.2] $$\text{Crd } 90° = \sqrt{R^2 + R^2} \qquad \text{Crd } 120° = 2\sqrt{R^2 - \left(\frac{R}{2}\right)^2} = R\sqrt{3}$$

78 Halma, 1813, pp.26–29, also Bulmer-Thomas, 1939, p.421. For sexagesimal multiplication, division and square rooting as performed by Greek mathematicians see Heath, 1921, Volume 1, pp.52–64, Bullmer-Thomas, 1939, pp.50–63 (includes also description of cube root extraction), Pedersen, 2011, pp.49–56, Ideler, 1812, pp.12–15. For square root extraction see also Flores, 2014.

From the known angles, other values can be calculated using the angle difference, at intervals of 6°, since the following applies:[79]

$$\text{Crd}(\beta - \alpha) = \frac{\text{Crd}\,\beta\,\text{Crd}(180° - \alpha) - \text{Crd}\,\alpha\,\text{Crd}(180° - \beta)}{120} \quad [0.3]$$

Using the following expression, we can calculate the chord function for every 3°:

$$\text{Crd}^2(180° - \alpha) = 120^2 - \text{Crd}^2\alpha \quad [0.4]$$

and the chords of the half-angles are calculated using the following expression:[80]

$$\text{Crd}^2\frac{\alpha}{2} = 60\big(120 - \text{Crd}(180° - \alpha)\big) \quad [0.5]$$

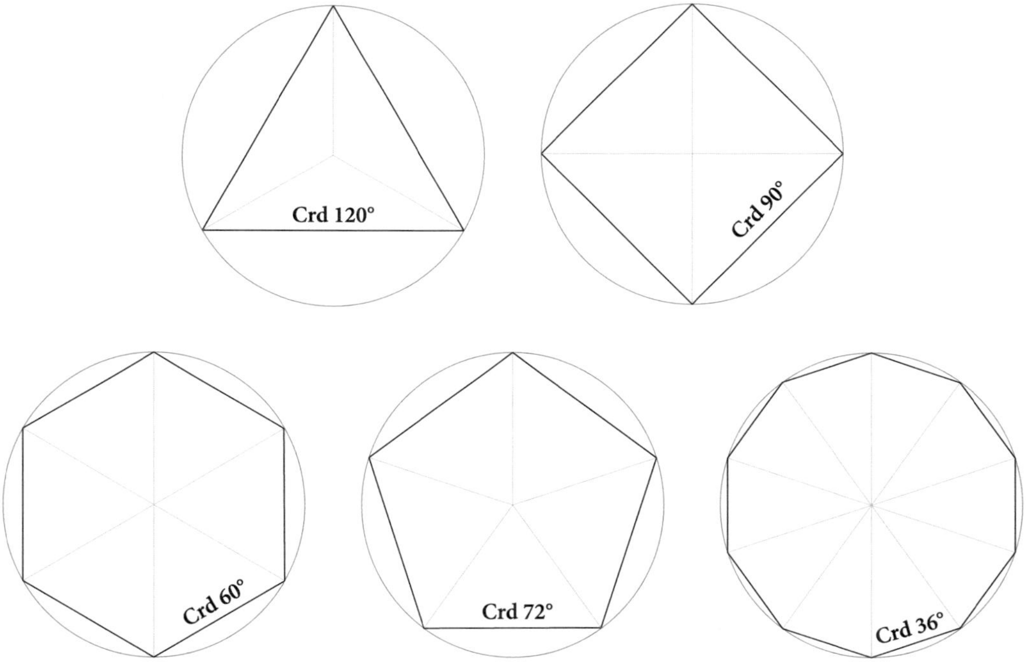

Figure 0.6 Basic chord functions derived by the rules of plane geometry or EUCLID's *Elements*.

79 Pedersen, 2011, pp.57–60.
80 The expression for the half-angle function, as written today using the sine and cosine functions, is:

$$\sin\frac{x}{2} = \pm\sqrt{\frac{1 - \cos x}{2}}$$

The algorithm for the sum and difference of the chord functions was probably already mastered by ARCHIMEDES. Tropfke, 1928, Archimedes und die Trigonometrie, Archiv für Geschichte der Mathematik und Naturwisenschaften und Technik, 10. Band, p.436, and Schoy, 1927, Die trigonometrischen Lehren des Al-Birūni, particularly pp.74–91. Also Toomer, 1974, The Chord Table of Hipparchus and the Early History of Greek Trigonometry, Centaurus, Volume 18, Issue 1, p.28, note 25.

In addition to the addition formulas for the difference (Equation 0.3) or sum of angles, which in modern notation are written as:

$$\sin(x - y) = \sin x \cos y - \cos x \sin y$$
$$\cos(x + y) = \cos x \cos y - \sin x \sin y$$

PTOLEMY also knew an identity, complex expressed in chords, that is equivalent to a well-known modern identity:[81]

$$\sin^2 x + \cos^2 x = 1$$

Using the Equations 0.3 and 0.5, we can calculate the chord function for 1½° and ¾°, but we cannot calculate it for 1°. PTOLEMY derived the value of Crd 1° by interpolation from the values Crd 1½° = 1;34,15 and Crd ¾° = 0;47,8. PTOLEMY's interpolation is based on a theorem that we assume to have been already known to ARISTARCHUS:[82]

[0.6] $$\frac{\mathrm{Crd}\,\alpha}{\mathrm{Crd}\,\beta} < \frac{\alpha}{\beta} \quad \text{or} \quad \frac{\sin\alpha}{\sin\beta} < \frac{\alpha}{\beta} \quad \text{applies to} \quad \frac{\pi}{2} > \alpha > \beta$$

The interpolation was carried out by PTOLEMY using the geometric construction shown in Figure 0.7.[83] Line BD bisects the angle $\widehat{ABG}$. The arc $\widehat{HET}$ is centred at point D.

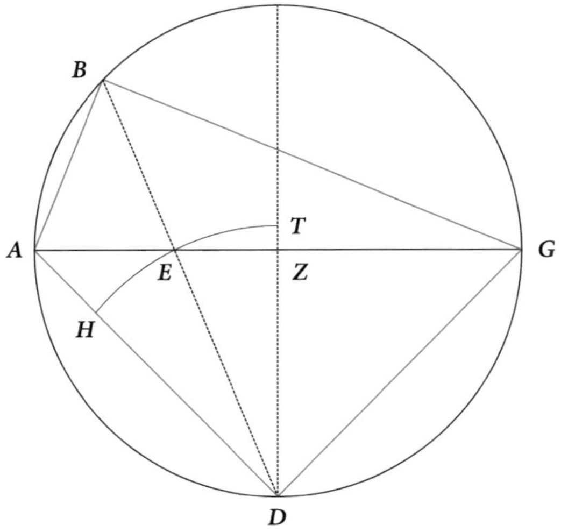

Figure 0.7 The geometric construction used by PTOLEMY to interpolate or prove Equation 0.6. Source: Halma, 1813, p.34.

81 Bulmer-Thomas, 1939, Selections illustrating the history of Greek mathematics, II, pp.420–423 and pp.431–434. The expression PTOLEMY calculated using the Chord function was still quite complex, but equivalent to the modern expression. The expression in its modern form dates back to the 6th century and was coined by VARĀHAMIHIRA.

82 Heath, 1921, II, p.281, also Bulmer-Thomas, 1939, II, pp.434–443, or Pedersen, 2011, pp.60–63.

83 Halma, 1813, p.34. For an explanation or demonstration, see Bulmer-Thomas, 1939, II, pp.435–439.

Using the rules of geometry, we can derive the following expression:[84]

[0.7] $$\frac{GB}{BA} < \frac{\overset{\frown}{GB}}{\overset{\frown}{BA}}$$

Equation 0.6 corresponds to the theorem above (Equation 0.7). Plugging in the values for Crd 1° and Crd 1½° and Crd ¾°, respectively, gives:

$$\frac{\text{Crd } 1^\circ}{\text{Crd } \frac{3}{4}^\circ} < \frac{1}{\frac{3}{4}} \qquad \text{and} \qquad \frac{\text{Crd } 1\frac{1}{2}^\circ}{\text{Crd } 1^\circ} < \frac{1\frac{1}{2}}{1}$$

After we rearrange the expression we get:

$$\frac{4}{3}\text{Crd}\frac{3}{4}^\circ > \text{Crd } 1^\circ > \frac{2}{3}\text{Crd } 1\frac{1}{2}^\circ \qquad \text{or} \qquad 1;2{,}50 > \text{Crd } 1^\circ > 1;2{,}50$$

The value 1;2,50 is a very good approximation for Crd 1°.[85] From this, PTOLEMY finally calculated Crd ½° (Equation 0.5). He then used his values for Crd 1° = 1;2,50 and Crd ½° = 0;31,25 for further calculations. With these values, we can calculate the values of the chord function for each ½°.[86]

Example 0.4 Using given values of Crd 90°, Crd 72° and Crd 1°, calculate Crd 8½° (solution: Crd 8½° = 8;53,35).

From Crd 90° and Crd 72°, we first calculate Crd 18° using Equation 0.3:

$$\text{Crd } 18^\circ = \frac{\text{Crd } 90^\circ\, \text{Crd}(180^\circ - 72^\circ) - \text{Crd } 72^\circ\, \text{Crd}(180^\circ - 90^\circ)}{120}$$

$$= \frac{\text{Crd } 90^\circ\, \text{Crd } 108^\circ - \text{Crd } 72^\circ\, \text{Crd } 90^\circ}{120} = 18;46{,}19$$

For this calculation, we also need Crd 108°, which is calculated using Equation 0.4:

$$\text{Crd}^2(180^\circ - 72^\circ) = 120^2 - \text{Crd}^2 72^\circ \quad \text{or} \quad \text{Crd } 108^\circ = \sqrt{120^2 - \text{Crd}^2 72^\circ} = 97;4{,}55$$

To calculate Crd 9°, we also need Crd 162°, which is calculated in the same way as Crd 108° above.

$$\text{Crd}^2(180^\circ - 18^\circ) = 120^2 - \text{Crd}^2 18^\circ \quad \text{or} \quad \text{Crd } 162^\circ = \sqrt{120^2 - \text{Crd}^2 18^\circ} = 118;31{,}21$$

Then, using Equation 0.5, calculate Crd 9°,

$$\text{Crd } 9^\circ = \sqrt{60(120 - \text{Crd}(180^\circ - 18^\circ))} = 9;24;54$$

The final value of Crd 8½° is calculated in the same way as above from Crd 9° and Crd ½° using Equation 0.3. The result is:

$$\text{Crd } 8\tfrac{1}{2}^\circ = 8;53;34$$

84 Heath, 1921, II, pp.281–282, also Bulmer-Thomas, 1939, II, pp.437–439.

85 The value 1;2,50 converted to decimal notation is precise to three decimal places =1.04722; Crd 1° = 1.04718.

86 For PTOLEMY's trigonometry, see also Heath, 1921, II, pp.276–286. For a summary of Greek trigonometry, see also Bulmer-Thomas, 1939, II, pp.406–463.

An example of some other values of the chord function that follow from the base angles (Figure 0.6) is shown in Table 0.1.

arc ⊳	Chord			arc ⊳	Chord		
36	37; 4,55	EUCLID XIII.9		54	54;28,44	Equation 0.5	$54° = {}^{108}/_2°$
60	60; 0, 0	= radius, R		126	106;55,15	Equation 0.4	$126° = 180° - 54°$
72	70;32, 3	EUCLID XIII.10		135	*110;51,56	Equation 0.4	$135° = 180° - 45°$
90	84;51,10	Equation 0.2		150	115;54,40	Equation 0.4	$150° = 180° - 30°$
120	103;55,23	Equation 0.2		162	*118;31, 2	Equation 0.4	$162° = 180° - 18°$
108	97; 4,55	Equation 0.4	$108° = 180° - 72°$	9	9;24,54	Equation 0.5	$9° = {}^{18}/_2°$
144	*114; 7,36	Equation 0.4	$144° = 180° - 36°$	15	15;39,47	Equation 0.5	$15° = {}^{30}/_2°$
18	*18;46,20	Equation 0.5	$18° = {}^{36}/_2°$	22½	23;24,39	Equation 0.5	$22\frac{1}{2}° = {}^{45}/_2°$
30	31; 3,30	Equation 0.5	$30° = {}^{60}/_2°$	27	28; 0,48	Equation 0.5	$27° = {}^{54}/_2°$
45	45;55,19	Equation 0.5	$45° = {}^{90}/_2°$				

Table 0.1 Example of the initial steps in the construction of a table with the values of the chord function, radius $R = 60$. * Values slightly differ from PTOLEMY's values due to rounding. For PTOLEMY's values see Toomer, 1998, pp.57–60 or *Almagest* I.11.

arc ⊳	Chord	arc ⊳	Chord	arc ⊳	Chord
6	6;16,49	66	65;21,24	126	106;55,15
12	12;32,36	72	70;32, 3	132	109;37,32
18	*18;46,20	78	*75;31, 6	138	112; 1,47
24	24;56,58	84	*80;17,44	144	*114; 7,36
30	31; 3,30	90	84;51,10	150	115;54,40
36	37; 4,55	96	89;10,39	156	117;22,40
42	43; 0,15	102	93;15,27	162	118;31,21
48	48;48,30	108	*97; 4,55	168	*119;20,33
54	54;28,44	114	100;38,26	174	119;50, 8
60	60; 0, 0	120	103;55,23	180	120; 0, 0

Table 0.2 Chord values for arcs multiples of 6°. Values are calculated by Equation 0.3 by using the values from Table 0.1, see Example 0.4. * Values slightly differ from PTOLEMY's values due to rounding. For PTOLEMY's values see Toomer, 1998, pp.57–60 or *Almagest* I.11. For an additional explanation see Bulmer-Thomas, 1939, pp.423–435 or Pedersen, 2011, pp.56–65.

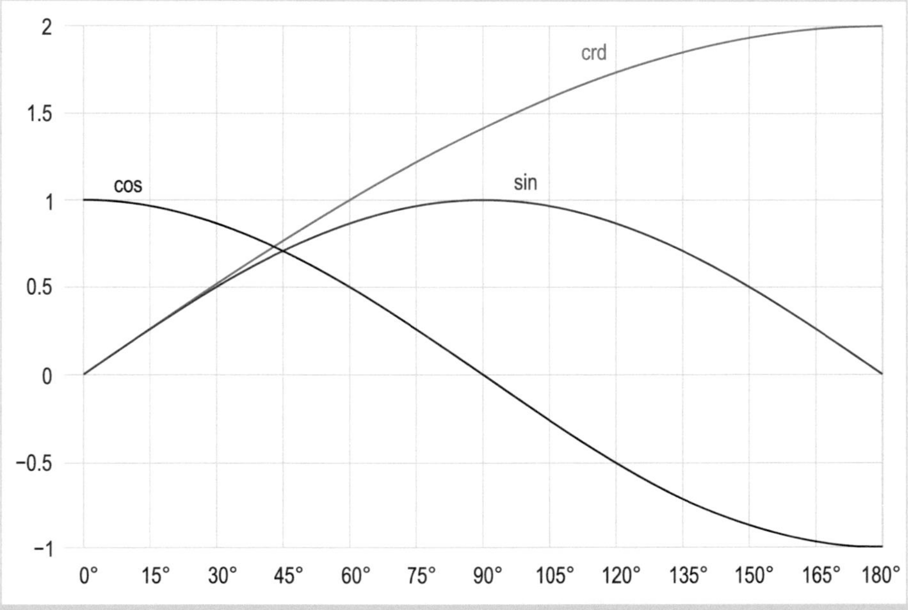

Figure 0.8 The sine and cosine functions and the chord function for the radius of a circle $R = 1$ at an interval from 0 to 180° or from 0 to π radians.

arc ⊳	Chord
3	3; 8,28
9	9;24,54
15	15;39,47
21	21;52, 6
27	28; 0,48
33	34; 4,55
39	40; 3,25
45	45;55,19
51	*51;39,41
57	57;15,33
63	*62;41,59
69	*67;58, 7
75	73; 3, 5
81	77;56, 2
87	82;36, 9

Table 0.3 Chord values for arcs multiples of 3°. Values are calculated by Equation 0.5 by using the values from Table 0.2, see Example 0.4. * Values slightly differ from PTOLEMY's values due to rounding. For PTOLEMY's values see Toomer, 1998, pp.57–60 or *Almagest* I.11. For an additional explanation see Pedersen, 2011, pp.56–65.

Theorem of Menelaus

Spherical trigonometry was finally established by MENELAUS OF ALEXANDRIA in the 1st century CE with the book *Sphaerics*, the original of which is lost, but manuscripts in Arabic have been preserved. In *Sphaerics*, he published and proved several theorems of great importance for the development of spherical astronomy, and we assume that much of the material published in *Sphaerics* was already well known at the time of the book's publication.[87] The work was an important source for PTOLEMY. It is likely that both MENELAUS and PTOLEMY covered knowledge from an older lost work, whose author may have been HIPPARCHUS.[88]

Most of the theorems of spherical trigonometry are summarised in the third book of *Sphaerics*.[89] Among the most important is statement III.1 or MENELAUS' theorem, which states that for a geometric construction of arcs on great circles[90] on a sphere where the two larger arcs $\overset{\frown}{ADB}$ and $\overset{\frown}{AEC}$ coincide with two other arcs $\overset{\frown}{DZC}$ and $\overset{\frown}{BCE}$ at point Z, and all four arcs are smaller than a semicircle, the following applies (Figure 0.9):[91]

$$\frac{\mathrm{Crd}\, 2\overset{\frown}{AC}}{\mathrm{Crd}\, 2\overset{\frown}{AE}} = \frac{\mathrm{Crd}\, 2\overset{\frown}{CD}}{\mathrm{Crd}\, 2\overset{\frown}{DZ}} \cdot \frac{\mathrm{Crd}\, 2\overset{\frown}{BZ}}{\mathrm{Crd}\, 2\overset{\frown}{BE}} \tag{0.8}$$

An equivalent modern record with the sine function is:[92]

$$\frac{\sin \overset{\frown}{AC}}{\sin \overset{\frown}{AE}} = \frac{\sin \overset{\frown}{CD}}{\sin \overset{\frown}{DZ}} \cdot \frac{\sin \overset{\frown}{BZ}}{\sin \overset{\frown}{BE}}$$

and the following also applies:

$$\frac{\mathrm{Crd}\, 2\overset{\frown}{ZC}}{\mathrm{Crd}\, 2\overset{\frown}{DZ}} = \frac{\mathrm{Crd}\, 2\overset{\frown}{CE}}{\mathrm{Crd}\, 2\overset{\frown}{AE}} \cdot \frac{\mathrm{Crd}\, 2\overset{\frown}{AB}}{\mathrm{Crd}\, 2\overset{\frown}{BD}} \tag{0.9}$$

or

$$\frac{\sin \overset{\frown}{ZC}}{\sin \overset{\frown}{DZ}} = \frac{\sin \overset{\frown}{CE}}{\sin \overset{\frown}{AE}} \cdot \frac{\sin \overset{\frown}{AB}}{\sin \overset{\frown}{BD}}$$

87 For some comments on earlier works see Berggren, Thomas, 1996, Euclid's Phaenomena, pp.1–18.

88 HIPPARCHUS knew the work of EUDOXUS and EUCLID. It is likely that he was also familiar with spherical trigonometry.

89 Van Brummelen, 2009, pp.56–63, and Neugebauer, 1975, A History of Ancient Mathematical Astronomy, pp.26–30.

90 Great circles are circles that lie on a plane passing through the centre of a sphere.

91 Björnbo, 1902, Studien über Menelaos' Sphärik, p.89. The original text of MENELAUS' theorem in *Sphaerics* is not known, since the texts in various preserved manuscripts do not match in this part. For PTOLEMY's calculation of the declination for a certain longitude, see Neugebauer, 1975, pp.30–31, or Toomer, 1998, pp.69–70 and 72, *Almagest* I.14 and I.15 (table of declination values). For the right ascension, see Neugebauer, 1975, pp.31–32, or Toomer, 1998, pp.71, 73–74, 100, *Almagest* I.16 and II.8 (table with right ascension values).

92 At the time of PTOLEMY the sine function was not known. Modern reference to the sine function is given for the reader's convenience only.

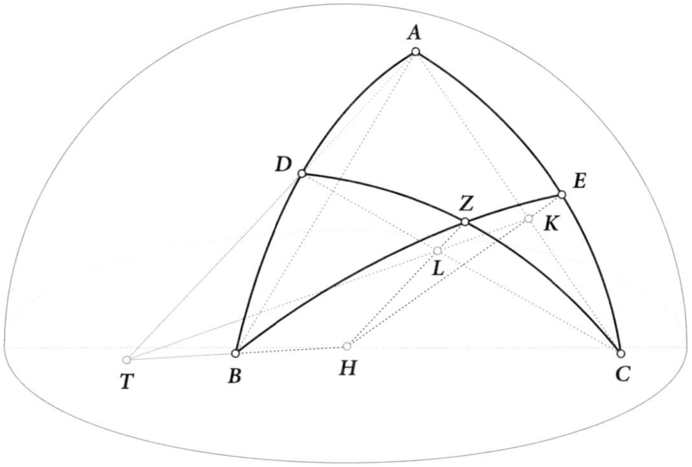

Figure 0.9 MENELAUS' geometric construction, where H is the centre of the sphere. Summarised from Björnbo, 1902, p.89, Figure 48.

The Latin translation by GERARD OF CREMONA reads:[93]

$$\frac{\text{Crd }2\overset{\frown}{CE}}{\text{Crd }2\overset{\frown}{EA}}=\frac{\text{Crd }2\overset{\frown}{CZ}}{\text{Crd }2\overset{\frown}{AB}}\cdot\frac{\text{Crd }2\overset{\frown}{DB}}{\text{Crd }2\overset{\frown}{ZD}}$$

MENELAUS' theorem also holds that if we replace the outer arc $\overset{\frown}{AC}$ with $\overset{\frown}{AB}$, we have to replace the inner arcs as well. In this case:[94]

[0.10]

$$\frac{\text{Crd }2\overset{\frown}{AB}}{\text{Crd }2\overset{\frown}{AD}}=\frac{\text{Crd }2\overset{\frown}{CZ}}{\text{Crd }2\overset{\frown}{CD}}\cdot\frac{\text{Crd }2\overset{\frown}{BE}}{\text{Crd }2\overset{\frown}{ZE}}$$

or

$$\frac{\sin\overset{\frown}{AB}}{\sin\overset{\frown}{AD}}=\frac{\sin\overset{\frown}{CZ}}{\sin\overset{\frown}{CD}}\cdot\frac{\sin\overset{\frown}{BE}}{\sin\overset{\frown}{ZE}}$$

and[95]

[0.11]

$$\frac{\text{Crd }2\overset{\frown}{BZ}}{\text{Crd }2\overset{\frown}{ZE}}=\frac{\text{Crd }2\overset{\frown}{AC}}{\text{Crd }2\overset{\frown}{CE}}\cdot\frac{\text{Crd }2\overset{\frown}{DB}}{\text{Crd }2\overset{\frown}{AD}}$$

or

$$\frac{\sin\overset{\frown}{BZ}}{\sin\overset{\frown}{ZE}}=\frac{\sin\overset{\frown}{AC}}{\sin\overset{\frown}{CE}}\cdot\frac{\sin\overset{\frown}{DB}}{\sin\overset{\frown}{AD}}$$

93 Björnbo, 1902, p.89, f.152a.
94 Neugebauer, 1975, p.28, MENELAUS' theorem I.
95 Neugebauer, 1975, p.28, MENELAUS' theorem II.

MENELAUS' theorem is particularly important because with it we can convert equatorial coordinates into ecliptic coordinates, or vice versa.[96] We can also use it to derive further expressions that are common in spherical astronomy.

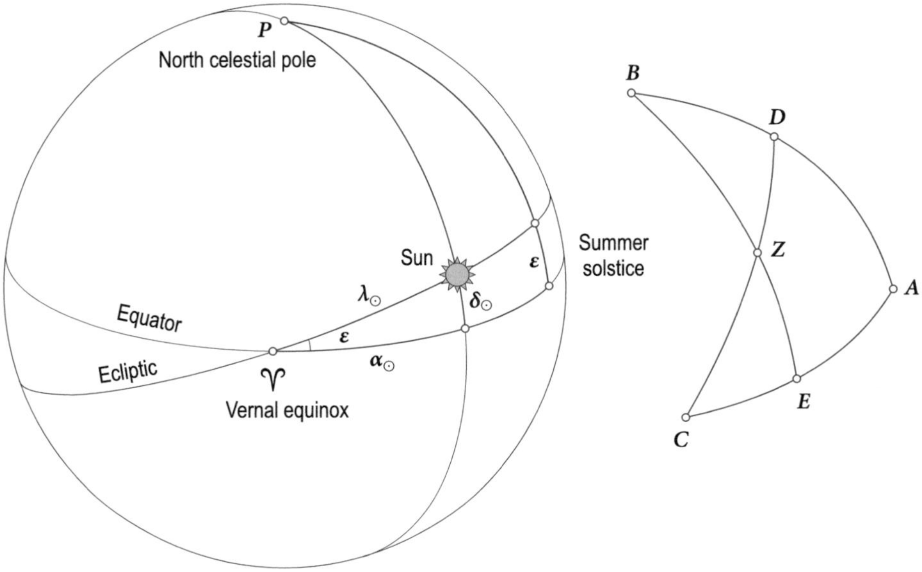

Figure 0.10 The use of MENELAUS' theorem in spherical astronomy.

Also interesting is proposition III.2 where, for two spherical triangles $\triangle ABC$ and $\triangle DEZ$ (Figure 0.11), if the conditions are satisfied:[97]

$$\widehat{BAC} = \widehat{EDZ} \quad \text{and} \quad \begin{cases} \widehat{BCA} = \widehat{EZA} \\ \widehat{BCA} + \widehat{EZA} = 180° \end{cases}$$

the following applies:

$$\frac{\text{Crd}\, 2\overset{\frown}{AB}}{\text{Crd}\, 2\overset{\frown}{BC}} = \frac{\text{Crd}\, 2\overset{\frown}{DE}}{\text{Crd}\, 2\overset{\frown}{EZ}} \quad \text{or} \quad \frac{\sin \overset{\frown}{AB}}{\sin \overset{\frown}{BC}} = \frac{\sin \overset{\frown}{DE}}{\sin \overset{\frown}{EZ}} \tag{0.12}$$

For example, this theorem can be used to determine the relationship between the declination, longitude and obliquity of the ecliptic more easily than MENELAUS' theorem.

96 Neugebauer, 1975, pp.26–32, and Nadal et al., 2004.

97 Björnbo, 1902, pp. 92–93 and 97–98. REGIOMONTANUS is said to have had considerable difficulty in understanding this theorem in his study of the *Sphaerics*, see pp.97–98.

Writing the Equation 0.12 as a sine function for the triangles $\triangle ABC$ and $\triangle DEZ$ (Figure 0.11) gives:

$$\frac{\sin \lambda_\odot}{\sin \delta_\odot} = \frac{\sin 90°}{\sin \varepsilon} \qquad \text{or} \qquad \sin \delta_\odot = \sin \lambda_\odot \sin \varepsilon$$

The same expression is obtained using NAPIER's rules for the triangle $\triangle \Upsilon BC$ defined by the vernal equinox, the Sun, and the Sun's projection on the equator (Appendix D).

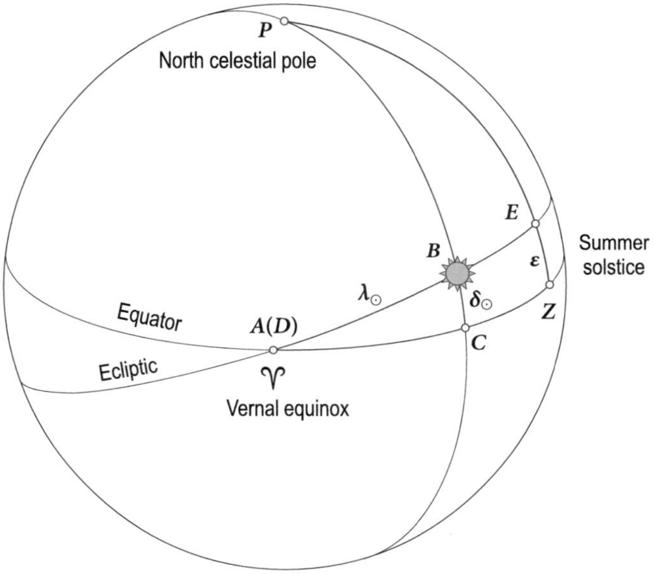

Figure 0.11 An example of the application of proposition III.2 of MENELAUS' *Sphaerics*, in spherical astronomy.

Theorem III.5 implies the now forgotten theorem of double (sine) ratios.[98] The theorem deals with the so-called anaharmonic property of four great circles intersecting at one point (for example, pole P) and intersected by two other great circles $\bigcirc ABCD$ and $\bigcirc EFGH$ (Figure 0.12), where:[99]

[0.13] $$\frac{\operatorname{Crd} 2\widehat{AD}}{\operatorname{Crd} 2\widehat{DC}} \cdot \frac{\operatorname{Crd} 2\widehat{BC}}{\operatorname{Crd} 2\widehat{AB}} = \frac{\operatorname{Crd} 2\widehat{EH}}{\operatorname{Crd} 2\widehat{HG}} \cdot \frac{\operatorname{Crd} 2\widehat{FG}}{\operatorname{Crd} 2\widehat{EF}}$$

or using the sine function:

$$\frac{\sin \widehat{AD}}{\sin \widehat{DC}} \cdot \frac{\sin \widehat{BC}}{\sin \widehat{AB}} = \frac{\sin \widehat{EH}}{\sin \widehat{HG}} \cdot \frac{\sin \widehat{FG}}{\sin \widehat{EF}}$$

98 Björnbo, 1902, pp.95–102.
99 Heath, 1921, II, pp.269–270.

Example 0.5 The longitude of the Sun $\lambda_\odot$ at point B is 15°, see Figure 0.12. Calculate the longitude of the Sun at point C if the arc $\widetilde{FG}$ or increment of right ascension $\Delta\alpha_\odot$ is 4 hours or 60°. The obliquity of the ecliptic is 23;15° (solution: $\lambda_\odot \approx 75°$).

From MENELAUS' theorem (Equation 0.9) and Figure 0.10:

$$\frac{\sin\lambda_\odot}{\sin(90° - \lambda_\odot)} = \frac{\sin\alpha_\odot}{\sin(90° - \alpha_\odot)} \cdot \frac{\sin 90°}{\sin(90° - \varepsilon)} \quad \text{or} \quad \frac{\sin\lambda_\odot}{\cos\lambda_\odot} = \frac{\sin\alpha_\odot}{\cos\alpha_\odot} \cdot \frac{1}{\cos\varepsilon}$$

Rearranging the above expression, the right ascension at a given longitude and the obliquity of the ecliptic is:

$$\begin{aligned} \tan\alpha_\odot &= \tan\lambda_\odot \cos\varepsilon = \tan 15° \cos 23°51' \\ \alpha_\odot &\approx 13°46' \end{aligned}$$

For the example from spherical astronomy shown in Figure 0.12:

$$\frac{\sin 90°}{\sin(90° - \lambda_\odot - \Delta\lambda_\odot)} \cdot \frac{\sin\Delta\lambda_\odot}{\sin\lambda_\odot} = \frac{\sin 90°}{\sin(90° - \Delta\alpha_\odot - \alpha_\odot)} \cdot \frac{\sin 60°}{\sin\alpha_\odot}$$

Inserting the values for longitude and right ascension gives:

$$\frac{\sin 90°}{\sin(75° - \Delta\lambda_\odot)} \cdot \frac{\sin\Delta\lambda_\odot}{\sin 15°} = \frac{\sin 90°}{\sin 16°14'} \cdot \frac{\sin 60°}{\sin 13°46'}$$

The denominator on the left is transformed using the addition theorem, then the expression is rearranged to give $\Delta\lambda_\odot \approx 60°$. The new longitude $\lambda_\odot$ is therefore approximately 75°.

It is not entirely clear whether MENELAUS' theorem was developed by MENELAUS or was known earlier, but the theorem of double ratios was very likely already widely known in MENELAUS' time. It is possible that the theorem was known and used by HIPPARCHUS, and according to some speculations, the theorem of double ratios may have been used by APOLLONIUS.[100]

100 Björnbo, 1902, p.99 and p.101, note f.173.

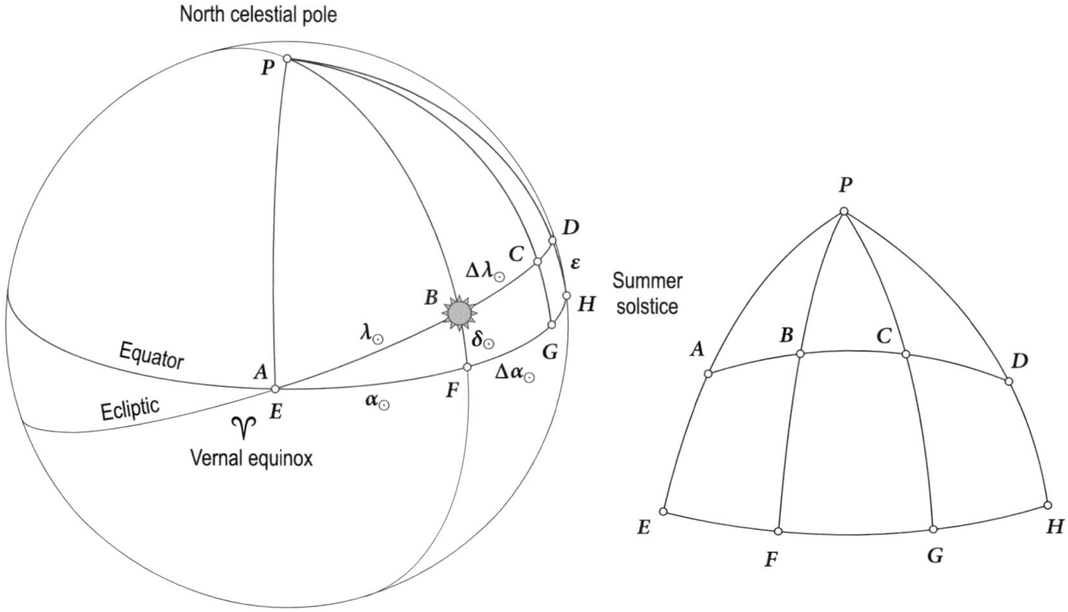

Figure 0.12 MENELAUS' proposition III.5, or the double-sine theorem, and a possible application to spherical astronomy. Based on Björnbo, 1902, p.99, Figure 56.

Oblique ascension and the length of the day

Among the most important tasks of ancient astronomy was determining the length of the day, being the time between sunrise and sunset. The oldest known calculation is the record on the *Ramesside papyrus*, in which the length of the day changes linearly.[101] The method used in Babylon was probably similar; it is seen on tables written around 400 BCE.[102] The first systematic treatise in which the calculation of the length of the day forms a good part of the book is EUCLID's *Phaenomena*, in which he deals with the topic in propositions 14 to 18.[103]

In antiquity, it was an important problem to calculate the rising time of the arc or the oblique ascension $\theta = \widehat{\Upsilon C}$ (Figure 0.13), which was used as a powerful tool by PTOLEMY who also tabulated the values of the oblique ascension in the *Almagest* and in the *Handy Tables* (Figures 0.15 and 0.16).[104]

101 Neugebauer, 1975, p.706. The *Ramesside papyrus* was probably written in the 12th century BCE.
102 Neugebauer, 1975, pp.709–710.
103 Berggren, Thomas, 1996, pp.97–113.
104 Halma, 1822, pp.2–65, Tihon, Mercier, 2011, 1b, pp.14–41 and Toomer, 1998, pp.100–103.

[0.14] $$\theta = \widehat{\Upsilon C} = \widehat{\Upsilon E} - n \qquad \text{or} \qquad \theta = \alpha_{\odot} - n$$

$\alpha_{\odot}$ right ascension
n ascensional difference

If the observer is not at the equator, the right ascension and oblique ascension differ by $n = \widehat{CE}$ (Figure 0.13). This difference is referred to in Arabic literature as the *equation of daylight*, and in medieval literature as the *ascensional difference*, and it depends on the observer's latitude (Figure 0.14):[105]

[0.15] $$\sin n = \tan\varphi \tan\delta_{\odot}$$

φ latitude
$\delta_{\odot}$ declination of the Sun

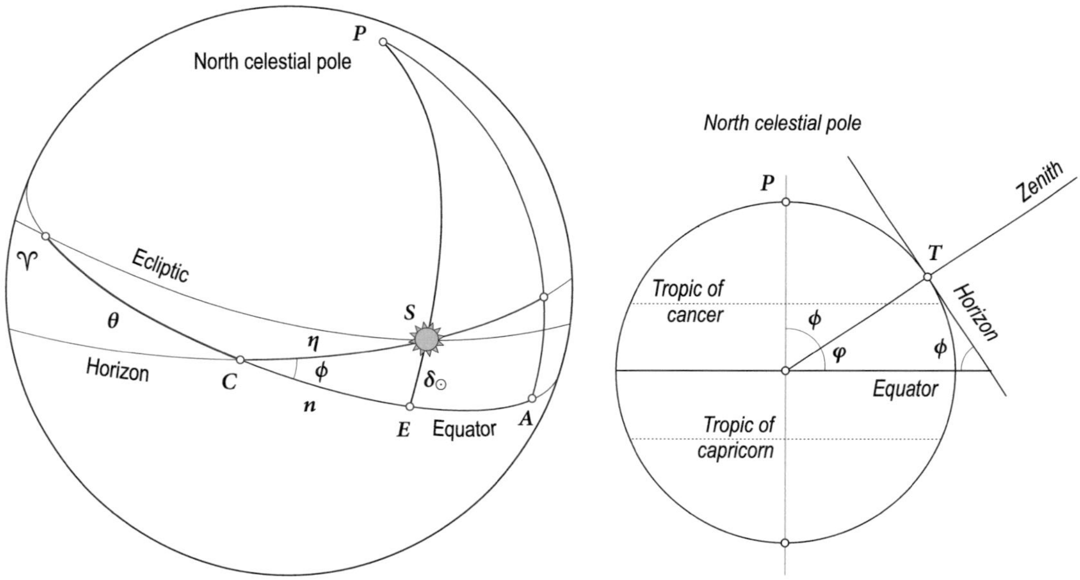

Figure 0.13 Oblique ascension θ, ascensional difference n and ortive amplitude η. Based on Evans, 1998, p.118, Figure 2.25 and Neugebauer, 1975, III, p.1215 Figure 30.

105 Neugebauer, 1975, p.36 also Pedersen, 2011, pp.110–113.

The expression for the ascensional difference is derived using MENELAUS' theorem (Equation 0.11 and Figure 0.13) where for the quadrilateral *CAPS*, if $\widehat{AC} = 90°$, we get:[106]

$$\frac{\sin(90° - \delta_\odot)}{\sin \delta_\odot} = \frac{\sin 90°}{\sin n} \cdot \frac{\sin(90° - \phi)}{\sin \phi} \qquad \text{or} \qquad \frac{\cos \delta_\odot}{\sin \delta_\odot} = \frac{1}{\sin n} \cdot \frac{\cos \phi}{\sin \phi}$$

Taking $\phi = 90° - \varphi$ into account, we obtain the following expression, which is equivalent to Equation 0.15:

$$\sin n = \frac{\sin \delta_\odot}{\cos \delta_\odot} \cdot \frac{\sin \varphi}{\cos \varphi}$$

PTOLEMY tabulated the values of oblique ascension for different latitudes or reference parallels, the so-called *climata*[107], in the *Handy Tables*. A geographical scheme using different reference parallels was also used by HIPPARCHUS and ERATOSTHENES.[108] PTOLEMY tabulated the values of oblique ascension for the following *climata*:[109]

1st clima	Meroë
2nd clima	Asuan, Gr. Syëne
3rd clima	Lower Egypt
4th clima	Rhodes
5th clima	Dardanelles, Gr. Hellespont
6th clima	Middle Pontus
7th clima	Scythia, mouth of the Dnieper, Gr. Borysthenes

He gives a comprehensive overview of the reference parallels in chapter 6 of the second book of the *Almagest*.[110] All the parallels given by PTOLEMY in the *Almagest*, for which he also gave the shadow length on the longest day of the year, are shown in Table 0.4. For a given parallel the duration in equinoctial hours of the longest day of the year is also given in the table.[111]

106 Neugebauer, 1975, pp.34–35.

107 The word *clima* Gr. *κλίμα* was first used, as far as we know, by HYPSICLES in his *Anaphoricos*, or *On the Rising and Setting of the Stars*, Hypsicles, 1966, De Falco, Krause (trans.), p.36, line 63, and in HIPPARCHUS' *Commentary on the Phainomena of Eudoxus and Aratus*, Shcheglov, 2003, p.2. The word *climata* could in a modern sense be called a typical or reference parallel.

108 Only fragments of ERATOSTHENES' work survive. Roller, 2010, Eratosthenes' Geography, pp.75–79. For more information on the parallels of ERATOSTHENES' work, see also Strabo, 2007, Geographica, 2.3.35–38. Also Shcheglov, 2003, and Shcheglov, 2007.

109 Tihon, Mercier, 2011, 1b, pp.14–41 also Halma, 1822, pp.2–65.

110 Halma, 1813, pp.76–89. For an overview in tabular form, see Pedersen, 2011, p.108. Also Toomer, 1998, pp.82–90. PTOLEMY also gives the lengths of the longest day of the year and, for locations up to and including 25, the lengths of the solstice shadows. See also Graßhoff et al., 2013, Of paths and places: the origin of Ptolemy's Geography.

111 For additional comments see Neugebauer, 1959b, Ptolemy's Geography, Book VII, Chapters 6 and 7.

1	Equator	0°	12h	Day length does not vary throughout the year.
2	Taprobane, Ceylon	4;15°	12;15h	The earliest known reference is in ERATOSTHENES' *Geographia*, but the island was little known to the Greeks at the time.
3	Avalite Bay, Gulf of Aden	8;25°	12;30h	Today Zeila Somalia at the mouth of the Red Sea near the border with Djibouti. Ancient Avalites was an important port on the shores of the Red Sea. Its inhabitants were the ancestors of today's Somalis, and their homeland stretched from Aswan to present-day Tanzania.
4	Adulis, Red Sea coast	12;30°	12;45h	Now Zula Bay, modern town of Zula near Massawa, Eritrea
5	Meroë, Sudan	16;27°	13h	Administrative centre of the Kingdom of Kush (1500–390 BCE), capital after the fall of Napata (c.590 BCE).
6	Napata, Sudan	20;14°	13;15h	For a time, the capital of the 25th dynasty in Egypt and, after its fall in 663 BCE, the capital of the kingdom of Kush.
7	Syene	23;51°	13;30h	Modern Aswan.
8	Thebaid, Hermeiou	27;12°	13;45h	Today the city of El-Mansha, Thebes, Egypt
9	Lower Egypt	30;22°	14h	
10	Phoenicia	33;18°	14;15h	
11	Rhodes	36;0°	14;30h	
12	Smyrna, Turkey	38;35°	14;45h	Today Izmir, Turkey. A port on the Aegean coast of Anatolia. Its favourable location led to a great boom in antiquity.
13	Hellespont	40;56°	15h	Dardanelles, Bosporus.
14	Massalia	43;1°	15;15h	A Greek colony, today's Marseille.
15	Middle of Pontus	45;1°	15;30h	Pontus was an ancient province in north-eastern Anatolia along the southern coast of the Black Sea.
16	Istros	46;51°	15;45h	The sources of the Danube.
17	Borysthenes	48;32°	16h	The mouth of the Dnieper River, present-day Ukraine.

Table 0.4/1 PTOLEMY's reference parallels or climata and length of the longest day of the year. Taken from Halma, 1813, reprinted 1927, pp.76–89 and Toomer, 1998, pp.82–90.

18	Lake Maiotis	50;4°	16;15 h	The Sea of Azov.
19	Southern Britannia	51;30°	16;30 h	
20	Mouth of Rhine	52;50°	16;45 h	
21	Tanais	54;1°	17 h	An ancient Greek city in the river Don delta. A meeting point of Greek culture and steppe peoples. It was protected by a large system of fortifications.
22	Brigantia, Brittania	55°	17;15 h	The region of Brigantia, inhabited in antiquity by British Celtic people. Now part of north-central England.
23	Middle of Brittania	56°	17;30 h	
24	Cataractonium, Brittania	57°	17;45 h	A Roman fort and settlement in North Yorkshire. Today the city of Catterick.
25	Little Brittania	58°	18 h	Ireland
26	Middle of Little Brittania	59;30°	18;30 h	
27	Northern Little Brittania	61°	19 h	
28	Eboudae	62°	19;30 h	Hebrides Islands
29	Thule	63°	20 h	PTOLEMY states that the Thule Islands were the Shetland Islands.
30	Unknown Scythian peoples	64;30°	21 h	
31		65;30°	22 h	
32		66°	23 h	
33	Arctic Circle	66;8;40°	24 h	
34		67°		
35		69;30°		
36		73;20°		
37		78;20°		
38		84°		
39	North Pole	90°		

Table 0.4/2 PTOLEMY's reference parallels or climata and length of the longest day of the year. Taken from Halma, 1813, reprinted 1927, pp.76–89 and Toomer, 1998, pp.82–90.

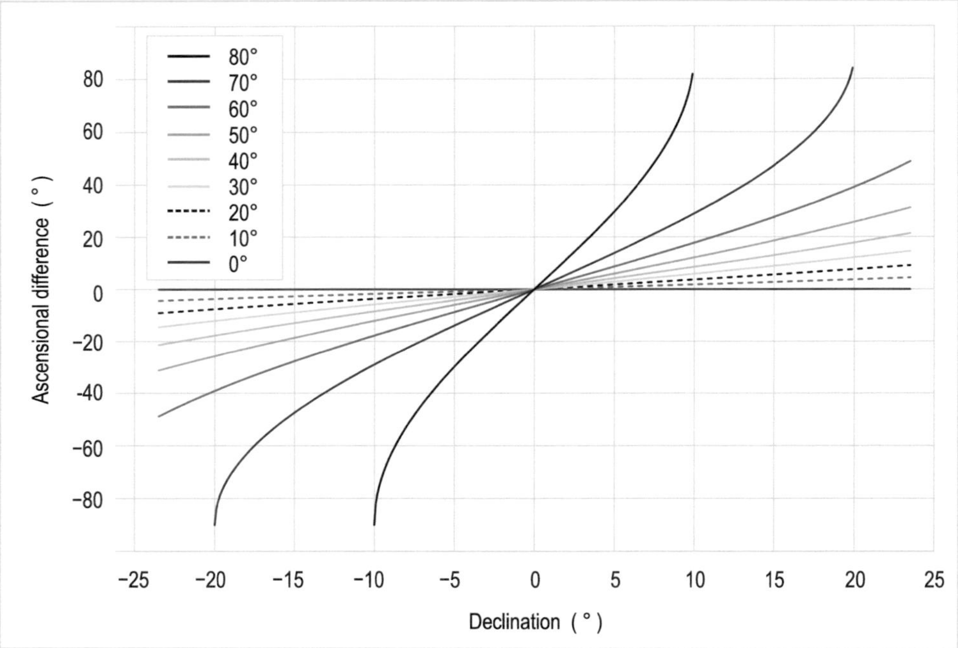

Figure 0.14 Ascensional difference as a function of declination for different latitudes. Calculated by Equation 0.15.

With the help of the oblique ascension, we can calculate the day length for locations in different latitudes, for which we need the values of oblique ascension at sunrise and sunset. Instructions for calculating the length of the day by means of the difference of ascensions were given by Ptolemy in the *Almagest* II.9 (H142), and he gave the following expression for the length of the day (Figures 0.17 and 0.18):[112]

[0.16] $$t = \theta_1(\lambda_{\odot} + 180°, \varphi) - \theta_2(\lambda_{\odot}, \varphi)$$

θ_1 oblique ascension for longitude $\lambda_{\odot} + 180°$ and latitude φ
θ_2 oblique ascension for longitude $\lambda_{\odot}$ and latitude φ

The calculated values of the length of the day are in degrees, which are converted into time units by a factor of $15° \equiv 1\,\text{h}$ or $1° \equiv 4\,\text{min}$, from which the term *"time-degree"* is also derived.[113]

112 Toomer, 1998, p.99. Pedersen, 2010, pp.102 and 113–114. The values of oblique ascension were tabulated in the *Almagest* and the *Handy Tables*. See Toomer, 1998, pp.100–103. The equation does not take into account the refraction of light in the atmosphere and is therefore approximate (applies to the equator).

113 Hypsicles, 1966, Die Aufgangszeiten der Gestirne, De Falco, Krause (trans.), p.79.

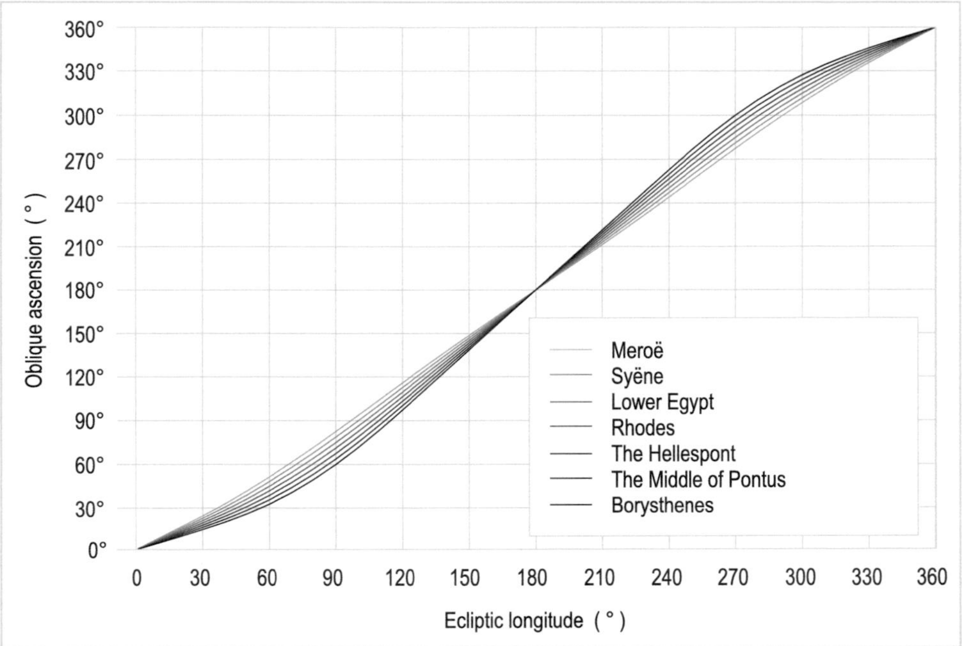

Figure 0.15 Oblique ascension for PTOLEMY's climatas as a function of solar longitude. Data source Tihon, Mercier, 2011, 1b, pp.14–41 with corrections. In addition, the value for the 5th clima at ♑ 30° and the 6th clima at ♎ 23° are corrected, where the true values are apparently 321;9° and 174;47°, see Halma, 1822, p.45.

The division of the day into formal parts was practiced in Ancient India, Mesopotamia and Egypt. With ALEXANDER's conquest of Egypt, the division of the day into 12 and 24 hours became a constant in Greek astronomy. Greek astronomers distinguished between seasonal hours, which had uneven lengths, and equinoctial hours. The light part of the day was also divided into 12 hours in antiquity. The ancient equinox hours, the so-called equinoctial hours, were equivalent in length to one hour as we understand it today, but as the days grew longer, the individual hours, the so-called seasonal hours, also grew longer.

The relationship between seasonal hours, which vary in length, and equinoctial hours, which are approximately equal to the modern hour, is:

[0.17] $$h_e = \frac{h_s t_s}{15°}$$

h_e equinoctial hours (h)
h_s seasonal hours (h)
t_s length of seasonal hours in time-degrees, 15° ≡ equinoctial hour

Example 0.6 Calculate the length of the day in equinoctial hours for the ancient Hellespont, today's Dardanelles, when the Sun has an ecliptic longitude of 90° or ♊ 30°. What is the length of the seasonal hour if the time between sunrise and sunset is divided into 6 hours? The following values for oblique ascension for longitudes of 90° and 270°, respectively, are taken from the *Handy Tables*:[114]

$$\theta_1 = 292;30° \quad \text{and} \quad \theta_2 = 67;30°$$

The length of the day is calculated using Equation 0.16 and, like the oblique ascension values, the calculated length of the day is in time-degrees:

$$t = \theta_1 - \theta_2 = 292;30° - 67;30° = 225°$$

Taking $15° \equiv 1\,\text{h}$ into account, the bright part of the day is 15 equinoctial hours long. That is 7 hours 30 minutes in the morning and 7 hours 30 minutes in the afternoon. Considering that the light part of the day was divided into 12 seasonal hours in antiquity, the length of an ancient seasonal hour in this case is 1 hour and 15 minutes.

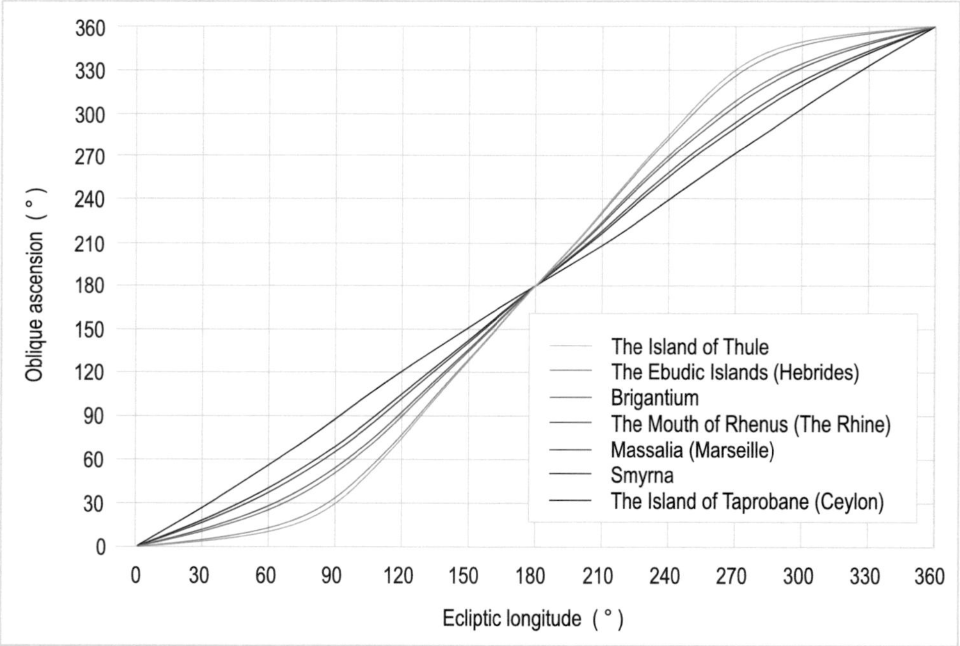

Figure 0.16 Oblique ascension for other locations or PTOLEMY's parallels versus solar longitude. Data source Pedersen, 2011, p.108. The values are calculated using the Equations 0.14 and 0.15.

114 Halma, 1822, pp.35 and 39.

Ortive amplitude

The ortive amplitude η is the angle or arc $\widehat{CS}$ on the horizon between the eastern point and the point of sunrise. In other words, it is the arc on the horizon between the intersection of the horizon with the celestial equator and the ecliptic (Figure 0.13).

Using the ortive amplitude, we can calculate the latitude if we know the length of the day. The expression for the ortive amplitude can also be derived using MENELAUS' theorem.[115] From Equation 0.10 and Figure 0.13:

$$\frac{\sin 90°}{\sin\phi} = \frac{\sin\eta}{\sin 90°}\cdot\frac{\sin 90°}{\sin\delta_\odot} \qquad \text{or} \qquad \frac{1}{\sin\phi} = \frac{\sin\eta}{\sin\delta_\odot}$$

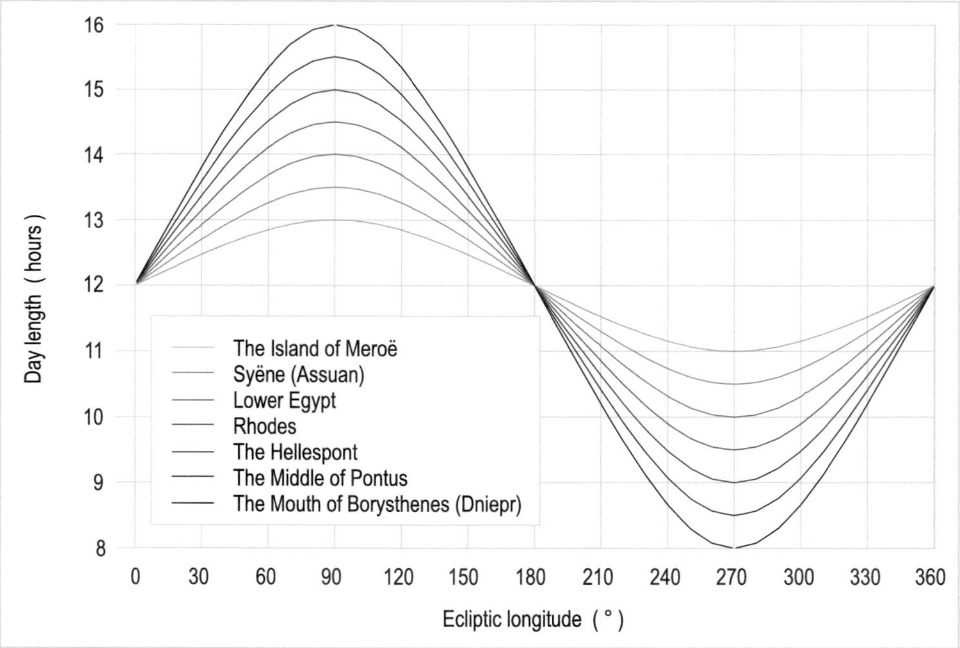

Figure 0.17 Day length for PTOLEMY's climatas as a function of solar longitude, calculated using the oblique ascension given in Tihon, Mercier, 2011, 1b, pp.14–41, with corrections as in Figure 0.15.

115 See Berggren, van Brummelen, 2001, pp.36–37.

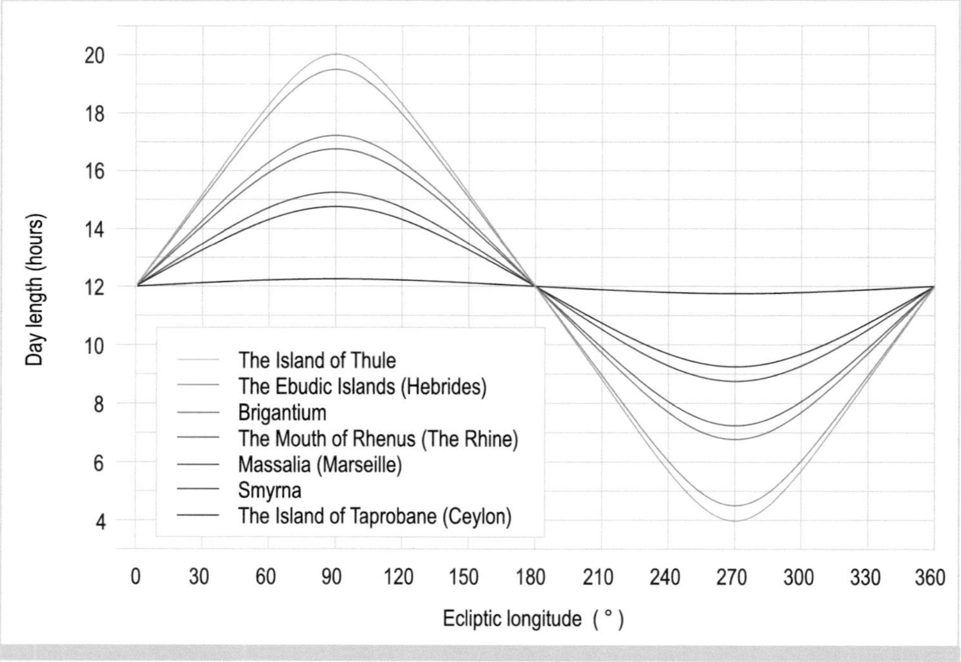

Figure 0.18 Day length for other locations or PTOLEMY's parallels vs. solar longitude. The values are calculated using the Equation 0.16.

Taking into account $\phi = 90° - \varphi$ or $\sin(90° - \varphi) = \cos\varphi$

We get an expression for the ortive amplitude (Figure 0.13):

[0.18] $$\sin\eta = \frac{\sin\delta_{\odot}}{\cos\varphi}$$

The relationship between the latitude and day length follows from MENELAUS' theorem, or Equation 0.9 and Figure 0.13:

$$\frac{\sin\eta}{\sin(90° - \eta)} = \frac{\sin n}{\sin(90° - n)} \cdot \frac{\sin 90°}{\sin(90° - \phi)} \quad \text{or} \quad \frac{\sin\eta}{\cos\eta} = \frac{\sin n}{\cos n \cos\phi}$$

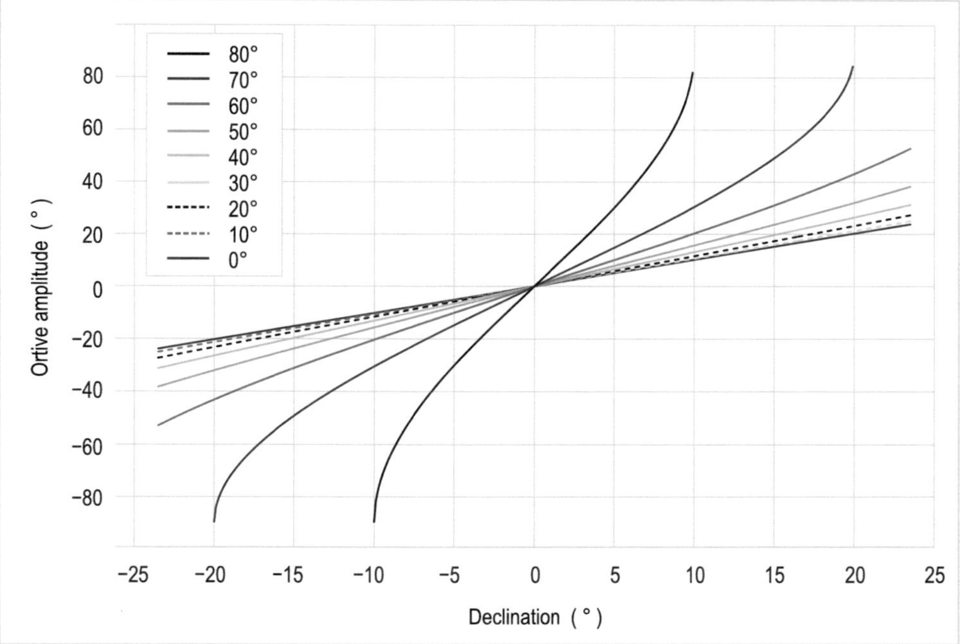

Figure 0.19 Ortive amplitude as a function of declination for different latitudes. Calculated by Equation 0.18.

Taking $n = 90° - \frac{t}{2}$ and $\phi = 90° - \varphi$ or $\sin(90° - \varphi) = \cos\varphi$, where t is the length of the day, gives an expression relating latitude and day length:[116]

[0.19] $$\sin\varphi = \frac{\cos\frac{t}{2}\cos\eta}{\sin\frac{t}{2}\sin\eta} = \cot\frac{t}{2}\cot\eta$$

In the same way, Equation 0.11 gives another equation relating to the length of day, latitude and declination:

$$\frac{\sin(90° - \delta_\odot)}{\sin\delta_\odot} = \frac{\sin 90°}{\sin n}\cdot\frac{\sin(90° - \phi)}{\sin\phi} \qquad \text{or} \qquad \frac{\cos\delta_\odot}{\sin\delta_\odot} = \frac{-\sin\varphi}{\sin n\cos\varphi}$$

Therefore:

[0.20] $$\cos\frac{t}{2} = -\tan\varphi\tan\delta_\odot$$

116 Neugebauer, 1975, p.37.

Example 0.7 Calculate the latitude if the maximum day length is 17 hours and 15 minutes, assuming a declination of 23°51′ (solution: 57°59′).

Latitude is calculated using Equation 0.20, which is rearranged to give:

$$\varphi = \arctan\left(-\frac{\cos\frac{t}{2}}{\tan\delta_\odot}\right) = \arctan\left(\frac{\cos\frac{270°}{2}}{\tan 23°51'}\right) \approx 57°59'$$

The result corresponds to the latitude of Ireland as given by Ptolemy for the 25th parallel or Little Brittania, see Table 0.4.

Library of Alexandria

The Library of Alexandria and its custodians are largely responsible for the preservation of much ancient material, including the sources mentioned in this book. The history of the library is scant, although it was one of the most important cultural institutions of ancient Greece. We know that there were various public and private archives in Egypt long before the founding of Alexandria, and papyrus production had long been perfected. The Library of Alexandria is thought to have been founded around 300 BCE or a little later. Its foundation is associated with a confidant of the pharaoh PTOLEMY I, DEMETRIUS OF PHALERON (c.350–282 BCE), a member of the Peripatetic school and former Athenian politician. After losing influence in Athens, he sought protection in Alexandria around 297 BCE. PTOLEMY I is said to have entrusted DEMETRIUS with the foundation of the Mouseion, a cultural centre for philosophical and intellectual lectures and debates, and which can be compared in importance to a modern university.[117] The Library of Alexandria was part of the *Mouseion* and made the city a popular destination for ancient scholars to acquire new knowledge.

> *... The Museum is also a part of the royal palaces; it has a public walk, an Exedra with seats, and a large house, in which is the common mess-hall of the men of learning who share the Museum.*
>
> *Strabo, Geographica, Book XVII, 1, 8 (Jones, trans., 1932, p.35).*

The *Mouseion* was dedicated to the Muses, the following patron saints of the arts:

CALLIOPE (epic poetry),
CLIO (history),

117 Eng. *Museum*. Ancient Gr. *Μοῦσαι*, Moûsai, Gr. *Μούσες*, Múses.

EUTERPE (flutes and lyric poetry),
THALIA (comedy and pastoral poetry),
MELPOMENE (tragedy),
TERPSICHORE (dance),
ERATO (love poetry),
POLYHYMNIA (sacred poetry),
URANIA (astronomy).

PTOLEMY I is said to have first asked THEOPHRASTUS OF LESBOS (372–287 BCE), ARISTOTLE's successor as head of the Lyceum of Athens, the Peripatetic school, to establish the *Mouseion*, but he did not take up the task. DEMETRIUS, who then took over, was also a member of the Lyceum. However, he later fell out of favour with PTOLEMY I and was forced to leave public life. The result of DEMETRIUS' efforts was the central library, one of the most important cultural institutions of the ancient Greek world, also used for philosophical meetings of scholars. Later, it was given a daughter library, the *Serapeum*, which is known to have existed in the time of pharaoh PTOLEMY III. The daughter library in the temple of the god Serapis was a public library, which indicates widespread literacy, at least among the Greek part of the population of Alexandria.

At its peak, the central library is said to have had around half a million scrolls, while the daughter library had just under 43,000, which were copies of works from the central library. It started acquiring material virtually from scratch. Greek books were available in Athens and Rhodes, but for a young city like Alexandria this was something new. The collection, which ranged from HOMERIC poetry to cookery books, was clearly deliberate. The staff responsible for collecting the material were supposed to have searched the ships that arrived in Alexandria. Any books found were to be confiscated and copies of the works returned to their owners. Foreign-language works were systematically translated into Greek. Books or written works were also acquired through purchases in Athens and Rhodes. The originals of AESCHYLUS, SOPHOCLES and EURIPIDES, which were kept in the Athenian state archives and were not available for loan, were allegedly borrowed by PTOLEMY I from the Athenians on payment of a large security, but copies were later returned to them and the security waived. The library's directors were the great names of ancient scholarship. In the golden age of the library, these were:[118]

ZENODOTUS OF EPHESUS (c.284–260BCE),
APOLLONIUS OF RHODES (c.260–245BCE),[119]
ERATOSTHENES OF CYRENE (245–205BCE),
ARISTOPHANES OF BYZANTIUM (c.205–185BCE),
ARISTARCHUS OF SAMOTHRACE (175–145BCE),
CYDAS.

118 See Hannak, 1867, Das Museum und die Bibliotheken in Alexandria, pp.26–35.
119 In some sources APOLLONIUS OF RHODES is named APOLLONIUS OF ALEXANDRIA. Ersch, Gruber (eds.), 1863, Allgemeine Encyclopädie der Wissenschaften und Künste, p.421.

The library developed thanks to the support of the ruling elite, who had an interest in culture, art and science. PTOLEMY I (reigned 305–282 BCE) was a historian. It was he who, when he asked EUCLID whether there was an easier way to mathematics, got the answer that "*there is no royal road to mathematics*". PTOLEMY II (reigned 285–246 BCE)[120] was a zoologist, PTOLEMY III (reigned 246–222 BCE) a patron of literature, and PTOLEMY IV (reigned 222–205 BCE) a playwright. At its peak, the library would have had a large staff, including copyists, translators, scroll correctors and similar professions, of which we have no information. Most likely, all these jobs were carried out mainly by slaves. The vastness of the material eventually led to the need for systematic filing or alphabetisation and library catalogues. ZENODOTUS was the first to address this issue, culminating in the bibliographical work of CALLIMACHUS OF CYRENE (active in the time of PTOLEMY I and II), entitled *Pinakes* or *Tables*, which is said to have comprised 120 books, listing works and authors, and was thus five times as voluminous as HOMER's *Iliad*.

The demise of the library was not sudden, but the result of several events. The first step towards the decline of the library as a cultural institution was taken by PTOLEMY VIII, who, apparently more concerned with internal dissensions among the ruling elite or family disagreements than with culture, expelled intellectuals from Alexandria in revenge when he came to power. Perhaps the then director, CYDAS, was acting on his instructions. Another near-fatal event for the library occurred in 48 BCE, when a besieged JULIUS CAESAR barricaded himself in part of the city and then fled after a fire that is said to have destroyed part of the main library. The final blow came in 391 and 392 CE with the decrees of THEODOSIUS I, which outlawed pagan practices. The *Serapeum*, like other pagan sanctuaries, was probably destroyed in 391 CE as a result of Christian fanaticism. What remained after the religious rampage was a pale shadow of the once magnificent library, but the mathematical tradition continued, at least to a limited extent, in Alexandria after the death of HYPATIA. The claim or allegation in some literature that the library was destroyed by Arab conquerors is not supported by historical sources.

> *It was a wise and useful provision of the ancients to transmit their thoughts to posterity by recording them in treatises, so that they should not be lost, but, being developed in succeeding generations through publication in books, should gradually attain in later times, to the highest refinement of learning. And so the ancients deserve no ordinary, but unending thanks, because they did not pass on in envious silence, but took care that their ideas of every kind should be transmitted to the future in their writings.*
>
> *Vitruvius, De Architectura, Book VII, 1*
> *(Hicky Morgan, trans., 1914, De Architectura, p.195)*

120 PTOLEMY II became co-regent around 285 BCE.

Bibliography

- Aaboe, Asger. 1955. On the Babylonian Origin of Some Hipparchian Parameters. In: Centaurus, 4, 2, 122–125. doi 10.1111/j.1600-0498.1955.tb00619.x
- Al-Rawi, Farouk N. H. and A. R. George. 1991/1992. Enūma Anu Enlil XIV and Other Early Astronomical Tables. In: Archiv für Orientforschung, Band 38/39, 52–73. Wien: Archiv für Orientforschung (AfO)/Institut für Orientalistik.
- Apuleius. 1998. The Golden Ass or Metamorphoses, E. J. Kenney (trans.). London: Penguin Books. e-isbn 978-0-141-90450-4. url: https://archive.org/details/TheGoldenAss_201509/page/n91/mode/2up (14 January 2024).
- Astronomical Phaenomena for the Year 2025. 2023. Washington DC: U.S. Government Printing Office; Taunton: The U.K. Hydrographic Office. isbn 978-0-16-094773-5.
- Berggren, J. L. and R. S. D. Thomas. 1996. Euclid's Phaenomena: A Translation and Study of a Hellenistic Treatise in Spherical Astronomy. New York, London: Garland Publishing. isbn 0-8153-0493-5.
- Björnbo, Axel Anthon. 1902. Studien über Menelaos' Sphärik. In: Abhandlungen zur Geschichte der Mathematischen Wissenschaften mit Einschluss ihrer Anwendungen, 14, 1–154. Leipzig: Teubner.
- Borchardt, Ludwig. 1920. Altägyptische Zeitmessung. Die Geschichte der Zeitmessung und der Uhren. Band I, Lieferung B. Berlin, Leipzig: Walter de Gruyter & Co.
- Brack-Bernsen, Lis and Hermann Hunger. 1999. The Babylonian Zodiac: Speculations on its invention and significance. In: Centaurus, 41, 4, 280–292. doi 10.1111/j.1600-0498.1999.tb00425.x.
- Brown, Robert. 1899–1900. Researches into the origin of Primitive Constellations of the Greeks, Phoenicians and Babylonians I-II. London: Williams and Norgate. url: https://archive.org/details/researchesintoor01browuoft/mode/2up, https://archive.org/details/researchesintoor02browuoft/mode/2up (6 November 2022).
- Bulmer-Thomas, Ivor. Theodosius of Bithynia. In: Encyclopedia.com [online source]. url: https://www.encyclopedia.com/science/dictionaries-thesauruses-pictures-and-press-releases/theodosius-bithynia (25 June 2023).
- Bulmer-Thomas, Ivor. 1939. Selections illustrating the history of Greek mathematics, Volume 1–2. Cambridge, MA: Harvard University Press. url: https://archive.org/details/selectionsillust01bulmuoft/mode/2up, https://archive.org/details/selectionsillust02bulmuoft/mode/2up (15 June 2024).
- Bulmer-Thomas, Ivor. 1971. Euclid. In: Dictionary of Scientific Biography, Gillispie, Charles Coulston (ed.), IV, 414–437. New York: Charles Scribner's Sons. sbn 684-10115-7.
- Bulmer-Thomas, Ivor. 1974. Menelaus of Alexandria. In: Dictionary of Scientific Biography, Gillispie, Charles Coulston (ed.), IX, 296–302. New York: Charles Scribner's Sons. isbn 0-684-10120-3.
- Bulmer-Thomas, Ivor. 1976. Theodosius of Bithynia. In: Dictionary of Scientific Biography, Gillispie, Charles Coulston (ed.), XIII, 319–321. New York: Charles Scribner's Sons. isbn 0-684-12925-6.
- Censorinus. 1887. De di Natali, Hultsch, Frederick Otto (ed.). Leipzig: Teubner. url: https://archive.org/details/dedienataliliber00cens/mode/2up (10 February 2024).
- Censorinus. 2011. Das Geburtstagsbuch, Brodersen, Kai (trans.). Darmstadt: Primus Verlag. isbn 978-3-89678-752-1.
- Chatterjee, Bina (trans., ed.). 1970. The Khaṇḍakhādyaka (an astronomical treatise) of Brahmagupta with the Commentary of Bhaṭṭotpala, I-II. New Delhi: Motilal Banarsidass.
- Clagett, Marshall. 1970. Archimedes. In: Dictionary of Scientific Biography, Gillispie, Charles Coulston (ed.), I, 213–231. New York: Charles Scribner's Sons.
- Couprie, Dirk L., Gerard Naddaf and Robert Hahn. 2002. Anaximander in Context: New Studies in the Origins of Greek Philosophy. New York: State University of New York Press. isbn 978-0791455388.
- Couprie, Dirk L. 2011. Heaven and Earth in Ancient Greek Cosmology, From Thales to Heraclides Ponticus. New York: Springer. isbn 978-1-4419-8115-8.
- Crowe, Michael J. 2001. Theories of the World from Antiquity to the Copernican Revolution. Second Edition. Mineola, NY: Dover Publications. isbn 0-486-41444-2.
- Czwalina, Arthur. 1931. Autolykos, Rotierende Kugel und Aufgang und Untergang der Gestirne. Theodosios von Tripolis, Sphaerik. Leipzig: Akademische Verlagsgesellschaft.
- Daressy, Georges. 1901. Catalogue général des antiquités égyptiennes du Musée du Caire / 25001-25385, Ostraca. Kairo. url: https://archive.org/details/DaressyOstraca1901/page/n1 (14 January 2024).
- Dewndney, Selwyn H. 1931. The Zodiac and Early Astronomy. In: Journal of the Royal Astronomical Society of Canada, 25, 400–408.
- Dicks, D. R. 1971. Euctemon. In: Dictionary of Scientific Biography, Gillispie, Charles Coulston (ed.), IV, 459–460. New York: Charles Scribner's Sons. sbn 684-10115-7.

- Drawnel, Henryk. 2011. The Aramaic Astronomical Book (4Q208–4Q211) from Qumran: Text, Translation, and Commentary. Oxford: Oxford University Press. isbn 978-0-19959043-8.
- Eastwood, Bruce Stansfield. 1992. Heraclides and Heliocentrism: Texts, Diagrams, and Interpretations. In: Journal for the History of Astronomy, 23, 4, 233–260. doi 10.1177/002182869202300401.
- Ebert, Max (ed.). 1927. Reallexikon der Vorgeschichte, Band 8. Berlin: Walter Gruyter & Co.
- Eggeling, Julius. 1897. The Śataphata Brāhmana According to the Text of the Mādhyamdina School, Part IV. Oxford: Clarendon Press. url: https://archive.org/details/satapathabrahman04egge/mode/2up (14 January 2024).
- Ersch, Johann Samuel and Johann Gottfried Gruber (eds.). 1863. Allgemeine Encyclopädie der Wissenschaften und Künste, Erste Section A–G. Leipzig: Brockhaus. (Google Books, 9 June 2024).
- Evans, James. 1998. The History and Practice of Ancient Astronomy. Oxford: Oxford University Press. isbn 978-0-19-509539-5.
- Evans, James and J. Lennart Berggren. 2006. Geminos's Introduction to Phaenomena. Princeton, NJ: Princeton University Press. isbn 9780691123394.
- Fitzgerald, Augustine. 1926. Letters of Synesius. Oxford University Press. London: Humphrey Milford.
- Flores, Alfinio. 2014. The Babylonian Method for Approximating Square Roots: Why Is It So Efficient? In: The Mathematics Teacher, 108, 3, 230–235. doi 10.5951/mathteacher.108.3.0230. url: https://www.jstor.org/stable/10.5951/mathteacher.108.3.0230.
- Friberg, Jöran, Hermann Hunger and Farouk Al-Rawi. 1990. Seed and Reeds - A Metro-mathematical Topic Text from Late Babylonian Uruk. In: Baghdader Mitteilungen. Deutsches Archäologisches Institut. Abteilung Baghdad. Bd 21, 483–557. Berlin: Gebr. Mann. isbn 3-7861-1592-3.
- Friberg, Jöran. 2007. Amazing traces of a Babylonian origin in Greek mathematics. Singapore: World Scientific. isbn 978-981-270-452-8.
- Friberg, Jöran. 2019. Three thousand years of sexagesimal numbers in Mesopotamian mathematical texts. In: Archive for History of Exact Sciences, 73, 1, 183–216. doi 10.1007/s00407-019-00221-3.
- Gibbs, Sharon L. 1976. Greek and Roman Sundials. New Haven, CT: Yale University Press.
- Glowatzki, Ernst and Helmut Göttsche. 1975. Die Sehnentafel des Klaudios Ptolemaios. München: R. Oldenbourg Verlag. isbn 3-486-20181-6.
- Gondhalekar, Prabhakar. 2013. The Time Keepers of the Vedas: History of the Calendar of the Vedic Period (from Ṛgveda to Vedāṅga Jyotiṣa). New Delhi: Manohar. isbn 978-81-7304-969-9.
- Graßhoff, Gerd, Florian Mittenhuber and Elisabeth Rinner. 2017. Of paths and places: the origin of Ptolemy's Geography. In: Archive for History of Exact Sciences, 71, 1, 483–508. doi 10.1007/s00407-017-0194-7.
- Gysembergh, Victor, Peter J. Williams and Emanuel Zingg. 2022. New evidence for Hipparchus' Star Catalogue revealed by multispectral imaging. In: Journal for the History of Astronomy, 53, 4, 383–393. doi 10.1177/00218286221128.
- Halma, Nicolas. 1813. Composition Mathematique I. Reprint 1927. Paris: Librarie Scientifique J. Hermann. Halma, Nicolas. 1813-1816. Composition Mathématique I-II. Translation with Comments of Delambre. Reprint 1988. Paris: Librairie A. Blanchard. isbn 2-85367-069-4.
- Halma, Nicolas. 1821. Commentaire De Théon d'Alexandrie sur le Premier & le Second Livres de la Composition Mathématique de Ptolemée. Translation and Comments. Reprint 1993. Paris: Librairie A. Blanchard. isbn 2-85-367-186-0.
- HalmaA, Nicolas. 1822. Commentaire De Theon d'Alexandrie sur le Livre III de l'Almaqeste cond Livres de la Composition de Ptolemee. Reprint 1990. Paris: Librairie A. Blanchard. Isbn 2-85-367-169-0.
- Hannak, E. 1867. Das Museum und die Bibliotheken in Alexandria. Wien: Leopoldstädter Realgymnasium. (Google Books, 9 June 2024).
- Haug, Martin. 1863. Aitareya Brahmanam of the Rigveda. Bombay: Government Central Book Depot, London: Trubner & Co. url: https://archive.org/details/in.ernet.dli.2015.24205 (14 January 2024).
- Heath, Thomas. 1897. The Works of Archimedes. Cambridge: Cambridge University Press.
- Heath, Thomas. 1913. Aristarchus of Samos, the Ancient Copernicus. Reprint 2007. Adamant Media. isbn 1-4021-4790-2.
- Heath, Thomas. 1921. A History of Greek Mathematics, Volume I. Reprint 2006. Adamant Media. isbn 0-543-97448-0.
- Heath, Thomas. 1921. A History of Greek Mathematics, Volume II. Reprint 2006. Adamant Media. isbn 0-543-97877-4.
- Heath, Thomas (transl.). 1956. The Thirteen Books of the Elements, Volume III (Books 10–13). New York: Dover Publ. isbn 978-0-486-60088-8.
- Heath, Thomas. 2009. The Works of Archimedes. Cambridge: Cambridge University Press. isbn 9780511695124.

- Hunger, Hermann and David Pingree. 1989. MUL.APIN. An Astronomical Compendium in Cuneiform. Archiv für Orientforschung Beiheft 24. Horn: Ferdinand Berger & Söhne.
- Huxley, G. L. 1970. Autolycus of Pitane. In: Dictionary of Scientific Biography, Gillispie, Charles Coulston (ed.), I, 338–339. New York: Charles Scribner's Sons.
- Huxley, G. L. 1971. Eudoxus of Cnidus. In: Dictionary of Scientific Biography, Gillispie, Charles Coulston (ed.), IV, 465–467. New York: Charles Scribner's Sons. sbn 684-10115-7.
- Hypsicles. 1966. Die Aufgangszeiten der Gestirne. De Falco, Vittorio and Max Krause (trans.), introduction Neugebauer, Otto. Abhandlungen der Akademie der Wissenschaften in Göttingen, Philogogisch-Historische Klasse, Dritte Folge, 62. Göttingen: Vandenhoeck & Ruprecht.
- Ideler, Christian Ludwig. 1812. Über die Trigonometrie der Alten. In: Monatliche Correspondenz zur Beförderung der Erd- und Himmelskunde, Franz Xaver von Zach (ed.), 26, 3–38. url: https://zs.thulb.uni-jena.de/receive/jportal_jpjournal_00001153?XSL.q=zach%20monatliche%20correspondenz (14 January 2024).
- Jacobus, Helen, R. 2014. Zodiac Calendars in the Dead Sea Scrolls and Their Reception: Ancient Astronomy and Astrology in Early Judaism. Lleiden: Brill. e-isbn 978-90-04-28406-7.
- Jacobus, Helen, R. 2018. Reconstructing 4Q208–4Q209 as an Astronomical Artefact. url: https://bibleinterp.arizona.edu/sites/bibleinterp.arizona.edu/files/docs/Aramaic%20calendars%20final%20(1).pdf (14 January 2024).
- Jacobus, Helen R. 2019. Reconstructing the Calendar of 4Q208–4Q209 (and a response to Eshbal Ratzon). In: Revue de Qumran, 31, 2, 114, 251–273. doi 10.2143/RQ.31.2.3287242.
- Jacobus, Helen R. 2022. How 4QAstronomical Enoch[a–b] (4Q208–209) Transformed Elements of Late Babylonian Magical Hemerological Texts into a Synchronistic Calendar. In: Science in Qumran Aramaic Texts, Fröhlich, Ida (ed.), 83–110. Tübingen: Mohr Siebeck. doi 10.1628/978-3-16-161388-3. isbn 978-3-16-161387-6.
- Jones, Alexander. 1991a. Hipparchus's Computation of Solar Longitudes. In: Journal for the History of Astronomy, 22, 68, 101–125. doi 10.1177/002182869102200201.
- Jones, Alexander. 1991b. The Adaptation of Babylonian Methods in Greek Numerical Astronomy.
- In: Isis, 82, 3, 440–453. doi 10.1086/355836.
- Jones, Alexander. 2002. Erathostenes, Hipparchus, and the Obliquity of the Ecliptic. In: Journal for the History of Astronomy, 33, 110, 15–19. doi 10.1177/002182860203300103.
- Joseph, George Gheverghese. 2011. The Crest of the Peacock. Princeton, NJ, Oxford: Princeton University Press. isbn 978-0-691-13526-7.
- Knorr, Wilbur R. 1990. Plato and Eudoxus on the Planetary Motions. In: Journal for the History of Astronomy, 21, 4, 313–329. doi 10.1177/002182869002100401.
- Krause, Max. 1936. Die Sphärik von Menelaos aus Alexandrien in der Verbesserung von Abū Naṣr Mansūr B. ʿAli B. ʿIrāq: mit Untersuchungen zur Geschichte des Textes bei den islamischen Mathematikern. In: Abhandlungen der Gesellschaft der Wissenschaften zu Göttingen, Philologisch-Historische Klasse, Neue folge 3, 17. Berlin: Weidmann.
- Kunitzsch, Paul and Richard Lorch. 2010. Theodosius Sphaerica. Stuttgart; Franz Steiner Verlag. isbn 978-3-515-09288-3.
- Machiela, David. 2023. A Handbook of the Aramaic Scrolls from the Qumran Caves. Leiden: Brill. e-isbn 978-90-04-51381-5.
- Malville, McKim J. 2008. The Astronomical Gnomon, A Series of Interactive Experiments in Archaeoastronomy. In: African Cultural Astronomy, Holbrook, J., R. Medupe and J. Urama (eds.). New York: Springer. isbn 978-1-4020-6638-2.
- Michel, Peter (ed.). 2008. Rig-Veda, das Heilige Wissen Indiens, Band I-II. Reprint of 1923 edition, Geldner Karl-Friedrich transl. Wiesbaden: Marix Verlag. isbn 978-3-86539-165-0.
- Mishra, Suresh Chandra. 2005. Vedanga Jyotisham. New Delhi: Ranjan Publications. isbn 81-88230-50-2.
- Murdoch, John. 1971. Euclid: Transmission of the Elements. In: Dictionary of Scientific Biography, Gillispie, Charles Coulston (ed.), IV, 437–459. New York: Charles Scribner's Sons. sbn 684-10115-7.
- Nadal, R. and J. P. Brunet. 1984. Le "Commentaire" d'Hipparque I. La sphère mobile. In: Archive for History of Exact Sciences, 29, 4, 201–236. doi 10.1007/BF00348621.
- Nadal, R. and J. P. Brunet. 1989. Le "Commentaire" d'Hipparque II. Position de 78 étoiles. In: Archive for History of Exact Sciences, 40, 4, 305–354. doi 10.1007/BF00329864.
- Nadal, Robert, Abdelkaddous Taha and Pierre Pinel. 2004. Le contenu astronomique des Sphèriques de Ménélaos. In: Archive for History of Exact Sciences, 58, 5, 381–436. doi 10.1007/s00407-003-0077-y.
- Nadal, Robert, Abdelkaddous Taha and Pierre Pinel. 2004. Erratum: "Le contenu astronomique des Sphériques de Ménélaos" In: Archive for History of Exact Sciences, 58, 5, 437. doi 10.1007/s00407-004-0084-7.

- Neuenschwander, E. 1973. Die ersten vier Bücher der Elemente Euklids. In: Archive for History of Exact Sciences, 9, 4–5, 325–380. doi 10.1007/BF00348367.
- Neugebauer, Otto. 1927. Zur Entstehung des Sexagesimalsystems. In: Abhandlungen der Gesselschaft der Wissenschaften zu Gottingen. Mathematisch Physikalische Klasse, Neue Folge, Bd. XIII, 1–50.
- Neugebauer, Otto. 1934. Vorlesungen über Geschichte der antiken mathematischen Wissenschaften. Berlin: Springer.
- Neugebauer, Otto. 1938. Untersuchungen zur Geschichte der antiken Astronomie III. In: Quellen und Studien zur Geschichte der Mathematik, Astronomie und Physik, Band 4. Berlin: Springer.
- Neugebauer, Otto. 1941. Some Fundamental Concepts in Ancient Astronomy. In: Studies in the History of Sciences. Reprint 1983, Astronomy and History, Selected Essays, 5–21. Springer. isbn 0-387-90844-7.
- Neugebauer, Otto. 1955a. Astronomical Cuneiform texts, vol. 1–3. Princeton, NJ: Institute for Advanced Study.
- Neugebauer, Otto. 1955b. Apollonius' Planetary Theory. In: Communications on Pure and Applied Mathematics, 8, 641–648. doi 10.1002/cpa.3160080413.
- Neugebauer, Otto. 1959a. The Equivalence of Eccentric and Epicyclic Motion According to Apollonius. In: Scripta Mathematica, 24, 5–21. Reprint 1983, Astronomy and History, Selected Essays, 335–351. Springer. doi 10.1007/978-1-4612-5559-8. isbn 0-387-90844-7.
- Neugebauer, Otto. 1959b. Ptolemy's Geography, Book VII, Chapters 6 and 7. In: Isis, 50, 1, 22–29. doi 10.1086/348723. url: https://www.jstor.org/stable/226535.
- Neugebauer, Otto. 1969. The Exact Sciences in Antiquity, 2nd Edition. New York: Dover. sbn 486-22332-9.
- Neugebauer, Otto. 1972. Planetary Motion in P. Mich. 149. In: The Bulletin of the American Society of Papyrologists, 9, 1–2, 19–22. url: https://hdl.handle.net/2027/spo.0599796.0009.001:05 (14 January 2024).
- Neugebauer, Otto. 1975. A History of Ancient Mathematical Astronomy I-III. New York: Springer. isbn 0-387-06995-X.
- Neugebauer, Otto. 1985. On Some Aspects of Early Greek Astronomy. In: Astronomy and History Selected Essays, 361–369. New York: Springer. isbn 0-387-90844-7. doi 10.1007/978-1-4612-5559-8.
- Ôhashi, Yukio. 1993. Development of Astronomical Observations in Vedic and post Vedic India. In: Indian Journal of History of Science, 28, 3, 185–251.
- Ôhashi, Yukio. 1994. Astronomical Instruments in Classical Siddhāntas. In: Indian Journal of History of Science, 29, 2, 155–313.
- Parker, R. A. 1974. Ancient Egyptian Astronomy. In: Philosophical Transactions of the Royal Society of London. Series A, Mathematical and Physical Sciences, 276, 1257, 51–65. doi 10.1098/rsta.1974.0009.
- Parpola, Asko. 2013. Beginnings of Indian Astronomy, with Reference to a Parallel Development in China. In: History of Science in South Asia 1, 21–78. doi 10.18732/H2VC7S.
- Pedersen, Olaf. 2011. A Survey of the Almagest. New York, Dordrecht, Heidelberg, London: Springer. isbn 978-0-378-84825-9. doi 10.1007/978-0-387-84826-6.
- Pedersen, Olaf and Mogens Pihl. 1974. Early Physics and Astronomy. London: MacDonald and Janes. New York: Elsevier. isbn 0444195750.
- Petersen, Viggo and Olaf Schmdt. 1968. The Determination of the Longitude of the Apogee of the Orbit of the Sun according to Hipparchus and Ptolemy. In: Centaurus, 12, 2, 73–96. doi 10.1111/j.1600-0498.1968.tb00080.x
- Pinches, Theophilus Goldrige, Johann Nepomuk Strassmaier and Abraham J. Sachs. 1955. Late Babylonian astronomical and related texts. Providence, RI: Brown University Press.
- Powell, Marwin A. 1976. Sumerian area measures and the alleged decimal substratum. In: Zeitschrift für Assyriologie und vorderasiatische Archäologie, Band 62, 165–221. doi 10.1515/zava.1972.62.2.165.
- Powell, Marvin A. 1990. Masse und Gewichte. In: Reallexikon der Assyriologie und vorderasiatischen Archäologie, Band 7. Ebeling Erich and Ernst Weidner (ed.). Berlin: De Gruyter. isbn 978-3-11-010437-0.
- Ratzon, Eshbal. 2019. Jewish time: First stages of seasonal hours in Judea. In: Studies in History and Philosophy of Science, 75, 23–33. doi 10.1016/j.shpsa.2018.11.003.
- Rehm, Albert. 1913. Griechische Kalender III. Das Parapegma des Euktemon. Sitzungsberichte der Heidelberger Akademie der Wissen-schaften. Philosophisch-Historische Klasse, Abhandlung 3, 3–38. doi 10.11588/diglit.33044#0036.
- Rehm, Albert. 1941. Parapegmastudien. In: Abhandlungen der Bayerischen Akademie der Wissenschaften, Philosophisch-Historische Abteilung, Neue Folge, Band 19, 5–145. url: https://publikationen.badw.de/en/006529515 (14 January 2024).
- Robson, Eleanor. 2008. Mathematics in Ancient Iraq. Princeton, NJ, Oxford: Princeton University Press. isbn 978-0-691-09182-2.

- Sachs, Abraham and Hermann Hunger. 1988. Astronomical Diaries and Related Texts. Volume I, Diaries from 652 B. C. to 262 B. C. Wien: Verlag der Österreichischen Akademie der Wissenschaften.
- Schaefer, Bradley E. 2004. The latitude and epoch for the origin of the astronomical lore of Eudoxus. In: Journal for the History of Astronomy, 35, part 2, 119, 161–223. doi 10.1177/002182860403500204.
- Schaefer, Bradley E. 2007. The Latitude and Epoch for the Origin of the Astronomical Lore in MUL.APIN. In: American Astronomical Society Meeting 210, Bulletin of the American Astronomical Society, 39, 157.
- Schaldach, Karlheinz. 2004. The Arachne of the Amphiareion and the Origin of Gnomonics in Greece. In: Journal for the History of Astronomy, 35, part 4, 121, 435–445. doi 10.1177/002182860403500404.
- Schodde, George Henry (trans.). 1882. The Book of Enoch. Andover, MA: Warren F. Draper. url: https://archive.org/details/bookenochtransl00schogoog/mode/2up (14 January 2024).
- Schoy, Carl. 1927. Die trigonometrischen Lehren des persischen Astronomen Abu'l'Raiḥân Muḥ. Ibn Aḥmad al-Bîrûnî dargestellt nach Al-Qânûn al-Mas'ûdî, Ruska, Julius and Heinrich Wieleitner (eds.). Hannover: Orient-Buchhandlung Heinz Lafaire.
- Shceglow, Dmitry A. 2003. Hipparchus' Table of Climata and Ptolemy's Geography. In: Orbis Terrarum, 9, 7, 159–192. url: https://www.academia.edu/214134/Hipparchus_Table_of_Climata_and_Ptolemy_s_Geography (author's text) (14 January 2024).
- Shceglow, Dmitry A. 2007. Ptolemy's Latitude of Thule and the Map Projection in the Pre-Ptolemaic Geography. In: AKAN, Antike Naturwissenschaft und ihre Rezeption, Band XVII. url: https://www.academia.edu/213001/Ptolemy_s_Latitude_of_Thule_and_the_Map_Projection_in_the_Pre_Ptolemaic_Geography (author's text) (14 January 2024).
- Sidoli, Nathan. 2004. Hipparchus and the Ancient Metrical Methods on the Sphere. In: Journal for the History of Astronomy, 35, part 1, 118, 71–84. doi 10.1177/002182860403500103.
- Stahl, William H. 1970. Aristarchus of Samos. In: Dictionary of Scientific Biography, Gillispie, Charles Coulston (ed.), I, 246–250. New York: Charles Scribner's Sons.
- Steele, John and Lis Brack-Bernsen. 2008. A Commentary Text to Enuma Anu Enlil 14. In: From the Banks of the Euphrates: Studies in Honor of Alice Louise Slotsky, Ross, Micah (ed.), 253–262. University Park, PA: Eisenbrauns. isbn 978-1-57506-144-3. url: https://www.academia.edu/2361093/A_Commentary_Text_to_Enuma_Anu_Enlil_14?email_work_card=view-paper (5 February 2023).
- Strabo. 1855–1898. Geographica. A. Forbiger (transl.). Reprint 2007. Wiesbaden: Marix Verlag. isbn 978-3-86539-051-6.
- Strabo. 1932. The Geography of Strabo, Jones, Horace Leonard (transl.). Reprint 1967. Cambridge, MA: Harvard University Press. London: William Heinemann. url: https://archive.org/details/Strabo08Geography17AndIndex/Strabo%2008%20Geography%2017%20and%20index/mode/2up (14 January 2024).
- Subbarayappa, B. V. and K. V. Sarma. 1985. Indian Astronomy: A Source Book. Bombay: Nehru Centre.
- Swerdlow, Noel. 1980. Hipparchus's Determination of the Length of the Tropical Year and the Rate of Precession. In: Archive for History of Exact Sciences, 21, 4, 291–309. doi 10.1007/BF00595374. url: https://www.jstor.org/stable/41133611.
- Szabó, Arpad. 1992. Das geozentrische Weltbild. München: Deutscher Taschenbuch Verlag. isbn 3-423-04490-X.
- The Astronomical Almanac for the Year 2022. 2021. Washington DC: U.S. Government Printing Office; Taunton: The U.K. Hydrographic Office. isbn 978-0-7077-46203.
- Thureau-Dangin, François. 1910. Inventaire des Tablets de Tello conservées au Musée Impérial Ottoman, Volume 1. Paris: Ernest Leroux. url: https://archive.org/details/inventairedestab01consuoft/mode/2up (14 January 2024).
- Thureau-Dangin, François. 1939. Sketch of a History of the Sexagesimal System. In: Osiris, 7, 95–141. doi 10.1086/368503.
- Tihon, Anne and Raymond Mercier. 2011. Ptolemaiou Procheiroi Kanones. Ptolemy's Handy Tables. Volume 1a, 1b: Tables A1-A2. Transcription and Commentary. Leuven: Peeters Publishers. isbn 978-90-429-2436-9.
- Toomer, Gerald J. 1970. Apollonius of Perga. In: Dictionary of Scientific Biography, Gillispie, Charles Coulston (ed.), I, 179–193. New York: Charles Scribner's Sons.
- Toomer, Gerald J. 1974. The Chord Table of Hipparchus and the Early History of Greek Trigonometry. In: Centaurus, 18, 1, 6–28. doi 10.1111/j.1600-0498.1974.tb00205.x.
- Toomer, Gerald J. 1978. Hipparchus. In: Dictionary of Scientific Biography, Gillispie, Charles Coulston (ed.), XV, Supplement I, 207–224. New York: Charles Scribner's Sons. isbn 0-684-14779-3.
- Toomer, Gerald J. 1998. Ptolemy's Almagest. Princeton, NJ: Princeton University Press. isbn 978-0-691-00260-6.
- Tropfke, Johannes. 1923. Geschichte der Elementar-Mathematik, Band 5, Ebene Trigonometrie, Sphärik und Sphärische Trigonometrie. Berlin, Leipzig: Walter de Gruyter.

- Tropfke, Johannes. 1928. Archimedes und die Trigonometrie. In: Archiv für Geschichte der Mathematik der Naturwissenschaften und der Technik, 10, 432–463.
- Unger, Eckhard. 1916. Zwei babylonische Antiken aus Nippur. Publikationen der Kaiserlich Osmanischen Museen 1.
- Van Brummelen, Glen. 2009. The Mathematics of the Heavens and the Earth. Princeton, NJ: Princeton University Press. ISBN 978-0-691-12973-0.
- Van der Waerden, Bartel Leendert. 1949. Babylonian Astronomy. II. The Thirty-Six Stars. In: Journal of Near Eastern Studies, 8, 1, 6–26. URL: https://www.jstor.org/stable/542436.
- Van der Waerden, Bartel Leendert. 1951a. Babylonian Astronomy. III. The Earliest Astronomical Computations. In: Journal of Near Eastern Studies, 10, 1, 20–34. URL: https://www.jstor.org/stable/542419.
- Van der Waerden, Bartel Leendert. 1951b. Die Astronomie der Pythagoreer. Amsterdam: North-Holland Publishing Company.
- Van der Waerden, Bartel Leendert. 1952–1953. History of the Zodiac. In: Archiv für Orientforschung, Band 16, 216–230. URL: https://www.jstor.org/stable/41635784.
- Van der Waerden, Bartel Leendert. 1968. Die Anfänge der Astronomie. Erwachende Wissenschaft Band 2. Basel: Birkhäuser.
- Van der Waerden, Bartel Leendert. 1970. Das Heliozentrische System in der griechischen, persichen und indischen Astronomie. Neujahrsblatt herausgegeben von der Naturforschenden Gesselschaft in Zürich. Zürich: Kommisionsverlag Leemann.
- Van der Waerden, Bartel Leendert. 1975. The Earliest Form of the Epicycle Theory. In: Journal for the History of Astronomy, 5, 175–185. DOI 10.1177/002182867400500303.
- Van der Waerden, Bartel Leendert. 1983. Geometry and Algebra in Ancient Civilizations. Berlin, Heidelberg: Springer. DOI 10.1007/978-3-642-61779-9. e-ISBN 978-3-642-61779-9.
- Van der Waerden, Bartel Leendert. 1984. Greek Astronomical Calendars: I. The Parapegma of Euctemon. In: Archive for History of Exact Sciences, 29, 2, 101–114. DOI 10.1007/BF00348241.
- Vitruvius. 1914a. The Ten Books on Architecture. Morris Hicky Morgan (trans.), Reprint 1960. New York: Dover. ISBN 978-0-486-20645-5.
- Vitruvius. 1914b. The Ten Books on Architecture. Morris Hicky Morgan (trans.). Cambridge, MA: Harvard University Press. URL: https://archive.org/details/vitruviustenbook00vitruoft/mode/2up (14 January 2024).
- Warmington, E. H. 1975. Posidonius. In: Dictionary of Scientific Biography, Gillispie, Charles Coulston (ed.), XI, 103–106. New York: Charles Scribner's Sons. ISBN 0-684-10122-X.
- Watson, Rita and Wayne Horowitz. 2011. Writing Science before the Greeks: A Naturalistic Analysis of the Babylonian Astronomical Treatise MUL.APIN. Leiden: Brill. ISBN 978 90 04 20230 6.
- Webster, Doris. 1940. The Origin of the Signs of the Zodiac: An Interpretation from the Psychological Viewpoint. In: American Imago, 1, 4, 31–47.
- Weidner, Ernst F. 1924. Ein Babylonisches Kompendium Der Himmelskunde. In: The American Journal of Semitic Languages and Literatures, 40, 3, 186–208. DOI 10.1086/370028.
- Weidner, Ernst F. 1931–1932. Der Tierkreis und die Wege am Himmel. In: Archiv für Orientforschung, Band 7, 170–178. URL: https://www.jstor.org/stable/41662003.
- Weinstock, Stefan. 1949. Lunar Mansions and Early Calendars. In: The Journal of Hellenic Studies, 69, 48–69. DOI 10.2307/629462.
- Wendorf, Fred and Romuald Schild. 2001. Holocene Settlement of the Egyptian Sahara: Volume 1: The Archaeology of Nabta Playa. New York: Springer. ISBN 978-0306466120.
- Winter, Jay (trans.). 2015. The Complete Book of Enoch. Winter Publications. ISBN 9781370207848. URL: https://archive.org/details/TheCompleteBookOfEnochStandardEnglishVersionJayWinter (14 January 2024).
- Yampolsky, Philip. 1950. The Origin of the Twenty-Eight Lunar Mansions. In: Osiris, 9, 62–83. DOI 10.1086/368524.

1

Equation of time in Ancient Greece

For the rotations of the cosmos are of equal duration, but the rising of the arcs that he Sun traverses during a rotation of the cosmos are not of equal duration. It is for this reason that every combined night and day is not equal to every [other] combined night and day.

Geminus of Rhodes, Introduction to the Phenomena VI, 4
(Evans, Berggren, trans., 2006, p.162)

1.1 Geminus

We know next to nothing about GEMINUS' life but historians place him, according to the citations in and of his works, between the years 150 BCE and 200 CE. He was a Stoic philosopher and it seems most likely that he lived in the first century BCE.[1] In Ancient Greek sources, he is often referred to as GEMINUS OR GEMINUS THE MATHEMATICAN.[2] His name is of Latin origin but, by judging from his education, way of thinking and work in science, he was a Greek, and in his works he cites the achievements of Babylonian astronomers.[3] The assumption that he worked on Rhodes is not supported by solid evidence, the only indication is that his calculations are often illustrated with Rhodes as an example. According to some interpretations, perhaps GEMINUS or POSIDONIUS, perhaps a friend or teacher of GEMINUS, could have designed the mechanism from Antikythera.[4] His book *Introduction to the Phenomena*[5], which he most probably wrote between 90 and 25 BCE, is preserved and is one of the best ancient astronomical works that have come down to us. It is of special significance since it contains detailed descriptions and references that are not available in other ancient sources. Designed for beginner astronomers, *Introduction to the Phenomena* provides a useful insight into astronomy in the era between HIPPARCHUS and PTOLEMY. Unlike many other ancient sources, it is written in understandable language and can be read through the eyes of historians as well as astronomers.

GEMINUS does not cite sources in the work, but it is apparent that he was familiar with the works of EUCLID, AUTOLYCUS, THEODOSIUS and possibly HYPSICLES OF ALEXANDRIA. The book also includes an important description of the Babylonian lunar theory and the 8 and 19-year lunisolar cycles, which are the most accurate among all ancient sources.[6] There is also a unique description of HIPPARCHUS' interpretation of constellations which cannot be found in other

1 For more information on the time frame of the life and work of GEMINUS, see Evans, Berggren, 2006, Geminos's Introduction to Phenomena, pp.15–22. See also Neugebauer, 1975, II, pp.579–581.

2 GEMINUS OF RHODES, Gr. *Γεμῖνος ὁ Ῥόδιος*, or GEMINOS, Lat. GEMINUS. For comments on his name see Heath, 1921, II, pp.222–223.

3 For his work see Neugebauer, 1975, II, pp.578–589.

4 See Voigt, 2010, Geminos von Rhodos und Oktaëteris, Acta Historica Astronomiae, vol. 37, p.11. Also, Kouremenos, 1994, Posidonius and Geminus on the Foundations of Mathematics. Hermes: Zeitschrift für laische Philologie 122, pp.437–450, and Evans, Carman, 2014, Mechanical Astronomy in Ancient Greece; From Alexandria Through Baghdad, Sidoli, van Brummelen (eds.), pp.166–171. For fragments and commentary, see Evans, Berggren, 2006, pp.243–289. For POSIDONIUS see Warmington, 1975, pp.103–106.

5 The title of the book in the original, as written by GEMINUS, is not known; today, we know the work under the title *Introduction to the Phenomena*, Gr. *Εἰσαγωγὴ εἰς τὰ Φαινόμενα*, *Eisagōgē eis ta phainomena*, often referred to as the *Isagoge*. Evans, Berggren, 2006, pp.1–3. GERARD OF CREMONA translated the work from Arabic in around 1170 CE, when PTOLEMY was wrongly believed to be the author. The first quality translation into Latin, *Gemini elementa astronomiae Graecè & Latinè*, was produced in 1590 by EDO HILDERICUS VON VAREL (1533–1599), a German historian, mathematician, philologist, and theologist. The *Phenomena* was also translated into French, Germaine Aujac (1975), and German, Karl Manitius (1898); Voigt, 2010, p.11.

6 Oktaëteris, Gr. *ὀκταετηρίς*, an eight-year cycle, a period of eight years, after which the moon phases occur on the same calendar day with a difference of a day or two. The period may have been designed by CLEOSTRATUS, see von Fritz, 1971, Grundprobleme der Geschichte der Antiken Wissenschaft, pp.139–140. A nineteen-year cycle *enneakaidekaēteris*, Gr. *εννεακαιδεκαετηρίς*, also the METONIC cycle. For a detailed description, see Evans, Berggren, 2006, pp.180–185, and for a possible correct interpretation of the work, see Voigt, 2010, pp.7–34.

ancient works. It is not unlikely that he could have extracted the data from an unknown work of HIPPARCHUS, since the information is not the same as in HIPPARCHUS' preserved works.[7] He also rejected the widespread belief in antiquity that changes in weather are caused by heliacal rising or setting of stars.[8] Other interesting topics of great pedagogical and didactic value are also found in GEMINUS' work, such as descriptions of ancient astronomical instruments and signs of the zodiac. From historical sources it is evident that he also wrote works in the fields of geometry and meteorology, which were lost.[9] His apparently very substantial but lost work on mathematics, known from the references in PROCLUS' commentaries on the first book of EUCLID's *Elements*, also contained a considerable number of definitions and may have been, at least in part, a kind of mathematical encyclopaedia.[10] GEMINUS himself contributed several mathematical definitions, and he also distinguished postulates and axioms. References to his mathematical achievements are also found in the notes of PAPPUS and EUTOCIUS.

It was already known in antiquity that the length of the day, measured as the time between two consecutive sunsets or noons, is not the same for all days of the year. Historical sources provide evidence for this. GEMINUS also commented in *Introduction to the Phenomena* on the different combined lengths of night and day – or the day as we understand it today – and explained this phenomenon. He also distinguished between the different notions of the day in chapter VI of *Introduction to the Phenomena*, where he separates[11]

- The day as the time interval between sunrise and sunset, which corresponds to the duration of daylight as we understand it today; and
- The day as the time interval between two consecutive sunrises, or the rotation of the cosmos and the Sun's arc rising time.

GEMINUS' term *rotation of the cosmos* corresponds to the concept of a sidereal day, and is equivalent to the modern notion of the time interval required for a single rotation of the Earth's axis relative to the fixed stars, and is about four minutes shorter than the 24-hour day.[12] Thus, in modern terms, the Earth rotates around its axis - relative to the fixed stars - in 23 hours, 56 minutes and 4.1 seconds. During this time, the Earth also travels about 1° along its orbit around

7 Evans, Berggren, 2006, pp.137–145. HIPPARCHUS' comments on the phenomena of EUDOXUS and ARATUS are preserved. Hipparchus, 1894, Hipparchi in Arati et Eudoxi Phaenomena Commentariorum, translation and commentary by K. Manitius. See also Graßhoff, 1990, The History of Ptolemy's Star Catalogue, pp.52–54. Neugebauer, 1975, vol. 1, pp.274–298. Dicks, 1960, The Geographical Fragments of Hipparchus, pp.16–17.

8 Evans, Berggren, 2006, p.219.

9 These may have been comments or an abbreviated version of POSIDONIUS' *Meteorology*. POSIDONIUS (around 135–51 BCE), Gr. *Ποσειδώνιος*, was a stoic philosopher who wrote several works in the fields of history, geography and astronomy, which are preserved only partially in fragments; Evans, Berggren, 2006, p.4, note 4.

10 For quotes of GEMINUS' work in PROCLUS' *Commentary* see Proclus, 1970, A Commentary on the First Book of Euclid's Elements, Morrow (trans.). For his mathematical works see also Heath, 1921, II, pp.222–234.

11 Evans, Berggren, 2006, Chapter VI.4, pp.161–162. Similar preserved works include THEON OF SMYRNA, *Mathematics Useful for Understanding Plato*, from the middle of the third century and CLEOMEDES' work *Meteōra*, written between the early third century and the middle of the fourth century. Lawlor, Lawlor, 1979, Theon of Smyrna, Mathematics Useful for Understanding Plato. Bowen, Todd, 2004, Cleomedes' Lectures on Astronomy.

12 In the fixed-Earth ontology of the ancients, the difference between the solar and sidereal day reflects the year-long rotation of the celestial vault and the Sun's progression along the ecliptic.

the Sun, which means it must turn around its axis by that same angle to bring the Sun to the zenith again, and this results in a 24-hour day. If the Sun rises at point *S*, simultaneously point *E* of the celestial equator also rises (Figure 1.1). The Sun once again reaches the horizon after one day. In one sidereal day, however, the Sun also moves eastward by about 1°, so the sunrise lags behind the rise of the celestial equator. After one day, the Sun reaches point *R*, while the arc $\widehat{RS}$ is an arc on which the Sun travels further to reach point *S*, where it rises again. Neither the angular velocity of the Sun nor the angle between the ecliptic and the horizon are constant. The arc $\widehat{RS}$ length depends on the Sun's angular velocity and on the angle that the ecliptic forms with the horizon, and thus changes from day to day.[13] The phenomenon, as described by GEMINUS, corresponds to the contribution to the equation of time due to the obliquity of the ecliptic.

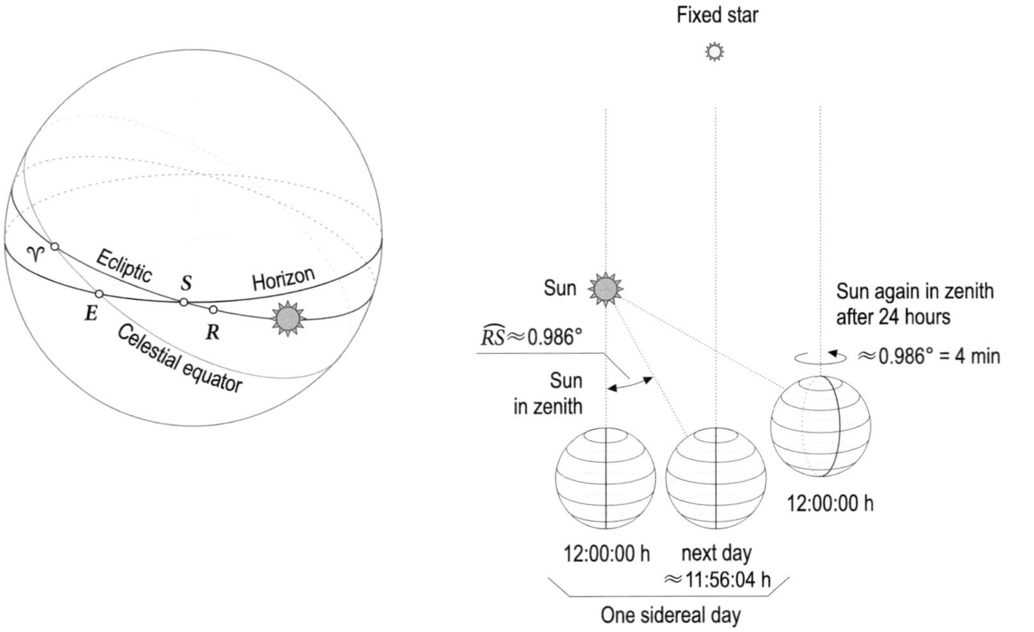

Figure 1.1 GEMINUS' explanation of the influence of the obliquity of the ecliptic on the equation of time. Adapted from Evans, Berggren, 2006, p.161, Figure 6.1, and Neugebauer, 1975, p.1215, Figure 27 (left). Brief sketch of sidereal day (right).

In chapter VI, GEMINUS also takes on related topics such as the longest day in the year and the changing length of the day depending on the seasons. According to GEMINUS, the days are longer than the nights in the zodiac signs of the ram ♈, the bull ♉, the twins ♊, the crab ♋, the lion ♌ and the virgin ♍, and the nights are longer than the days in the signs of the scales ♎, the scorpio ♏, the archer ♐, the sea goat ♑, the water bearer ♒ and the fish ♓.

13 Evans, Berggren, 2006, p.161, note 1, Figure 6.1 and p.162, note 2.

1.2 Claudius Ptolemy

1.2.1 Ptolemy's life and work

As with GEMINUS, very little is known about the life of CLAUDIUS PTOLEMY, Gr. *Κλαύδιος Πτολεμαῖος*, Lat. CLAUDIUS PTOLEMAEUS. The results of his astronomical observations place him between the years 125 CE – the ninth year of the reign of the Roman Emperor HADRIAN – and 141, the fourth year of the reign of ANTONINUS PIUS, the beginning of which was chosen by PTOLEMY as the epoch in the preparation of his star catalogue.[14] We assume that he worked in Alexandria or at least in the immediate vicinity, and was closely associated with Alexandria's library, which provided access to the works of some of his predecessors.[15] In the *Almagest*, written some time during the reign of Emperor ANTONINUS PIUS around 150 CE, PTOLEMY devoted a section of the chapter at the end of the third book to the equation of time.[16] He also dedicated several works to SYRUS, about whom we have no information.[17] In the *Almagest*, he also mentions the results of observations by THEŌN OF SMYRNA, who observed Mercury and Venus between 127 and 132 CE and may have been PTOLEMY's teacher.[18] We know that PTOLEMY had access to the works of HIPPARCHUS. Unfortunately, most of the work from the period between HIPPARCHUS and PTOLEMY was later lost, mainly because it was deemed obsolete due to the professional recognition and importance of PTOLEMY's work. References to HIPPARCHUS' works are also found in the works of PTOLEMY's contemporaries, GALEN and VETTIUS VALENS. With PAPPUS, the *Almagest* became the dominant astronomical work for the next thousand years.[19]

Of all the preserved ancient works, the expertise of the *Almagest* is only surpassed by EUCLID's *Elements*.

14 HADRIAN (76–138), Lat. PUBLIUS AELIUS HADRIANUS, Roman emperor in the period 117–138. Succeeded by ANTONINUS PIUS (86–161), Lat. ANTONINUS PIUS, who ruled between 138 and 161.

15 Riley, 1996, Ptolemy's Use of His Predecessor's Data, Transactions of the American Philological Association vol. 125, pp.221–250. For more information on PTOLEMY's Star Catalogue, see also Grasshoff, 1990. For more information on the operation of the library in Alexandria and the historical circumstances, see El-Abbadi et al. (eds.), 2008, What Happened to the Ancient Library of Alexandria? See also Haas, 2006, Alexandria in Late Antiquity.

16 Toomer, 1998, III.9, H263, p.171. The title in Greek was the *Systematic Mathematical Treatise*, Gr. *Μαθηματικὴ Σύνταξις*. The word *almagest* derives from the Greek word *μεγίστη*, the largest, large (treatise). But later, possibly due to incorrect pronunciation, it was converted into the *al-majastī* form and then into the Latinized form of *almagesti* or *Almagestum*. For explanations and Latin words of Arab origin, see Latham, 1975, Arabic into Medieval Latin, p.49.

17 SYRUS is also mentioned at the beginning of the *Almagest*. See also Toomer, 1998, p.35, n.5.

18 For citations of THEON's observations, see Toomer, 1998, Ptolemy's Almagest IX.9, p.456 and X.1, p.469, X.2, p.471. THEON OF SMYRNA (c.70–135), also called THEON THE MATHEMATICIAN or OLD THEON. The value of his works lies above all in the numerous references to the works of older authors. Neugebauer, 1975, pp.951–952. For more data, see Heath, 1921, II, pp.238–244.

19 GALEN (129?–216?), Lat. AELIUS GALENUS, Gr. *Γαληνός*, GALENOS, also GALEN OF PERGAMON, who was better known as a doctor, was also a philosopher. For more information about his philosophical work, see Lloyd, 2006, Principles and Practices in Ancient Greek and Chinese Science. VETTIUS VALENS (120–175), Lat. VETTIUS VALENS, Greek astronomer and astrologer. See Riley, 1996, A Survey of Vettius Valens. Dicks, 1960, p.16. Kroll, 1908, Vettii Valentis Anthologiarum Libri, I, 19. PAPPUS, Gr. *Πάππος*, a Greek mathematician, known for commenting on EUCLID's *Elements*, and other works. He worked at the end of the 3rd century CE at the time of Emperor DIOCLETIAN and probably a little later. Almost nothing is known about his life. For more on his work see Sefrin-Weis, 2010, Pappus of Alexandria: Book 4 of the Collection, Jones, 1986, Pappus of Alexandria, Book of the Collection and Heath, 1921, A History of Greek Mathematics, vol. II, pp.355–439. PAPPUS also commented on the *Almagest*, but his commentary was only partially preserved, and it is unclear whether his comments, unlike THEON's a century later, covered the whole book or only some of the chapters. Also ISIDORE OF SEVILLE mentions PTOLEMY – falsely as a king – in an encyclopaedia from the 7th century. *The Etymologies*, Book III, XXVI, Barney et al., 2014, p.99 or Möller, 2008, p.140.

1.2.2 Equation of time

PTOLEMY, who described the unequal length of the day in the *Almagest*, considered the day as the time interval between two successive positions of the Sun at the same points of the horizon or meridian, which is basically similar to the present definition.[20] The word equation had several meanings in ancient literature; among others, PTOLEMY used it to describe the value to be added or taken away from some other value – in modern language, this would be called a correction. PTOLEMY described the phenomenon mathematically in chapter 9 of book III of the *Almagest*, where he quoted both the influence of the uneven motion of the Sun and the influence of the obliquity of the ecliptic, which is the oldest known mathematical treatment of the concept that we presently refer to as the equation of time.[21] For the argument of the true ecliptic longitude, he tabulated the values of the equation of time in *Handy Tables*, which are the oldest known values of the equation written in tabular form.[22] The mathematical reconstruction of the equation of time shows that PTOLEMY knew linear interpolation, and the sources that he used in his works are also partly known.[23] PTOLEMY's equation of time has a value of zero at ♏ 0° (0° of the scorpio), and the epoch of the values tabulated in *Handy Tables* is the first day of the era of PHILIP.[24] It is evident from the *Almagest* that PTOLEMY used the values of two different, temporally separate, astronomical observations when discussing the equation of time. For PTOLEMY's equation, the following expression applies:[25]

[1.1] $$E = \Delta t_{\odot} - \Delta t_{\mathrm{M}}$$

$\Delta t_{\odot}$ the difference between the passages of the true Sun across the meridian at moments A and B at midday

Δt_{M} the difference between the passages of the mean Sun across the meridian at moments A and B at midday

20 Toomer, 1998, Ptolemy's Almagest, III.9, pp.169–170.

21 Toomer, 1998, pp.169–172. For the graphical description of PTOLEMY's text, see also Neugebauer, 1975, vol. 3, p.1222, Figure 57.

22 Delambre, 1817, Histoire de l'Astronomie Ancienne, reprint 2005, 2nd book, p.632. Halma, 1822, Commentaire de Théon d'Alexandrie, sur le Livre III de l'Almageste de Ptolémée; Tables Manuelles des Mouvemens des Astres, traduites pour la première fois du grec en français, sur les manuscrits de la bibliothèque du Roi, par M. L'Abbé Halma, reprint 1990, I, pp.148–155. Tihon, Mercier, 2011, Ptolemy's Handy Tables, volume 1b, table A1, pp.10–13.

23 THEON's *Great commentary on Ptolemy's Handy Tables* is preserved, where THEON also commented on the equation of time in chapter 2. Mogenet and Tihon, 1985, Le "Grand Commentaire" de Théon d'Alexandrie aux Tables Faciles de Ptolémée, pp.237–244. Also, Halma, 1822, Commentaire de Théon d'Alexandrie sur les Tables manuelles Astronomiques de Ptolémée. The text of Handy Tables was also published in Heiberg, 1907, Claudii Ptolemaei opera quae exstant omnia/2, Opera astronomica minora, pp.157–185. For more information on resources see Rome, 1939, Le problème de l'équation du temps chez Ptolémée, Annales de la Société Scientifique de Bruxelles 59, pp.211–224.

24 12 November 324 BCE, i.e. Thoth 1, 324 BCE or 425th year of NABONASSAR at midday in Alexandria. Thoth was the first month of the Egyptian calendar, and also the first month of floods. For a more detailed description of the periods and a list of rulers, see Tihon, Mercier, 2011, 1b, pp.51–78.

25 Pedersen, 2011, A Survey of the Almagest, p.156.

Using modern mathematical notation, PTOLEMY's equation of time $(\lambda_\odot, \Gamma)$, as a function of the longitude of the Sun and the apogee, can be written as:[26]

[1.2] $$E(\lambda_\odot, \Gamma) = \alpha(\lambda_\odot) - \lambda_M(\lambda_\odot, \Gamma) - (\alpha(\lambda_0) - \lambda_M(\lambda_0, \Gamma))$$

$\lambda_\odot$	true longitude
λ_0	true longitude for 0° ♏, or $\lambda_0 = 210°$, when $E = 0$[27]
α	right ascension
λ_M	mean longitude of ecliptical mean Sun
Γ	apogee

Unlike in modern astronomy, ancient astronomers understood the term 'mean Sun' as the Sun moving along the ecliptic at a steady pace and coinciding with the true Sun in the apogee and perigee. In modern astronomy, the mean Sun moves along the equator and coincides with the true Sun at longitude $\lambda_\odot = 0°$, or at the vernal equinox.[28] The equation of the centre, or in the historical context the solar equation, is the difference between the true and the mean anomaly, i.e. the difference between the angular positions of the true Sun, as it is observed to move along the ecliptic, and the mean Sun, which would move evenly along the ecliptic. The values of the equation of the centre depend on the parameters of the circle, and especially on the position of the apogee. whether the equation is positive or negative sign depends on the quadrant, or on whether the mean Sun overtakes or falls behind the true Sun. The values of the solar equation as a function of the mean anomaly are tabulated in *Almagest* III.6, but can also be calculated using the equations below.[29] The expression for the equation of the centre can be obtained by means of the sine law, in which case it applies for the triangle $\triangle SCO$ (Figure 1.3):

♈	0	22;54
	10	18;42
	20	14;41
♉	30	11; 4
	40	8;17
	50	6;34
♊	60	6; 2
	70	6;38
	80	8; 8
♋	90	10;13
	100	12;25
	110	14;17
♌	120	15;29
	130	15;45
	140	15; 1
♍	150	13;24
	160	11; 5
	170	8;23
♎	180	5;36
	190	3; 5
	200	1;10
♏	210	0; 7
	220	0;13
	230	1;38
♐	240	4;22
	250	8;18
	260	13; 5
♑	270	18;17
	280	23;22
	290	27;47
♒	300	31; 8
	310	33; 3
	320	33;28
♓	330	32;25
	340	30; 6
	350	26;50
	360	22;54

Table 1.1 Excerpt from the equation of time table from *Handy Tables*. Source: Delambre, 1817, Histoire de l'Astronomie Ancienne, II, p.632.

26 Tihon, Mercier, 2011, 1b, pp.89–96 and Appendix A, pp.158–159, Mogenet, Tihon, 1985, pp.240–242. Van Dalen, 1994, pp.100–104. For general information, see Evans, 1998, The History and Practice of Ancient Astronomy, pp.235–245.

27 The equation of time as given in the *Handy Tables*, with the epoch of the era of PHILIP as a reference moment, has the value 0 at 210°. Tihon, Mercier, 2011, p.13 or Halma, 1822, p.155. Equation of time from Delambre, 1817 has value 0 at 214° (Table 1.1).

28 Smart, 1990, Textbook on Spherical Astronomy, p.42.

29 In ancient and medieval literature, two types of solar equations can be found: one for the case of dependence on the ecliptic longitude of the mean Sun travelling on the ecliptic, and the other for the case of dependence on the ecliptic longitude of the true Sun, depending on how the equation of time was used in the old sources. For an explanation, see van Dalen, 1994, pp.101–103.

[1.3] $$\frac{e}{\sin c} = \frac{R}{\sin(\lambda_\odot - \lambda_\Gamma)} \quad \text{or} \quad c = \arcsin\left(\frac{e}{R} \cdot \sin(\lambda_\odot - \lambda_\Gamma)\right)$$

The maximum value of the solar equation c_{max} occurs when the true anomaly $\kappa = 90°$ or $\sin\kappa = 1$. By taking into account $\kappa = \lambda_\odot - \lambda_\Gamma$ (Equation 1.5) we get:

[1.4] $$c = \arcsin\left(\frac{e \sin\kappa}{R}\right) \quad \text{or} \quad \sin c_{max} = \frac{e}{R}$$

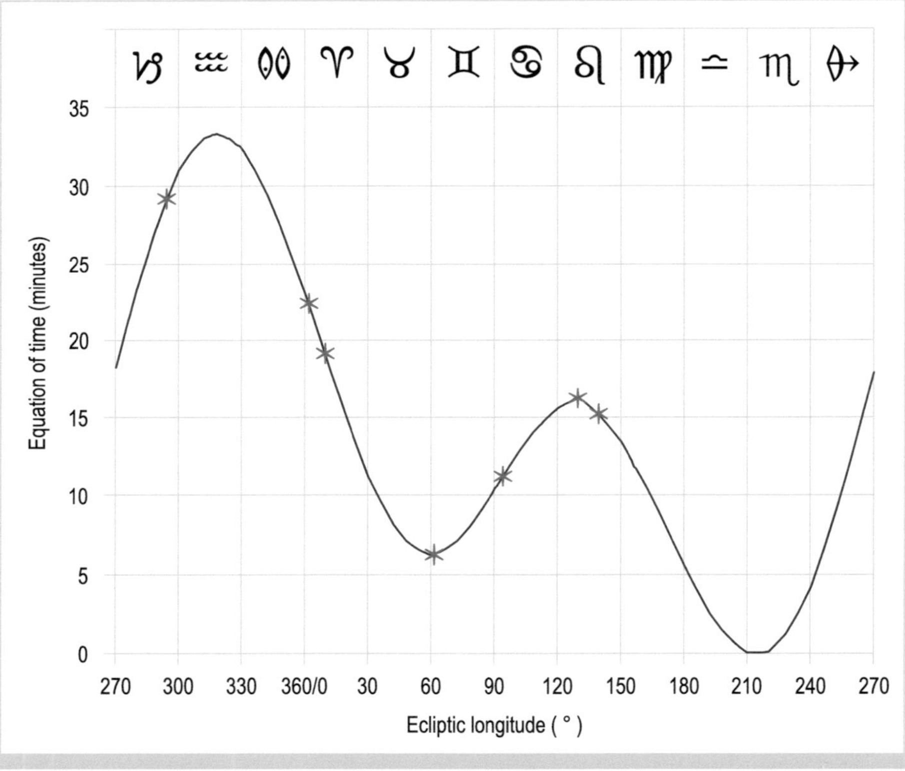

Figure 1.2 The course of PTOLEMY's equation of time in units of time (minutes), tabulated in *Handy Tables*, for the longitude of the Sun in steps of 1°. Data source for the diagram: Halma, 1822, part I, pp.148–155, the places marked with the symbol (✱) contain obvious (scribal?) errors. For data on PTOLEMY's equation of time, see also Tables 1.2 and 1.3.

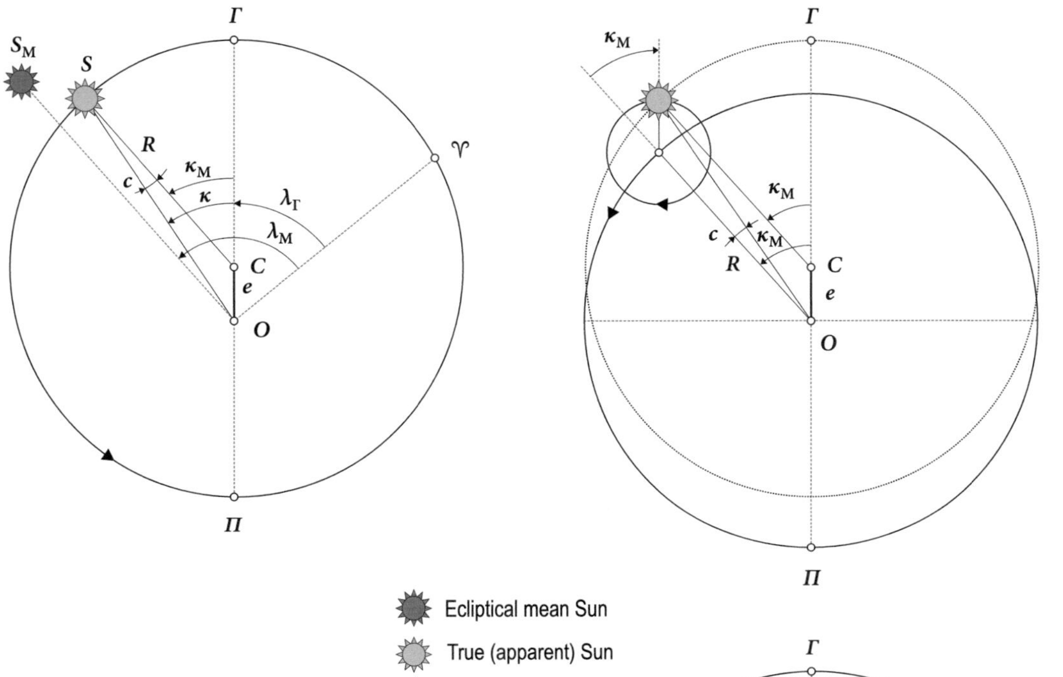

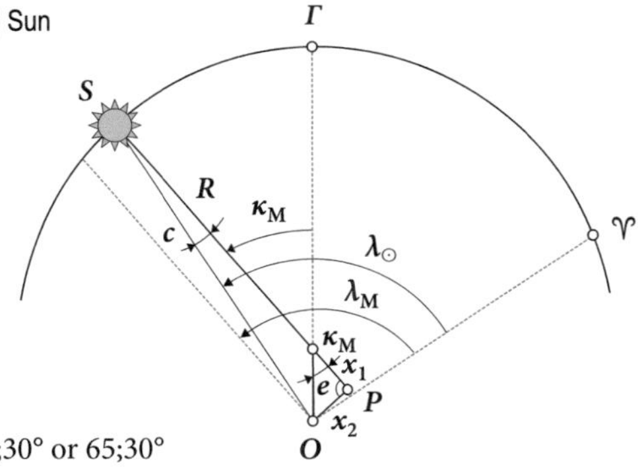

♈	vernal equinox
$\lambda_{\odot}$	true longitude
Γ, Π	apogee and perigee
κ	true anomaly
κ_{M}	mean anomaly
λ_{M}	mean longitude of ecliptical mean Sun
R	radius, Ptolemy's value = 60
e	eccentricity = OC, Ptolemy's value = 2;30
λ_{Γ}	the longitude of the apogee, Ptolemy's value = ♊ 5;30° or 65;30°
c	the equation of the centre, in the historical context also the solar equation
O, C	the observer (Earth) and the centre of the eccentric circle
OS_{M}	direction of the ecliptical mean Sun

Figure 1.3 Ptolemy's model of eccentric movement and epicycles. The figure on the right is important for understanding the calculation of the solar equation or the equation of the centre, see the text below. Adapted from: Neugebauer, 1975, p.1220, Figures 49–52, and Chabás, Goldstein, 2011, pp.63–64, Figure 10. See also Evans, 1998, p.227, Figure 5.21.

The following expressions, valid for all values of κ and $\lambda_{\odot}$, are also apparent from Figure 1.3:[30]

$$\lambda_{\odot} = \lambda_M - c \qquad \kappa_M = \lambda_M - \lambda_\Gamma = \lambda_{\odot} - \lambda_\Gamma + c \qquad \lambda_{\odot} = \lambda_\Gamma + \kappa \qquad \kappa = \kappa_M - c \tag{1.5}$$

The sign of the equation of the centre depends on the quadrant where the Sun is currently located. In the literature, the equation of the centre is sometimes given as a value to be added to the mean longitude:

$$\lambda_{\odot} = \lambda_M + c \qquad \kappa_M = \lambda_{\odot} - \lambda_\Gamma - c \qquad \kappa = \kappa_M + c \tag{1.6}$$

In this case:[31]

$$c(\kappa) < 0 \quad \text{when} \quad 0° < \kappa < 180°$$
$$c(\kappa) > 0 \quad \text{when} \quad 180° < \kappa < 360°$$

Ptolemy's value $c_{max} = 2;23°$ is valid for radius $R = 60$, eccentricity $e = 2;30$, apogee $\lambda_\Gamma = 65;30°$ or ♊ 5;30° and mean anomaly $\kappa_M = 92;23°$.[32] The expression for the equation of the centre or the solar equation c can also be obtained by solving the right triangle ΔSPO with sides (catheti) $R + x_1$ and x_2. See Figure 1.3, where $x_1 = e \cos \kappa_M$ and $x_2 = e \sin \kappa_M$ from which follows:

$$\tan c = \frac{x_2}{R + x_1} = \pm \frac{e \sin \kappa_M}{R + e \cos \kappa_M} \qquad \begin{cases} + \text{ if Equation 1.5 is applied} \\ - \text{ if Equation 1.6 is applied} \end{cases} \tag{1.7}$$

By applying (see Figure 1.3):

$$OS = \sqrt{(R + e \cos \kappa_M)^2 + (e \sin \kappa_M)^2} = \sqrt{R^2 + 2R \cdot e \cos \kappa_M + e^2}$$

from the triangle ΔSPO, another useful expression also follows:[33]

$$\sin c = \frac{x_2}{OS} = \pm \frac{e \sin \kappa_M}{\sqrt{R^2 + 2R \cdot e \cos \kappa_M + e^2}} \qquad \begin{cases} + \text{ if Equation 1.5 is applied} \\ - \text{ if Equation 1.6 is applied} \end{cases} \tag{1.8}$$

30 Evans, 1998, p.226 or Van Dalen, 1994, p.101.

31 For further explanation, see Pedersen, 2011, pp.138–141.

32 Neugebauer, 1975, pp.59–60. For apogee, the value 65;30 was supposedly already used by Hipparchus. In historical sources, the value 65;30° is given but based on mathematical analysis it is more likely that Ptolemy used the value 66°. Van Dalen, 1994, p.126 and Tihon, Mercier, 2011, 1b, pp.90–96. For the calculation, see Neugebauer, 1975, pp.57–58.

33 Evans, 1998, pp.232–233.

By taking into account the relations $\sin c_{max} = \frac{e}{R}$ and $\kappa_M = \lambda_M - \lambda_\Gamma$ (Equation 1.5), it follows from the Equation 1.7:

$$c = \arctan\left(\frac{\sin(\lambda_M - \lambda_\Gamma)\sin c_{max}}{1 + \cos(\lambda_M - \lambda_\Gamma)\sin c_{max}} \right) \quad [1.9]$$

If we insert this expression into the equation for the longitude of the true Sun $\lambda_\odot = \lambda_M - c$ (Equation 1.5), and if we consider trigonometric addition formula:

$$\tan(x \pm y) = \frac{\tan x \pm \tan y}{1 \mp \tan x \tan y}$$

we get:[34]

$$\begin{aligned}
\tan\lambda_\odot &= \tan(\lambda_M - c) \\
&= \frac{\tan\lambda_M - \tan c}{1 + \tan\lambda_M \tan c} \\
&= \frac{\tan\lambda_M(1 + \cos(\lambda_M - \lambda_\Gamma)\sin c_{max}) - \sin(\lambda_M - \lambda_\Gamma)\sin c_{max}}{1 + \cos(\lambda_M - \lambda_\Gamma)\sin c_{max} + \tan\lambda_M \sin(\lambda_M - \lambda_\Gamma)\sin c_{max}} \\
&= \frac{\sin\lambda_M + \sin c_{max}(\sin\lambda_M\cos(\lambda_M - \lambda_\Gamma) - \cos\lambda_M \sin(\lambda_M - \lambda_\Gamma))}{\cos\lambda_M + \sin c_{max}(\cos\lambda_M\cos(\lambda_M - \lambda_\Gamma) + \sin\lambda_M \sin(\lambda_M - \lambda_\Gamma))} \\
&= \frac{\sin\lambda_M + \sin c_{max}\sin\lambda_\Gamma}{\cos\lambda_M + \sin c_{max}\cos\lambda_\Gamma}
\end{aligned}$$

The equation can be rearranged and in the following we obtain the expressions:

$$\begin{aligned}
&\tan\lambda_\odot(\cos\lambda_M + \sin c_{max}\cos\lambda_\Gamma) = \sin\lambda_M + \sin c_{max}\sin\lambda_\Gamma \\
&\sin\lambda_\odot(\cos\lambda_M + \sin c_{max}\cos\lambda_\Gamma) = \cos\lambda_\odot(\sin\lambda_M + \sin c_{max}\sin\lambda_\Gamma) \\
&\sin c_{max}(\sin\lambda_\odot\cos\lambda_\Gamma - \cos\lambda_\odot\sin\lambda_\Gamma) = \cos\lambda_\odot\sin\lambda_M - \sin\lambda_\odot\cos\lambda_M \\
&\sin c_{max}\sin(\lambda_\odot - \lambda_\Gamma) = \sin(\lambda_M - \lambda_\odot)
\end{aligned}$$

The final result gives a relation between the longitude of the mean Sun and the true Sun:[35]

$$\lambda_M = \lambda_\odot + \arcsin(\sin c_{max}\sin(\lambda_\odot - \lambda_\Gamma)) \quad [1.10]$$

$\lambda_\odot$ true longitude
c_{max} the maximum of the solar equation
λ_Γ longitude of apogee

34 Tihon, Mercier, 2011, 1b, Appendix A, pp.158–159. See also Pedersen, 2011, pp.138–143.
35 Tihon, Mercier, 2011, 1b, pp.158–159. For further explanation, see Neugebauer, 1975, vol. I, pp.61–68, also Pedersen, 2011, Chapter 5 and Toomer, 1998, III.7.

From the data in the *Almagest* and *Handy Tables*, the main contributions to the equation of time can be calculated: the contribution due to the obliquity of the ecliptic $\lambda_{\odot} - \alpha_{\odot}$ and the contribution due to eccentricity, the equation of the centre or the solar equation *c*. Right ascension is tabulated in *Almagest* II.8 and in the *Handy Tables*, and the solar equation as a function of the mean anomaly in *Almagest* III.6. PTOLEMY's equation of time, as a function of longitude, is written in modern notation as:[36]

[1.11] $$E = \lambda_{M} - \lambda_{M0} - (\alpha_{\odot} - \alpha_{0})$$

By taking $\lambda_M = \lambda_{\odot} + c$ into account (Equation 1.5) and rearranging the above equation, we obtain the equation that is the basis for calculating the equation of time, if we know the values of the position of the Sun at a given moment (t_0) or within a given time interval.[37]

[1.12] $$\begin{aligned} E &= \lambda_{\odot} + c - \lambda_{M0} - (\alpha_{\odot} - \alpha_{0}) \\ &= \underbrace{\alpha_0 - \lambda_{M0}}_{k} + (\lambda_{\odot} - \alpha_{\odot}) + c \end{aligned}$$

$\lambda_{\odot}$ true longitude
λ_{M0} mean longitude, for reference moment (♓ 0;45° or 330;45°)
$\alpha_{\odot}$ right ascension
α_0 right ascension, for reference moment (335;8°)
c equation of the centre, or solar equation;
the sign as in the Equation 1.5 should be applied
$\alpha_0 - \lambda_{M0}$ constant due to epoch of NABONASSAR; $k = +4;23°$

The above values for α_0 and λ_{M0} apply to the epoch of the era of NABONASSAR, which is on Thoth the 1st in the first year of NABONASSAR's rule, at midday in Alexandria.[38] This epoch, which corresponds to February 26, 747 BCE with the Julian day number *JDN* = 1,448,638, was used by PTOLEMY in the *Almagest*, whereas the absolute precise epoch constant used by PTOLEMY in the *Almagest* or in the *Handy Tables* cannot be determined on the basis of the investigated sources.[39] In the *Handy Tables*, he apparently used the era of PHILIP as the epoch, beginning on Thoth 1, 425th year of NABONASSAR at midday in Alexandria; however, he mentions the era of PHILIP only in the *Instructions for the Use of Handy Tables*.[40]

36 Neugebauer, 1975, vol. 1, p.66.

37 Neugebauer, 1975, vol. 1, p.66. See also Equation 1.2, the sign of the equation of time is not uniquely determined.

38 Neugebauer, 1975, vol. 1, p.67.

39 Van Dalen, 1994, p.114, epoch constant is −3;18. For the values from Halma's tables under $\lambda_0 = 210°$ and $\lambda_\Gamma = 65;30°$, the value of the epoch constant is −3;33, see Example 1.2 or Example 14.2. Tihon, Mercier, 2011, 1b, p.93. See also Manitius, 1912, Ptolemäus Handbuch der Astronomie, Band I, III.9, pp.186–191. Toomer, 1998, pp.169–172, also note 71, p.172.

40 12 November 324 BCE, with the Julian day number *JDN* = 1,603,398. Tihon, Mercier, 2011, 1b, p.52, note 1 and Van Dalen, 1994, p.114. Heiberg, 1907, Volume 2. Opera astronomica minora, p.160, lines 20–5. The *Instructions for Use of Handy Tables* are only available in Greek. The translation, Halma, 1822, I, pp.1–26 is merely a framework, see van der Waerden, 1958, Die Handlichen Tafeln des Ptolemaios, Osiris, 13, pp.54–78. For a more detailed description of the periods and a list of rulers, see Tihon, Mercier, 2011, 1b, pp.51–78.

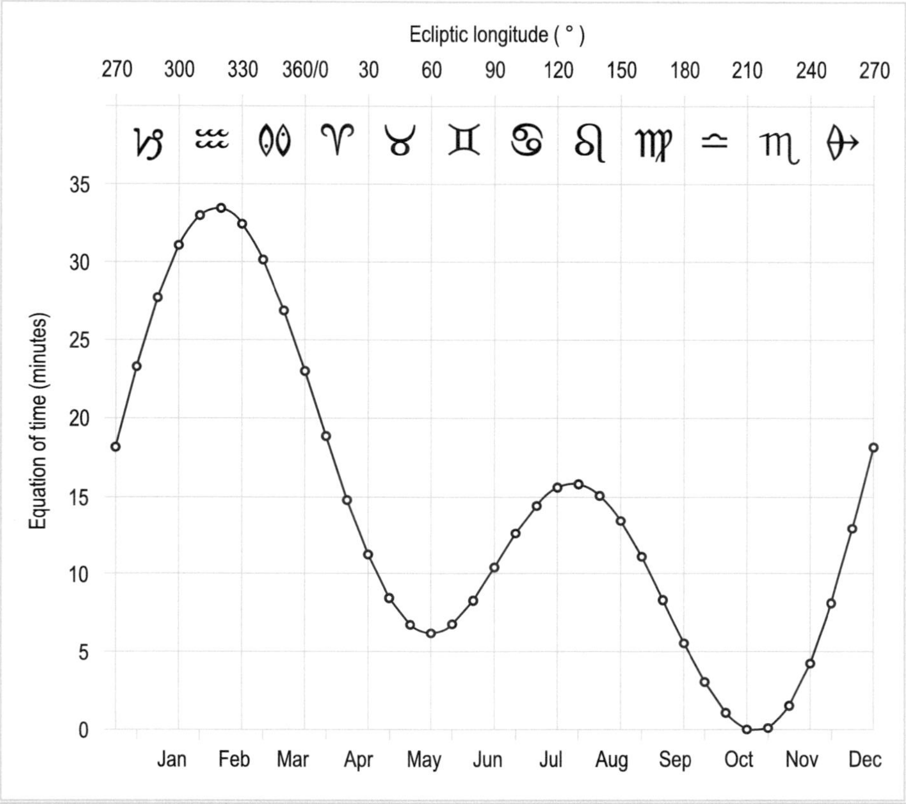

Figure 1.4 The course of PTOLEMY's equation of time in units of time (minutes) as a function of the ecliptic longitude of the Sun in steps of 10°. Data source for the diagram: Delambre, 1817, p.632 or Table 1.1. For the calculation, see Example 1.3.

The oldest known fragments of the equation of time with the lowest zero value are contained in *Papyrus P. London 1278*. Values from this papyrus are calculated for the epoch of the era of NABONASSAR, as in the *Almagest*, and not for the era of PHILIP, as in the *Handy Tables*.[41] The tables are very likely to have been calculated with parameters similar to those in the *Handy Tables*, such as $e = 2;30$ and $\lambda_\Gamma = 66°$. The value of obliquity of the ecliptic also differs from the one in the *Handy Tables*, which may be due to an error.

41 Neugebauer, 1962, p.108. Tables with a minimum zero value are also published in *Handy Tables* and works of THEON. Neugebauer, Skeat, 1958, The Astronomical Tables P. Lond. 1278. With a Note on the Palaeography of the Fragments. Osiris, vol. 13, pp.93–113, especially pp.101 and 109. Also Van Dalen, 1994, pp.138–143. Heiberg, 1907, pp.162.23–163.6, and Halma, 1822, reprint 1990, I, pp.148–155, column 3 of right ascension table.

Example 1.1 Calculate the reference longitude and the right ascension for the epoch of NABONASSAR on 26 February 747 BCE at midday in Alexandria. The results of PTOLEMY's Sun observations on 25 September 132 CE two hours after midday in Alexandria are known, when he measured the longitude of the Sun as ≏ 0° or 180°.

When we take into consideration that $R=60$, $\lambda_\Gamma=65;30°$ and the eccentricity e, which is $OC=2;30$, we obtain the result $c=2;10°$ for the equation of the centre (Equation 1.4) and for the mean anomaly $\kappa_M=\lambda_\odot-\lambda_\Gamma+c=116;40°$ (Equation 1.5).[42]

The number of Julian days *JDN* or the Julian day number can be obtained from the tables, the best source being Schram, 1908, pp.36 and 182–183.[43] The result from the Schram's tables is obtained by taking the contributions for the year −746 (p.182) and the month Thoth, and adding the number of days into the month (1 Thoth, hence one day).[44] We get (Schram, 1908, pp.182–183):

Era of NABONASSAR		contributions from the table
Year −746		1 448 000
Thoth		637
1 day		1
	Sum	$JD=1\,448\,638.0$

PTOLEMY's observation, 25 September 132 CE two hours after noon (Schram, 1908, p.36):

Year 132 CE		1 769 000
September		514
25 days		25
2 hours		0.083
	Sum	$JD=1\,769\,539.083$

If we know the calendar date, we can get the same result by calculating the Julian day number *JDN* or the Julian date *JD* for the epoch and moment of the PTOLEMY's observation.

Next, we need to consider that the Egyptian year is 365 days long. The difference between the epoch of NABONASSAR, 26 February 747 BCE, and the moment of PTOLEMY's observation on 25 September 132 CE, $\Delta JD=320\,901.083$ or $\Delta t=879$ years, 66 days and 2 hours.

For each time period, Table III.2 (Toomer, 1998, pp.142–143) gives the following values for the change in the mean anomaly $\Delta\kappa_M$:

	°	′	″	‴	⁗	′′′′′	′′′′′′
810 years	163	4	12	15	25	52	30 = 163.07007
18 years × 3	355	37	25	36	20	34	30 = … × 3
15 years	356	21	11	20	17	8	45 = …
30 days × 2 (2 months)	29	34	8	36	36	15	30 = … × 2
6 days	5	54	49	43	19	15	6 = …
2 hours	0	4	55	41	26	6	2 = …
						Sum	= 1651.4286

42 Adapted from Neugebauer, 1975, pp.59–60. The example follows PTOLEMY's procedure for calculating a true longitude which is summarised in *Almagest* III.8. See Toomer, 1998, p.169. For additional explanation see also Pedersen, 2011, pp.153–154.

43 See also Appendix A.

44 PTOLEMY uses Egyptian year or calendar, Toomer, 1998, pp.9–10. Egyptian year is 365 days long, for an explanation see Schram, 1908, pp.XXII–XXIII.

From Table III.2 in the *Almagest*, we obtain the change of the mean anomaly from the epoch (initial moment) until the moment of PTOLEMY's observations, for which we must know the length of the time interval between both temporal moments. For the epoch of NABONASSAR on 26 February 747 BCE, $\Delta t = 879$ years, 66 days and 2 hours (Schram, 1908, pp.182–183 and 36).[45] The result for the change of mean anomaly $\Delta\kappa_M$ in the given time period Δt is obtained by taking into consideration the contributions from Table III.2 in the *Almagest*, for $\Delta t = 810$ years + 3 × 18 years + 15 years + 2 months + 6 days + 2 hours (Toomer, 1998, pp.142–143). The sum of the contributions $\Delta\kappa_M = 1651;25°$. We reduce this to a more meaningful form by subtracting multiples of 360° to achieve the final result, 211;25°. We can then calculate the values of the mean anomaly, mean longitude and true longitude at the epoch (Equation 1.5):

Mean anomaly at the epoch: $\kappa_{M0} = \kappa_M - \Delta\kappa_M = 116;40° - 211;25° = 265;15°$

Mean longitude at the epoch: $\lambda_{M0} = \lambda_\Gamma + \kappa_{M0} = 65;30° + 265;15° = 330;45°$ or ♓ 0;45°

True longitude at the epoch: $\lambda_0 = \lambda_{M0} - c_{max} = 330;45° + 2;23° = 333;8°$

The right ascension α_0 obtained from *Almagest* II.8 or from the *Handy Tables* by means of interpolation is 335;8°.

Longitude, λ	right ascension, α
333°	335; 1°
334°	335;58°

For the calculated longitude $\lambda_0 = 335;8°$, $\alpha_0 = 333;8°$ follows by interpolation for right ascension.
For the era of NABONASSAR, we can calculate or verify the known reference moment constant k:

$$\kappa = \alpha_0 - \lambda_{M0} = 335;8° - 330;45° = +4;23°$$

Example 1.2 Calculate the value of PTOLEMY's equation of time as a function of the mean longitude for a day with a Sun longitude of $\lambda_\odot = 290°$ or ♑ 20°. When the equation of time has a value of zero, the reference longitude is $\lambda_{M0} = 210°$ or ♎ 30°. From the *Handy Tables*, the values of $\sin c_{max} = 1/24$ and the longitude of the apogee $\lambda_\Gamma = 66°$ (Van Dalen, 1994, p.126, Tihon, Mercier, 2011, 1b, pp.90–96). The right ascension values used by PTOLEMY are listed in *Almagest* II.8 (Toomer, 1998, p.100), (solution: $E = 27;42$ or 27 minutes 42 seconds).

Mean longitude can be calculated with Equation 1.10:

$$\lambda_{M0} = 210° + \arcsin(\sin c_{max} \sin(210° - 66°)) = 211;24,12°$$

$$\lambda_M = 290° + \arcsin(\sin c_{max} \sin(290° - 66°)) = 288;20,29°$$

The values of right ascension $\boldsymbol{\alpha}_\odot$ and $\boldsymbol{\alpha}_0$ are taken from *Almagest* II.8: $\alpha_0 = 207;50°$ for $\lambda_{M0} = 210°$ or ♎ 30° and $\alpha_\odot = 291;42°$ for $\lambda_\odot = 290°$ or ♑ 20°. The final result for the equation of time can be calculated with Equation 1.11. PTOLEMY did not use negative values, so with the intention to get the positive result, the sign of Equation 1.11 is reversed:

$$E = \alpha_\odot - \lambda_M - (\alpha_0 - \lambda_{M0})$$

$$= 291;42° - 288;20,29° - (\underbrace{207;50° - 211;24,12°}_{\approx -3;34°\text{, see footnote 39}}) = 6;55,43° \text{ (time-degrees) or 27 min 42 s}$$

45 Schram, 1908, pp.XXII–XXIII.

$\lambda_{\odot}$ (°)	$\alpha_{\odot}$ (°)[a]	λ_M (°)[b]	$E(\lambda)$[c]	$E(\lambda)$[d]	$\lambda_{\odot}$ (°)	$\alpha_{\odot}$ (°)[a]	λ_M (°)[b]	$E(\lambda)$[c]	$E(\lambda)$[d]
Ram ♈					**Scale ♎**				
10	9;10	8; 1,14	0;18,51	0;18,54	190	189;10	191;58,46	0; 3, 1	0; 3, 0
20	18;25	18;16,57	0;14,49	0;14,54	200	198;25	201;43, 3	0; 1, 4	0; 1, 9
30	27;50	28;35,48	0;11,13	0;11,13	210	207;50	211;24,12	0; 0, 0	0; 0, 0
Bull ♉					**Scorpio ♏**				
40	37;30	38;57,12	0; 8,27	0; 8,37	220	217;30	221; 2,48	0; 0, 5	0; 0, 6
50	47;28	49;20,31	0; 6,46	0; 6,55	230	227;28	230;39,29	0; 1,30	0; 1,40
60	57;44	59;45, 2	0; 6,12	0; 6,12	240	237;44	240;14,58	0; 4,12	0; 4,13
Twins ♊					**Archer ♐**				
70	68;18	70;10, 0	0; 6,48	0; 6,55	250	248;18	249;50, 0	0; 8, 8	0; 8,15
80	79; 5	80;34,39	0; 8,18	0; 8,22	260	259; 5	259;25,21	0;12,55	0;12,59
90	90; 0	90;58,16	0;10,23	0;10,28	270	270; 0	269; 1,44	0;18, 9	0;18, 9
Crab ♋					**Sea goat ♑**				
100	100;55	101;20, 6	0;12,36	0;12,32	280	280;55	278;39,54	0;23,17	0;23,16
110	111;42	111;39,31	0;14,26	0;14,20	290	291;42	288;20,29	0;27,42	0;27,35
120	122;16	121;55,54	0;15,37	0;15,37	300	302;16	298; 4, 6	0;31, 4	0;31, 4
Lion ♌					**Water bearer ♒**				
130	132;32	132; 8,46	0;15,49	0;16,15	310	312;32	307;51,14	0;32,59	0;32,49
140	142;30	142;17,44	0;15, 5	0;15,40	320	322;30	317;42,16	0;33,27	0;33,17
150	152;10	152;22,30	0;13,26	0;13,27	330	332;10	327;37,30	0;32,26	0;32,26
Virgin ♍					**Fish ♓**				
160	161;35	162;22,56	0;11, 5	0;11, 1	340	341;35	337;37, 4	0;30, 8	0;30, 2
170	170;50	172;19, 1	0; 8,20	0; 8,19	350	350;50	347;40,59	0;26,52	0;26,49
180	180; 0	182;10,53	0; 5,33	0; 5,33	360	360; 0	357;49, 7	0;23, 0	0;23, 0

Table 1.2 The equation of time $E(\lambda)$ in units of time (h; min, sec) as a function of the Sun's true longitude $\lambda_{\odot}$, calculated by the procedure as in Example 1.2 (c) and summarised by Halma, 1822, pp.148–155 (d). For the calculated values, the calculation procedure is summarised by Tihon, Mercier, 2011. For an example of calculation and values in shorter steps, see Tihon, Mercier, 2011, Ptolemy's Handy Tables, Table A1, pp.10–13, and comments on Table A1, pp.89–96, and Van Dalen, 1994, On Ptolemy's Table for the Equation of Time, Table 5, p.134.

Notes

a The values of right ascension are summarised by Halma, 1822, reprint 1990, I, pp.148–153. For every 10° of longitude, they are also listed in Toomer, 1998, Ptolemy's Almagest II.8, p.100, Sphaera Recta.
b The longitude values of the mean Sun are rounded and calculated by the procedure explained in Example 1.2.
c The equation of time is calculated taking into account $\sin c_{max} = 1/24$ and $\lambda_{\Gamma} = 66°$. The final values are rounded down to the nearest whole second. The newly calculated values may slightly deviate from the tabulated ones, and some of the values in the source tables are clearly incorrect. For the original values, more data and additional explanation, see Halma, 1822, I, pp.148–153, van Dalen, 1994, Tihon, Mercier, 1b, 2011, pp.10–13 and 89–107, and Pedersen, 2011, Chapter 5, pp.122–158.

$\lambda_\odot$ (°)	$\lambda_\odot - \alpha_\odot$ (°)	c (°)	E (°)	$\lambda_\odot$ (°)	$\lambda_\odot - \alpha_\odot$ (°)	c (°)	E (°)
Ram ♈				**Scales ♎**			
10	0;50	−1;58,46	−4;25,53	190	0;50	1;58,46	−0;28,20
20	1;35	−1;43, 3	−3;25,10	200	1;35	1;43, 3	0; 0,56
30	2;10	−1;24,12	−2;31,19	210	2;10	1;24,12	0;17, 5
Bull ♉				**Scorpio ♏**			
40	2;30	−1; 2,48	−1;49,54	220	2;30	1; 2,48	0;15,41
50	2;32	−0;39,29	−1;24,36	230	2;32	0;39,29	−0; 5,38
60	2;16	−0;14,58	−1;16, 5	240	2;16	0;14,58	−0;46, 8
Twins ♊				**Archer ♐**			
70	1;42	0; 9,59	−1;25, 7	250	1;42	−0; 9,59	−1;45, 6
80	0;55	0;34,39	−1;47,27	260	0;55	−0;34,39	−2;56,46
90	0; 0	0;58,16	−2;18,51	270	0; 0	−0;58,16	−4;15,22
Crab ♋				**Sea goat ♑**			
100	−0;55	1;20, 6	−2;52, 0	280	−0;55	−1;20, 6	−5;32,13
110	−1;42	1;39,31	−3;19,36	290	−1;42	−1;39,31	−6;38,38
120	−2;16	1;55,54	−3;37,12	300	−2;16	−1;55,54	−7;29, 1
Lion ♌				**Water bearer ♒**			
130	−2;32	2; 8,46	−3;40,20	310	−2;32	−2; 8,46	−7;57,53
140	−2;30	2;17,44	−3;29,23	320	−2;30	−2;17,43	−8; 4,50
150	−2;10	2;22,30	−3; 4,37	330	−2;10	−2;22,30	−7;49,36
Virgin ♍				**Fish ♓**			
160	−1;35	2;22,56	−2;29,11	340	−1;35	−2;22,56	−7;15, 3
170	−0;50	2;19, 1	−1;48, 5	350	−0;50	−2;19, 1	−6;26, 8
180	0; 0	2;10,53	−1; 6,13	360	0; 0	−2;10,53	−5;28, 0

Table 1.3 The contributions to PTOLEMY'S equation of time due to the obliquity of the ecliptic ($\lambda_\odot - \alpha_\odot$), eccentricity ($c$, equation of the centre) and equation of time for the epoch of the era of PHILIP. Values are stated in time-degrees (15° ≡ 1 hour or 1° ≡ 4 minutes) and calculated using the procedure in Example 1.3 but calculated with $\lambda_\Gamma = 66°$.

Notes

d The values of the equation of time from the *Handy Tables*, summarised by Halma, 1822, I, pp.148–155. Values that differ slightly from the ones in the table are also available in Delambre, 1817, Histoire de l'Astronomie Ancienne, II, p.632. For these values, according to DELAMBRE, $\sin c_{max} = 1/24$ and $\lambda_\Gamma = 65°$. Halma's tables contain values in several places that are clearly inconsistent with the function of the equation of time. The cause may be errors in the original manuscripts available to Halma, or mistakes in Halma's work. To the left is the value from Halma's work and to the right is the author's correction as used in the equation of time plot in Figure 1.2:
♈ 1° – 4°: 21;36/22;36, 21;11/22;11, 22;47/21;47, 22;22/21;22; ♈ 9°: 9;19/19;19; ♊ 1°: 6;6/6;16; ♋ 3°: 11;40/11;4

e For the given sign of the equation of centre, Equation 1.5 applies.

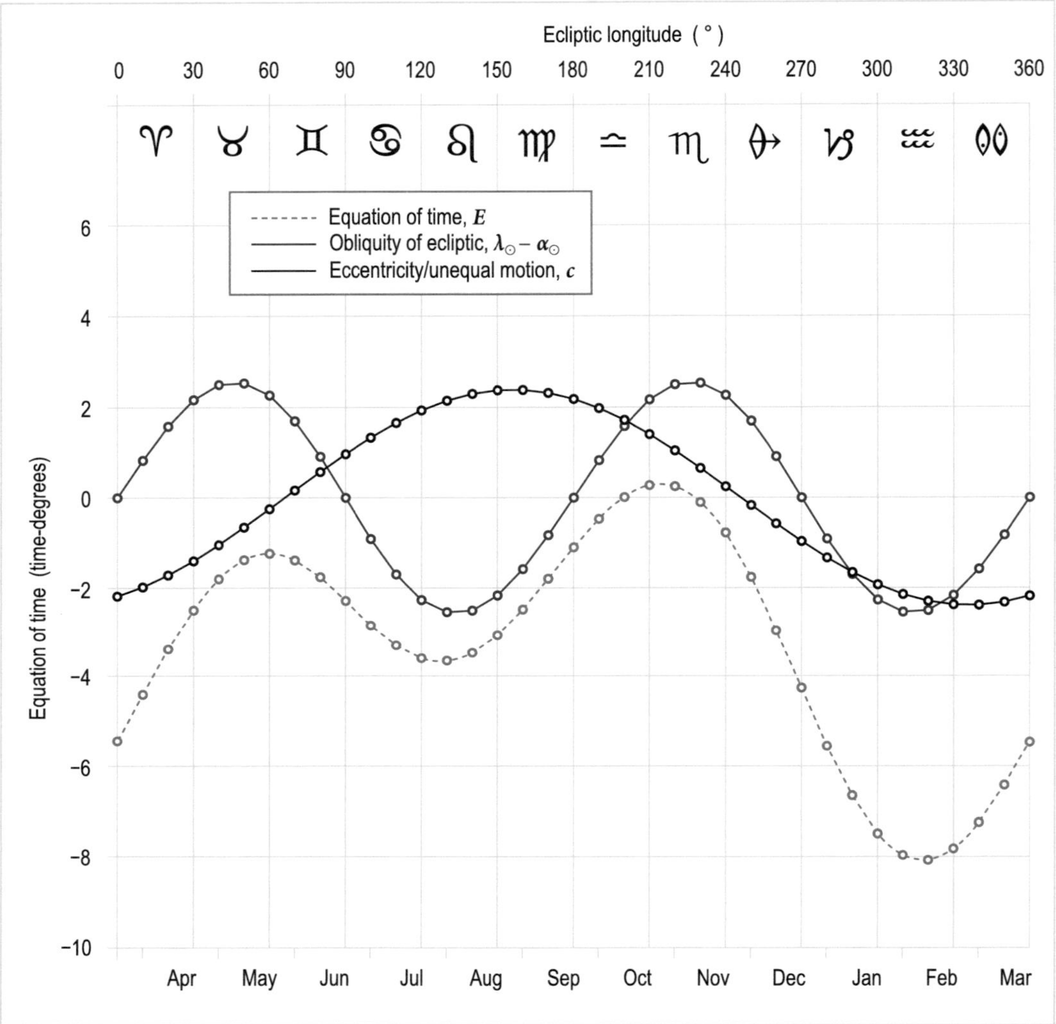

Figure 1.5 The influence of the obliquity of the ecliptic and the uneven revolution or eccentric orbit on the equation of time. The values are in time-degrees, and the conversion is as follows: 1° ≡ 4 minutes or 15° ≡ 60 minutes. The dashed grey line applies to the equation of time for the era of PHILIP with an epoch constant of −3;18°. The values are calculated by the procedure in Example 1.3. By applying a negative sign, we get very similar results as calculated in the literature for the epoch of NABONASSAR; see Neugebauer, 1975, III, p.1222, Figure 57.

Example 1.3 Calculate the equation of time as a function of the true longitude and the contributions to PTOLEMY's equation of time due to the obliquity of the ecliptic $\lambda_{\odot} - \alpha_{\odot}$ and the eccentricity c, for the longitude $\lambda_{\odot} = 290°$. Also, calculate the value of the equation of time with respect to the epoch of the era of PHILIP, Thoth 1, i.e. November 12, 324 BCE, at midday in Alexandria, as used by PTOLEMY in the *Handy Tables* (solution: $\lambda_{\odot} - \alpha_{\odot} = -1;42°$, $E = -6;39°$ or $E = 0;26,36$ h).

In this case, we also consider the results of PTOLEMY's observation on September 25, 132 CE, two hours after midday in Alexandria. In the calculation, the values $\lambda_{\odot} = 180°$, $\sin c_{max} = 1/24$, $R = 60$ and $\lambda_{\Gamma} = 65;30°$ are applied.

Similar to Example 1.1, we obtain the result $c = 2;10°$ for the equation of the centre (Equation 1.4) and for the mean anomaly $\kappa_M = \lambda_{\odot} - \lambda_{\Gamma} + c = 116;40°$ (Equation 1.5).[46]

From Table III.2 in the *Almagest*, we obtain the change $\Delta\kappa_M$ of the mean anomaly from the epoch (initial moment) until the moment of PTOLEMY's observations, for which we must know the length of the time interval between both temporal moments.

Era of PHILIP (Schram, 1908, pp.182–183)		contributions from the table
Year −323		1 603 000
Thoth		397
1 day		1
	Sum	$JD = 1\,603\,398.0$

PTOLEMY's observation, 25 September 132 CE two hours after noon (see Example 1.1) $JD = 1\,769\,539.083$.

Considering that the Egyptian year is 365 days long for the epoch of the era of PHILIP on 12 November 324 BCE, or 1 Thoth, in the 425th year of the era of NABONASSAR, $\Delta JD = 166\,141.083$ or $\Delta t = 455$ years, 66 days and 2 hours (Schram, 1908, pp.36, 182–183).

The mean anomaly is obtained by consulting the table for $\Delta t = 450$ years + 5 years + 2 × 30 days + 6 days + 2 hours (Toomer, 1998, pp.142–143). The change of mean anomaly $\Delta\kappa_M = 674;30°$, after subtracting the redundant multiples of 360°, we get the final result 314;30° for the given time period Δt.

As in Example 1.1, we can then calculate the mean anomaly and mean longitude at the epoch (Equation 1.5):

Mean anomaly at the epoch: $\kappa_{M0} = \kappa_M - \Delta\kappa_M = 116;40° - 314;30° = 162;10°$

Mean longitude at the epoch: $\lambda_{M0} = \lambda_{\Gamma} + \kappa_{M0} = 65;30° + 162;10° = 227;40°$ or ♏ 17;40°,

The true longitude at the epoch, incorporating the value of the equation of the centre that can be found or interpolated from Table III.6 in the *Almagest* is (Equation 1.5):

$$\lambda_0 = \lambda_{M0} - c_{max} = 227;40° - 0;45° = 226;55°$$

The right ascensions $\alpha_{\odot}$ and α_0 can be found or interpolated from *Almagest* II.8 or the *Handy Tables*. For the right ascension (Toomer, 1998, p.100), interpolation gives the values of $\alpha_0 = 224;23°$ or $\alpha_{\odot} = 291;42°$.

The contribution due to the obliquity of the ecliptic is:

$$\lambda_{\odot} - \alpha_{\odot} = 290° - 291;42° = -1;42°$$

The mean longitude can be calculated using Equation 1.10, in the same way as in Example 1.2. The value of the equation of time (Equation 1.11) is:

$$E = (\lambda_M - \lambda_{M0}) - (\alpha_{\odot} - \alpha_0)$$
$$= (288;20° - 227;40°) - (291;42° - 224;23°) = 60;40° - 67;19° = -6;39°$$

46 Based on the example for the epoch of NABONASSAR, Neugebauer, 1975, pp.59–60. See also Example 1.1. For a further explanation of the calculation, see Neugebauer, 1975, I, pp.61–68, and III, p.1081 or p.1222, Figure 57.

or by means of constant $k = \alpha_0 - \lambda_{M0} \approx -3;17°$ for the era of PHILIP, whereas the equation of the centre is calculated using Equation 1.4, and taking into account $e = 2;30°$ and $R = 60$. By Equation 1.12:

$$\begin{aligned} E &= (\alpha_0 - \lambda_{M0}) + (\lambda_{\odot} - \alpha_{\odot}) + c \\ &= (224;23 - 227;40) + (290 - 291;42) + (288;20 - 290) \\ &= -3;17° - 1;42° - 1;40° = -6;39° \end{aligned}$$

For the epoch of the era of PHILIP, the value of the equation of time for longitude $\lambda_{\odot} = 290°$ is $E = -6;39°$ expressed in time-degrees or −26 min and 36 sec. The equation of time is presented as a plot in Figure 1.5.

1.2.3 Spherical astronomy

CLAUDIUS PTOLEMY presented the mathematical procedures for calculating the main parameters of spherical astronomy by applying MENELAUS' theorem and chord trigonometric function in the first three books of the *Almagest*:

- Declination - *Almagest* I.14, I.15
- Right ascension - *Almagest* I.16
- Mean anomaly and epicycles - *Almagest* III.3
- True anomaly - *Almagest* III.4
- True longitude - *Almagest* III.8.

PTOLEMY calculated the true longitude using the following procedure:[47]

- First, we calculate the time interval between the epoch and the moment when the parameters are calculated. The result should be in the form *number of years* + *number of months* + *number of days* + *number of hours*.[48]
- Table III.2 shows the mean anomaly from the calculated time. The mean anomaly is the sum of the contributions for each time period (years, months, days, hours). Table III.2 gives the values for 18-year periods, individual years, months, days and hours of the day.[49]
- Add the mean anomaly at the epoch to the mean anomaly value in Table III.2.[50]
- Using Equation 1.6, calculate the mean longitude. The apogee value is 65;30° or ♊ 5;30°.
- From Table III.6, the calculated value of the mean anomaly, we determine the value of the equation of the centre or solar equation, and the sign of the centre equation is determined according to the quadrant in which the result is located. The intermediate values are interpolated.[51]
- Using Equation 1.6, we then calculate the true longitude.

47 Toomer, 1998, III.8, p.169 (H258), for additional explanation see also Pedersen, 2011, pp.153–154. See also Examples 1.1 and 1.3.

48 For more information about epochs see also Chapter 14.

49 Toomer, 1998, III.2, pp.142–143.

50 For the NABONASSAR epoch, the value of the mean anomaly is 265;15°. Toomer, 1998, III.2, p.142.

51 The equation of centre or solar equation Gr.*prosthaphairesis* angle. Toomer, 1998, III.6, p.167.

The calculation of right ascension given by PTOLEMY in *Almagest* I.16 follows from MENELAUS' theorem:[52]

[1.13] $$\frac{\text{Crd}\,2\widehat{BD}}{\text{Crd}\,2\widehat{AD}} = \frac{\text{Crd}\,2\widehat{BZ}}{\text{Crd}\,2\widehat{EZ}} \cdot \frac{\text{Crd}\,2\widehat{CE}}{\text{Crd}\,2\widehat{CA}}$$

The following values apply (Figure 1.6):

$2\widehat{BD} = 2(90° - \varepsilon)$

$2\widehat{AD} = 2\varepsilon$

$2\widehat{BZ} = 2(90° - \delta_\odot)$

$2\widehat{ZE} = 2\delta_\odot$

$2\widehat{CE} = 2\alpha_\odot$

$2\widehat{CA} = 180°$

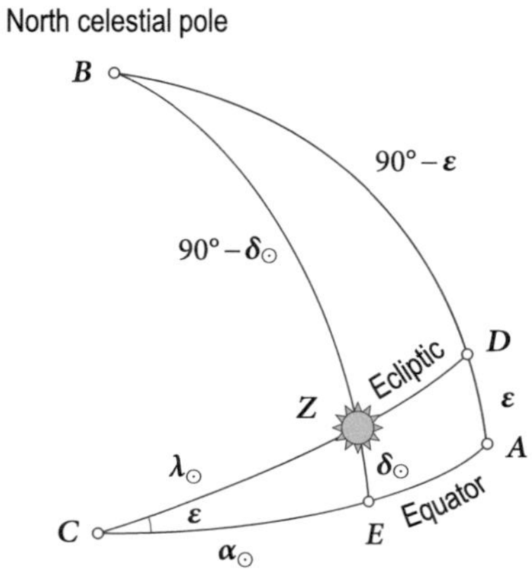

Figure 1.6 Application of the theorem of MENELAUS in PTOLEMY's spherical astronomy. See also Figure 0.10.

By using PTOLEMY's method, one can get the result for the right ascension with the help of the chord values given in *Almagest* I.11.[53] For modern convenience, by rearranging the expression of MENELAUS' theorem or Equation 1.13 and by applying the sine function, taking $R = 60$ into account, we get (Figure 1.6):[54]

[1.14] $$\text{Sin}\,\alpha_\odot = \frac{\text{Sin}(90° - \varepsilon)}{\text{Sin}\,\varepsilon} \cdot \frac{\text{Sin}\,\delta_\odot}{\text{Sin}(90° - \delta_\odot)} \cdot 60$$

or, in modern mathematical notation, with $R = 1$:[55]

$$\sin \alpha_\odot = \cot \varepsilon \tan \delta_\odot$$

52 Toomer, 1998, Almagest I.16, p.71, explanation pp.71–74. Halma, 1821, Commentaire De Théon d'Alexandrie sur le Premier & le Second Livres de la Composition Mathématique de Ptolemée, pp.257–269. See also Pedersen, 2011, p.70. For more detailed explanation and mathematical background of MENELAUS' theorem see also Björnbo, 1902, pp.88–92 or Toomer, 1998, Almagest I.14–I.16, pp.64–74.

53 For calculation of the result by applying the chord function all necessary procedures and tables are given in Toomer, 1998, pp.48–56 or tables on pp.57–60. Also Halma, 1813, pp.26–48. For additional information see also Toomer, 1974 or Pedersen, 2011.

54 See Introduction, note 75.

55 See Appendix D or Equation D.37.

1.3 The Almagest and the Handy Tables

1.3.1 Translations of the Almagest

The *Almagest* is mentioned in PTOLEMY's *Tetrabiblos, Handy Tables, Planetary Hypotheses* and *Geography*, which proves that PTOLEMY wrote it earlier than these other important works.[56] In spite of incorrect assumptions, the entries in the book were matched effectively with observations. PTOLEMY's *Almagest* and *Handy Tables* were later used in Constantinople, and were also appreciated among the Arabs who continued a considerable part of the rich Greek astronomical tradition. It is very likely that the *Almagest* was known to the Persians under the rule of SHAPUR I in the third century CE, and perhaps it was even translated into Pahlavi. Later on, the *Almagest* was translated into Syriac, and then several times into Arabic.[57] The first Arabic version is the work of an unknown translator(s) in the second half of the eighth, possibly at the beginning of the ninth century.[58] The translations by AL-ḤASSAN IBN QURAYSH and AL-ḤAJJĀJ IBN YŪSUF IBN MATAR, who was assisted by HLYL IBN SRGWN or SERGIUS, followed in the early ninth century. Of these translations, two original manuscripts are preserved, one entirely and one in fragments. Half a century later, a translation was undertaken by ISHAQ IBN ḤUNAYN, which was later revised by THĀBIT IBN QURRA. In the West, after the collapse of the Roman Empire, the *Almagest* sank into oblivion for centuries.

The *Almagest*'s route to Latin was long. It was revived only by a translation of the Arabic edition into Latin by GERARD OF CREMONA, with the help of GALIPPUS[59], in Toledo between 1150 and 1187.[60] The translations by AL-ḤAJJĀJ and the translation revised by THĀBIT reached Spain, and GERARD used them as the basis for the preparation of the Latin translation.[61] It is less well known

56 *Tetrabiblos* contains topics on physical astronomy and is PTOLEMY's attempt to establish a scientific foundation of astrology. Riley, 1988, Science and Tradition in Tetrabiblos, Proceedings of the American Philosophical Society 132.1, pp.67–84. For PTOLEMY's work before the *Almagest*, see also Hamilton, Swerdlow, Toomer, 1987, The Canobic Inscription: Ptolemy's Earliest Work; From Ancient Omens to Statistical Mechanics, Berggren, Goldstein (eds.). Jones, 2005, Ptolemy's Canobic Inscription and Heliodorus Observation Report, SCIAMVS 6.

57 Lorch, 2001, Greek-Arabic-Latin: The Transmission of Mathematical Texts in the Middle Ages. In addition to Arabic translations, the Syriac translation was used by IBN AL-ṢALĀH in the 12th century. He contributed some important comments and corrections to the *Almagest*, and many of his works have not yet been studied. For more information on Almagest translations, see Kunitzsch, 1974, Der Almagest: Die Syntaxis Mathematica des Claudius Ptolemäus in arabisch-lateinischer Überlieferung. Kunitzsch, 1975, Ibn as-Salah. Zur Kritik der Koordinatenüberlieferung im Sternkatalog des Almagest, pp.15–112. Kunitzsch, 1977, Über einige Spurren der Syrischen Almagestübersetzung. For a brief review also in Glowatzki, Göttsche, 1976, pp.11–20, Saliba, 1995, A History of Arabic Astronomy: Planetary Theories During the Golden Age of Islam, p.143, notes 1–3, and Toomer, 1998, pp.2–3.

58 Kunitzsch, 1974, p.23.

59 GHALIB THE MOZARAB, Lat. GALIPPUS MIXTARABUS, translated from Arabic into the local dialect, and GERARD wrote it down in Latin. The Mozarabs were Iberian Catholics who partly assumed elements of Arabic culture and language.

60 The *Almagest* was copied in 1175 by the THADEUS THE HUNGARIAN. The widely accepted year of translation, 1175, is not necessarily correct. Juste, 2021, Ptolemy, Almagesti, http://ptolemaeus.badw.de/work/3.

61 Precise information on the operation of GERARD and the financing of his translations are not known, Fletcher, 2006, Moorish Spain, p.151, for his translations see McVaugh, 1974, The Translation of Greek and Arabic Science into Latin. A Source Book in Medieval Science, Grant (ed.), pp.35–38. It is very likely that Christian rulers sought the knowledge of astronomers from Alexandria, mainly to determine the precise date of Easter.

than another translation of the *Almagest* from Greek into Latin which was done presumably around 1160 in Sicily by an unknown translator; most likely the translation was done or at least reviewed by EUGENE OF SICILY, also known as EUGENE OF PALERMO, but it remains more or less unknown.[62] This copy was brought from Constantinople to Sicily by HENRY ARISTIPPUS.[63] At the beginning of the 13th century, the *Almagest* was also translated from Arabic to Hebrew by JACOB ANATOLI.[64] Two other translations by unknown authors are also known, i.e. a translation probably done from Arabic in Spain in the 13th century that has only been preserved in fragments, and another translation of four of the *Almagest*'s books probably done before 1300 which seems to be based on another original translation or compilation of THĀBIT IBN QURRA. The presumed translator of this work is ʿABD AL-MASĪḤ WITTONIENSIS, about whom nothing is known.[65]

1.3.2 Handy Tables

PTOLEMY published the tabulated values of the equation of time in the *Handy Tables*.[66] Of all the works of antiquity, these tables came closest to fulfilling the practical needs of astronomers. From the citations of SYNCELLUS[67] it is known that the *Handy Tables* were already familiar to PANODORUS.[68] SEVERUS SEBOKHT[69] also knew the tables, and they were known to other Syriac speaking astronomers who had a great influence on the development of astronomy and science, representing a bridge between Greek, Persian, and Indian mathematical and astronomical knowledge. Certainly they were also translated into Arabic, although there is no exact information on the translators and translations. The original Greek manuscript has not been preserved; however, THEON's *Small Commentary on the Handy Tables* and *Great Commentary on the Handy Tables* have been. A separate part of the *Handy Tables* is a manual with the title *Instructions for the Use of the Handy Tables*, which is only in Greek.

62 Haskins, Lockwood, 1910, The Sicilian Translators of the Twelfth Century and the First Latin Version of Ptolemy's Almagest. Several manuscripts of the translation are known, see Haskins, 1924, Studies in the History of Mediaeval Science, pp.153–193 in particular pp.157–159. EUGENE OF SICILY, Lat. EUGENIUS SICULUS (about 1130–1202) was an admiral, poet and translator. He also translated PTOLEMY's *Optics*. For his life and works see Jamison, 1957, Admiral Eugenius of Sicily: His life and work, and the authorship of the Epistola ad Petrum, and the Historia Hugonis Falcandi Siculi.

63 Haskins, 1924, p.53. HENRY ARISTIPPUS (?1105/1110–1162) also HENRICUS ARISTIPPUS was archdeacon, translator and royal administrator in Sicily.

64 JACOB ANATOLI (?1194–1256) was born in Provence and was a philosopher, preacher, physician and translator. His most important translations include works of AVERROËS or IBN RUSHD.

65 ʿABD AL-MASĪḤ WITTONIENSIS also EBDELMESSIE OF WINCHESTER seems to be closely related to STEPHAN OF PISA (12th century), also known as STEPHAN OF ANTIOCH. Translation was most likely done in Antioch and copied later. Only the first four books without tables are preserved. ʿABD AL-MASIḤ seems to be a local expert practised at such translations. Grupe, 2019, Stephen of Pisa and Antioch: Liber Mamonis, pp.10–13.

66 *Handy Tables*, Gr. *Πρόχειροι Κανόνες*. See Tihon, Mercier, 2011, Ptolemaiou Procheiroi Kanones. Ptolemy's Handy Tables. Volume 1a: Tables A1-A2: Introduction. Édition critique. Volume 1b: Tables A1-A2: Transcription and Commentary. The manuscripts are kept, for example, by the National Library, Bibliothèque Nationale in Paris and the University Library in Leiden - the manuscript of Biblioteca Publica Graeca (BPG) 78, ff.50v–155r.

67 Tihon, Mercier, 2011, 1b, p.3 and p.71. GEORGE SYNCELLUS, Gr. Γεώργιος Σύγκελλος (8th century - died after the year 810), was a Byzantine monk and historian who lived for a long time in Palestine.

68 PANODORUS OF ALEXANDRIA (worked around the year 400), historian and writer.

69 SEVERUS SEBOKHT (575?–666/667), a Syrian philosopher and bishop in Qēnneshre near Aleppo, also dealt with astronomy. His most important work is a treatise on the astrolabe, based on the lost work of THEON. We know that he was acquainted with the other works of PTOLEMY.

1.3.3 Theon of Alexandria and Hypatia

THEON OF ALEXANDRIA (supposedly 335–405?) Gr. Θέων ὁ Ἀλεξανδρεύς, was one of the last – if not the very last – directors of the Alexandria Mouseion or library. From references in the *Suda encyclopaedia* it appears that he flourished during the reign of Emperor THEODOSIUS I, who ruled from 379 to 395. From his notes it is evident that on the 16th June 364 he observed a solar eclipse in Alexandria, and on the 25th of November of the same year he also observed a lunar eclipse. We also know that he compiled a list of Roman consuls, which he revised until 372. Several works are attributed to him, including some that are lost.[70] His preserved works are *Comments on Almagest*, a five-volume *Great Commentary on Handy Tables* and a one-volume *Small Commentary on Handy Tables*. Of the *Great Commentary*, the fifth volume is lost and the preserved fourth volume is not complete. THEON's comments are the only preserved ancient work on the structure of PTOLEMY's *Handy Tables*.[71] Apparently, THEON did not know much more about the tables than we know today, as PTOLEMY obviously did not leave other records on the background of the tables, except for a short treatise. In interpreting the tables, THEON obviously had many difficulties since, as the sources show, he did not fully understand all of PTOLEMY's procedures.[72] From the preserved fragments, it is evident that the *Handy Tables* were already commented on before THEON, but the authors of the comments are not known.[73] He also commented on EUCLID's *Elements* and *Optics*. Commenting on *Almagest*, he was helped by his daughter HYPATIA (about 355?–415).[74] According to the *Suda encyclopaedia*, HYPATIA also contributed commentaries on the works of APOLLONIUS and DIOPHANTUS, but these were not preserved.[75]

This learned lady is said to have been mistress of the whole of the pagan science, especially of philosophy and medicine...

Heath, 1921, A History of Greek Mathematics, II, p.528

70 *Suidae Lexicon* II, Adler (ed.), p.702, cited in Tihon, 1978, Le "Petit Commentaire" de Théon d'Alexandrie aux Tables faciles de Ptolémée, p.2, and in Tihon, 1999, Theon of Alexandria and Ptolemy's Handy Tables. Ancient Astronomy and Celestial Divination, p.357, note 1.

71 For a chronological sequence of manuscript copies of the Small Commentary on Handy Tables, see Tihon, 1978, Le Petit Commentaire de Théon d'Alexandrie aux Tables faciles de Ptolémée; Tableau récapitulatif de l'histoire du texte, a brochure in the appendix. For the Small Commentary, see Halma, 1822, Commentaire de Theon d'Alexandrie sur les tables manuelles astronomiques de Ptolemée, II, pp.27–108.

72 Tihon, 1999, p.358.

73 Jones, 1990, Ptolemy's First Commentator. Sources mention commentators from the beginning of the 3rd century. The author may have been SERAPION(?). Mogenet, Tihon, 1985, Le "Grand commentaire" de Théon d'Alexandrie aux Tables faciles de Ptolémée, I, pp.288–290. The most likely author of the commentary from a later period is STEPHANOS OF ALEXANDRIA. Lempire, 2016, Le commentaire astronomique aux Tables Facile de Ptolémée attribué à Stéphanos d'Alexandrie. For more information on his life and work, see Papathanassiou, 2006, Stephanos of Alexandria: A Famous Byzantine Scholar, Alchemist and Astrologer. The occult sciences in Byzantium.

74 Heath, 1921, II, p.528.

75 *Suidae Lexicon*, Adler (ed.), IV, p.644, 4–5 as cited in Dzielska, 1996, Hypatia of Alexandria, pp.70–73 or p.140 note 35. See also *Suda On Line*. For comments see Heath, 1921, II, p.528.

We know almost nothing about Hypatia's life and even her year of birth is unknown. It is very likely that she worked all her life in Alexandria; the only reliable source on her work and activities are Synesius' surviving letters.[76] Like her father, she taught in Alexandria, but there is no preserved information about her students. She died a violent death in 415.[77]

76 For Synesius' letters see Fitzgerald, 1926, The Letters of Synesius of Cyrene. Garzya, Roques, 2003, Synésios de Cyrene; Tome II et III: Correspondance: Lettres I-CLVI. Fritz, 1898, Die Briefe des Bischofs Synesius von Kyrene, reprint 1982. For more information about his life and works see Crawford, 1901, Synesius the Hellene. For commentaries on Hypatia's mathematical achievements, see Knorr, 1989, Textual Studies in Ancient and Medieval Geometry, pp.753–804.

77 Historical sources on the reasons for Hypatia's death are very scarce. Some interesting insights are given in the commentaries of Socrates of Constantinople, from the 5th century. In relation to Hypatia's death the Christian religious fanaticism is not directly stated in available ancient sources but it might be used indirectly as a way to remove the respected teacher or philosopher. The main motive seems to be envy supported by doctrinal or political disagreements in the city, whereas Bishop Cyril of Alexandria seems to be the key person. For ancient sources and their possible interpretation see Schwarzbauer, 2011, in particular pp.215–216 and also Haase, 2020, Hypatia's Death According to Socrates, Hist. eccl. 7.15: A Textual Commentary. Religious fanaticism but without detailed background is mentioned, for instance in Heath, 1921, A History of Greek Mathematics, II, p.528. For more information about historical circumstances, see Dzielska, 1996, Hypatia of Alexandria and Deakin, 2007, Hypatia of Alexandria: Mathematician and Martyr.

Bibliography

- BARNEY, Stephen A., W. J. LEWIS, J. A. BEACH and Oliver BERGHOF (trans.). 2014. The Etymologies of Isidore of Seville. 4th printing. Cambridge: Cambridge University Press. ISBN 978-0-521-14591-6.
- BJÖRNBO, Axel Anthon. 1902. Studien über Menelaos' Sphärik. In: Abhandlungen zur Geschichte der Mathematischen Wissenschaften mit Einschluss ihrer Anwendungen, 14. Heft, 1–154. Leipzig: Teubner.
- BOLL, Franz. 1894. Studien über Claudius Ptolemaeus: Ein Beitrag zur Geschichte der griechischen Philosophie und Astrologie. Besonderer Abdruck aus dem Einundzwanzigsten Supplementbande der Jahrbücher für classische Philologie. Leipzig: Teubner.
- BOWEN, Alan C. and Robert B. TODD. 2004. Cleomedes' Lectures on Astronomy. Oakland, CA: University of California Press. ISBN 978-0520233256.
- CRAWFORD, William Saunders. 1901. Synesius: The Hellene. London: Rivingtons. URL: https://archive.org/details/synesiushellene00crawiala/page/n6/mode/2up (20 January 2024).
- CZWALINA, Arthur. 1928. Die Trigonometrie des Ptolemäus von Alexandria. In: Archiv für Geschichte der Mathematik der Naturwissenschaften und der Technik, 10, 241–249.
- DEAKIN, Michael A. B. 1994. Hypatia and Her Mathematics. In: The American Mathematical Monthly, 101, 3, 234–243. DOI 10.2307/2975600.
- DEAKIN, Michael A. B. 2007. Hypatia of Alexandria: Mathematician and Martyr. Amherst, NY: Prometheus Books. ISBN 978-1591025207.
- DELAMBRE, Jean-Baptiste. 1817. Histoire de l'Astronomie Ancienne, Book 2. Paris: Courcier. URL: https://gdz.sub.uni-goettingen.de/id/PPN618023852 (20 January 2024).
- DELAMBRE, Jean-Baptiste. 1817. Histoire de l'Astronomie Ancienne, Book 2. Reprint 2005. Paris: Jacques Gabay. ISBN 2-87647-250-3.
- DICKS, D. R. 1960. The Geographical Fragments of Hipparchus. London: University of London.
- DICKS, D. R. 1972. Geminus. In: Dictionary of Scientific Biography, Gillispie, Charles Coulston (ed.), V, 344–347. New York: Charles Scribner's Sons. SBN 684-10116-5.
- DZIELSKA, Maria. 1996. Hypatia of Alexandria. Cambridge, MA: Harvard University Press. ISBN 9780674437760.
- EL-ABBADI, Mostafa, Omnia FATHALLAH and Ismail SERAGELDIN (eds.). 2008. What Happened to the Ancient Library of Alexandria? Leiden: Brill. DOI 10.1163/ej.9789004165458.i-259. e-ISBN 978-90-47-43302-6.
- EL-ABBADI, Mostafa. 2020. Library of Alexandria. In: Encyclopædia Britannica [online source]. URL: https://www.britannica.com/topic/Library-of-Alexandria (15 December 2020).
- EVANS, James. 1998. The History and Practice of Ancient Astronomy. Oxford: Oxford University Press. ISBN 978-0-19-509539-5.
- EVANS, James and Lennart BERGGREN. 2006. Geminos's Introduction to the Phenomena. Princeton, NJ: Princeton University Press. ISBN 978-0-691-12339-4.
- EVANS, James and Christián Carlos CARMAN. 2014. Mechanical Astronomy in Ancient Greece. In: From Alexandria Through Baghdad, Sidoli, Nathan and Glen van Brummelen (eds.), 145–174. Berlin: Springer. ISBN 978-3-642-36735-9.
- FITZGERALD, Augustine. 1926. The Letters of Synesius of Cyrene. Oxford: Oxford University Press; London: Humphrey Milford.
- FLETCHER, Richard. 2006. Moorish Spain. Berkeley, CA: University of California Press. ISBN 9780520248403.
- FRITZ, Wilhelm. 1898. Die Briefe des Bischofs Synesius von Kyrene. Reprint 1982. Aalen: Scientia Verlag. ISBN 3511091500.
- GARCZYA, Antonio and Denis ROQUES. 2003. Synésios de Cyrène. Tome II et III: Correspondance: Lettres I-CLVI. Paris: Les Belles Lettres. ISBN 9782251004792.
- GINZEL, Friedrich Karl. 1906. Handbuch der mathematischen und technischen Chronologie, Band 1. Leipzig: J. C. Hinrichs'sche Buchhandlung. Reprint ALO Arbeitsgemeinschaft. ISBN 3-226-00478-6.
- GLOWATZKI, Ernst and Helmut GÖTTSCHE. 1975. Die Sehnentafel des Klaudios Ptolemaios. München: R. Oldenbourg Verlag. ISBN 3-486-20181-6.
- GRAβHOFF, Gerd. 1990. The History of Ptolemy's Star Catalogue. New York, Berlin, Heidelberg: Springer. ISBN 0-387-97181-5.
- GRUPE, Dirk. 2019. Stephen of Pisa and Antioch: Liber Mamonis. An Introduction to Ptolemaic Cosmology and Astronomy from the Early Crusader States. Cham: Springer. DOI 10.1007/978-3-030-19234-1. e-ISBN 978-3-030-19234-1.

- Haas, Christopher. 2006. Alexandria in Late Antiquity. Baltimore, MD: Johns Hopkins University Press. SBN 978-0801885419.
- Haase, Mareile. 2020. Hypatia's Death According to Socrates, Hist. eccl. 7.15: A Textual Commentary. In: Hypatia of Alexandria, Her Context and Legacy, LaValle Norman, Dawn and Alex Petkas (eds.), 255–284. Tübingen: Mohr Siebeck. DOI 10.1628 / 978-3-16-158954-6. eISBN 978-3-16-158954-6.
- Halma, Nicolas. 1813–1816. Composition Mathématique I-II. Translation with Comments of Delambre. Reprint 1988. Paris: Librairie A. Blanchard. ISBN 2-85367-069-4.
- Halma, Nicolas. 1821. Commentaire De Théon d'Alexandrie sur le Premier & le Second Livres de la Composition Mathématique de Ptolemée. Translation and Comments. Reprint 1993. Paris: Librairie
- A. Blanchard. ISBN 2-85-367-186-0.
- Halma, Nicolas. 1821. Commentaire De Théon d'Alexandrie sur le Premier & le Second Livres de la Composition Mathématique de Ptolemée, Tome I. Paris: Merlin. URL: https://gallica.bnf.fr/ark:/12148/bpt6k6211371c (20 January 2024).
- Halma, Nicolas. 1822. Commentaire de Théon d'Alexandrie, sur le Livre III de l'Almagestede Ptolemée; Tables Manuelles des Mouvements des Astres, traduites pour la première foisdu grec en français, sur les manuscrits de la bibliothèque du Roi, par M. L'Abbé Halma. Paris: A. Bobée. URL: https://gallica.bnf.fr/ark:/12148/bpt6k3974x?rk=21459;2 (20 January 2024).
- Halma, Nicolas. 1822. Commentaire de Théon d'Alexandrie, sur le Livre III de l'Almagestede Ptolemée; Tables Manuelles des Mouvements des Astres, traduites pour la première foisdu grec en français, sur les manuscrits de la bibliothèque du Roi, par M. L'Abbé Halma. Reprint 1990. Paris: Librarie A. Blanchard. ISBN 2-85367-169-0.
- Hamilton, N. T., Noel M. Swerdlow and Gerald J. Toomer. 1987. The Canobic Inscription: Ptolemy's Earliest Work. In: From Ancient Omens to Statistical Mechanics, Berggren, Lennart and Bernard Goldstein (eds.), 55–73. Copenhagen: University Library. ISBN 978-8777090028.
- Harich-Schwarzbauer, Henriette. 2011. Hypatia; Die spätantiken Quellen: Eingeleitet, kommentiert und interpretiert. Bern: Peter Lang. DOI 10.3726/978-3-0351-0244-4. eISBN 978-3-0351-0244-4.
- Haskins, Charles Homer. 1924. Studies in the History of Mediaeval Science. Cambridge: Harvard University Press. URL: https://archive.org/details/studiesinhistory00hask (10 February 2024).
- Haskins, Charles Homer and Dean Putnam Lockwood. 1910. The Sicilian Translators of the Twelfth Century and the First Latin Version of Ptolemy's Almagest. In: Harvard Studies in Classical Philology, 21, 75–102. DOI 10.2307/310505.
- Heath, Thomas. 1921. A history of Greek mathematics, I. Oxford: Clarendon Press. Reprint 2006. Adamant Media. ISBN 0-543-97448-0.
- Heath, Thomas. 1921. A history of Greek mathematics, II. Oxford: Clarendon Press. Reprint 2006. Adamant Media. ISBN 0-543-96877-4.
- Heiberg, Johan Ludvig (ed.). 1907. Claudii Ptolemaei opera quae exstant omnia. Volume 2. Opera astronomica minora. Leipzig: Teubner.
- Hypsicles. 1966. Die Aufgangszeiten der Gestirne. De Falco, Vittorio and Max Krause (trans.), introduction Neugebauer, Otto. Abhandlungen der Akademie der Wissenschaften in Göttingen, Philogogisch-Historische Klasse, Dritte Folge, 62. Göttingen: Vandenhoeck & Ruprecht.
- Jamison, Evelyn Mary. 1957. Admiral Eugenius of Sicily: His life and work, and the authorship of the Epistola ad Petrum, and the Historia Hugonis Falcandi Siculi. London: Oxford University Press. URL: https://archive.org/details/admiraleugeniuso0000jami (10 February 2024).
- Jones, Alexander. 1986. Pappus of Alexandria Book 7 of the Collection. New York: Springer. ISBN 3-540-96257-3.
- Jones, Alexander. 1990. Ptolemy's First Commentator. In: Transactions of the American Philosophical Society, 80, 7, 1–61. DOI 10.2307/1006577.
- Jones, Alexander. 2005. Ptolemy's Canobic Inscription and Heliodorus' Observation Report. In: SCIAMVS, Sources and Commentaries in Exact Sciences, 6, 53–98. URL: https://www.sciamvs.org/2005.html (31 March 2024).
- Juste, David. 2021. Ptolemy, Almagesti (tr. Gerard of Cremona), (update 07 October 2021), Ptolemaeus Arabus et Latinus. Works. URL: http://ptolemaeus.badw.de/work/3 (9 October 2021).
- Knorr, Wilbur Richard. 1989. Textual Studies in Ancient and Medieval Geometry. Basel: Birkhäuser. ISBN 0817633871.
- Kouremenos, Theokritos. 1994. Posidonius and Geminus on the foundations of mathematics. In: Hermes, Zeitschrift für laische Philologie, 122, 4, 437–450. URL: https://www.jstor.org/stable/4477044.
- Kramer, Edna. 1972. Hypatia. In: Dictionary of Scientific Biography, Gillispie, Charles Coulston (ed.), VI, 615–616. New York: Charles Scribner's Sons. ISBN 0-684-10117-3.

- Kroll, Wilhelm (ed.). 1908. Vettii Valentis Anthologiarum. Leipzig: Weidmannsche Buchhandlung. URL: https://archive.org/details/bub_gb_OQ_gAAAAMAAJ (10 February 2024).
- Kunitzsch, Paul. 1974. Der Almagest: Die Syntaxis Mathematica des Claudius Ptolemäus in arabisch-lateinischer Überlieferung. Wiesbaden: Harrassowitz. ISBN 978-3-447-01517-2.
- Kunitzsch, Paul. 1975. Ibn as-Salah. Zur Kritik der Koordinatenüberlieferung im Sternkatalog des Almagest. Abhandlungen der Akademie der Wissenschaften zu Göttingen. Philologisch-Historische Klasse. Dritte Folge, 94. Göttingen: Vandenhoeck & Ruprecht. ISBN 9783525823705.
- Kunitzsch, Paul. 1977. Über einige Spurren der Syrischen Almagestübersetzung. In: Prismata: Festschrift für Willy Hartner, Maeyama, Y. and W. G. Salzer (eds.), 203–210. Wiesbaden: Franz Steiner Verlag. ISBN 3-515-02683-5.
- Latham, John Derek. 1975. Arabic into Medieval Latin. In: Journal of Semitic Studies, 17, 1, 30–67. DOI 10.1093/jss/17.1.30.
- Lawlor, R. and D. Lawlor (trans.). 1979. Theon of Smyrna, Mathematics Useful for Understanding Plato or, Pythagorean Arithmetic, Music, Astronomy, Spiritual Disciplines. Mecosta, MI: Wizards Bookshelf. ISBN 0913510246.
- Lemay, Richard. 1978. Gerard of Cremona. In: Dictionary of Scientific Biography, Gillispie, Charles Coulston (ed.), XV, Supplement I, 173–192. New York: Charles Scribner's Sons. ISBN 0-684-14779-3.
- Lempire, Jean. 2016. Le commentaire astronomique aux Tables Faciles de Ptolémée attribué a Stéphanos d'Alexandrie. Louvain-la-Neuve: Institut orientaliste de l'Université catholique de Louvain & Peteers Press. ISBN 978-90-429-3074-2.
- Lloyd, Geoffrey E. R. 2006. Principles and Practices in Ancient Greek and Chinese Science. Abingdon: Routledge. ISBN 978-0860789932.
- Lorch, Richard. 2001. Greek-Arabic-Latin: The Transmission of Mathematical Texts in the Middle Ages. In: Science in Context, 14, 1–2, 313–331. DOI 10.1017/S0269889701000114.
- Manitius, Karl (trans.). 1894. Hipparchi in Arati et Eudoxi Phaenomena Commentariorum. Leipzig: Teubner.
- Manitius, Karl (trans.). 1898. Gemini Elementa Astronomiae. Leipzig: Teubner.
- Manitius, Karl (trans.). 1912. Ptolemäus Handbuch der Astronomie. Leipzig: Teubner. URL: https://archive.org/details/desclaudiusptole01ptoluoft/mode/2up (20 January 2024).
- McVaugh, Michael. 1974. The Translation of Greek and Arabic Science into Latin. In: A Source Book in Medieval Science, Grant, Edward (ed.), 35–38. Cambridge, MA: Harvard University Press. ISBN 674-82360-5.
- Mogenet, Joseph and Anne Tihon. 1985. Le "Grand commentaire" de Théon d'Alexandrieaux Tables faciles de Ptolémée, Livre I. Studi e Testi 315. Città del Vaticano: Biblioteca Apostolica Vaticana. ISBN 88-210-0527-5.
- Möller, Lenelotte. 2008. Die Enzyklopädie des Isidor von Sevilla. Wiesbaden: Marix Verlag. ISBN 978-3-86539-177-3.
- Mueller, Ian. 1987. Hypatia. In: Women of Mathematics: A Bibliographic Sourcebook, Grinstein, Louise S. and Paul J. Campbell (eds.), 74–79. New York: Greenwood. ISBN 0-313-24849-4.
- Neugebauer, Otto. 1962. The astronomical tables of al-Khwārizmī, Translation with Commentaries on the Latin Version, Suter, Heinrich (ed.), Supplemented by Corpus Christi College MS 283 by O. Neugebauer. København: Ejnar Munksgaard.
- Neugebauer, Otto. 1975. A History of Ancient Mathematical Astronomy. Volumes I-III. Berlin, Heidelberg, New York: Springer. ISBN 978-3-642-61912-0.
- Neugebauer, Otto and T. C. Skeat. 1958. The Astronomical Tables P. Lond. 1278. With a Note on the Palaeography of the Fragments. In: Osiris, 13, 93–113. DOI 10.1086/368611.
- Papathanassiou, Maria. 2006. Stephanos of Alexandria: A Famous Byzantine Scholar, Alchemist and Astrologer. In: The Occult Sciences in Byzantium, Magdalino, Paul and Maria Mavroudi (eds.). Geneva: La Pomme d'or. ISBN 987-954-8446-04-4.
- Pedersen, Olaf. 2011. A Survey of the Almagest. New York, Dordrecht, Heidelberg, London: Springer. ISBN 978-0-378-84825-9. DOI 10.1007/978-0-387-84826-6.
- Petersen, Viggo M. and Olaf Schmidt. 1968. The Determination of the Longitude of the Apogee of the Orbit of the Sun according to Hipparchus and Ptolemy. In: Centaurus, 12, 2, 73–96. DOI 10.1111/j.1600-0498.1968.tb00080.x.
- Proclus. 1970. A Commentary on the First Book of Euclid's Elements, Morrow, Glenn R. (trans.). Princeton, NJ: Princeton University Press. ISBN 0-691-07160-8.
- Riley, Mark T. 1995. Ptolemy's Use of His Predecessors' Data. In: Transactions of the American Philological Association, 125, 221–250. DOI 10.2307/284353.

- Riley, Mark. 1988. Science and Tradition in the Tetrabiblos. In: Proceedings of the American Philosophical Society, 132, 1, 67–84. URL: https://www.jstor.org/stable/3143825.
- Riley, Mark. 2004. A Survey of Vettius Valens. URL: https://www.academia.edu/10706681/A_Survey_of_Vettius_Valens (28 April 2018).
- Rome, Adolphe. 1939. Le problème de l'équation du temps chez Ptolémée. In: Annales de la Société Scientifique de Bruxelles, 59, série 1, 211–224.
- Saliba, George. 1994. A History of Arabic Astronomy: Planetary Theories During the Golden Age of Islam. New York: New York University Press. ISBN 0-8147-7962-X.
- Sarton, George. 1931. Introduction to the History of Science, Volume II. London: Baillière, Tindall and Cox.
- Schram, Robert. 1908. Kalendariographische und chronologische Tafeln. Leipzig: J. C. Hinrichs'sche Buchhandlung.
- Sefrin-Weis, Heike. 2010. Pappus of Alexandria: Book 4 of the Collection. London: Springer. ISBN 978-1-84996-004-5. DOI 10.1007/978-1-84996-005-2.
- Smart, William Marshall. 1990. Textbook on Spherical Astronomy. Reprint. Cambridge: Cambridge University Press. ISBN 0 521 29180 1.
- Tihon, Anne. 1978. Le "Petit Commentaire" de Théon d'Alexandrie aux Tables faciles de Ptolémée. Studi e Testi 282. Città del Vaticano: Biblioteca Apostolica Vaticana. ISBN 88-210-0503-8.
- Tihon, Anne. 1999. Theon of Alexandria and Ptolemy's Handy Tables. In: Ancient Astronomy and Celestial Divination, Swerdlow, Noel (ed.), 357–369. Cambridge, MA: MIT Press. ISBN 9780262194228.
- Tihon, Anne and Raymond Mercier. 2011. Ptolemaiou Procheiroi Kanones. Ptolemy's Handy Tables. Volume 1a, 1b: Tables A1-A2. Transcription and Commentary. Leuven: Peeters Publishers. ISBN 978-90-429-2436-9.
- Toomer, Gerald J. 1974. The Chord Table of Hipparchus and the Early History of Greek Trigonometry. In: Centaurus, 18, 1, 6–28. DOI 10.1111/j.1600-0498.1974.tb00205.x.
- Toomer, Gerald J. 1975. Ptolemy. In: Dictionary of Scientific Biography, Gillispie, Charles Coulston (ed.), XI, 186–206. New York: Charles Scribner's Sons. ISBN 0-684-10122-X.
- Toomer, Gerald J. 1976. Theon of Alexandria. In: Dictionary of Scientific Biography, Gillispie, Charles Coulston (ed.), XIII, 321–325. New York: Charles Scribner's Sons. ISBN 0-684-12925-6.
- Toomer, Gerald J. 1998. Ptolemy's Almagest. Princeton: Princeton University Press. ISBN 978-0-691-00260-6.
- Van Dalen, Benno. 1994. On Ptolemy's Table for the Equation of Time. In: Centaurus, 37, 2, 97–153. DOI 10.1111/j.1600-0498.1994.tb00005.x.
- Van Der Waerden, Bartel Leendert. 1958. Die Handlichen Tafeln des Ptolemaios. In: Osiris, 13, 54–78. DOI 10.1086/368609.
- Von Fritz, Kurt. 1971. Grundprobleme der Geschichte der antiken Wissenschaft. Berlin, New York: Walter de Gruyter. ISBN 3110018055.
- Voigt, Ulrich. 2010. Geminos von Rhodos und Oktaēteris. In: Acta Historica Astronomiae, 37, Dick, Wolfgang R., Hilmar W. Duerbeck and Jürgen Hamel (eds.), 7–34. Frankfurt am Main: Harri Deutsch. ISBN 978-3-8171-1863-2.
- Warmington, Eric Herbert. 1975. Posidonius. In: Dictionary of Scientific Biography, Gillispie, Charles Coulston (ed.), XI, 103–106. New York: Charles Scribner's Sons. ISBN 0-684-10122-X.

Internet links

- Almagest Ephemeris Calculator, https://webspace.science.uu.nl/~gent0113/astro/almagestephemeris_main.htm (20 January 2024)
- Suda On Line: Byzantine Lexicography, https://www.cs.uky.edu/~raphael/sol/sol-html/ (20 January 2024)

This page is intentionally left blank.

2

Equation of time in Ancient India

Multiply the daily motion (bhukti) of a planet by the sun's result from the base-sine (bāhuphala), and divide by the number of minutes in a circle (bhacakra); the result, in minutes, apply to the planet's true place, in the same direction as the equation was applied to the sun.

Sūrya Siddhānta II.46 (Burgess, 1860, p.76)

2.1 Trigonometry and epicycles

2.1.1 Indian trigonometry

Between late antiquity and the Middle Ages astronomy is marked by significant contributions from Indian and Arab mathematicians and astronomers. In India, mathematics and astronomy advanced greatly during the GUPTA dynasty, and several important extant astronomical treatises, the *Siddhāntas*, date from this period.[1] The earliest extant Indian work with the values of the trigonometric function given is the work of ĀRYABHAṬA I (470–550). No doubt the mathematical procedures for calculating trigonometric functions were known to his predecessors, but their works are lost. The existence of older sources is indicated by VARĀHAMIHIRA's *Pañcasiddhāntikā*, which also contains references from older, lost *Siddhāntas*.[2] Indian trigonometric functions are functions of the arc, not of the angle as we know it today.

The Indian or Hindu trigonometric functions of *Jyā*, *Koṭjyā* and *Utkramajyā* are identical with the sine, cosine and versine trigonometric functions, and their values can be obtained by multiplying the sine, cose, or versine by the radius of a circle.[3] The geometric model for determining the value of the *Jyā*, *Koṭjyā* and *Utkramajyā* functions (Equations 2.1 – 2.3) can be seen in Figure 2.1.

[2.1] $$\text{Jyā}\,\alpha \equiv AC = R \cdot \sin\theta$$

[2.2] $$\text{Koṭjyā}\,\alpha \equiv OA = R \cdot \cos\theta$$

[2.3] $$\text{Utkramajyā}\,\alpha \equiv AB = R \cdot \text{versin}\,\theta = R(1 - \cos\theta)$$

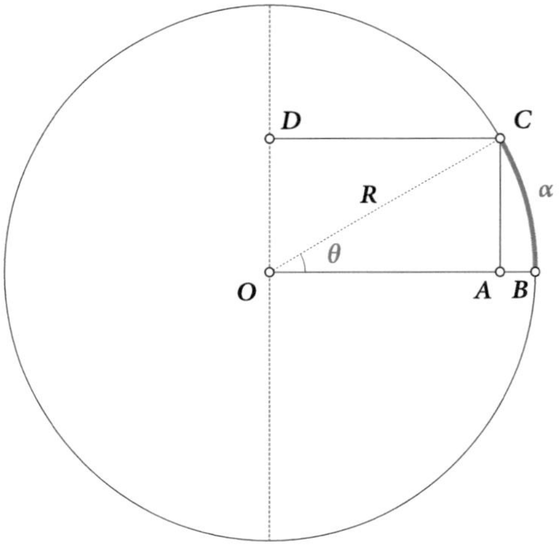

Figure 2.1 Indian trigonometric functions *Jyā*, *Koṭjyā* and *Utkramajyā*.

1 The GUPTA dynasty existed from the early 4th century to the late 6th century.

2 The *Pañcasiddhāntikā* comprises five treatises. Of these, the *Sūrya-*, *Romaka-* and *Pauliśa Siddhānta* are mathematically more demanding, while *Vāsiṣṭha-* and *Paitāmaha Siddhānta* are simpler and indicate a connection with Babylonian astronomy. For *Romaka Siddhānta* see van der Waerden, 1988, On the Romaka-Siddhānta.

3 In the literature on Hindu trigonometric functions, Hsin, Hcos and Hversin descriptions can also be found.

The Sanskrit word *jyā* means the string of the bow or the chord, the word *koṭi* means the curved end of the bow, and the word *utkrama* means reverse.[4]

Different astronomers have used different values for circle radius R.[5] All three functions were known in India in the fifth century. The relationship between the chord, Indian and modern trigonometric functions is:[6]

[2.4] $$R \cdot \sin\theta \equiv \mathrm{Sin}\,\alpha = \frac{1}{2}\mathrm{Crd}\,2\alpha \qquad \text{or} \qquad 2R \cdot \sin\frac{\theta}{2} \equiv 2\,\mathrm{Sin}\frac{\alpha}{2} = \mathrm{Crd}\,\alpha$$

ĀRYABHAṬA I in his treatise *Āryabhaṭīya* gives the values of the *Jyā* function in steps of 225′ or 3¾°. The values are given indirectly as differences between sines; see Table 2.1. The function corresponds to the sine function for a circle of radius 3438. This radius can be calculated as in Equation 2.5, where the value 21600 is the number of arc minutes in a circle:[7]

[2.5] $$\frac{21600}{2\pi} = 3437.7$$

Given that ĀRYABHAṬA's description is rather sparse, we assume that the mathematical procedures were already well known by the time *Āryabhaṭīya* was written.[8]

ĀRYABHAṬA was already familiar with the following expressions, which were later used by BRAHMAGUPTA:[9]

$$\sin\left(\frac{\pi}{2}+\theta\right) = \sin\frac{\pi}{2} - \mathrm{versin}\,\theta$$

$$\sin(\pi+\theta) = \sin\frac{\pi}{2} - \mathrm{versin}\frac{\pi}{2} - \sin\theta$$

$$\sin\left(\frac{3\pi}{2}+\theta\right) = \sin\frac{\pi}{2} - \mathrm{versin}\frac{\pi}{2} - \sin\frac{\pi}{2} + \mathrm{versin}\,\theta$$

ĀRYABHAṬA is also known for the formula relating the circumference and diameter of a circle, the result of which is of course π:[10]

[2.6] $$\frac{\text{circumference}}{\text{diameter}} = \frac{62832}{20000} = 3.1416$$

4 Datta, Singh, 1983, Hindu Trigonometry, p.40.
5 For a table of the radii used by different authors, see Datta, Singh, 1983, p.89.
6 Gupta, 1978, Indian Values of the Sinus Totus, p.125.
7 By calculating the values of the chord function, HIPPARCHUS supposedly also used the value 3438. Toomer, 1974, The Chord Table of Hipparchus and the Early History of Greek Trigonometry.
8 Shukla, Sarma, 1976, Āryabhaṭīya, xxviii or pp.44-54. For applications see also Madhavan, 2011, Jyā: Concept, Properties and Applications.
9 Shukla, Sarma, 1976, xxviii.
10 ĀRYABHAṬA, *Ganita*, verse 10. Shukla, Sarma, 1976, p.45, ĀRYABHAṬA states that it is an approximation. CLAUDIUS PTOLEMY used the value 3.141666. Heath, A History of Greek Mathematics, I, p.233.

θ (°)	Jyā		Δ	Koṭjyā	Utkramajyā	
0	0			3438	0	
3.75	225		225	3431	7	
7.5	449		224	3409	29	
11.25	671		222	3372	66	
15	890		219	3321	117	
18.75	1105		215	3256	182	
22.5	1316	*1315	210	3176	262	*261
26.25	1521	*1520	205	3083	355	*354
30	1719		199	2977	461	*460
33.75	1910		183	2859	579	
37.5	2093		174	2728	710	
41.25	2267		164	2585	853	
45	2431		154	2431	1007	
48.75	2585		143	2267	1171	
52.5	2728		131	2093	1345	
56.25	2859		119	1910	1528	
60	2977	*2978	106	1719	1719	
63.75	3083	*3084	93	1521	1917	*1918
67.5	3176	*3177	79	1316	2122	
71.25	3256		65	1105	2333	
75	3321		51	890	2548	
78.75	3372		37	671	2767	
82.5	3409		22	449	2989	
86.25	3431		7	225	3213	
90	3438			0	3438	

Table 2.1 Indian trigonometric functions of the arc length of a circle of radius 3438, in steps of $3\,^3/_4$° calculated using Equations 2.1, 2.2 and 2.3. *ARYABHAṬA's values. The values differ from those used by ARYABHAṬA due to rounding. The calculation of the grey-shaded values based on the geometric model is given below. Differences summarised from: Clark, 1930, Aryabhatiya, p.19.

ĀRYABHAṬA's geometric construction for calculating the value of the Jyā function can be seen in Figure 2.2:[11]

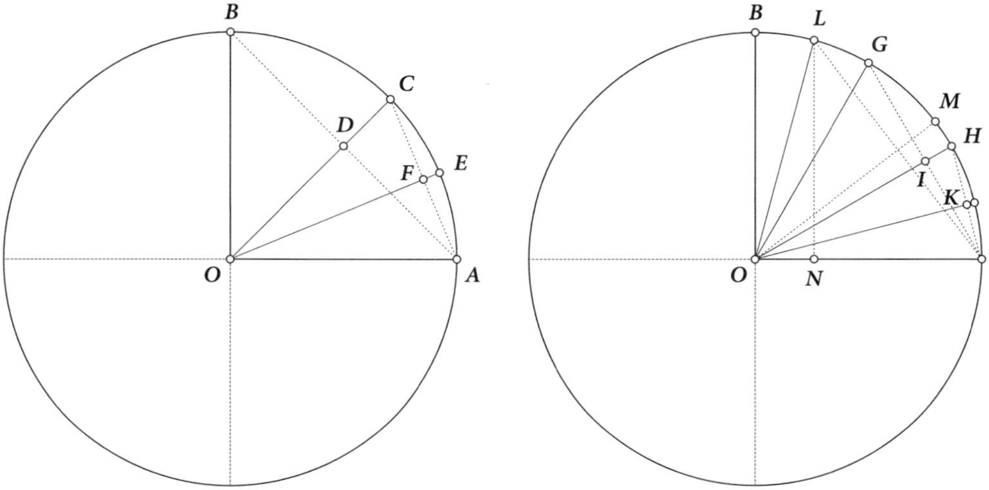

Figure 2.2 Example of geometric constructions to determine the value of the function Jyā (Sine), $R=3438$. Adopted from Chatterjee, 1970, I, p.190, Figure 3, see also Shukla, Sarma, 1976, pp.46–47.

Divide the circle into angles of 30° (Figure 2.2, right) and 45° (Figure 2.2, left), then halve them and repeat the process. The values of the Jyā function in steps of 15° can be calculated using the procedure below. For the right-angled triangle ΔOIA, Figure 2.2 right, $AI = R/2$ and $OA = R$, resulting in Jyā of 60°:

$$OI = \sqrt{R^2 - \left(\frac{R}{2}\right)^2} = \frac{\sqrt{3}}{2}\mathrm{R} = 2978 = \text{Jyā}\,60°$$

The following also applies:

$$AI = \sqrt{R^2 - \frac{3}{4}R^2} = \frac{R}{2} = 1719 = \text{Jyā}\,30°$$

For the triangle ΔAIH applies $IH = R - OI$ (=Utkramajyā 30°, Figure 2.2 right), which implies:

$$AH = \sqrt{AI^2 + IH^2} = \frac{1}{2}\sqrt{R^2 + (2 - \sqrt{3})^2 R^2} = 1780 = \text{Crd}\,30°$$

From Equation 2.4, the expression for Jyā 15° follows:

$$AK = \frac{R}{2}\sqrt{2 - \sqrt{3}} = \frac{1}{2}\text{Crd}\,30° = 890 = \text{Jyā}\,15°$$

11 Chatterjee, 1970, The Khaṇḍakhādyaka, Vol. I, pp.190–191. Shukla, Sarma, 1976, Āryabhaṭīya; Ganita 11, p.45 also Elfering, 1975, p.105.

From the triangle $\triangle ODA$, Figure 2.2 left, calculate the value for Jyā 45°:

$$OA = OB = R = 3438 \qquad \text{and} \qquad AB = \sqrt{2OA^2} = \sqrt{2}R = 4862$$

The result is the half value: $AD = \frac{\sqrt{2}}{2}R = 2431 = \text{Jyā } 45°$.

The value for Jyā 75° follows from the triangle $\measuredangle AOK$, Figure 2.2 right:

$$OK = \sqrt{R^2 - AK^2} = \sqrt{R^2 - \frac{2-\sqrt{3}}{4}R^2} = \frac{R}{2}\sqrt{2+\sqrt{3}} = 3321 = \text{Jyā } 75°$$

Then continue calculating the value in increments of 7°30′. For the triangle $\triangle AOD$ (Figure 2.2, left), $OD = AD$, follows:

$$DC = R - OD = R - \frac{\sqrt{2}}{2}R = 1007$$

Then calculate the AC and AE values:

$$AC = \sqrt{AD^2 + CD^2} = R\sqrt{2-\sqrt{2}} = 2631 \qquad \text{and} \qquad AE = 1315 = \text{Jyā } 22.5°$$

The value for Jyā 7°30′ follows from the triangle $\triangle AJK$, where $JK = R - OK$:

$$AJ = \sqrt{AK^2 + JK^2} = \frac{R}{2}\sqrt{(2-\sqrt{3}) + (2-\sqrt{2+\sqrt{3}})^2} = R\sqrt{2-\sqrt{2+\sqrt{3}}} = 897$$

or

$$\frac{AJ}{2} = 448 = \text{Jyā } 7°30'$$

The value of Jyā 37°30′ is obtained using the triangles $\measuredangle LNO$ and $\measuredangle LNA$ (Figure 2.2, left), where $LN = \text{Jyā } 75°$, $NO = R - NA$ and $LA = \text{Crd } 75°$.

$$NO = \sqrt{R^2 - LN^2} = \sqrt{R^2 - \frac{R^2}{4}(2+\sqrt{3})} = \frac{R}{2}\sqrt{(2-\sqrt{3})} = 889$$

or $$LA = \sqrt{LN^2 + NA^2} = \sqrt{LN^2 + (R - NO)^2}$$

$$= \frac{R}{2}\sqrt{(2+\sqrt{3}) + \left(2-\sqrt{2-\sqrt{3}}\right)^2} = R\sqrt{2-\sqrt{2-\sqrt{3}}} = 4185$$

resulting in a Jyā of 37°30′:

$$\frac{LA}{2} = 2093 = \text{Jyā } 37°30'$$

Repeat the process in the same way to obtain the function values for other angles from Table 2.1. The radius or *Trijyā* is not constant in Indian astronomical literature.[12] Different Indian astronomers have used different radii. VARĀHAMHIRA, BRAHMAGUPTA and BHĀSKARA II also used the value of 120 as the radius; BRAHMAGUPTA used 3270 and ŚRĪPATI used 3415.[13] The radius 120 indicates a connection with, or knowledge of, Greek astronomy. If we use the half argument in PTOLEMY's tables and multiply the radius by 2, we obtain the same values for the sine function as PTOLEMY's chord.[14] Indian astronomers knew trigonometry well, but apparently not MENELAUS' theorem. However, they developed their own procedures for solving problems of spherical trigonometry or spherical astronomy.[15]

2.1.2 Indian model of the epicycles

Indian astronomers also used an epicycle model, which we assume was adapted from the Greek model and then transferred to the Arabs. The first descriptions of a simple model of epicycles in Indian literature are found in the work of VARĀHAMHIRA (505–587), the *Pañcasiddhāntikā*, which was probably written around 575.[16] VARĀHAMHIRA mentions the original, the now lost edition of *Sūrya Siddhānta* as the source for determining the longitude of the Sun.[17] A simple Indian epicycle model with the Earth at the centre is shown in Figure 2.3.

For the rectangular triangle $\Delta A'DO$, the following relations apply:

[2.7] $$\sin \widehat{\Gamma OA'} = \sin \widehat{DOA'} = \frac{A'D}{A'O}$$

Using this model, we can determine or calculate the contribution to the equation of time due to the eccentricity of the Earth's orbit.[18] It follows from the Figure 2.3 that $A'D = R \sin \widehat{\Gamma OA'}$, where the radius $R = A'O$, for similar right-angled triangles $\Delta SCA'$ and $\Delta A'DO$ applies:

[2.8] $$\frac{SC}{SA'} = \frac{A'D}{A'O} \quad \text{or} \quad SC = A'D \cdot \frac{SA'}{A'O}$$

12 In Medieval sources also called *Sinus Totus*. The term *Sinus Totus*, or greatest sinus, is said to have been first used by GERARD OF CREMONA in his translation of AL-ZARQALI's work.

13 Somayaji, 1971, A Critical Study of the Ancient Hindu Astronomy, pp.4–5. For an extensive overview of different radii used in Indian astronomy see Gupta, 1978, Indian Values of the Sinus Totus.

14 Van Brummelen, 2009, pp.96–97.

15 For a comparison of Greek and Hindu methods in spherical astronomy see Subbarayappa (ed.), 2008, The Tradition of Astronomy in India *Jyotihśāstra*, pp.321–342.

16 Sen, 1974, Epicyclic Eccentric Planetary Theories in Ancient and Medieval Indian Astronomy. Van der Waerden, 1974, Science Awakening II, pp.299–300. Abraham, 1980, The Motion of the Sun in the Vasiṣṭha Siddhānta. Archive for History of Exact Sciences, 22, 1/2, pp.1–3.

17 This is the old edition of *Sūrya Siddhānta*, which is no longer preserved in its original form. Neugebauer, Pingree, 1970, The Pañcasiddhāntikā of Varāhamihira, Part I, IX.1, p.91.

18 For the calculation procedure and additional explanation, see Rao, 2000, Ancient Indian Astronomy, pp.55–63. For geometry in Ancient India, see Amma, 1979, Geometry in Ancient & Medieval India or Prakash, 1987, Geometry in Ancient India.

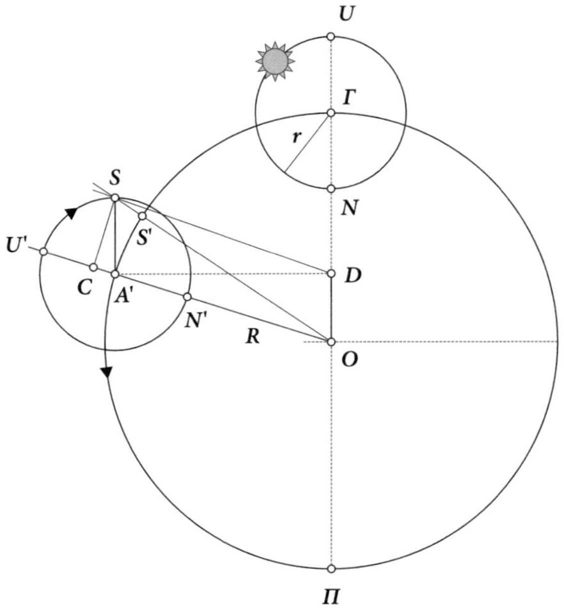

ΓA'Π	orbit, *kakṣavṛtta*,
ΓOΠ	apse line, apsides, *nīcoccaṛekhā*,
U	apogee, *mandocca*,
U'SN'	epicycle with radius *r*, *nīcoccavṛtta*,
N	perigee, *mandanicā*,
A'	mean Sun, *madhya*,
ΓO	radius of the orbit, *trijyā*,
A'O	radius *R*, *mandakendrajyā*,
A'S'	equation of the centre, *mandaphala*,
ΓOA'	mean anomaly, *mandakendra*

Figure 2.3 A model of epicycles with the basic concepts of Indian astronomy. Summarised after Shukla, 1960, p.111, Figure 11. See also Neugebauer, Pingree, 1971, volume II, p.69, Figure 34. For the terminology, see Rao, 2000, pp.55–56.

SA' and *A'O* are the radii of epicycles and orbits, and are proportional to their diameters or circumferences:

$$\frac{SA'}{A'O} = \frac{\text{circumference of the epicycle}}{\text{circumference of the orbit}} \quad \text{or} \quad SC = \frac{\text{circumference of the epicycle}}{\text{circumference of the orbit}} \cdot A'D$$

From Figure 2.3 it follows that:

$$A'D = R \sin \kappa_M$$

κ_M mean anomaly

Bearing in mind that the orbit circumference is 360 units:[19]

[2.9] $$SC = r \frac{R \sin \kappa_M}{360}$$

19 The assumption is that the circle or orbit circumference is 360 units, which follows the ratio of the radii of the epicycles. In the literature, the radii of the orbits and epicycles are often expressed in degrees (°), for example 360° for the orbit and 14° for the epicycle. The values in degrees are incorrect and misleading in this case.

In most Indian astronomical literature, the epicycle radius r is variable. According to VARĀHAMHIRA, in the original edition of *Sūrya Siddhānta*, and in *Khaṇḍakhādyaka*, it is constant. Assuming that SC is approximately the same as $A'S'$, we get the expression for the equation of the centre c or *mandaphala*. The calculation shown is based on simplifications; the actual result is an angle and not a distance:[20]

$$c = A'S' \approx r\frac{R\sin\kappa_{\mathrm{M}}}{360} = \frac{r}{R}\mathrm{Sin}\,\kappa_{\mathrm{M}} \qquad [2.10]$$

2.2 Periods in Indian astronomy and ahargaṇa

2.2.1 Periods in Indian astronomy

Indian astronomy is characterised by long periods that commence with a certain position of celestial bodies. The scheme, cumbersome at first glance, has some good properties, since it enables simple calculation. For practical calculations in Indian astronomy, it is important to understand its periods and epochs:

- *Kaliyuga* - a period of 432,000 years starting at midnight between the 17th and 18th February 3102 BCE. ĀRYABHAṬA I used February 18th 3102 BCE at sunrise in Laṅkā as the epoch, which was also the case with BHĀSKARA I.[21]
- *Mahāyuga* - a period of 432×10^4 (or 4.32 million) years.
- *Kalpa* - a period equal to 1,000 mahāyuga, or 432×10^7 years; the total number of 360° revolutions of the Sun along the ecliptic during this period is 4,320,000,000. From the beginning of the *kalpa* until the beginning of the *kaliyuga*, BRAHMAGUPTA claims there were 1,972,944,000 years, ĀRYABHAṬA I claims there were 1,986,120,000 years, and *Sūrya Siddhānta* claims a period of 1,955,880,000 years.[22]

Treatises that take into account the beginning of the *kalpa* or the *mahāyuga* periods for the epoch are generally classified as *siddhāntas*, treatises with the epoch of the *kaliyuga* are classified as *tantra*, and treatises with an arbitrary epoch chosen by the author are *karaṇa*.[23]

20 Neugebauer, Pingree, 1971, Part II, pp.69–70, Rao, 2000, pp.55–56. See also PTOLEMY's calculation, Equations 1.3 to 1.6 and explanation.

21 Laṅkā is an imaginary place on the equator where the Indian primordial meridian, the meridian through Ujjayinī, 23°11′ N, 75°47′ E, intersects with the equator. Shukla, 1960, Bhāskara I and his works, Part II, Mahā-Bhāskarīya, p.3, Rao, 2000, p.7 and 42.

22 Singh, Chand, 1992, Ahargaṇa and its Applications, p.25. Plofker, 2009, Mathematics in India, p.77. BRAHMAGUPTA's value was also used by BHĀSKARA II.

23 Chatterjee, 1981, Śiṣyadhīvṛddhida Tantra of Lalla with the commentary of Mallikārjuna Sūri, p.3.

Author, era	Number of days in *mahāyuga*	Source
ĀRYABHAṬA I, BHĀSKARA I[a]	1 577 917 500	*Mahā-Bhāskarīya*
BRAHMAGUPTA[b]	1 577 917 800	*Khaṇḍakhādyaka*
Unknown	1 577 917 828	*Sūrya Siddhānta*[c]
BHĀSKARA II[d]	1 577 916 450	*Siddhānta Śiromaṇi*

Table 2.2 Number of days in the *mahāyuga* period according to various authors.

Notes

a Rao, 2000, Ancient Indian Astronomy, p.43, Table 6.1. Shukla, 1960, pp.22–23.
b Chatterjee, 1970, The Khaṇḍakhādyaka, p.91.
c Sastri, Wilkinson (trans.), 1974, The Surya-Siddhānta, or an Ancient System of Hindu Astronomy. English Translation with Notes, p.6. *Sūrya Siddhānta* is an astronomical treatise whose content has been supplemented over time. The origins of the treatise, which deals with the motion of the planets, solar and lunar eclipses, conjunctions of planets, and in Chapters 3 and 13 also the use of the gnomon and other astronomical themes, may have originated from the time before the common era.
d Siddhānta Śiromaṇi, Madhyādhikāra, Bhagaṇādhyāya, Verse 9. Arkasomayaji, 1980, Siddhānta Śiromaṇi of Bhāskarācārya, p.24. Arkasomayaji, 2000, p.25.

2.2.2 Ahargaṇa

The number of days from the selected starting moment or epoch in Indian astronomy is called the *ahargaṇa*.[24] In today's terms, the term *ahargaṇa* is similar to the Julian day number, whose initial moment, unlike the *ahargaṇa*, has no strict astronomical background. The number of days or *ahargaṇa* is generally determined using the following procedure:[25]

- Convert the number of elapsed solar years since the epoch into lunar months.
- Add the number of additional lunar months or *adhimāsas* to get the actual number of lunar months until the beginning of the given year. *Adhimāsas* are extra lunar months, beyond the number of solar months.
- Add the number of lunar months in the observed year to the observed epoch and multiply them by 30. By doing this, we turn them into lunar days, or *tithis*. A *tithi* is the time period when the longitude of the moon gains 12° in comparison with the longitude of the Sun. The length of the lunar days, *tithis*, is not constant; it varies between 22 and 26 hours.
- Add the number of lunar days, *tithis*, in the observed lunar month, and finally convert the number of *tithis* to days.

Individual astronomers also defined the *ahargaṇa* with various procedures. Some notable examples are described below.

24 *Ahan* = day, *gaṇa* = group. Schmidt, 1952, On the Computation of the Ahargana. Bhattacharyya, Sen, 1969, Ahargana in Hindu Astronomy. Singh, Chand, 1992. A summary of the procedure Brahmagupta uses to calculate the number of days of ahargaṇa: Chatterjee, 1970, The Khaṇḍakhādyaka (an astronomical treatise) of Brahmagupta, translation, I.3–5, pp.48–49, comments pp.92–93. Rao, 2000, Chapter 5. For an explanation of BHĀSAKARA I see Shukla, 1960, pp.4–5. The *ahargaṇa* can be calculated by the Julian day number (*JDN*) if we know the calendar date according to the Gregorian or Julian calendar.
25 Rao, 2000, p.25. For comments on possible errors in the calculation of *ahargaṇa* see Ramasubramanian, Sriram, 2011, pp.32–33.

Arbitrary epoch – Khaṇḍakhādyaka

In the *Khaṇḍakhādyaka*, BRAHMAGUPTA gives a procedure for calculating the *ahargaṇa* since BRAHMAGUPTA's epoch.[26]

- Subtract the number 587 from the *śaka year*.[27] Multiply the result by 12 and add to the product the number of lunar or synodic months elapsed since the light half or *śukla pakṣa* of the first month or *Caitra*.[28]
- Multiply the sum by 30 and add to the product the number of *tithis* elapsed since the last *amāvāsyā*.[29] The result is the total number of solar days, *sauradinas*, elapsed. Add 5 to the result.
- Write down the sum in two places. At one place divide it by 14,945 and subtract the obtained result from the sum in the other place. Divide the remainder by 976. The result is the number of intercalary months, *adhimāsas*. Convert it into days and add the result to the number of solar days, *sauradinas*, as obtained above.
- The result is the number of lunar days or *cāndradinas* elapsed. Multiply the result by 11 and add 497. Again write down the sum in two places. In the first place divide it by 111,573 and subtract the result from the sum in the other place. Divide the reminder by 703. The quotient gives the number of omitted lunar days or *avamarātras*.
- Subtract the quotient from the number of lunar days or *cāndradinas*. The result is the *ahargaṇa*.

Example 2.1 Calculate the *ahargaṇa* at the end of the 5th tithi and one lunar month from the *śukla pakṣa* of the *Caitra* in the *śaka year* 672 (750), for BRAHMAGUPTA's epoch. See explanation in text for the calculation of *ahargaṇa* since BRAHMAGUPTA's epoch (solution: *ahargaṇa* = 31071, for a similar example see Rao, 2000, pp.28–29):

$$(\textit{śaka year} - 587)\cdot 12 + \textit{elapsed lunar months} = (672 - 587)\cdot 12 + 1 = 1021$$

$$1021 \cdot 30 + \textit{elapsed tithis} = 1021 \cdot 30 + 5 = 30635 \ \textit{solar days}$$

$$30635 + 5 = 30640$$

$$30640{:}14945 = 2.050184 \qquad 30640 - 2.050184 = 30637.95$$

$$30637.95{:}976 = 31.39134$$

$$\text{INT}\ (31.39134) = 31 \ \textit{intercalary months or adhimāsas}$$

$$31 \ \textit{intercalary months} = 930 \ \textit{days}$$

$$30635 \ \textit{solar days} + 930 \ \textit{days} = 31565 \ \textit{lunar days or cāndradinas}$$

$$\textit{lunar days} \cdot 11 + 497 = 31565 \cdot 11 + 497 = 347712$$

$$347712{:}111573 = 3.116453 \qquad 347712 - 3.116435 = 347708.9$$

$$347708.9{:}703 = 494.6072 \ \textit{omitted lunar days or avamarātras}$$

26 Chatterjee, 1970, I.3–5, p.48, comment, pp.92–93. Rao, 2000, pp.28–29. Epoch in the *Khaṇḍakhādyaka* starts at midnight between March 22nd and 23rd, 665.

27 *Śaka year* is the actual year – 78. Year 78 is initial year of the *Śaka era*. The epoch of the *Śaka era* is 15th March 78 with *JDN* = 1,749,621. Ginzel, 1906, I, reprint, pp.390–391.

28 Sen, Shukla, 2000, pp.309–311. *Caitra* is the first month in a year. *Śukla pakṣa* is a fortnight period, the so-called light half of the lunar month, the period of brightening or waxing moon. *Kṛṣṇa pakṣa* is the period of darkening or waning moon.

29 *Amāvāsyā* is a new moon day.

Subtract the number of lunar days omitted from the total number of lunar days to get the *ahargaṇa*:

$$31565 \text{ lunar days} - 494 \text{ omitted lunar days} = 31071 \text{ days}$$

Day with *ahargaṇa* number = 31,071 corresponds to 16th of April 750 at midnight. Considering the new moon on 11th of April 750, this is indeed 5 *tithi* days from the new moon.[30]

Kaliyuga epoch

The following or a very similar procedure for calculating *ahargaṇa* since the *kaliyuga* epoch can be found in many Indian astronomic treatises, including works of BRAHMAGUPTA, BHĀSKARA I, LALLA, BHĀSKARA II, ŚRĪPATI et al.:[31]

- Sum of a *śaka year* and 3179 gives the number of solar years elapsed since the beginning of the *kaliyuga*.
- Multiply the result by 12 and add the months since the beginning of *Caitra* elapsed in the current year.[32]
- Multiply the result by 30 and add the number of *tithis* elapsed since the last *amāvāsyā*.
- Write down the result in two places. In one place multiply the result by the number of intercalary months in a *yuga* = 159,330 and divide the result by the number of solar days in a yuga = 155,520,000. The result is the corresponding number of intercalary months, *adhimāsas*. Convert it into days by multiplying by 30 and add to the result in the second place.
- Write down the sum in two places again. In the first place multiply it by the number of omitted *tithis* in a *yuga* = 2,508,255 and divide by the number of tithis in the same period = 160,299,900. The result is the number of omitted *tithis*.
- Subtract the result from the sum in the second place. The remainder is the number of civil days or *ahargaṇa* elapsed since the beginning of the kaliyuga until the given day.

Example 2.2 Calculate the *ahargaṇa* from the beginning of *kaliyuga* until February 10 2005. The new moon occured on 8th February 2005.[33] Consider that, in a *kaliyuga* there are 159,330 intercalary months, 155,520,000 solar days, 160,299,900 *tithis* and 2,508,255 omitted *tithis* (solution: *ahargaṇa* = 1864947).[34]

Because the *śaka year* begins on 22nd March 2005, the year 2004 is selected as the final year for calculation.

30 Adopted from Fred Espenak's website: http://astropixels.com/ephemeris/phasescat/phasescat.html (19 August 2019).

31 Chatterjee, 1981, pp.8–12. Shukla, 1960, 4–6, pp.2–6. Singh, Chand, 1992, p.30.

32 This definition adopted from the literature is not very clear, see Shukla, 1960, p.2, Chatterjee, 1981, p.8. For a better understanding see Example 2.2. Month *Caitra* has 30 days and begins on 22nd March. On leap years it has 31 days and begins on 21st March.

33 Adopted from Fred Espenak's website http://astropixels.com/ephemeris/phasescat/phasescat.html (19 August 2019).

34 Singh, Chand, 1992, p.30.

$$3\,179 + \textit{śaka year} = 3\,179 + 1\,926 = 5\,105 \textit{ elapsed solar years}$$

$$5\,105 \cdot 12 + \textit{elapsed lunar months} = 61\,260 + 10 = 61\,270$$

$$61\,270 \cdot 30 + \textit{elapsed tithis} = 1\,838\,100 + 2 = 1\,838\,102 \textit{ solar days}$$

$$1\,838\,102 \cdot \frac{159\,330}{155\,520\,000} = 1\,883.133 = 1\,883 \quad \textit{intercalary months or adhimāsas}$$

$$1\,883 \cdot 30 + 1\,838\,102 = 56\,490 + 1\,838\,102 = 1\,894\,592$$

$$1\,894\,592 \cdot \frac{2\,508\,255}{160\,299\,900} = 29\,645.18 \textit{ omitted tithis}$$

$$1\,894\,592 - 29\,645.18 = 1\,864\,947 \textit{ days}$$

The calculation of Julian date *JD* gives a result of 1 864 946.5 for the desired date at noon. If the average lunar month length is taken into account, an *ahargaṇa* calculation error of ±1 day may occur. For an explanation see Rao, 2000, p.26.

Bhāskara II - Siddhānta Śiroman'i

To calculate the number of days of *ahargaṇa* from the beginning of the *kalpa* period until the beginning of the *kaliyuga*, it follows from the work of BHĀSKARA II that:[35]

- The number of solar years from the beginning of the *kalpa* period is multiplied by 12, thus obtaining the number of months (BRAHMAGUPTA's number of years from the beginning of the *kalpa* to the beginning of the *kaliyuga* is 1,972,944,000 years).[36]
- The resulting value is multiplied by 30, so we get the total number of solar days d from the beginning of the kalpa to the beginning of the *kaliyuga*.
- By means of a partial result, we calculate the number of additional lunar months m_l or *adhimāsas*, compared to solar months:[37]

[2.11] $$m_l = \frac{d \times b}{s} = \frac{(d \times 1\,593\,300\,000)}{1\,555\,200\,000\,000}$$

b number of additional lunar months, *adhimāsās* in *kalpa* = 1,593,300,000
s number of solar days in *kalpa* = 1,555,200,000,000
d number of solar days from the beginning of the *kalpa* to the beginning of the *kaliyuga*

- The result is multiplied by 30 and added to the number of solar days d from the beginning of the *kalpa* and we get the number of lunar days or *tithis*, d_t.

35 *Siddhānta Śiromaṇi*, *Grahānayanādhyāya*, Verse 1. Arkasomayaji, 1980, pp.29–31. Arkasomayaji, 2000, pp.30–32. Singh, Chand, 1992, p.25. For the ahargaṇa of BHĀSKARA I see Shukla, 1960, pp.2–6, or *Mahā-Bhāskarīya* I.4-6.

36 Miśra (ed.), 1947, The Siddhānta Śekhara of Śrīpati, Part II, p.XX. Singh, Chand, 1992, p.25.

37 Arkasomayaji, 1980, p.23 Verse 8, p.24 Verse 10. Arkasomayaji, 2000, p.24 Verse 8, p.25 Verse 10.

[2.12] $$d_t = m_l \cdot 30 + d$$

d number of solar days from the beginning of the *kalpa*
m_l number of additional lunar months or *adhimāsās*

- Then we calculate the number of redundant lunar days d_0, thus obtaining the final result, or the sought *ahargaṇa* number from the beginning of the *kalpa* period to the start of the observed epoch.[38]

[2.13] $$d_0 = \frac{d_t \times k_d}{t} = \frac{(d_t \times 25\,082\,550\,000)}{1\,602\,999\,000\,000}$$

k_d difference between solar and lunar days in *kalpa*, *kshayās* = 25,082,550,000
d_t number of lunar days since the beginning of *kalpa*
t number of lunar days in *kalpa*, *tithis* = 1,602,999,000,000

Omitted days or *kshayās* are days when, for example, the *tithi* begins after sunrise and ends before the next sunrise. In this case, the *tithi* day is superfluous in the calculation of the number of days and is omitted.

- *Ahargaṇa* is the difference between the number of lunar days and redundant lunar days:

[2.14] $$a = d_t - d_0$$

38 Arkasomayaji, 1980, p.23 Verse 8, p.24 Verse 10. Arkasomayaji, 2000, p.24 Verse 8, p.25 Verse 10.

2.3 Mean Sun

2.3.1 Mean Sun in Indian astronomy

The longitude of the mean Sun is generally given using the following expression:[39]

[2.15] $$\lambda_M = \lambda_{M0} + (a \times d_\odot)$$

a *ahargaṇa*
$d_\odot$ diurnal motion of the Sun
λ_{M0} mean longitude at an epoch

The daily motion of the Sun $d_\odot$ in degrees is determined using:

[2.16] $$d_\odot = \frac{\text{number of revolutions in mahāyuga}}{\text{number of days in mahāyuga}}$$

The integer part of the result, which means the number of revolutions around the Sun, is neglected and only the decimal part is multiplied, by 360° or by 2π, thus obtaining the ecliptic longitude in either degrees or radians.

2.3.2 Varāhamihira

In the *Pañcasiddhāntikā*, which is one of the oldest preserved Indian written sources, VARĀHAMIHIRA gives an expression for calculating the longitude of the mean Sun, which, according to the statements, is derived from the initial edition of *Sūrya Siddhānta*:[40]

[2.17] $$\lambda_M = \frac{(a \times 150) - 65}{54\,787}$$

a *ahargaṇa*, number of days since the epoch

VARĀHAMIHIRA's epoch starts on 22 March, 505, which roughly corresponds to 25;54° in Scorpio.[41]

39 Shukla, 1960, pp.6–7. Rao, 2000, pp.42–44.
40 Neugebauer, Pingree, 1971, Part II, IX.1, p.59 and Part I, p.85. The expression is valid for a tropical year. For a sidereal year see Part II, IX.1, p.65 and Part I, p.91.
41 Modern value, VARĀHAMIHIRA's value is 25;59°. Neugebauer, Pingree, 1971, Part II, III.28–29, p.35.

2.3.3 Brahmagupta

In the work *Khaṇḍakhādyaka*, BRAHMAGUPTA presents the following expression for the calculation of the longitude of the mean Sun, where the epoch from that work starts at midnight between March 22nd and 23rd, 665:[42]

[2.18] $$\lambda_M = \frac{(a \times 800) + 438}{292\,207}$$

a *ahargaṇa*, number of days since the epoch

2.3.4 Bhāskara I

We know next to nothing about the life of the Indian astronomer and mathematician BHĀSKARA I. Perhaps he came from the district of Aśmak, which he mentions several times in his works. Other sources suggest that he worked for at least some time in the capital city of Surāstra, Valabhī, on the west bank of Khambat Bay (Cambay).[43] He worked in the footsteps of ĀRYABHAṬA I, but much later, most likely at the beginning of the 7th century. More is known about his work than about his life. He was one of the greatest names of the ĀRYABHAṬA astronomy school. Three of his works are known: *Mahā-Bhāskarīya* and *Laghu-Bhāskarīya*, the abbreviated and processed edition of *Mahā-Bhāskarīya*, and partially preserved comments on the work of ĀRYABHAṬA I, *Āryabhaṭīyabhāṣya*.[44] In the commentary, he considers the method of solving Diophantine equations and also worked on trigonometry. His astronomical works were in use as textbooks until the 16th century. Among other achievements, he calculated the sines of angles larger than 90° and cubic roots.[45]

The expression for the mean longitude given below follows from his work. The epoch starts on Friday, February 18th, 3102 BCE at sunrise at Laṅkā.[46]

[2.19] $$\lambda_M = \frac{(a \times 4\,320\,000)}{1\,577\,917\,500}$$

a *ahargaṇa*, number of days since beginning of the kalpa

42 Chatterjee, 1970, I.8, p.49, comment, p.96. Rao, 2000, pp.27–28, and Pingree, 1981a, Census of the Exact Sciences in Sanskrit; series A, vol. 4, p.256.

43 Rao, 1998, pp.59–62.

44 For comments on the work of ĀRYABHAṬA I, see Keller, 2006, I, II. Shukla, 1960; Shukla, 1963, Bhāskara I and His Works Part III, Laghu-Bhāskarīya.

45 Shukla, 1971/72, Hindu mathematics in the seventh century as found in Bhāskara I's commentary on the Āryabhaṭīya, Ganita, vol. 22, no. 1, pp.115–130, Ganita, vol. 22, no. 2, pp.61–78, Ganita, vol. 23, no. 1, pp.57–79, and Ganita, vol. 23, no. 2, pp.41–50.

46 Shukla, 1960, pp.3 and 6–7.

The mean anomaly, *mandakendra*, is determined using Equation 1.5:

$$\kappa_{\mathrm{M}} = \lambda_{\mathrm{M}} - \lambda_{\Gamma}$$

λ_{Γ} apogee, BRAHMAGUPTA's value = 80°
λ_{M} mean longitude
κ_{M} mean anomaly

The equation of centre c, *mandaphala*, in Indian astronomy is:[47]

[2.20] $$c = \frac{r}{R} \operatorname{Sin} \kappa_{\mathrm{M}}$$

R radius of orbit ≈ 3438 (circumference of orbit = 360), value of ĀRYABHAṬA I
r circumference of epicycle = 14
κ_{M} mean anomaly
$\operatorname{Sin} \kappa_{\mathrm{M}}$ sine of the mean anomaly, capitalised to denote the Indian or Hindu sine = $R \sin \kappa_{\mathrm{M}}$

The true longitude of the Sun is (Equation 1.5): $\lambda_{\odot} = \lambda_{\mathrm{M}} - c$

Daily motion of the Sun:[48]

[2.21] $$n + \Delta n = 59'.133 + \Delta n$$

where $$\Delta n = \pm \frac{r}{R} \cos \kappa_{\mathrm{M}} \frac{\Delta \kappa_{\mathrm{M}}}{\Delta t}$$

R radius/circumference of the orbit
r radius/circumference of the epicycle
n daily motion of the mean Sun[49]
Δn correction of daily motion of the Sun
κ_{M} mean anomaly
$\Delta \kappa_{\mathrm{M}} / \Delta t$ daily change of the mean anomaly = 59'.133

The correction is positive if the anomaly of the Sun is in the second and third quadrants, or between 90° and 270°, and negative if the anomaly is in the first or fourth quadrants, or between 0° and 90° or between 270° and 360°.[50]

47 Rao, 2000, p.56, see also Equation 2.10. For guidance on calculating the equation of centre in Indian astronomy, see also Shukla, 1960, pp.110–111.

48 Rao, 2000, p.69.

49 Value from the *Khaṇḍakhādyaka*. BHĀSKARA II used the value $0°59'8''10'''21''''$ ≈ 59'.136. Arkasomayaji, 1980, p.64, Verse 15. Arkasomayaji, 2000, p.69. For other typical values, see Rao, 2000, p.44, Table 6.2.

50 Rao, 2000, p.67, note.

Example 2.3 Calculate the mean and true longitude of the Sun and equation of the centre on 23 March 665 at noon. Also calculate the daily motion of the Sun. Apogee is 80° (solution: $\lambda_M = 1°1'56''$, $\lambda_\odot = 3°13'9''$, $n + \Delta n = 58'.6929$):

The mean longitude is calculated using Equation 2.18 and the mean anomaly using Equation 1.5:

$$\lambda_M = \frac{(a \times 800) + 438}{292\,207} = \frac{(0.5 \times 800) + 438}{292\,207} = 0.002868$$

$$\lambda_M = 0.002868 \cdot 360° = 1°1'56''$$

$$\kappa_M = \lambda_M - \lambda_\Gamma = 1°1'56'' - 80° = -78°58'4''$$

$$c = \frac{14°}{360°} \sin \kappa_M \text{ rad} \quad \text{or in degrees} \quad c = \frac{14}{360} \sin \kappa_M \cdot \frac{180}{\pi} = -2°11'13''$$

$$\lambda_\odot = \lambda_M - c = 1°1'56'' + 2°11'13'' = 3°13'9''$$

The anomaly is in the fourth quadrant, so the correction is negative.

$$\Delta n = \frac{-r}{R} \cos \kappa_M \underbrace{\frac{\Delta \kappa_M}{\Delta t}}_{=59'.133} = -0'.4406$$

The daily motion of the Sun, calculated using Equation 2.21, is:

$$n + \Delta n = 59'.133 + \Delta n = 59'.133 - 0'.4406 = 58'.6929$$

2.3.5 Sūrya Siddhānta

The *Sūrya Siddhānta* uses the epoch of the *kaliyuga* starting at midnight between 17th and 18th February 3102 BCE. The number of full revolutions of the Sun during the *mahāyuga* period $= 432 \times 10^4$ years is 4,320,000, and the *kalpa* period is $1000 \times$ *mahāyuga* $= 432 \times 10^7$ years.[51] The number of days in the *mahāyuga* period is 1,577,917,828.

[2.22] $$\lambda_M = \frac{a \times \text{number of revolutions}}{\text{number of days}} = \frac{a \times 4\,320\,000}{1\,577\,917\,828}$$

According to the claims in *Sūrya Siddhānta*, the Sun's apogee completes 387 revolutions during the *kalpa* period, or 432×10^7 years.[52] At the beginning of the *kaliyuga* period, at midnight between February 17th and February 18th, 3102 BCE, the apogee of the Sun is at $77°7'48''$.[53] *Sūrya Siddhānta* cites a solution to the equation of the centre in a variable epicycle.

51 For periods and days, see Rao, 2000, pp.43–45.

52 Sastri, Wilkinson (trans.), 1861, The Surya-Siddhānta, or an Ancient System of Hindu Astronomy. English translation with explanatory notes, reprint 1974, chapter I, 41&42, p.7.

53 Rao, 2000, p.45, Table 6.3.

The mean anomaly, *mandakendra* or κ_M, is determined by Equation 1.5, and the general expression for the circle circumference is:

$$2\pi R = 360° = 21\,000' \qquad \text{or} \qquad R = \frac{21\,000'}{2\pi\,\text{rad}} = 3437.7468 \approx 3438$$

According to the references in the sources, the circumference of the epicycle varies within the limits of 13°40′ to 14°, where the correction of the radius of the epicycle r_k is:[54]

$$r_k = 14° - (20' \cdot |\sin \kappa_M|)$$

The variable radius of the epicycle $\boldsymbol{r}$ is thus:

$$r = \frac{3438'}{360°} \cdot (14° - 20' \cdot |\sin \kappa_M|) \qquad [2.23]$$

κ_M mean anomaly

The circumference of the solar epicycle is determined from the above equation.[55]

2.3.6 Bhāskara II

Bhāskara II provides the following expression for calculating the mean longitude:[56]

$$\lambda_M = \frac{(a \times 4\,320\,000\,000)}{1\,577\,916\,450\,000} \qquad [2.24]$$

a *ahargaṇa*, number of days since the beginning of the kalpa

54 Sastri, Wilkinson (trans.), 1861, reprint 1974, chapter II, 34, p.17. See also Rao, 2000, pp.57–58.
55 Rao, 2000, p.56, maximum of the equation of centre of Bhāskara II is 2°11′36″ = 131′.5.
56 Arkasomayaji, 1980, pp.12, 24, 32 or Arkasomayaji, 2000, pp.13, 25, 33.

2.4 Correction due to eccentricity - bhujāntara

2.4.1 Bhāskara I and Brahmagupta

The mean longitudes of the Sun, moon and planets, calculated using *ahargaṇa*, are given for the mean Sun at midnight. The true longitude of the Sun is obtained by corrections added or subtracted from the mean longitude value. The correction of the mean longitude due to the orbit's eccentricity for the position of the true Sun at midnight, as indicated by Indian astronomical sources, is called *bhujāntara*.[57] BRAHMAGUPTA refers to the correction of *bhujāntara* in *Khaṇḍakhādyaka*, written around 665, and this rule is also mentioned in the *Brāhmasphuṭasiddhānta*, written by BRAHMAGUPTA in 628.[58] BHĀSKARA I also knew the correction in the 7th century, as can be seen from his work *Mahā-Bhāskarīya*.[59]

The correction of longitude or *bhujāntara*, according to the references in the literature, is obtained by multiplying the equation of centre or *mandaphala* with the daily motion of the Sun and dividing by the number of degrees or time units in a circle. The correction is positive if the equation of centre is positive, and negative if it is negative. If the equation of centre and the path of the Sun are given in arc minutes, both BHĀSKARA I and BRAHMAGUPTA cite the following expression for calculating the correction of *bhujāntara*:

$$E_{\mathrm{b}} = \frac{c \times \text{(mean) daily motion of the Sun}}{21600'} \qquad [2.25]$$

If the equation of centre and the path of the Sun are given in degrees:

$$E_{\mathrm{b}} = \frac{c \times \text{(mean) daily motion of the Sun}}{360^\circ}$$

If the result is in *ghaṭikā* time units:

$$E_{\mathrm{b}} = \frac{\frac{c}{15} \times \text{(mean) daily motion of the Sun}}{24\ \text{hrs}}$$

Part of the *bhujāntara* correction is the equation of centre *c*, *mandaphala*, which is the contribution to the equation of time due to eccentricity, and whose values correspond to time units. The degrees are converted into time units at a factor of 15° per hour or 6° per *ghaṭika*.[60]

57 *Bhujāntara* also means a correction of the position of the planets or the moon.

58 Chatterjee, 1970, pp.73–74, comment, p.138. *Brāhmasphuṭasiddhānta* XIV.19; Shukla, 1960, p.126, note 2.

59 Shukla, 1960, p.129; Chatterjee, 1970, pp.73–74 and 137–138; Subbarayappa, Sarma, 1985, Indian Astronomy: A Source Book, p.161. For a summary of the equation of time in Indian astronomy, see Das, 1928, The Equation of Time and Hindu Astronomy, The American Mathematical Monthly, vol. 35, no. 10, pp.540–543.

60 *Ghaṭikā*, old Indian time unit, 24 hours = 60 *ghaṭikās*.

The correction of longitude due to eccentricity, *bhujāntara*, as mathematically described by BHĀSKARA I and BRAHMAGUPTA, is thus determined in modern notation by the following expression:[61]

$$E_b = \frac{(\lambda_M - \lambda_\odot) \times (n + \Delta n)}{360°} \qquad [2.26]$$

$\lambda_\odot$ true longitude
λ_M mean longitude of the ecliptical mean Sun
n mean daily motion of the Sun[62]
Δn correction of daily motion of the Sun
$\lambda_M - \lambda_\odot$ equation of centre

Later, the *bhujāntara* correction also mentions the *Sūrya Siddhānta*, whose author is unknown.[63]

Example 2.4 Calculate the equation of centre and *bhujāntara* correction on 25 March 665 at noon. Apogee is 80° (solution: $E_b = -0'.3535$):

The first part of the calculation is identical to Example 2.3:

$$\lambda_M = \frac{(a \times 800) + 438}{292\,207} = \frac{(2.5 \times 800) + 438}{292\,207} = 0.008343$$

$$\lambda_M = 0.008343 \cdot 360° = 3°0'13''$$

$$\kappa_M = \lambda_M - \lambda_\Gamma = 3°0'13'' - 80° = -76°59'47''$$

$$c = \frac{14°}{360°} \sin \kappa_M = \frac{14°}{360°} \sin(-76°.9964) \frac{180}{\pi} = -2°10'15''$$

$$\Delta n = -0'.51744$$

The mean daily motion of the Sun is 59′.133; the true daily motion is 58′.6155. For a calculation, see Example 2.3. The correction of *bhujāntara* is thus:

$$E_b = \frac{c \times n}{21\,600'} = \frac{-130'.262 \times 58'.6155}{21\,600'} = -0'.3535$$

It is a correction to the ecliptic longitude, which can be converted into units of time by taking into account 24 h ≡ 59′.133. For the given example, $E_b = -8$ min 36 sec.

61 See Rao, 2000, p.62.
62 For a correction of the mean daily motion, see Equation 2.21.
63 Subbarayappa, Sarma, 1985, pp.161–162. Sastri, Wilkinson (trans.), 1974, p.20 and 109, note †. Rao, 2000, pp.61–62.

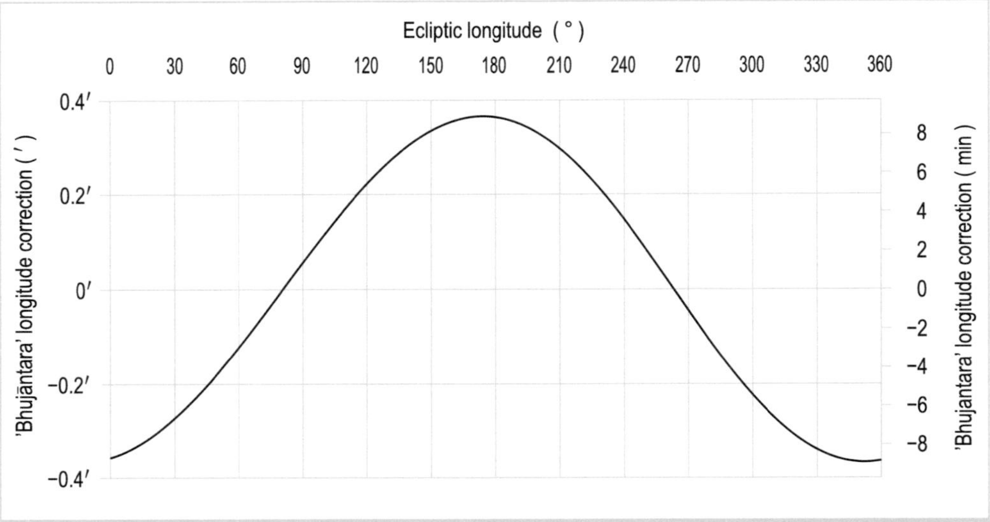

Figure 2.4 The correction of the longitude of the Sun due to eccentricity, or *bhujāntara*. Values are calculated in the same way as in Example 2.3 for the BRAHMAGUPTA epoch for the period from March 665 to March 666. Indian values in arc minutes, left scale and minutes of time, right scale.

2.4.2 Bhāskara II

BHĀSKARA II (1114 – c.1185), considered one of the greatest Indian names in mathematics, is also known as *Bhāskarācārya*, or MASTER BHĀSKARA. In his works, he collected much of the knowledge of his predecessors ĀRYABHAṬA, BRAHMAGUPTA, SRĪDHARA and PADMANĀBHE, and he was obviously influenced by the works of MAHĀVIRA. He solved quadratic equations with one or more variables, as well as Diophantine equations.[64] Among his important astronomical works is the treatise *Siddhānta Śiromaṇi*, written at the age of 36, which enjoyed a great reputation among Indian astronomers, and *Karaṇa Kutūhala*, written around 1183. *Siddhānta Śiromaṇi* has two parts: the first refers to the motion of planets, time in astronomy, eclipses and conjunctions, and the other deals with the firmament. The *Brahmasphuṭasiddhānta* treatise is one of the few preserved documents that gives an insight into the development of Indian mathematics in the period between BRAHMAGUPTA and BHĀSKARA II. It is based on the work of BRAHMAGUPTA and also contains two chapters on mathematics.

64 The method of solving Diophantine equations had already been developed by ĀRYABHAṬA. Acharya, 2015, Bhaskaracharya's Method for Diophantine Equations; International Journal of Mathematics and Statistics Invention, vol. 3, no. 1, pp.2321–4759.

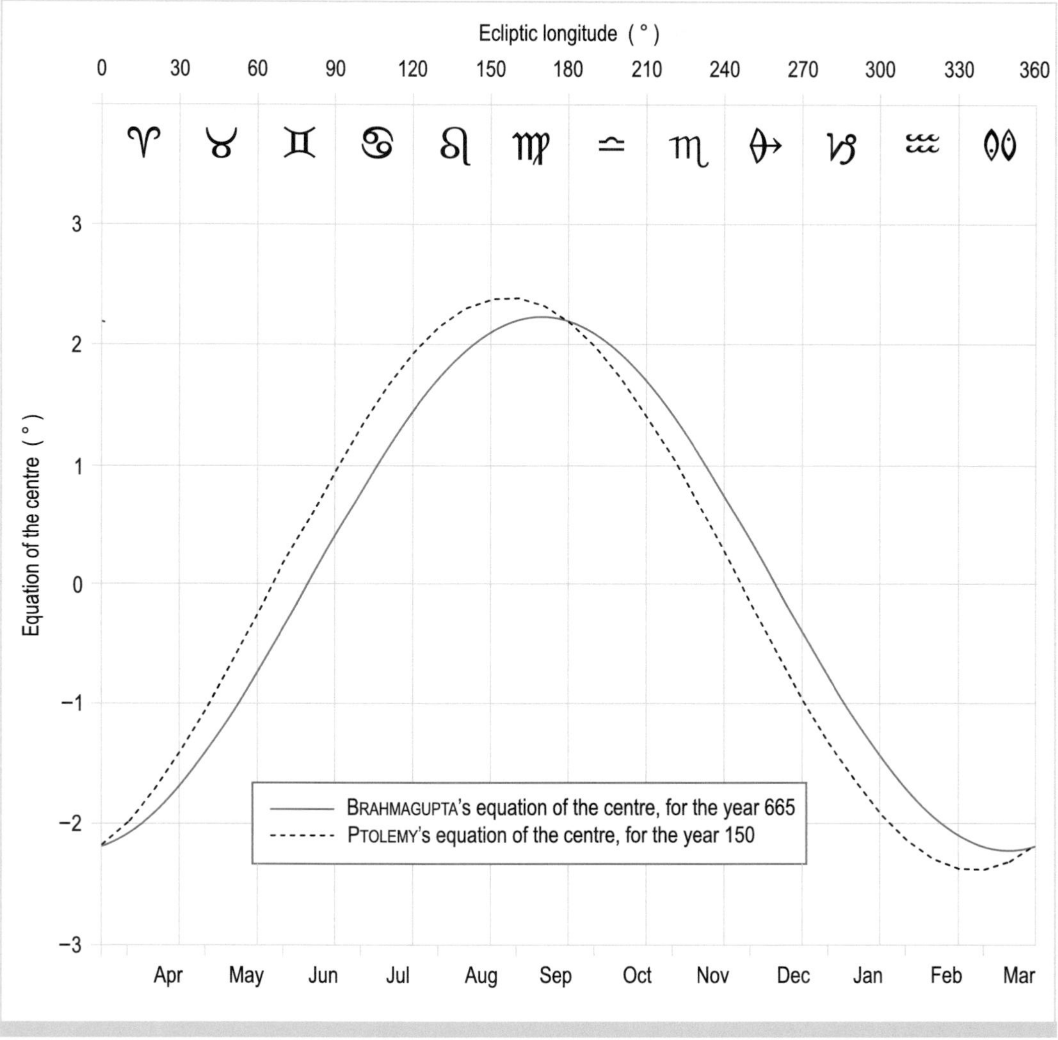

Figure 2.5 The equation of centre, or the contribution to the equation of time due to eccentricity in time-degrees. Comparison of PTOLEMY's value, with its origin from about the year 150, calculated in the same way as in Example 1.2, and the values calculated for the BRAHMAGUPTA epoch, for the period from March 665 to February 666.

BHĀSKARA II gave a more precise definition of the *bhujāntara* correction:[65]

[2.27] $$E_b = \frac{c \times \alpha_M}{30 \cdot 60} \cdot \frac{n + \Delta n}{60 \cdot 60 \cdot 6} = \frac{c \times \alpha_M}{1800'} \cdot \frac{n + \Delta n}{21659'}$$

α_M right ascension of the mean Sun
c equation of centre
n daily motion of the mean Sun
Δn correction of daily motion of the Sun

The average diurnal motion of the Sun according to BHĀSKARA II is $0°59'8''10'''21''''$.[66] The highest value of the equation of centre, as stated by BHĀSKARA II, is $2°11'30''$, which is also the value of radius r.[67] This knowledge was soon assumed by other astronomers.

The right ascension, which often occurs in Hindu literature, is defined as the time of the equatorial sunrise in a certain sign of the zodiac circle.[68] It can be defined by the following expression (where the capitalised functions denote their Indian form):[69]

[2.28] $$\text{Sin}\,\alpha_\odot = \frac{\text{Sin}\,\lambda_\odot \cdot \text{Cos}\,\varepsilon}{\text{Cos}\,\delta_\odot}$$

$\alpha_\odot$ right ascension
$\lambda_\odot$ true longitude
ε obliquity of ecliptic, Indian or Hindu value = 24°
$\delta_\odot$ declination

The above term follows from the equations of spherical astronomy (see Appendix D, Equations D.37):

$$\cos\varepsilon = \frac{\tan\alpha_\odot}{\tan\lambda_\odot} \qquad \cos\lambda_\odot = \cos\alpha_\odot \cos\delta_\odot$$

If we multiply and rearrange the above equations, we obtain an expression that is identical to Equation 2.28:

$$\sin\alpha_\odot = \frac{\sin\lambda_\odot \cdot \cos\varepsilon}{\cos\delta_\odot}$$

65 *Siddhānta Śiromaṇi*, *Spaṣṭādhikāra*, Verse 61. Arkasomayaji, 1980, pp.203–204. $21{,}659' = 21{,}600' + 59'$, where $59'$ = mean diurnal motion of the Sun. VAṬEŚVARA also used the value $21{,}659'$. Shukla (ed.), 1985, p.186. See also Arkasomayaji, 2000, pp.216–217 and Somayaji, 1971, pp.117–118.

66 Somayaji, 1971, p.71, Table II, see also Rao, 2000, p.44, Table 6.2.

67 Somayaji, 1971, p.76.

68 For a comparison of Greek and Indian models in spherical trigonometry see Sengupta, 2008, Greek and Indian Methods in Spherical Astronomy. In: The Tradition of Astronomy in India Jyotiḥśāstra, Subbarayappa, B. V. (ed.), pp.321–342.

69 Arkasomayaji, 1980, pp.205–207, Arkasomayaji, 2000, pp.219–221, Somayaji, 1971, pp.112–113.

2.5 Correction due to obliquity of ecliptic - udayāntara

2.5.1 Śrīpati and Bhāskara II

Unlike the *bhujāntara* longitude correction due to the eccentricity of the orbit, which was known in Indian astronomy as early as the 7th century, the longitude correction due to the obliquity of the ecliptic was first mentioned in the work of *Siddhānta Śekhara*, which was written in the first half of the 11th century, probably around 1030, by the mathematician and astronomer Śrīpati, who initially called the phenomenon *yātāsava*.[70]

Śrīpati is the most important Indian mathematician of the 11th century. He followed Lalla's astronomical principles. Between 1039 and 1056, he wrote three astronomical treatises: *Dhīkoṭidakaraṇa* (1039), a work in twenty verses about solar and lunar eclipses, *Dhruvamānasa* (1056), a work in one hundred and five verses on calculating the longitudes and transits of planets and eclipses, and *Siddhānta Śekhara*, a work in twenty chapters with content in the fields of geometry, arithmetic, algebra (chapters XIII, XIV), astronomy and astronomic instruments (chapter XIX). The contribution to the equation of time due to the obliquity of the ecliptic, or *udayāntara*, is given in section 3.46.[71] He is one of the few Indian astronomers to also be involved in astrology, with the works *Śrīpatipaddhati* and *Jyotiṣaratnamālā*. His mathematical records include, among other things, algorithms for solving quadratic equations and indeterminate equations of the first degree.

A more precise explanation was later provided by Bhāskara II, who named the correction *udayāntara-samskāra*.[72]

[2.29]
$$E_u = \frac{(\lambda_M - \alpha_M) \times \text{(mean) daily motion of the Sun}}{21\,600'}$$

λ_M	longitude of the ecliptical mean Sun
α_M	right ascension of the mean Sun
$\lambda_M - \alpha_M$	contribution to the equation of time due to the obliquity of the ecliptic[73] $\approx \lambda_\odot - \alpha_\odot$

70 Rao, 1998, Indian Mathematics and Astronomy, p.139. Pingree, 1981b, Jyotihśāstra: Astral and Mathematical Literature, p.25. Subbarayappa, Sarma, 1985, p.162. Somayaji, 1971, pp.118–123.

71 Pingree, 1981b, p.25. Sastri, Wilkinson (trans.), 1974, pp.132–133.

72 Arkasomayaji, 1980, pp.204–208 or Arkasomayaji, 2000, pp.217–222. The *udayāntara* correction is mentioned in the *Siddhánta Śiromaṇi*. Sastri, Wilkinson (trans.), 1974, pp.132–133. Das, 1928, pp.540–543.

73 For an approximation, see Somayaji, 1971, p.120.

The part of the equation of time due to the obliquity of the ecliptic can be calculated using the equation:[74]

[2.30] $$\lambda_M - \alpha_M = \lambda_M - \arcsin\left(\frac{\sin\lambda_M \cdot \cos\varepsilon}{\cos\delta_\odot}\right)$$

or $$\alpha_M = \arcsin\left(\frac{\sin\lambda_M \cdot \cos\varepsilon}{\cos\delta_\odot}\right)$$

For the correction of *udayāntara*, Bhāskara II also mentions another expression, where the Indian sine is calculated for a radius of 120 and not for 3438, as is most common in Indian astronomy:[75]

[2.31] $$E_u = \frac{\text{mean daily motion of the Sun} \times \text{Sin } 2\lambda_M}{270}$$

λ_M longitude of the ecliptical mean Sun

If mean daily motion of the Sun $(n + \Delta n)$ is given in degress ° result is in arc minutes ′ and if it is given in minutes ′ result is in arc seconds ″.

2.6 Nīlakaṇṭha Somayājī

A modern concept of the equation of time is evident from the Indian astronomical literature of the late Middle Ages. Nīlakaṇṭha Somayājī, one of the most important Indian astronomers of the 15th century, thus basically describes the equation of time in the way we understand it today:[76]

[2.32] $$E = (\alpha_\odot - \lambda_\odot) + (\lambda_\odot - \lambda_M)$$

$\alpha_\odot - \lambda_\odot$ *prāṇakalāntara* or *prāṇaliptā*, the difference between the right ascension and the longitude of the Sun[77]

$\lambda_\odot - \lambda_M$ *doḥphala*, the difference between the longitude of the true Sun and the longitude of the equatorial mean Sun

Prāṇakalāntara or *prāṇaliptā* is a correction due to the obliquity of the ecliptic, and *doḥphala* is a correction due to eccentricity.

74 Somayaji, 1971, pp.118–119.

75 Somayaji, 1971, pp.121–122. Arkasomayaji, 1980, pp.209–211 or Arkasomayaji, 2000, pp.222–224.

76 Ramasubramanian, Sriram, 2011, Tantrasaṅgraha of Nīlakaṇṭha Somayājī, pp.79, 81–82. For mathematics see Delire, 2009, Mathematics of the Tantrasaṅgraha and its Successors. Joseph (ed.), 2009, Kerala Mathematics, pp.137–151.

77 The concepts of *prāṇa* and *kalā* refer to the difference between the right ascension and the longitude of the Sun. Ramasubramanian, Sriram, 2011, p.79, note 10. For a mathematical derivation and comments about *prāṇakalāntara* see Ramasubramanian, Sriram, 2011, pp.249–252 or Pai et al., 2018, Karaṇapaddhati of Putumana Somayājī, pp.249–251.

Example 2.5 Calculate the *ahargaṇa* from the beginning of the *kalpa* to the beginning of the *kaliyuga*.[78] Also calculate the longitude correction due to the obliquity of the ecliptic, *udayāntara*, for April 10 in the year 1150:

- the number of years from the beginning of the *kalpa* to the beginning of the *kaliyuga*; the data was used by both BRAHMAGUPTA and BHĀSKARA II: 1,972,944,000[79]
- the number of additional (intercalary) lunar months, *b adhimāsās* in the *kalpa*, due to differences in the length of the lunar and sidereal years: 1,593,300,000
- the number of solar days *s* in the *kalpa*: 1,555,200,000,000
- the number of lunar days *t*, *tithis*, in the *kalpa*: 1,602,999,000,000
- the difference between solar and lunar days in the *kalpa*, or *kshayās*: 25,082,550,000

First we calculate the number of days, *ahargaṇa*, from the beginning of the *kalpa* to the beginning of the *kaliyuga*; an example of the calculation for BRAHMAGUPTA's number of years from the beginning of the *kalpa* to the beginning of the *kaliyuga*:

Number of days *d* from the beginning of the *kalpa* to the beginning of the *kaliyuga*:

$$d = 1\,972\,944\,000 \cdot 12 \cdot 30 = 710\,259\,840\,000$$

Number of additional lunar months m_l *adhimāsās* in the *kalpa*:

$$m_l = \frac{d \times b}{s} = 1\,593\,300\,000 \cdot \frac{710\,259\,840\,000}{1\,555\,200\,000\,000} = 727\,660\,110$$

Number of lunar days d_t *tithis* in the *kalpa*:

$$d_t = m_l \cdot 30 + d = 727\,660\,110 \cdot 30 + 710\,259\,840\,000 = 732\,089\,643\,300$$

Number of redundant lunar days d_0 tithis in the kalpa:

$$d_0 = \frac{d_t \times k}{t} = 25\,082\,550\,000 \cdot \frac{732\,089\,643\,300}{1\,602\,999\,000\,000} = 11\,455\,200\,585$$

The *Ahargaṇa* is the difference between the number of lunar days d_t and redundant lunar days d_0:

$$a = 732\,089\,643\,300 - 11\,455\,200\,585 = 720\,634\,442\,715$$

The *Ahargaṇa* since the beginning of the *kaliyuga* can be calculated using the same procedure as in Example 2.2. The new moon day *amāvāsyā* in the year 1150 is March 30.[80] The number of months elapsed since the beginning of the *Caitra* is 1 and the number of *tithis* elapsed from the new moon day *amāvāsyā* is 11. Following the same procedure used in Example 2.2 gives the result *ahargaṇa* = 1,552,729.

We can obtain a similar result by calculating the Julian date or Julian day number. The Julian date, *JD*, at the beginning of the *kaliyuga* (17th/18th February 3102 BCE midnight) is 588 465.5 and the Julian day number *JDN* for 10 April 1150 midday is 2 141 195.

Number of days from the beginning of the *kaliyuga* until 10 April 1150 is:

Number of days $= 2\,141\,195 - 588\,466 = 1\,552\,729$

Total number of days or *ahargaṇa* $= 720\,634\,442\,715 + 1\,552\,729 = 720\,635\,995\,444$

78 Adopted from Singh, Chand, 1992, pp.27–28.

79 Also until the end of Dvāparayuga which precedes the Kaliyuga.

80 Adopted from Fred Espenak's website: http://astropixels.com/ephemeris/phasescat/phasescat.html (19 August 2019).

Calculation of the mean longitude:

$$\lambda_{\mathrm{M}} = \frac{a \times 4\,320\,000\,000}{1\,577\,916\,450\,000} = \frac{720\,635\,995\,444 \times 4\,320\,000\,000}{1\,577\,916\,450\,000}$$

$$\lambda_{\mathrm{M}} = 0.0421132 \cdot 360° = 15{.}°1607$$

The true daily motion of the Sun is calculated in the same way as in Example 2.3:[81]

$$n + \Delta n = 58{.}'1583$$

The *udayāntara* correction then follows from Equation 2.31, calculated for the radius of the circle 120:[82]

$$E_{\mathrm{u}} = \frac{(n + \Delta n) \times \mathrm{Sin}\, 2\lambda_{\mathrm{M}}}{270} = \frac{\frac{58{.}'1583}{60} \times 120 \cdot \sin(2 \cdot 15{.}°1607)}{270} = 0{.}'21749$$

The result converted to time units is 5 min 17 sec.

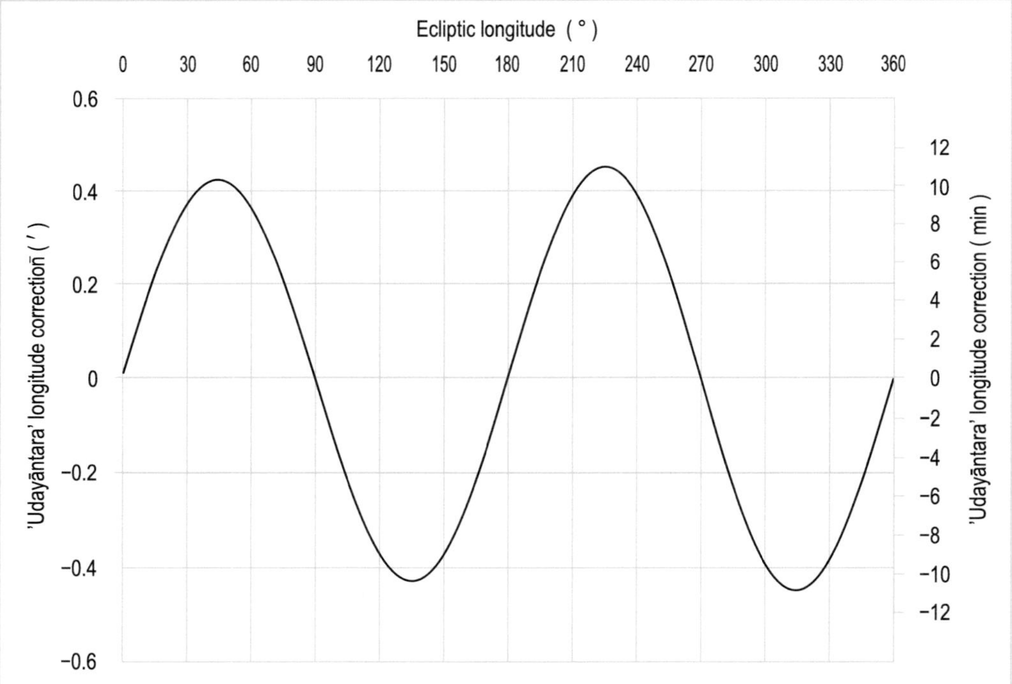

Figure 2.6 Correction of the longitude of the Sun due to the obliquity of the ecliptic, or *udayāntara*. Indian values in arc minutes, left scale and minutes of time, right scale. The values are calculated for the period from March 1150 to February 1151, which roughly corresponds to the time of BHĀSKARA II. Calculation of the *udayāntara* correction according to Equations 2.29 or 2.31. For the calculation of the *ahargaṇa* and the calculation of the daily motion of the Sun, see Example 2.3 or 2.5.

81 The calculation in Example 2.3 is based on *Khaṇḍakhādyaka*. For BHĀSKARA's method, see Arkasomayaji, 1980 or Arkasomayaji, 2000. For a detailed explanation, see Somayaji, 1971, chapter VI.

82 The maximal value of the *udayāntara* correction of BHĀSKARA II was 624 seconds. Somayaji, 1971, pp.122–123.

Author	Correction	Sources and era
Unknown [a]	Bhujāntara	*Sūrya Siddhānta*
Bhāskara I [b]	Bhujāntara	*Mahā-Bhāskarīya* } 7th century *Laghu-Bhāskarīya* }
Brahmagupta [c]	Bhujāntara	*Khaṇḍakhādyaka*, about 665
Lalla [d]	Bhujāntara	*Śiṣyadhīvṛddhida Tantra*, 8th century
Vaṭeśvara [e]	Bhujāntara	*Vaṭeśvara-Siddhānta*, 10th century
Śrīpati [f]	Bhujāntara, Udajāntara, Yatasava	*Siddhānta Śekhara*, 11th century
Bhāskara II [g]	Bhujāntara, Udajāntara-samskāra	*Siddhānta Śiromaṇi*, 12th century
Nīlakaṇṭha [h]	Prāṇakalāntara or Prāṇaliptā, Doḥphala	*Tantrasaṅgraha*, 14th/15th century
Jñānarāja [i]	Bhujāntara	*Siddhānta Sundara*, 15th century
Putumana Somayājī [j]	Prāṇakalāntara	*Karaṇapaddhati*, 15th–18th (?) century

Table 2.3 Significant mentions of corrections due to eccentricity and the obliquity of the ecliptic, or equations of time in Indian literature. The sources are listed in the notes.

Notes

a Subbarayappa, Sarma, 1985, Indian Astronomy: A Source Book, pp.161–162. Sastri, Wilkinson (trans.), 1974, The Surya-Siddhānta, or an ancient system of Hindu astronomy. English translation with notes, pp.20 and 109, note †.

b Shukla, 1960, Bhāskara I and his works, Mahā-Bhāskarīya, p.114. Shukla, 1960, Bhāskara I and his works, Laghu-Bhāskarīya, pp.18–19.

c Chatterjee (trans., ed.), 1970, The Khaṇḍakhādyaka (an astronomical treatise) of Brahmagupta, translation I.18, p.51, comment, p.138.

d Chatterjee (trans.), 1985, Śiṣyadhīvṛddhida Tantra of Lalla with the Commentary of Mallikārjuna Sūri, Part two, pp.18–19.

e Vaṭeśvara used an improved *bhujāntara* correction, similar to Bhāskara II. Shukla (trans., ed.), 1985, Vaṭeśvara-Siddhānta and Gola of Vaṭeśvara, Part two, translation and commentary, pp.XLVII, 185–187.

f Miśra (ed.), 1947, The Siddhānta Śekhara of Śrīpati, Part II. Reprint 2017, pp.XXXIV–XXXVIII. Pingree, 1981b, Jyotiḥśāstra: Astral and Mathematical Literature, p.25.

g Sastri, Wilkinson (trans.), 1974, Siddhānta Śiromaṇi, *udayāntara* IV.19–22, pp.132–133, *bhujāntara* V.43, p.150. Siddhānta Śiromaṇi, Spaṣṭādhikāra, Verses 61–65. Arkasomayaji, 1980, pp.203–211. Arkasomayaji, 2000, pp.216–224.

h Ramasubramanian, Sriram, 2011, Tantrasaṅgraha of Nīlakaṇṭha Somayājī, pp.79, 81–82, 249–252.

i Knudsen, 2014, The Siddhāntasundara of Jñānarāja: An English Translation with Commentary, p.199.

j Pai et al., 2018, Karaṇapaddhati of Putumana Somayājī, pp.249–251.

Bibliography

- Abraham, George. 1980. The Motion of the Sun in the Vasiṣṭha Siddhānta. In: Archive for History of Exact Sciences, 22, 1/2, 1–3. doi 10.1007/BF00327867. url: https://www.jstor.org/stable/41133571.
- Acharya, Manisha M. 2015. Bhaskaracharya's Method for Diophantine Equation. In: International Journal of Mathematics and Statistics Invention, 3, 1, 01–06. url: https://www.ijmsi.org/Volume 3 - Issue1.html (21 January 2024).
- Amma, Sarasvati T. A. 1979. Geometry in Ancient & Medieval India. New Delhi: Motilal Banarsidass. isbn 978-8120813441.
- Arkasomayaji, D. 1980. Siddhānta Śiromaṇi of Bhāskarācārya. Tirupati: Rashtriya Sanskrit Sansthan.
- Arkasomayaji, D. 2000. Siddhāntaśiromaṇi of Bhāskarācārya. Tirupati: Rashtriya Sanskrit Vidyapeetha.
- Bag, A. K. 2003. Luni-Solar Calendar, Kali Ahargaṇa and Julian Days. In: Indian Journal of History of Science, 38, 1, 17–37. url: http://repository.ias.ac.in/74673/ (21 January 2024).
- Bhattacharyya, S. P. and S. N. Sen. 1969. Ahargaṇa in Hindu Astronomy. In: Indian Journal of History of Science, 4, 1&2, 144–155. url: http://www.insa.ndl.iitkgp.ac.in/xmlui/handle/1234567/899 (21 January 2024).
- Burgess, Ebenezer (trans.). 1860. Sûrya-Siddhânta, a Text-Book of Hindu Astronomy. New Haven, CT: American Oriental Society. (Google Books, 12 May 2022).
- Chatterjee, Bina (trans., ed.). 1970. The Khaṇḍakhādyaka (an astronomical treatise) of Brahmagupta with the Commentary of Bhaṭṭotpala, I-II. New Delhi: Motilal Banarsidass.
- Chatterjee, Bina (trans., ed.). 1981. Śiṣyadhīvṛddhida Tantra of Lalla with the commentary of Mallikārjuna Sūri. New Delhi: Indian National Science Academy.
- Clark, Walter Eugene (trans., ed.). 1930. The Aryabhatiya of Aryabhata. Chicago, IL: University of Chicago Press. Reprint Kessinger Publishing. isbn 1425485995.
- Das, Sukumar Ranjan. 1928. The Equation of Time in Hindu Astronomy. In: The American Mathematical Monthly, 35, 10, 540–543. doi 10.2307/2298168.
- Datta, Bibhutibhusan and Avandesh Narayan Singh. 1983. Hindu Trigonometry. In: Indian Journal of History of Science, 18, 1, 39–108. url: http://www.insa.ndl.iitkgp.ac.in/xmlui/handle/1234567/1125 (21 January 2024).
- Delire, Jean Michel. 2009. Mathematics of the Tantrasaṅgraha and its successors. In: Kerala Mathematics, History and its Possible Transmission to Europe, Joseph, George Gheverghese (ed.), 137–151. New Delhi: B. R. Publishing. isbn 978-81-7646-662-2.
- Elfering, Kurt. 1975. Die Mathematik des Āryabhaṭa I. München: Wilhelm Fink Verlag. isbn 3-7705-1326-6.
- Ginzel, Friedrich Karl. 1906. Handbuch der mathematischen und technischen Chronologie, Band 1. Leipzig: J. C. Hinrichs'sche Buchhandlung. Reprint ALO Arbeitsgemeinschaft. isbn 3-226-00478-6.
- Gupta, Radha Charan. 1978. Indian Values of the Sinus Totus. In: Indian Journal of History of Science, 13, 2, 125–143. url: http://www.insa.ndl.iitkgp.ac.in/xmlui/handle/1234567/1087 (21 January 2024).
- Hayashi, Takao. 1997. Āryabhata's Rule and Table for Sine-Differences. In: Historia Mathematica, 24, 4, 396–406. doi 10.1006/hmat.1997.2160.
- Joseph, George Gheverghese (ed.). 2009. Kerala mathematics: history and its possible transmission to Europe. New Delhi: B.R. Publishing Corp. isbn 9788176466622.
- Kak, Subhash C. 1989. The Brahmagupta algorithm for square rooting. In: Ganita Bhārati 11, 1–4, 27–29.
- Keller, Agathe. 2006. Expounding the Mathematical Seed, Volume 1: The Translation. Basel: Birkhauser. isbn 978-3-7643-7291-0.
- Keller, Agathe. 2006. Expounding the Mathematical Seed, Volume 2: The Supplements. Basel: Birkhauser. isbn 978-3-7643-7291-7.
- Khadilkar, S. D. 2003. Katyayana Sulba Sutra (With English Translation, Explanatory Notes, Diagrams and Articles). Pune: Vaidika Samsodhana Mandala.
- Knudsen, Toke Lindegaard. 2014. The Siddhāntasundara of Jñānarāja: An English Translation with Commentary, Baltimore: Johns Hopkins University Press. isbn 978-1-4214-1442-3.
- Kulkarni, R. P. 1983. Geometry According to Sulba Sutra. Pune: Vaidika Samsodhana Mandala.
- Madhavan, S. 2011. Jyā: Concept, Properties and Applications. In: Proceedings of the National Workshop on Ancient Indian Mathematics with Special Reference to Vedic Mathematics and Astronomy, Satapathy Harekrishna et al. (eds.), 205–227. Rashtriya Sanskrit Vidyapeetha Publication Series. Tirupati: Rashtriya Sanskrit Vidyapeetha.
- Mercier, Raymond. 1984. The Astronomical Tables of Rajah Jaj Singh Sawa'i. In: Indian Journal of History of Science, 19, 2, 143–171.

- Miśra, Babuāji (ed.). 1947. The Siddhānta Śekhara of Śrīpati, Part II. University of Calcutta. Reprint 2017. Delhi: Facsimile Publisher. ISBN 8888006854948.
- Neugebauer, Otto and David Pingree. 1970. The Pañcasiddhāntikā of Varāhamihira. Part I, Text and translation. København: Munksgaard.
- Neugebauer, Otto and David Pingree. 1971. The Pañcasiddhāntikā of Varāhamihira. Part II, Commentary. København: Munksgaard.
- Pai, Venketeswara, K. Ramasubramanian, M. S. Sriram and M. D. Srinivas. 2018. Karaṇapaddhati of Putumana Somayājī. Singapore: Springer Nature. New Delhi: Hindustan Book Agency. ISBN 978-981-10-6814-0. DOI 10.1007/978-981-10-6814-0.
- Pingree, David. 1970. Bhāskara I. In: Dictionary of Scientific Biography, Gillispie, Charles Coulston (ed.), II, 114–115. New York: Charles Scribner's Sons.
- Pingree, David. 1970. Bhāskara II. In: Dictionary of Scientific Biography, Gillispie, Charles Coulston (ed.), II, 115–120. New York: Charles Scribner's Sons.
- Pingree, David. 1970. Brahmagupta. In: Dictionary of Scientific Biography, Gillispie, Charles Coulston (ed.), II, 416–418. New York: Charles Scribner's Sons.
- Pingree, David. 1970. Āryabhaṭa I. In: Dictionary of Scientific Biography, Gillispie, Charles Coulston (ed.), I, 308–309. New York: Charles Scribner's Sons.
- Pingree, David. 1973. Lalla. In: Dictionary of Scientific Biography, Gillispie, Charles Coulston (ed.), VII, 582–583. New York: Charles Scribner's Sons. SBN 684-10118-1.
- Pingree, David. 1975. Śrīpati. In: Dictionary of Scientific Biography, Gillispie, Charles Coulston (ed.), XII, 598–599. New York: Charles Scribner's Sons. ISBN 0-684-12924-8.
- Pingree, David. 1976. Varāhamihira. In: Dictionary of Scientific Biography, Gillispie, Charles Coulston (ed.), XIII, 581–583. New York: Charles Scribner's Sons. SBN 0-684-12925-6.
- Pingree, David. 1978. History of Mathematical Astronomy in India. In: Dictionary of Scientific Biography, Gillispie, Charles Coulston (ed.), XV, Supplement I, 533–633. New York: Charles Scribner's Sons. ISBN 0-684-14779-3.
- Pingree, David. 1981a. Census of the Exact Sciences in Sanskrit; Series A. Philadelphia: American Philosophical Society. ISBN 978-0-87169-213-9.
- Pingree, David. 1981b. Jyotihśāstra: Astral and Mathematical Literature. Wiesbaden: Otto Harrassowitz. ISBN 3-447-02165-9.
- Plofker, Kim. 2007. Lalla. In: Biographical Encyclopedia of Astronomers, Hockey, Thomas, Virginia Trimble and Thomas R. Williams (eds.), 669–670. New York: Springer. ISBN 9780387310220.
- Plofker, Kim. 2009. Mathematics in India. Princeton, NJ: Princeton University Press. ISBN 9781400834075.
- Prakash, Satya. 1987. Geometry in ancient India. New Delhi: Govindram Hasanand. ISBN 978-8170770121.
- Rai, R. N. 1972. Computation of Ahargaṇa in Vaṭeśvara Siddhānta. In: Indian Journal of History of Science, 7, 1, 27–37. URL: http://www.insa.ndl.iitkgp.ac.in/xmlui/handle/1234567/949 (21 January 2024).
- Ramasubramanian, K. and M. S. Sriram. 2011. Tantrasaṅgraha of Nīlakaṇṭha Somayājī. New Delhi: Hindustan Book Agency. ISBN 978-93-80250-09-0.
- Rao, Balachandra S. 2000. Ancient Indian Astronomy. New Delhi: B.R. Publishing. ISBN 81-7646-162-8.
- Rao, Balachandra S. 1998. Indian Mathematics and Astronomy. Bangalore: Jnana Deep Publications. ISBN 81-900962-0-6.
- Sastri, Pundit Bapu Deva and Lancelot Wilkinson (trans.). 1861. The Surya-Siddhānta, or an ancient system of Hindu astronomy. English translation with notes. Reprint 1974. Amsterdam: Philo Press. ISBN 90 6022 662 3.
- Schmidt, Olaf. 1952. On the Computation of the Ahargana. In: Centaurus, 2, 2, 140–180. DOI 10.1111/j.1600-0498.1952.tb00402.x.
- Seidenberg, Abraham. 1962. The Ritual Origin of Geometry. In: Archive for History of Exact Sciences, 1, 5, 488–527. DOI 10.1007/BF00327767. URL: https://www.jstor.org/stable/41133224.
- Seidenberg, Abraham. 1981. The Ritual Origin of Circle and Square. In: Archive for History of Exact Sciences, 25, 4, 269–327. DOI 10.1007/BF01395658. URL: https://www.jstor.org/stable/41133635.
- Sen, Samarendra Nath. 1974. Epicyclic Eccentric Planetary Theories in Ancient and Medieval Indian Astronomy. In: Indian Journal of History of Science, 9, 1, 107–121. URL: http://www.insa.ndl.iitkgp.ac.in/xmlui/handle/1234567/980 (21 January 2024).
- Sen, Samarendra Nath and Kripa Shankar Shukla. 2000. History of Astronomy in India. New Delhi: Indian National Science Academy.

- Shukla, Kripa Shankar. 1960. Bhāskara I and his works, Part II, Mahā-Bhāskarīya. Lucknow: Dept. of Mathematics and Astronomy Lucknow University.
- Shukla, Kripa Shankar. 1963. Bhāskara I and his works, Part III, Laghu-Bhāskariya. Lucknow: Dept. of Mathematics and Astronomy Lucknow University.
- Shukla, Kripa Shankar. 1971. Hindu mathematics in the seventh century as found in Bhāskara I's commentary on the Āryabhaṭīya, Part I. In: Gaṇita, 22, 1, 115–130.
- Shukla, Kripa Shankar. 1971. Hindu mathematics in the seventh century as found in Bhāskara I's commentary on the Āryabhaṭīya, Part II. In: Gaṇita, 22, 2, 61–78.
- Shukla, Kripa Shankar. 1972. Hindu mathematics in the seventh century as found in Bhāskara I's commentary on the Āryabhaṭīya, Part III. In: Gaṇita, 23, 1, 57–79.
- Shukla, Kripa Shankar. 1972. Hindu mathematics in the seventh century as found in Bhāskara I's commentary on the Āryabhaṭīya, Part IV. In: Gaṇita, 23, 2, 41–50.
- Shukla, Kripa Shankar (trans., ed.). 1985. Vaṭeśvara-Siddhānta and Gola of Vaṭeśvara. Part two. New Delhi: Indian National Science Academy.
- Shukla, Kripa Shankar and K. V. Sarma. 1976. Āryabhaṭīya of Āryabhaṭa. New Delhi: Indian National Science Academy. URL: https://archive.org/details/Aryabhatiya1976/Aryabhatiya%20v1%201976/mode/2up (27 January 2022).
- Singh, S. L. and Ramesh Chand. 1992. Ahargaṇa and its Applications. In: Gaṇita Bhāratī, 14, 1–4, 25–36.
- Somayaji, D. A. 1971. A Critical Study of the Ancient Hindu Astronomy. Dharwar: Karnatak University.
- Subbarayappa, B. V. (ed.). 2008. The Tradition of Astronomy in India Jyotiḥśāstra. History of Science, Philosophy and Culture in Indian Civilization, Volume IV, Part 4. New Delhi: Centre for Studies in Civilizations. ISBN 81-87586-32-X.
- Subbarayappa, B. V. and K. V. Sarma. 1985. Indian Astronomy: A Source Book. Bombay: Nehru Centre.
- Toomer, Gerald J. 1974. The Chord Table of Hipparchus and the Early History of Greek Trigonometry. In: Centaurus, 18, 1, 6–28. DOI 10.1111/j.1600-0498.1974.tb00205.x.
- Van Brummelen, Glen. 2009. The Mathematics of the Heavens and the Earth. Princeton: Princeton University Press. ISBN 978-0-691-12973-0.
- Van der Waerden, Bartel Leendert. 1968. Erwachende Wissenschaft, Band 2. Basel: Birkhäuser Verlag.
- Van der Waerden, Bartel Leendert. 1988. On the Romaka-Siddhānta. In: Archive for History of Exact Sciences, 38, 1, 1–11. DOI 10.1007/BF00329976. URL: https://www.jstor.org/stable/41133822.

3

Equation of time in Arabic astronomy

The beginning of this equation is from four degrees of Scorpio, but if it is at twenty-one degrees of Libra up to four degrees of Scorpio there will be no equation for it, and the greatest it can be in this direction is thirty-four minutes, and that is if the sun is in Aquarius…

Al-Hāshimī, Book of the Reasons Behind Astronomical Tables
(Kennedy, Haddad, trans., 1981, p.157)

3.1 Early beginnings of Arabic astronomy

3.1.1 Access to ancient knowledge

We know very little about the beginnings of Arabic astronomy since almost no sources have been preserved from the first century after the Hijra in 622 CE. However, the Arabs must have been highly familiar with the stars, as they used them for navigation when they led nomadic lives before establishing larger cities.[1] The Arabs took ancient astronomical knowledge from several sources; Indian texts were among the first foreign translations. A delegation from the province of Sindh visited Baghdad around the year 770 and brought a written source, a *siddhānta*, possibly a work by BRAHMAGUPTA, and the Caliph AL-MANṢŪR commissioned its translation, which was overseen by an Indian astronomer who accompanied the delegation.[2] Other ancient, especially Greek texts, reached the Arabs by different routes, in part also through the Syriac and Pahlavi languages. There is very little information about the translations of astronomical works into Syriac and Pahlavi, however there is no doubt that some works from antiquity were translated into those languages and later also became available to the Arabs.

It is unknown why there was an increase in the translation of works from antiquity into Arabic around the middle of the 8th century during the late UMAYYAD Dynasty or the early ABBĀSID Dynasty. Most probably, the reasons included administrative goals – in order to harmonise the language in Arab-controlled areas where various nations lived and society was diverse. The intensity and quantity of translations were so expressive and systematic that the period is still considered one of the most concentrated in the history of mankind when it comes to translation. Among the biggest names in science in antiquity, whose works were translated into Arabic and also left quite a mark on the Arabs, were ARISTOTLE, GALEN, PLUTARCH and PTOLEMY. The Arabs not only summarised foreign knowledge, but also contributed their own achievements.[3]

One of the oldest Islamic astronomical works, or *zījes*,[4] that also includes a brief section on the equation of time is *al-Zīj al-Mumtaḥan*. It is based on Indian and Persian sources, and on data collected from astronomical observations at the request of Caliph AL-MA'MŪN in Baghdad in

1 Kunitzsch, 1961, Untersuchungen zur Sternnomenklatur der Araber. Henninger, 1954, Über Sternkunde und Sternkult in Nord- und Zentralarabien.

2 Most likely it was BRAHMAGUPTA's work *Brāhmasphuṭasiddhānta*, which was translated into Arabic in 771? by AL-FAZĀRĪ and YA'QŪB IBN ṬĀRIQ. For the circumstances and explanation of the historical circumstances that led to the transfer of knowledge, see Saliba, 1994, A History of Arabic Astronomy: Planetary Theories During the Golden Age of Islam, pp.51–81.

3 For Arabic comments on the *Almagest*, see Saliba, 1994, pp.143–160; Sezgin, 1978, Geschichte des Arabischen Schrifttums, VI, pp.88–94; de Vaux, Journal Asiatique, 8e 189, 1892, pp.408–471. Also, Graßhoff, 1990, The History of Ptolemy's Star Catalogue, pp.17–22.

4 The word *zīj* is of Persian origin, most likely derived from the Persian word *zīg*, which means an astronomical manual with tables and explanations; Schoy, 1927, p.1, Kennedy, 1956, pp.123–124, van Dalen, 1996, p.198, note 3. The word *zīg* in Persian also means a cord. See Pingree, 1985, The Astronomical Works of Gregory Chioniades, I, p.37. For more etymological data see Mercier, 2000, From Tantra to Zīj, In: Sic Itur Ad Astra: Studien Zur Geschichte Der Mathematik und Naturwissenschaften, pp.451–460. For an overview of astronomical tables see Kennedy, 1956, A Survey of Islamic Astronomical Tables. Transaction of the American Philosophical Society, vol. 46, 2, pp.123–177.

828 and 829. In Latin, these tables are known as the *Tabulae Probatae*. YAḤYĀ IBN ABĪ MANṢŪR is considered to be the author of at least part of this *zīj*, though some parts of the preserved manuscripts are the work of other unknown author(s) and were added later.[5]

YAḤYĀ (died 830) originated from Persia. We assume that he spent his youth in Baghdad, where his father, ABŪ MANṢŪR ABĀN, was an astrologer in the service of the second ABBĀSID Caliph AL-MANṢŪR (754–775). First, as an astrologer, he served the vizier of Caliph AL-MA'MŪN. MA'MŪN supported scientific work and translated works from Greek and Syriac to Arabic. Following MA'MŪN's orders, YAḤYĀ carried out astronomical observations in Baghdad with the astronomers JAWHARĪ, SANAD IBN 'ALĪ and MARWARRŪDHĪ in 828 and 829. He is a co-author or author of the *Mumtaḥan zīj*, a famous and frequently cited source historically in astronomy. He wrote several other works but very little is known about them. THĀBIT IBN QURRA wrote a treatise about the differences between the *Mumtaḥan zīj* and PTOLEMY's tables, which, unfortunately, was lost.

'ALĪ IBN SULAYMĀN AL-HĀSHIMĪ, about whose life we know nothing, is the author of *Kitāb fī'ilal al-zījāt* (*Book of the Reasons Behind Astronomical Tables*).[6] His legacy, which, according to the sources, must be read with reservations because of its author's uncritical nature, appears to date from the 9th century. It is important because it contains many descriptions of events that had a significant influence on the creation of astronomical tables in the Islamic world. The work was influenced by Indian and Persian astronomy, rather than Greek astronomy, so was probably written at a time before the Arabs had fully embraced Greek astronomy. In his *Book of the Reasons Behind Astronomical Tables*, AL-HĀSHIMĪ also describes the equation of time and the reasons for it, and cites an equation of time from the 9th century work of ABŪ MA'SHĀR AL-BALKHI.[7] ABŪ MA'SHĀR grew up in Khorazan, the city of Balkh, a meeting point of ancient cultures that gave the new civilisation its intellectual elites. It is these intellectual circles that are largely responsible for the large-scale systematic translations of ancient foreign works into Arabic during the ABBĀSID Caliphate. ABŪ MA'SHĀR was a highly respected author of astrological works.

ABŪ MA'SHĀR's equation of time is said to have a maximum value of 34 minutes in Aquarius ♒, and a value of 0 between 21°♎ (Libra) and 4°♏ (Scorpio), where the equation is supposed to begin.[8] The sources suggest that ABŪ MA'SHĀR's work is based either directly or indirectly on the *Shāh zīj* or on a related, possibly Indian, astronomical treatise.[9] While ABŪ MA'SHĀR's

5 Van Dalen, 2004, A Second Manuscript of the Mumtaḥan Zīj, Suhayl 4, pp.9–44. Vernet, 1954, Un antiguo tratado sobre el calendario judio en las "Tabulae Probatae". In: Sefarad 14, pp.59–78. Reprinted in Vernet, 1979, pp.213–232. Vernet, 1956, Las Tabulae probatae. In: omenaje a Millás-Vallicrosa, vol. II, pp.501–522. Reprinted in Vernet, 1979, pp.191–212. Vernet, 1979. Estudios sobre historia de la ciencia medieval.

6 Sezgin, 1978, Geschichte des Arabischen Schrifttums, VI, p.176.

7 Lat. ALBUMASAR (787–886), early Persian-Islamic astrologer, reputedly the foremost astrologer of the ABBĀSID court in Baghdad. His work later influenced the development of astrology in Europe. In addition to ABŪ MA'SHĀR, HĀSHIMĪ also mentions ḤABASH in the work.

8 Perhaps the start of the values in the table? Pingree, Kennedy (trans., eds.), 1981, pp.155–157.

9 Van der Warden, 1987, The Heliocentric system in Greek, Persian and Hindu Astronomy, pp.535–536. Both in the *Shāh zīj* and in the work of ABŪ MA'SHĀR and ARYABHAṬA we find the same date 17/18 February 3102 BCE in relation to the epoch.

astronomical works are lost and known only through references in sources, we know more about his astrological works. Several of his works were translated into Latin and exerted significant influence on the development of medieval philosophy in Europe. His works have been translated or studied by ADELARD OF BATH, ROGER BACON, ALBERT THE GREAT and others.[10]

3.1.2 Al-Khwārizmī

Computed on the basis of PTOLEMY's works, as well as other, especially Indian or Persian sources, the Arab tables were highly valued in the Middle Ages. Thus, Arabic astronomical tables were widely used, for example those by AL-KHWĀRIZMĪ[11] and AL-BATTĀNĪ[12], and later became the basis for the *Toledan Tables*, which transferred the rich Arab astronomical knowledge to Islamic Spain. The works of AL-KHWĀRIZMĪ were based on Persian and Hindu sources, and his main work, *Zīj al-Sindhind*[13] (*Sindhind zīj*), is known from citations in other sources. In addition to the *Brāhmasphuṭasiddhānta*, AL-KHWĀRIZMĪ also used as sources *Khaṇḍakhādyaka*, another work by BRAHMAGUPTA, as well as the lost Persian astronomical work of *Shāh zīj* from the 6th century. The Latin translation of AL-MAJRĪṬĪ's partial review of AL-KHWĀRIZMĪ's *zījes* is preserved, while an equally important source of information on this period can also be found in the *Toledan Tables*.[14] The latter contain several tables from the original Arab version of AL-KHWĀRIZMĪ's *zīj*.

AL-KHWĀRIZMĪ (c.780–850) worked primarily as a mathematician in the Baghdad court under the rule of the ABBĀSID caliphs AL-MA'MŪN (813–833), AL-MU'TAṢIM (833–842) and AL-WĀTHIQ (842–847). Among his astronomical works, the most important are the astronomical manual *Sindhind zīj*, based on Hindu sources, a treatise on sundials, and two astrological treatises. His astronomical tables were used by astronomers from Central Asia to the Iberian Peninsula. Among others, the values from his manuals are also an integral part of European medieval astronomical tables. Also important is his contribution to the development of mathematics; his arithmetic treatises, in which he used Hindu numbers, was felt considerably in the development of European arithmetic in the 12th and 13th centuries. The translations of his works thus brought to Europe, among other things, the decimal system as we know it today, and techniques for

10 ROGER BACON (around 1220–1292), also known by his Latin nickname DOCTOR MIRABILIS (Eng. Wonderful Teacher), Franciscan, advocate of experimental science, education reformer. ALBERT THE GREAT, Lat. ALBERTUS MAGNUS (c.1200–1280), born in Germany, one of the greatest names in medieval philosophy.

11 The equation of time table from the *Sindhind zīj* is available in the Chartres Bibliothèque Publique (No. 214, foil 80r–80v) and Bodleian Library (cod. Auct. F. I. 9, foil 137r–137v). For translation, see Suter, 1914, Muḥammad ibn Mūsā al-Khwārizmī in der Bearbeitung des Maslama ibn Ahmed al-Madjrītī und der lateinischen Übersetzung des Adelard von Bath, as well as Al-Khwarizmi, Neugebauer, 1962, The astronomical tables of al-Khwārizmī, Translation with Commentaries on the Latin Version. For more information on tables with the equation of time, see van Dalen, 1996, Al-Khwārizmī's astronomical tables revisited: analysis of the equation of time in Casulleras and Samsó (eds.), From Baghdad to Barcelona: Studies in the Islamic Exact Sciences in Honour of Prof. Juan Vernet, vol. 1, pp.195–252. In Table 4b on page 243, there is a small mistake at longitude 127°. Instead of 14,0 the value should be 14,40. (Anonymous expert, personal communication 23 July 2017).

12 Pedersen, 2002, The Toledan Tables, p.970, tabulated equation of time from *Toledan Tables*, whose values are equal to AL-BATTĀNĪ's, pp.972–975.

13 The word *sindhind* originates from the word *siddhānta*, which in Sanskrit means doctrine or learning.

14 ABŪ MASLAMA MUḤAMMAD AL-MAJRĪṬĪ (circa first half of the 10th century–1007), an Andalusian mathematician and astronomer who worked in Córdoba, was the undeniable authority of his time.

solving linear and quadratic equations. A little less known, but no less important, is his contribution to the development of geography.

An equation of time, the alleged author of which is AL-KHWĀRIZMĪ, is preserved in AL-MAJRĪṬĪ's comments that have come down to us in nine manuscripts, some of which only contain fragments.[15] AL-KHWĀRIZMĪ is not specifically listed in sources as the author of this equation of time, but we assume that it is him. The values of the equation of time in AL-MAJRĪṬĪ's comments are displaced by 2°. The actual value of the equation of time for 0° is thus tabulated at longitude 2°, the actual value for 1° is tabulated at longitude 3° and so on, and the reasons for the offset are not identified in the sources.[16] The equation of time obviously follows the Ptolemaic model with the obliquity of the ecliptic 23°51′, the maximum value of the solar equation being 2°14′, and the longitude of the apogee being 82°39′. The equation of time is of zero at ♒ 22° or at 322° longitude, and the epoch is 4;30°.[17] AL-KHWĀRIZMĪ's equation of time (Table 3.1) as a function of the displaced true longitude of the Sun can be written as:[18]

[3.1] $$E_s(\lambda_{\odot s}) = E(\lambda_{\odot s} + \Delta)$$

$$= \frac{1}{15} \cdot \left(\lambda_{\odot s} + \Delta + c(\lambda_{\odot s} + \Delta) - \alpha(\lambda_{\odot s} + \Delta) + k\right)$$

E_S equation of time as function of shifted longitude
E equation of time as given in historical sources
$\lambda_{\odot S}$ shifted true longitude = $\lambda_\odot - \Delta$ or $\lambda_{\odot S} + \Delta = \lambda_\odot$
Δ longitude shift = −2°
c equation of centre
k epoch constant = 4;30

♈	0	7;48
	10	11;28
	20	15; 8
♉	30	18;28
	40	21; 4
	50	22;48
♊	60	23;28
	70	23; 0
	80	21;32
♋	90	19;40
	100	17;32
	110	15;52
♌	120	14;44
	130	14;44
	140	15;44
♍	150	17;36
	160	20;24
	170	23;32
♎	180	26;52
	190	29;52
	200	32;24
♏	210	34; 0
	220	34;28
	230	33;36
♐	240	31;24
	250	27;44
	260	23; 4
♑	270	17;52
	280	12;32
	290	7;44
♒	300	3;48
	310	1;12
	320	0; 2
♓	330	0;20
	340	1;52
	350	4;28
	360	7;48

Table 3.1 The values of AL-KHWĀRIZMĪ's equation of time from the Latin translation of AL-MAJRĪṬĪ's comments. The equation has the value zero at ♒ 22°. Source: Suter, 1914, pp.181–182, Table 67–68. See also Neugebauer, 1962, pp.61–62 and van Dalen, 1996, pp.242–244.

15 Van Dalen, 1996, p.199. AL-MUTHANNĀ also commented on AL-KHWĀRIZMĪ's work, but he did not address the equation of time therein. Goldstein, 1967, Ibn al-Muthannā's Commentary on the Astronomical Tables of al-Khwārizmī, two Hebrew translations with commentary, and Millás Vendrell, 1963, El commentario de Ibn al-Mutannā' a las Tablas Astronomicas de al-Jwārizmī, Estudio y edicion critica del texto latino, en la version de Hugo Sanctallensis, Latin translation with commentary.

16 For an overview of the parameters for calculating the equation of time and a detailed explanation, see Van Dalen, 1996, pp.245–246.

17 The value of the longitude of the apogee is based on observations by the order of Caliph AL-MA'MŪN. These values were also used by YAḤYĀ IBN ABĪ MANṢŪR and ḤABASH AL-ḤĀSIB in their *zījes*. It cannot be reliably asserted whether the value of the apogee's longitude was 82°39′, or perhaps 82°40′, which was used by ḤABASH in his *zīj*. Debarnot, 1987, The Zīj of Habash al-Hāsib: A Survey of MS II Istanbul Yeni Cami 784/2. In: From Deferent to Equant, King, Saliba (eds.). The equation of time in MAJRĪṬĪ's comments is of Indian-Persian origin with the maximum of the solar equation at 2°14′. PTOLEMY's value is 2°23′. Van Dalen, 1996, pp.240, 245.

18 Van Dalen, 1996, pp.238–239, reprinted in Van Dalen, 2014, Islamic Astronomical Tables, Mathematical Analysis and Historical Investigation.

3.1.3 Al-Battānī

Important tables with the equation of time were also those by AL-BATTĀNĪ, who used the *Almagest* as the basis, and his values for the equation were later used in the *Toledan Tables*. AL-BATTĀNĪ (c.858–929), also known under the Latin name ALBATENIUS, was one of the most influential astronomers of the early Islamic period. He was probably the son of a maker of mathematical instruments, and thus he learned the basics of their use. He is known for his precise astronomical observations and the accuracy of his observations of the autumn and spring equinoxes, ande his determination of the length of the year was no worse than calculated by TYCHO BRAHE some 700 years later. He also calculated the obliquity of the ecliptic precisely. His most important work is the astronomical manual, which continued the tradition of PTOLEMY's *Handy Tables*. Among his mathematical works, his contribution to trigonometry is worth mentioning; around the year 900, he calculated sine tables at half-degree intervals.[19]

3.1.4 Al-Bīrūnī and Abū al-Wafā'

AL-BĪRŪNĪ, who is considered one of the greatest minds of the Islamic period, has also left several exceptional works.[20] He dedicated his encyclopaedic astronomical treatise *Al-Qānūn al-Mas'ūdī*, which covers the equation of time in section six, to the eldest son of Sultan MAḤMŪD, MAS'ŪD.

AL-BĪRŪNĪ's *Exhaustive Treatise on Shadows* shows that he had settled on the convention of using 1 as the radius of the circle,[21] as also used by ABŪ AL-WAFĀ' and AL-BĪRŪNĪ's teacher ABŪ NAṢR MANṢUR.[22] This practice did not become widespread in trigonometry until later in the Middle Ages, when decimal arithmetic replaced sexagesimal arithmetic.

I added to the operations in the zījes the merit of (additional) simplicity by making both/the total sine and the gnomon one part, removing the need for depression and elevation/completely.

Al-Bīrūnī, The Exhaustive Treatise on Shadows, I, 33:6
(Kennedy, trans., 1976, p.72)

19 For more information on his life and bibliography, see Hartner, 1970, Al-Battānī, Dictionary of Scientific Biography, I, pp.507–516, also Hartner, 1984, Oriens, Occidens, II, pp.201–210.

20 AL-BĪRŪNĪ came into contact with Indian culture at the time of the Muslim conquests; his book, India, written from his own experience around 1030, is considered to be the fundamental work of Indology in the modern sense of the word.

21 Al-Bīrūnī, Kennedy (trans.), 1976, I, 33:6, p.72.

22 Von Braunmühl, 1900, Vorlesungen uber Geschichte der Trigonometrie, 1, p.58 and Van Brummelen, 2009, p.145. ABŪ AL-WAFĀ' AL-BŪZJĀNĪ (940–997/998?), one of the most important astronomers of his time. For more information about his approach to spherical astronomy see Jensen, 1972, Abū Nasr Mansūr's Approach to Spherical Astronomy as Developed in His Treatise "The Table of Minutes".

Abū al-Wafā', mathematician and astronomer, was active in the 10th century. His contribution to trigonometry is very important. We know from sources that he wrote, among other works, a *Kitab al-kamil* (*Complete Book*), which is closely related in theme to the *Almagest* but is unfortunately lost. He was also the first to publish all six modern trigonometric functions, without names but in a modern form, in his *Kitāb al-Majisṭī* around 980.[23] Abū al-Wafā' calculated trigonometric function tables to half-degree accuracy, using a similar method to Ptolemy (Equation 0.6), but using the following theorem:[24]

$$\sin\frac{15^\circ}{32}+\frac{1}{3}\left(\sin\frac{18^\circ}{32}-\sin\frac{15^\circ}{32}\right)<\sin 30'<\sin\frac{15^\circ}{32}+\frac{1}{3}\left(\sin\frac{15^\circ}{32}-\sin\frac{12^\circ}{32}\right)$$

The required initial values of $\sin {}^{15}/_{32}$ and $\sin {}^{18}/_{32}$ can be calculated using the angles 60° and 72° and arithmetic. The value of $\sin {}^{12}/_{32}$ is obtained as the sine of the difference ${}^{72}/_{32} - {}^{60}/_{32}$. Other mathematicians, such as Theon of Alexandria and later Regiomontanus, also used similar theorems.[25]

A summary of Arab astronomers' major works and their calculations of the equation of time are given in Table 3.5.

3.2 Spherical trigonometry

Arabic works on spherical trigonometry are based on Greek developments but also contain original contributions. Menelaus' *Spherica* was commented on in the 10th century by Abū Naṣr Manṣur.[26] Many of Abū Naṣr's works survive, including his *Treatise on the Determination of Spherical Arcs*, where he used the concept of the polar triangle to solve problems.[27] His works show his knowledge of the sine law for spherical triangles:[28]

[3.2] $$\frac{\operatorname{Sin} a}{\operatorname{Sin} A}=\frac{\operatorname{Sin} b}{\operatorname{Sin} B}=\frac{\operatorname{Sin} c}{\operatorname{Sin} C}$$

23 These are the trigonometric functions sin, cos, tan, cot, sec and cosec. The *Kitāb al-majisṭī* treatise includes chapters on trigonometry, the application of spherical trigonometry to astronomy and planetary theory. For more information, see Von Braunmühl, 1900, vol. 1, p.57, or Cantor, 1907, vol. 1, p.748, also van Brummelen, 2009, p.155, footnote 50.

24 Youschkevitsch, 1970, p.42.

25 Bond, 1921, The Development of Trigonometric Methods down to the Close of the XVth Century, p.304.

26 For German translation see: Krause, 1936, Die Sphärik von Menelaos aus Alexandrien in der Verbesserung von Abū Nasr Mansūr b. Ali ibn Iraq. In: Abhandlungen der Gesselschaft der Wissenschaften zu Göttingen, Philologisch Historische Klasse, vol. 3, no. 17, pp.1–254.

27 Debarnot, 1978, Introduction du triangle polaire par Abu Nasr b. Iraq. In: Journal for the History of Arabic Sciences, vol. 2, no. 1, pp.126–136.

28 Van Brummelen, 2009, The Mathematics of the Heavens and the Earth, p.183.

It is not clear whether the sine law was discovered by ABŪ NAṢR or perhaps by ABŪ AL-WAFĀ', who was active in the same period.

Among the more important works on spherical trigonometry is AL-JAYYĀNĪ's[29] 11th century *Book on Unknown Arcs of a Sphere*,[30] which is also very similar in content to a manuscript by an unknown author from Isfahan, the *Collection of Rules of Astronomy*, from the same period. AL-JAYYĀNĪ cites MENELAUS' theorem for the entire spherical quadrilateral *ACFBA* (see Figure 3.1), albeit without proof:[31]

[3.3] $$\frac{\operatorname{Sin}\widehat{BA}}{\operatorname{Sin}\widehat{EA}}=\frac{\operatorname{Sin}\widehat{BD}}{\operatorname{Sin}\widehat{FD}}\cdot\frac{\operatorname{Sin}\widehat{CF}}{\operatorname{Sin}\widehat{CE}}\quad\equiv \text{Equation } 0.8$$

[3.4] $$\frac{\operatorname{Sin}\widehat{BE}}{\operatorname{Sin}\widehat{EA}}=\frac{\operatorname{Sin}\widehat{BF}}{\operatorname{Sin}\widehat{FD}}\cdot\frac{\operatorname{Sin}\widehat{CD}}{\operatorname{Sin}\widehat{CA}}\quad\equiv \text{Equation } 0.9$$

[3.5] $$\frac{\operatorname{Sin}\widehat{CA}}{\operatorname{Sin}\widehat{DA}}=\frac{\operatorname{Sin}\widehat{CE}}{\operatorname{Sin}\widehat{FE}}\cdot\frac{\operatorname{Sin}\widehat{BF}}{\operatorname{Sin}\widehat{BD}}\quad\equiv \text{Equation } 0.10$$

[3.6] $$\frac{\operatorname{Sin}\widehat{DC}}{\operatorname{Sin}\widehat{DA}}=\frac{\operatorname{Sin}\widehat{CF}}{\operatorname{Sin}\widehat{FE}}\cdot\frac{\operatorname{Sin}\widehat{BE}}{\operatorname{Sin}\widehat{BA}}\quad\equiv \text{Equation } 0.11$$

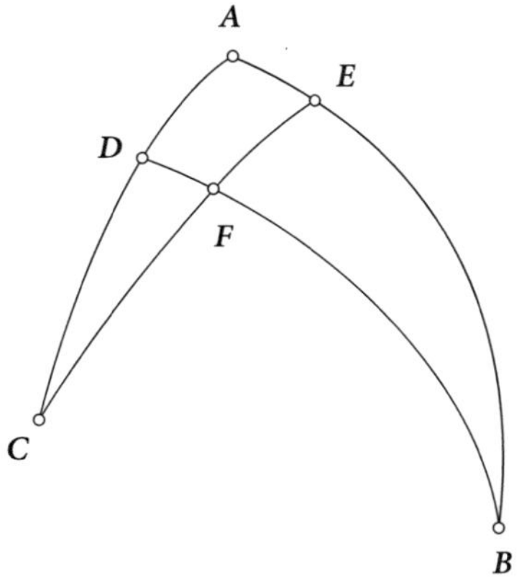

Figure 3.1 AL-JAYYĀNĪ's spherical quadrilateral. Based on Hairetdinova, 1986, p.137, Figure 1.

29 AL-JAYYĀNĪ, also IBN MU'ĀDH (active in the 11th century, died after 1079), served as a judge in Jaén. He was educated in Egypt and was probably a pupil of IBN AL-HAYTHAM, Lat. ALHAZEN. He is the author of several astronomical works, and his astronomical tables were translated into Latin by GERARD OF CREMONA.

30 *Kitāb majhūlāt qisiy al-kurah.*

31 Hairetdinova, 1986, On Spherical Trigonometry in the Medieval Near East and in Europe, p.138.

His work was one of REGIOMONTANUS's key sources for writing *De triangulis omnimodis libri quinque*, the book that brought spherical trigonometry to medieval Europe.[32]

AL-JAYYĀNĪ, too, studied the problems of spherical trigonometry using his conception of the polar triangle, which was later refined by AL-ṬŪSĪ in his *Treatise on the Complete Quadrilateral*, written in 1260.[33] AL-ṬŪSĪ is one of the greatest names in mathematics and astronomy of his time, although he is less well known for his mathematical advances. Among his most important achievements were new planetary models in which the equant was replaced by eccentric orbits, and in which he sought to improve on the Ptolemaic system then in use. His models were highly influential, very likely known to COPERNICUS, and were also known to Jewish astronomers as evidenced by Hebrew astronomical works of the time.[34]

3.3 Kūshyār's equation of time

The values of the equation of time are also preserved in the *zījes* of some other astronomers, for example in the *Jāmiʿ zīj* by KŪSHYĀR IBN LABBĀN. Almost nothing is known about KŪSHYĀR's life, but we do have some information on his work.[35] His extensive astronomical treatise *Bāligh zīj* was lost, but the manuscripts of *Jāmiʿ zīj* and the *the Basics of Hindu Reckoning*, which deal with arithmetic algorithms on a decimal and hexadecimal basis, have been preserved and translated into English, Persian, Russian and French. His most important work, the *Jāmiʿ zīj* is based on AL-BATTĀNĪ's *zīj* and the *Almagest*. It contains tables of trigonometric and astronomical functions together with a star catalogue and the geographic coordinates of cities.[36] KŪSHYĀR also wrote on astrology and astrolabes. KŪSHYĀR's original tables contained values for every six degrees of longitude, but were later supplemented. His equation of time argument is the longitude of the mean Sun λ_{Ms}, displaced by 2°. This calculation of the equation of time is an anomaly due to an offset mean longitude, which was already known earlier; for example, this method was also used by ḤABASH AL-ḤĀSIB in tables on the movement of the moon.[37]

32 REGIOMONTANUS may have used the work directly or indirectly through the work of IBN AFLAḤ (Lat. GEBER), whose work was crucial for the transmission of knowledge to Europe. For a brief description of his life and work, see Lorch, 1975, pp.85–87.

33 Papadopoulos, 2019, Nasir al-Din al-Tusi's treatise on the quadrilateral: The art of being exhaustive. Juschkewitsch, 1964, Mathematik im Mittelalter, pp.304–308.

34 Nosonovsky, 2018, Abner of Burgos: The Missing Link between Nasir al-Din al-Tusi and Nicolaus Copernicus? Zutot, vol. 15, pp.25–30.

35 KŪSHYĀR IBN LABBĀN (latter half of 10th/11th century), a Persian astronomer, known for his tables. For some data about his life and work see Bagheri, 2006, pp.XIII–XVI.

36 Van Dalen, 2021, Ptolemaic Tradition and Islamic Inovation: The Astronomical Tables of Kūshyār ibn Labbān, pp.120–122.

37 ḤABASH AL-ḤĀSIB (9th century) was one of the most important astronomers of the early period of Islamic astronomy. He was associated with the ABBĀSID court, although the sources do not place him in the narrower group of astronomers led by YAḤYĀ IBN ABĪ MANṢŪR. According to sources, he was engaged in astronomical observations in Baghdad in the years 829/830 and was also an author of *zījes*. See Salam, Kennedy, 1967, Solar and Lunar Tables in Early Islamic Astronomy, Journal of the American Oriental Society, vol. 87, no. 4, pp.496–497.

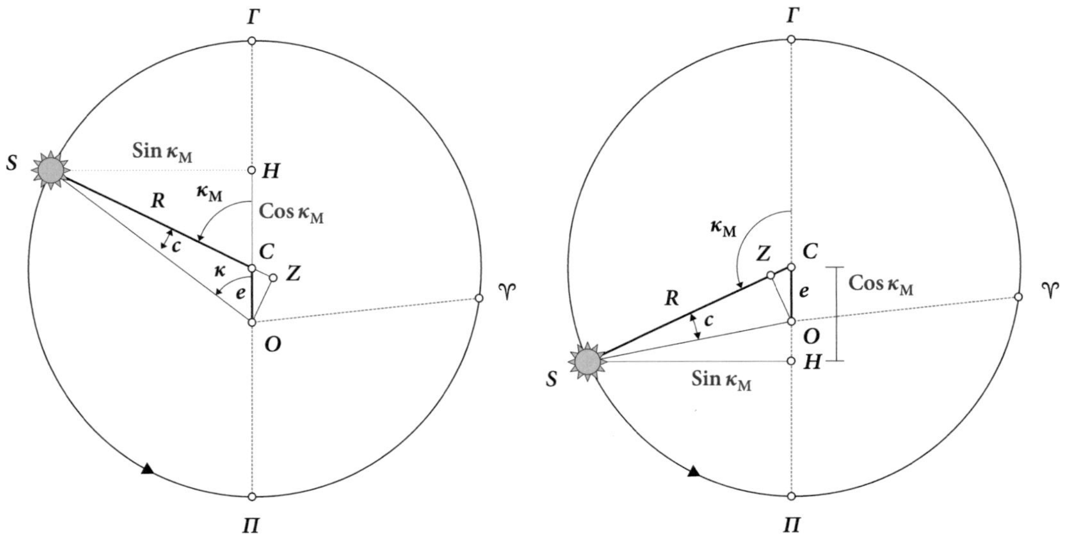

Figure 3.2 KŪSHYĀR's model for calculation of the solar equation or equation of centre. The model is obviously adapted from the *Almagest*. KŪSHYĀR calculated the result with the help of the sine function and not the tangent, which was already known in his time. Based on Bagheri, 2006, pp.140–141, see also van Brummelen, 2009, pp.156–158.

KŪSHYĀR's equation of time, as stated by the source, is written in modern mathematical notation as:[38]

[3.7] $$E = 0;4\left(\left(\lambda_{Ms} - 10^S 16°\right) - \left(\alpha(\lambda_\odot) - 10^S 22;4°\right)\right)$$

0;4 conversion factor from time-degrees to time units (hours)
10^S 10th sign of the Zodiac or ♑

The above equation can be rearranged into:[39]

[3.8] $$E = 0;4\left(\lambda_{Ms} + 6;4° - \alpha(\lambda_\odot)\right)$$

The following relations are valid for the calculation:[40]

[3.9] $$\lambda_{Ms} = \lambda_M - 2 \qquad \kappa_M = \lambda_M - \lambda_\Gamma \qquad \kappa_{Ms} = 2 - \kappa_M$$

38 Kennedy, 1988, Two Medieval Approaches to the Equation of Time, Centaurus, vol. 31, no. 1, p.2. For KŪSHYĀR's description of the equation of time, see Bagheri, 2006, Books I and IV of Kūshyār ibn Labbān's Jāmiʿ zīj: An Arabic Astronomical Handbook by an Eleventh-Century Iranian Scholar, PhD, Utrecht University, pp.139–140. For the reconstruction of the calculation see van Dalen, 1993, Ancient and Mediaeval Astronomical Tables: Mathematical structure and parameter values, pp.134–141, and van Dalen, 1996, pp.236–241.

39 Kennedy, 1988, pp.2–3. Constant should be 4;4° and not 6;4°. For an explanation, see Kennedy, 1988, p.3 and van Dalen, 1993, p.138.

40 Van Dalen, 1996, pp.236–239.

or: $\lambda_{Ms} + \kappa_{Ms} = (\lambda_M - 2) + (2 - \kappa_M) = \lambda_M - \kappa_M = \lambda_\odot$

λ_M mean longitude of ecliptical mean Sun
λ_{Ms} mean longitude shifted for 2°, argument of equation of time
$\lambda_\odot$ true longitude
λ_Γ longitude of apogee, KŪSHYĀR's value = ♊ 24° or 84°
α right ascension
κ_M mean anomaly
c, c_s equation of centre also solar equation, shifted equation of centre

The shifted solar equation $c(\kappa_M)$ is always added in the equation of time calculation process. KŪSHYĀR gives the instructions for calculating the equation in the first book of the *Jāmiʿ zīj*. From Figure 3.2 it is evident that the angles $\widehat{HCS}$ and $\widehat{ZCO}$ are equal, and considering that $R = 60$, $SH \equiv \sin \kappa_M$ and $HC \equiv \cos \kappa_M$, the following relations obtain:[41]

$$\frac{CS}{SH} = \frac{CO}{OZ} \quad \text{or} \quad \frac{R}{\text{Sin}\,\kappa_M} = \frac{e}{OZ} \quad \text{thus} \quad OZ = e \sin \kappa_M$$

$$\frac{CS}{HC} = \frac{CO}{CZ} \quad \text{or} \quad \frac{R}{\text{Cos}\,\kappa_M} = \frac{e}{CZ} \quad \text{thus} \quad CZ = e \cos \kappa_M$$

By applying the sine law on the triangle $\triangle COZ$:

$$\frac{\sin c}{e} = \frac{\sin \kappa}{R}$$

we get the equation of centre as a function of the true anomaly (Equation 1.4):

[3.10] $$c = \arcsin\left(\frac{e \sin \kappa}{R}\right)$$

The distance between SZ and SO, as in the *Almagest*, is determined by the following terms:

$$SZ = R + CZ \qquad SO = \sqrt{SZ^2 + OZ^2}$$

41 Bagheri, 2006, pp.140–141. See also van Brummelen, 2009, p.157.

λ_{Ms}	0	9;45
	10	13;16
	20	16;30
	30	19;10
	40	21; 1
	50	21;53
	60	21;43
	70	20;38
	80	18;50
λ_{Ms}	90	16;42
	100	14;36
	110	12;57
	120	12; 3
	130	12; 6
	140	13; 8
	150	15; 6
	160	17;47
	170	20;54
λ_{Ms}	180	24;10
	190	27;12
	200	29;39
	210	31;11
	220	31;31
	230	30;27
	240	27;57
	250	24;10
	260	19;27
λ_{Ms}	270	14;17
	280	9;17
	290	5; 1
	300	1;55
	310	0;14
	320	0; 0
	330	1; 7
	340	3;19
	350	6;19
	360	9;45

Table 3.2 The values of KŪSHYĀR's equation of time for each 10° shifted mean longitude λ_{Ms}. The values are recalculated as in Example 3.1. For the calculation see van Dalen, 1996, pp.236–241 and van Dalen, 1993, pp.138–139. For the data see Kennedy, 1988, p.4 and van Dalen, 1993, p.136.

Solar equation c or the equation of centre, as a function of the mean anomaly, is:[42]

[3.11] $$c = \arcsin\frac{OZ}{SO} = \arcsin\frac{e\sin\kappa_{M}}{\sqrt{(R + e\cos\kappa_{M})^2 + (e\sin\kappa_{M})^2}}$$

Example 3.1 Calculate the value of KŪSHYĀR's equation of time for longitudes $\lambda_{\odot} = 210°$ and $\lambda_{Ms} = 105°$. What is the mean shifted longitude λ_{Ms} at $\lambda_{\odot} = 210°$ and what is the true longitude $\lambda_{\odot}$ at $\lambda_{Ms} = 105°$? KŪSHYĀR's values are: radius $R = 60$, longitude of apogee $\lambda_{\Gamma} = 84°$, eccentricity $e = 2;4,45 = 2.079166$, obliquity of ecliptic $\varepsilon = 23;35°$ (solution: $E = 31$ min 9 s, at $\lambda_{Ms} = 209°36'23''$ and $E = 13$ min 42 s, at $\lambda_{\odot} = 106°14'53''$).

We calculate mean longitude by Equation 1.10 and right ascension by Equation D.37:

$$\lambda_{M} = \lambda_{\odot} + \arcsin\left(\frac{e}{R}\sin(\lambda_{\odot} - \lambda_{\Gamma})\right)$$

$$= 210° + \arcsin\left(\frac{2.0791\overline{6}}{60}\sin(210° - 84°)\right) = 211°36'23''$$

$$\alpha_{\odot} = \arctan(\cos\varepsilon\tan\lambda_{\odot}) = \arctan(\cos 23^{\circ}\!.58\overline{3}\tan 210°) = 207°53'4''$$

We calculate shifted mean longitude by Equation 3.9:

$$\lambda_{Ms} = \lambda_{M} - 2 = 209°36'23''$$

Then we calculate the equation of time in time-degrees (Equation 3.8):

$$E = 4\cdot(\lambda_{Ms} + 6;4° - \alpha_{\odot})$$

$$= 4\cdot(209°36'23'' + 6°4' - 207°53'4'') = 9°14'32''$$

After converting time-degrees into minutes and seconds ($1° \equiv 4$ min) we get:

$$E = 31 \text{ min } 9 \text{ s}$$

In calculating the equation of time for the longitude $\lambda_{Ms} = 105°$ we first calculate the mean longitude and solar equation.

We calculate mean longitude by Equation 3.9:

$$\lambda_{M} = \lambda_{Ms} + 2 = 105° + 2° = 107°$$

42 For the sign see Chapter 1, Equations 1.6, 1.7, 1.8 and explanation.

By algebraic or geometrical derivation as above we get:

$$OZ = e \sin \kappa_M = 2.0791666 \cdot \sin \overbrace{(107° - 84°)}^{\text{Equation 1.5}} = 0.812395$$

$$CZ = e \cos \kappa_M = 2.0791666 \cdot \cos(107° - 84°) = 1.91388$$

$$SZ = R + CZ = 60 + 1.91388 = 61.91388$$

$$SO = \sqrt{OZ^2 + SZ^2} = \sqrt{0.812395^2 + 61.91388^2} = 61.91921$$

We calculate the equation of the centre by Equation 3.11:

$$c = \arcsin \frac{OZ}{SO} = \arcsin \frac{0.812395}{61.91921} = 0\overset{\circ}{.}751756$$

The true longitude and right ascension can be calculated using Equation 1.5 or Equation D.37:[43]

$$\lambda_{\odot} = \lambda_M - c = 107° - 0\overset{\circ}{.}751756 = 106°14'53''$$

$$\alpha_{\odot} = \arctan\left(\cos \varepsilon \tan \lambda_{\odot}\right) = \arctan\left(\cos 23\overset{\circ}{.}58\overline{3} \tan 106\overset{\circ}{.}2482\right) = 107°38'26''$$

Therefore, the equation of time is (Equation 3.8):

$$E = 4 \cdot (\lambda_{Ms} - \alpha_{\odot} + 6;4°)$$
$$= 4 \cdot (105° - 107°38'26'' + 6°4') = 3°25'33'' \quad \text{or} \quad 13 \text{ min } 42 \text{ s}$$

43 The right quadrant needs to be considered. The atan2 (arctan2) function is useful in such cases. The common syntax in a spreadsheet program is ATAN2(x, y), whereas at least one variable or coordinate x or $y \neq 0$.

3.4 Al-Kāshī and Ulugh Beg

3.4.1 Jamshīd al-Kāshī

AL-KĀSHĪ (died about 1429), who worked at the end of the 14th and early 15th century, originated in Persia and was one of the most active astronomers of his time.[44] He wrote treatises in Arabic and also several mathematical works in addition to astronomy. Before finding a patron, he worked alone. Later he travelled to Samarkand at the invitation of ULUGH BEG, probably around 1420, where he took over a newly built observatory, that he also planned and supervised the construction of. He was ULUGH BEG's closest associate and consultant. The results of the observations were later used to prepare ULUGH BEG's zīj. AL-KĀSHĪ is regarded as one of the greatest masters of calculation of his time; his calculation of the sine of 1° is accurate to eighteen decimal places, and the number 2π to sixteen.

AL-KĀSHĪ's tables, published in the *Khāqānī zīj*, with the epoch at 712th Yazdegerd (December 21, 1342), were mathematically reconstructed, and the results of subsequent calculations matched the original values to a high degree of precision, which shows the carefulness of AL-KĀSHĪ's work. AL-KĀSHĪ's equation of time (Table 3.3) is the difference between the change of mean solar longitude and the change in the right ascension of the true Sun from epoch to the moment of observation:[45]

[3.12] $$E = \Delta\lambda_{M} - \Delta\alpha_{\odot} = (\lambda_{M} - \lambda_{M0}) - (\alpha_{\odot} - \alpha_{0})$$

or

[3.13] $$E = \frac{\lambda_{M} - \alpha_{\odot} - \overbrace{(\lambda_{M0} - \alpha_{0})}^{\text{epoch constant, see Chapter 1}}}{15;2,28} = \frac{\lambda_{M} - \alpha_{\odot} + 3;57,34^{\circ}}{15;2,28}$$

$\alpha_{\odot}$ right ascension
α_{0} right ascension of the Sun, at AL-KĀSHĪ's epoch = 322;25,17°
λ_{M} mean longitude of ecliptical mean Sun
λ_{M0} mean longitude, at AL-KĀSHĪ's epoch = 318;27,42,37°

44 For information about his life and mathematical works see Luckey, 1951, Die Rechnenkunst bei Ğamšīd B. Mascūd al-Kāšī, pp.4–17. Also Kennedy, 1960, pp.1–9.

45 Kennedy, 1988, p.5. See also Kennedy, 1960, The Planetary Equatorium of Jamshīd Ghiyāth al-Dīn al-Kāshī, pp.157 and 238–240. AL-KĀSHĪ has obviously also used the value 3;57,30°.

The relationship between the longitude of the true and the mean Sun is given by the following equation (see Equations 1.6 and 1.7):

[3.14] $$\lambda_{\odot} = \lambda_M + \arctan\left(\frac{e \sin \kappa_M}{60 + e \cos \kappa_M}\right) \qquad \text{where (Equation 1.5)} \qquad \kappa_M = \lambda_M - \lambda_\Gamma$$

Equation 3.14 is not simply solvable. However, the longitude of the mean Sun can be approximated numerically, which is repeated until the difference between two successive approximations for λ_M is sufficiently small or negligible:[46]

[3.15] $$\lambda_{M1} = \lambda_{\odot} \qquad \lambda_{Mn+1} = \lambda_{\odot} - c(\kappa_{M1}) \qquad \kappa_{M1} = \lambda_{Mn} - \lambda_\Gamma \qquad n = 1, 2, 3, \ldots$$

$\lambda_{\odot}$ true longitude
λ_M mean longitude of ecliptical mean Sun
λ_Γ longitude of apogee, 90°
e eccentricity, 2;6,9
ε obliquity of ecliptic, 23;30°
κ_M mean anomaly
c equation of centre or solar equation

Example 3.2 Calculate AL-KĀSHĪ's equation of time when the Sun is at $\lambda_{\odot} = 30°$. The parameters are: right ascension of the mean Sun for the epoch $\alpha_0 = 322;25,17°$; longitude of the mean Sun for the reference moment, $\lambda_{M0} = 318;27,42,37°$; longitude of the apogee, $\lambda_\Gamma = 90°$; eccentricity, $e = 2;6,9 = 2.1025$; obliquity of ecliptic, $\varepsilon = 23;30°$ (solution: $E = 17$ min 14 s).

The longitude of the mean Sun is calculated numerically where the sign of $c(\kappa_M) < 0$ for $0° < \kappa_M < 180°$ is taken into account with respect to Equation 3.15, which is iterated until the difference between the two successive approximations for λ_M is negligible:

$$\lambda_{M1} = 30° \qquad \kappa_M = \lambda_{M1} - \lambda_\Gamma = 30° - 90° = -60°$$

$$c(\kappa_{M1}) = -\arctan\frac{e \sin \kappa_{M1}}{60 + e \cos \kappa_{M1}} = -\arctan\frac{2.1025 \sin(-60°)}{60 + 2.1025 \cos(-60°)} = -1.^\circ708307$$

After a few repetitions, we get the following value for the mean Sun: $\lambda_M = 28;15,39°$.

46 Here, too, we take note of the sign; see Pedersen, 2011, A Survey of the Almagest, p.141, Equations 5.15 and 5.16.

$$\lambda_{M2} = 30° - c(\kappa_{M1}) = 30° - 1{.}^{\circ}708307 = 28{.}^{\circ}29169$$

$$\kappa_{M2} = \lambda_{M2} - \lambda_{\Gamma} = 28{.}^{\circ}29169 - 90° = -61{.}^{\circ}708307$$

$$c(\kappa_{M2}) = -\arctan\frac{e\sin\kappa_{M2}}{60 + e\cos\kappa_{M2}}$$

$$= -\arctan\frac{2.1025\sin(-61{.}^{\circ}708307)}{60 + 2.1025\cos(-61{.}^{\circ}708307)} = 1{.}^{\circ}738491$$

$$\lambda_{M3} = 30° - c(\kappa_{M2}) = 30° - 1{.}^{\circ}738491 = 28{.}^{\circ}2615$$

The right ascension $\alpha_{\odot}$ is calculated as in the previous example:

$$\alpha_{\odot} = \arctan(\cos\varepsilon\tan\lambda_{\odot})$$
$$= \arctan(\cos 23;30° \tan 30°) = 27;53,58°$$

The value of the equation of time is therefore (Equation 3.13):

$$E = \frac{\lambda_M - \alpha_{\odot} + 3;57,34°}{15;2,28} = \frac{28;15,39° - 27;53,58° + 3;57,34°}{15;2,28}$$
$$= 4;18,32° \equiv 0.2872644\ \text{h} = 17\ \text{min}\ 14\ \text{s}$$

which corresponds to the value in the original tables.[47]

It is interesting to note AL-KĀSHĪ's calculation of the circumference of the circle, which AL-KĀSHĪ approached in a similar way to ARCHIMEDES - using a circle with the radius $R = 60$ and inscribing regular polygons. AL-KĀSHĪ applied the following theorem:[48]

$$\sqrt{R(2R + \text{Crd}\,\alpha)} = \text{Crd}\left(\alpha + \frac{180° - \alpha}{2}\right)$$

Considering Crd 60° = R, we can calculate Crd 120° and then proceed to calculate the chord function for angles of 150°, 165°, 172°30′, etc. Taking $\alpha = 2\varphi$ and $R = 1$, the above equation can be written in a modern form as:

$$\sin\left(45° + \frac{\varphi}{2}\right) = \sqrt{\frac{1 + \sin\varphi}{2}}$$

which corresponds to the expression in LAMBERT's work of 1770.[49]

♈	4	9; 7
	14	12;29
	24	15;36
♉	34	18;10
	44	19;53
	54	20;36
♊	64	20;17
	74	19; 1
	84	17; 6
♋	94	14;54
	104	12;54
	114	11;29
♌	124	10;57
	134	11;28
	144	12;59
♍	154	15;24
	164	18;26
	174	21;47
♎	184	25; 6
	194	28; 2
	204	30;15
♏	214	31;27
	224	31;25
	234	30; 1
♐	244	27;18
	254	23;26
	264	18;46
♑	274	13;47
	284	9; 1
	294	4;58
♒	304	2; 0
	314	0;20
	324	0; 2
♓	334	1; 0
	344	3; 2
	354	5;51

Table 3.3 The recomputed values of AL-KĀSHĪ's equation of time, calculated with the procedure as in Example 3.2. For other recomputed values see Kennedy, 1988, p.6.

47 Kennedy, 1988, p.6.
48 Youschkevitsch, Rosenfeld, 1973, Al-Kāshī, Dictionary of Scientific Biography, p.258.
49 Lambert, 1770, pp.138–139. For LAMBERT's work, see Chapter 11.

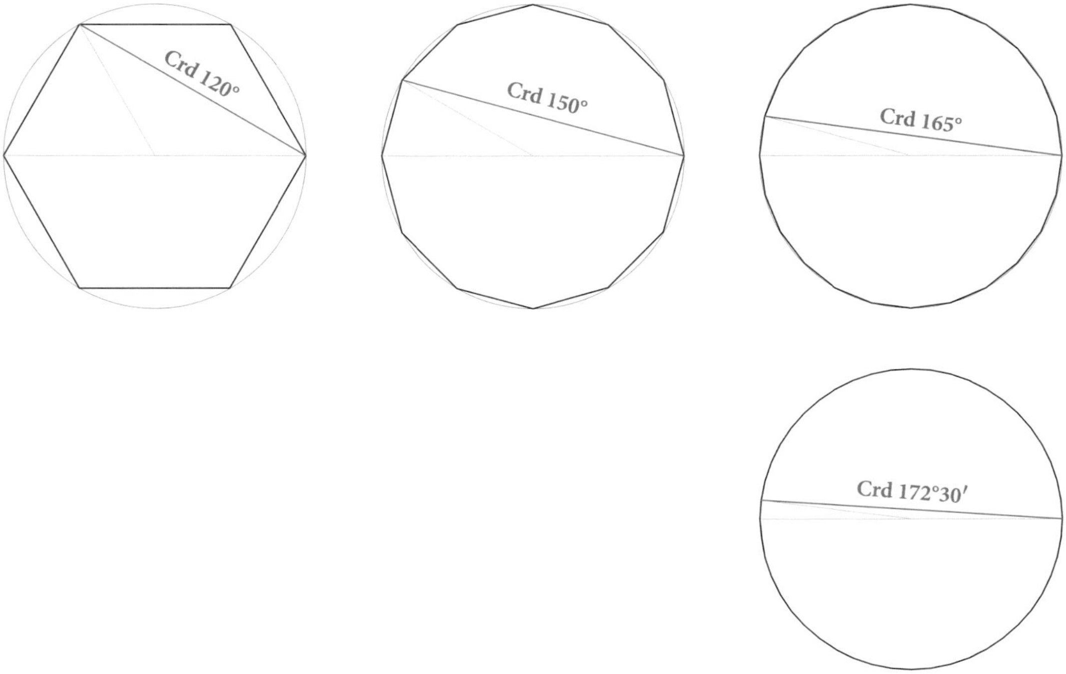

Figure 3.3 Examples of the sides of circles of regular polygons as drawn by AL-KĀSHĪ in determining the value of π using the values of Crd 120°, Crd 150°, Crd 165° and Crd 172°30′. The values correspond to the sides of a hexagon, a dodecagon, an icositetragon and a tetracontaoctagon, see also Figure 0.6.

3.4.2 Ulugh Beg

ULUGH BEG (1394–1449), which in Turkish means Grand Prince, TIMUR's grandson, was the governor of Transoxania and Turkestan, with Samarkand and Bukhara as cultural centres. He had apparently mastered astronomical knowledge and, as AL-KĀSHĪ's correspondence shows, was also involved in recruiting astronomers and actively participated in astronomical seminars. ULUGH BEG is known for accurate calculations that went beyond those of his precursors. The sines in his manual for every 5″ from 0 to 87° are accurate to nine decimal places, and between 87° and 90° to eleven.[50] The Observatory in Samarkand, founded in 1420, carried out many observations in a few decades of operation. It was one of the best observatories in the Islamic world. ULUGH BEG was its head, and AL-KĀSHĪ was in charge of observations. After his death, QĀDĪZĀDE became the head of the observatory, and the observations were completed by QŪSHJĪ.[51] The observatory was destroyed in the 16th century and today its remains are part of

50 For ULUGH BEG's sine tables from 0° to 44°59′, see Schoy, 1927, pp.92–100.

51 QADI ZADA AL-RUMI (1364–1436), astronomer of Turkish origin working in Samarkand. QŪSHJĪ (died 1474) was probably born in Samarkand, where he used to work.

the world's cultural heritage.[52] The *Zīj-i Ulugh Beg*, also *Zīj-i Jadīd-i Sulṭānī*, *Zīj-i Gūrgānī*, which was based on the results of observations and was among the most widely used Islamic astronomical textbooks of the Middle Ages, has been translated into Arabic, Turkish, Latin and, in the 17th century, also into English.[53]

The true longitude is also an argument in ULUGH BEG's equation of time (Table 3.4), with the epoch of July 5 1437, or the beginning of the 841th Islamic year.[54] The calculation was repeated by GAUSS by the method of least squares that he developed in 1799, and his approach was not historical but mathematical (Table 3.5). In calculating, he used the following expression to calculate the contribution due to eccentricity:[55]

[3.16] $$\sin c = e \sin\left(\lambda_{\odot} - \lambda_{\Pi}\right)$$

$\lambda_{\odot}$ true longitude
λ_{Π} longitude of perigee; longitude of apogee = 93;0,18°
e eccentricity = 0.0337037
c equation of centre or solar equation

How GAUSS calculated the contribution to the equation of time due to the obliquity of the ecliptic is not known since his calculations for this part have not been preserved. He may have used the following expression:[56]

[3.17] $$E = \left(\lambda_{\odot} - \alpha_{\odot}\right) - c + k$$

$\lambda_{\odot}$ true longitude
$\alpha_{\odot}$ right ascension
c equation of centre or solar equation
k constant due to epoch, 55114‴

♈	0	7;16,54
	10	10;54,46
	20	14; 5,40
♉	30	16;49;35
	40	18;49,12
	50	19;51,54
♊	60	19;51,56
	70	18;52,27
	80	17; 6,10
♋	90	14;54,21
	100	12;53,16
	110	10;59,10
♌	120	10; 3,14
	130	10; 8, 7
	140	11;16,46
♍	150	13;23,18
	160	16;14,27
	170	19;33,55
♎	*180	23; 0,11
	190	26;12,27
	200	28;49,50
♏	210	30;33,21
	220	31; 7,34
	230	30;22,25
♐	240	28;15,28
	250	24;53,42
	260	20;34,14
♑	270	15;42,49
	280	10;50,42
	290	6;28,59
♒	300	3; 3,38
	310	0;51,51
	320	0; 0,43
♓	330	0;28, 0
	340	2; 3,52
	350	4;33, 5
	360	7;16,54

Table 3.4 ULUGH BEG's equation of time. Source: Delambre, 1819, p.210. * For 0° of the scale, two different values are given in Delambre's table: 23′0″15‴ and 23′0″11‴.

52 The remains of the observatory were discovered during archaeological excavations in 1908 and are now under the protection of UNESCO.

53 More information and comments can be found in Sédillot, 1839, Tables astronomiques d'Oloug Beg, commentées et publiées avec le texte en regard, Tome I, 1 fascicule. Sédillot, 1847, Prolégomènes des Tables astronomiques d'Oloug Beg, publiées avec Notes et Variantes, et précédés d'une Introduction. Sédillot, 1853, Prolégomènes des Tables astronomies d'Oloug Beg, traduction et commentaire.

54 Akhmedov, 2002, Astronomy, Astrology, Observatories and Calendars, History of Civilizations of Central Asia, Bosworth, Asimov (eds.), vol. IV, part 2, p.204.

55 Gauss, 1929, Werke 12, pp.64–66.

56 Gauss, 1929, Werke 12, p.66. The sign of the equation of the centre in Equation 3.17 is the opposite as in Equation 1.11. For GAUSS' calculation procedure and equation of time, see Gauss, 1929, Werke 12, pp.64–68. For ULUGH BEG's values of the equation of time, see Delambre, 1819, Histoire de l'Astronomie du Moyen Age, reprint 2006, pp.209–210. Also Hyde, 1665, Tabulæ longitudinis et latitudinis Stellarum fixarum, ex observatione Ulugh Beghi, Tamerlani magni nepotis, regionum ultra citra que Gjihun (Oxum) principis potentissimi. Minor deviations are due to rounding and possible errors in transcribing tables.

Example 3.3 Calculate the value of ULUGH BEG's equation of time for longitude $\lambda_{\odot} = 75°$. The reference values are: longitude of apogee $\lambda_{\Gamma} = 93;0,18°$, eccentricity $\boldsymbol{e} = 0.0337037$, obliquity of ecliptic $\varepsilon = 23;30,17°$. The constant that is taken into account in the calculation according to the epoch is $k = 55114'''$ or 15.309 minutes (1 as = 4‴).[57] For more precise explanations, see GAUSS, *Werke*, pp.64–68 (solution: $E = 18$ min 3 s or 65038.‴68).

We calculate the equation of the centre with Equation 3.16:

$$c = \arcsin(e \cdot \sin(\lambda_{\odot} - \lambda_{\Gamma}))$$
$$= \arcsin(0.0337037 \cdot \sin(75° - 93;0,18°)) = -0;35,48°$$

Right ascension $\alpha_{\odot}$ is calculated in the same way as in the previous two examples:

$$\alpha_{\odot} = \arctan(\cos\varepsilon \tan\lambda_{\odot})$$
$$= \arctan(\cos 23;30,17° \tan 75°) = 73;42,43°$$

The equation of time is defined with a known expression, where the constant employed by GAUSS is used instead of the value of the longitude at the epoch. See GAUSS, *Werke*, pp.64–65. The result is then converted into sixtieths of a second, taking into account the more precise expression $1° \equiv 4 \cdot {}^{36525}/_{36625}$ min used by GAUSS, instead of $15° \equiv 60$ min or $1° \equiv 4$ min.

$$\lambda_{\odot} - \alpha_{\odot} = 75° - 73;42,43° = 1;17,17° = 1.^{\circ}288008$$

$$1.^{\circ}288008 \cdot \frac{4 \cdot 36525}{36625} \cdot 3600 = 18496.'''67$$

Similarly, the equation of the centre is converted into sixtieths of a second (=−8572‴) and, taking into account the constant $k = 55114'''$ and the right sign of the equation of the centre, we get the final result (Equation 3.17):

$$E = (\lambda_{\odot} - \alpha_{\odot}) - c + k = 18496.'''67 - 8572''' + 55114'''$$
$$= 65038.'''67 \approx 18 \text{ min } 3 \text{ s}$$

♈	0*	7;36,55
	10*	10;54,47
	20	14; 5,40
♉	30	16;49,35
	40	18;49,12
	50*	19;51,53
♊	60	19;51,56
	70*	18;52,26
	80*	17; 6, 9
♋	90*	14;54,20
	100*	12;43,16
	110	10;59,10
♌	120	10; 3,14
	130	10; 8, 7
	140*	11;16,47
♍	150*	13;23,19
	160*	16;14,53
	170*	19;33,56
♎	180*	23; 0,13
	190*	26;12,28
	200*	28;49,51
♏	210*	30;33,23
	220*	31; 7,35
	230	30;22,25
♐	240*	28;15,29
	250	24;53,42
	260*	20;34,10
♑	270*	15;42,48
	280*	10;50,41
	290*	6;28,58
♒	300*	3; 3,37
	310*	0;51,50
	320	0; 0,43
♓	330	0;28, 0
	340	2; 3,52
	350	4;33, 5
	360*	7;36,55

Table 3.5 The values of ULUGH BEG's equation of time, calculated with the procedure as used by GAUSS; sixtieths of seconds are rounded down to the nearest integer value (Example 3.3). See also Gauss, 1929, Werke 12, pp.64–68. * The values differ slightly from the values as given by Delambre, 1819, p.210.

57 The sixtieth of an (arc) second is often referred to in older German literature by the Latin word tertia or German word Tertie. Schultz, 1797, Kurzer Lehrbegriff der Mathematik. Zum Gebrauch der Vorlesungen und für Schulen, Königsberg, p.185.

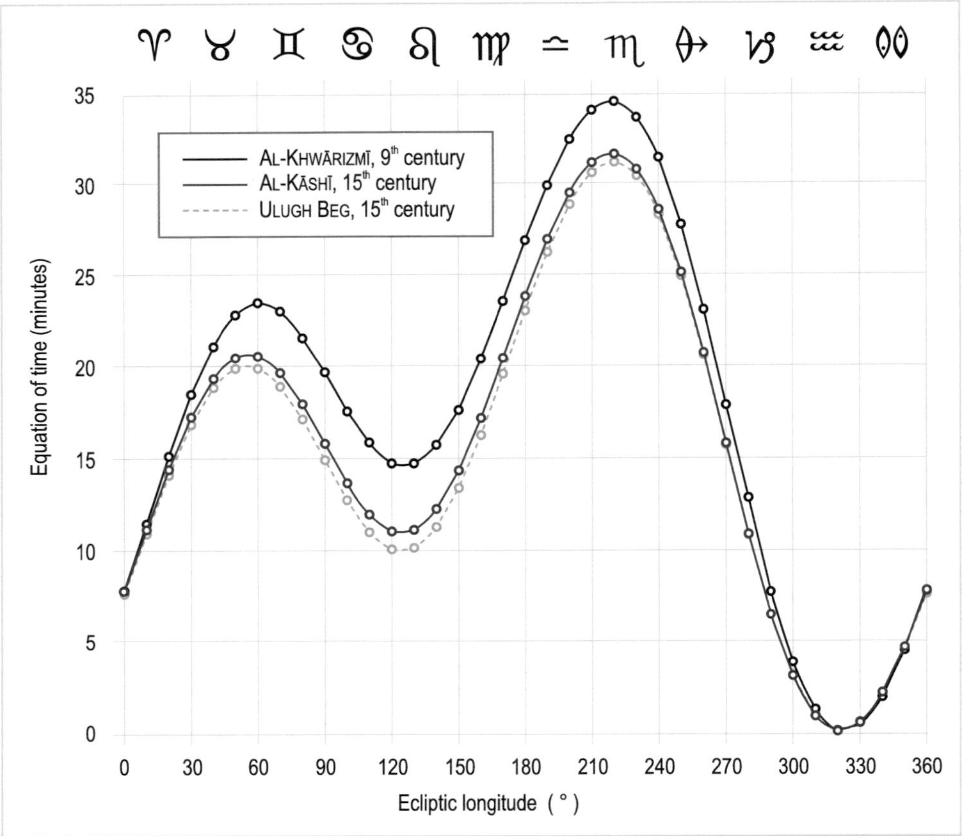

Figure 3.4 Comparison of AL-KHWĀRIZMĪ's values of the equation of time from the 9th century and AL-KĀSHĪ's and ULUGH BEG's values from the 15th century. The values are partially evident from Tables 3.1 and 3.3 or 3.5. Source of the data: Suter, 1914, pp.181–182, Table 67–68, Kennedy, 1988, p.6, Table 2 and Gauss, 1927, pp.64–67. See also van Dalen, 1996, pp.242–244, Table 4. For a comparison of the equation of time from PTOLEMY's *Handy Tables* and AL-KHWĀRIZMĪ's equation of time see Al-Khwarizmi, Neugebauer, 1962, p.108, Figure 32.

Mohammed Taragâi, surnommé Ulugh Beigh, … Il fit construire à Samarcande un gymnase qui passa pour une des merveilles du monde. Il eut d'abord pour cooperateur Gijath Eddin Gjimschïd, qui mourut avant que les observations ne fussent achevées, et fut enterré non loin de l'observatoire.

Delambre, 1819, Histoire de l'Astronomie du Moyen Age, p.204

Author	Year	Zīj	Equation of Time, precision, argument
Yaḥyā ibn Abī Manṣūr [a]	~ 820	*al-Zīj al-Mumtaḥan* Lat. *Tabulae probatae*	seconds for each integer degree of the argument
Abū Ma'shār [b]	9th century	*Kitāb fī'ilal al-zījāt*	time minutes
Al-Khwārizmī [c]	~ 840	*Zīj al-Sindhind*	two places for each integer degree of the true longitude. In al-Majrīṭī's recension, values are tabulated with a shift of $\lambda_{\odot} + 2°$
Al-Battānī	~ 900	*az-Zīj aṣ-Ṣābi*	arc minutes for each integer degree of the argument
Kūshyār ibn Labbān [d]	~ 1000	*az-Zīj al-Jami* also *Jamic zīj*	time seconds for 2° shifted mean longitude, additional tables for each six degrees of the mean longitude
Al-Bīrūnī	1030	*al-Qānūn al-Mas'ūdi*	equation of time described in *Treatise* VI
Al-Khāzinī [e]	1120	*al-Zīj al-Sanjarī*	time seconds for each integer degree of the argument
Naṣīr al-Dīn al-Ṭūsī [f]	1240	*Zīj-i Īlkhānī*	two places for each integer degree of the longitude
Ibn al-Raqqām [g]	1280/90	*al-Zīj al-Mustawfī*	two tables for each integer degree of the longitude, one in time-degrees
Al-Baghdādī [h]	~ 1285	*Baghdādī zīj*	time minutes and seconds for each integer degree of the true longitude given from 1° to 30° for each sign of the ecliptic
Ibn al-Shāṭir [i]	1350	*al-Zīj al-Jadīd*	seconds for each integer degree of the argument
Jamshīd al-Kāshi [j]	1420	*Zīj-i Khaqānī*	two places for each integer degree of the argument
Ulugh Beg	1440	*Zīj-i Sultānī*	three places for each integer degree of the argument

Table 3.6 Important *zījes* with equation of time tables included. Based on Kennedy, 1956, A Survey of Islamic Astronomical Tables, pp.145–166 and Samsó, 2014, p.320. For other sources see notes below.

Notes

a YAḤYĀ IBN ABĪ MANṢŪR was the chief astronomer/astrologer at the court of the ABBĀSID CALIPH AL-MA'MŪN. It is not clear whether the *al-Zīj al-Mumtaḥan* is his own work or the work of a group of astronomers under his leadership. Two manuscripts of the *zīj* are known; Van Dalen, 2004, A Second Manuscript of the Mumtahtan Zīj, Suhayl 4, 9–44.

b Pingree, Kennedy (trans., eds.), 1981, The Book of the Reasons behind the Astronomical Tables, pp.155–157.

c Van Dalen, 1996, pp.245–246.

d For translations and comments on Books I and IV of *Jāmiʿ zīj* see Bagheri, 2006, Books I and IV of Kūshyār ibn Labbān's Jāmiʿ zīj: An Arabic Astronomical Handbook by an Eleventh-Century Iranian Scholar, PhD dissertation, Utrecht University. For mathematical reconstruction see Van Dalen, 1993, Ancient and Mediaeval Astronomical Tables: Mathematical structure and parameter values, PhD dissertation, Utrecht University.

e ABŪ AL-FATḤ ʿABD AL-RAḤMĀN AL-KHĀZINĪ, the *zīj* contains a number of chronological and calendar tables, as well as tables for calculating holidays, a table of eclipses and more. Two incomplete manuscripts and an abridged version of the *zīj* translated by GREGOR CHIONIADES at the end of the 13th century in Greek were preserved.

f MUḤYI AL-DĪN AL-MAGHRIBĪ was also credited with the authorship of the treatise in the past, but this is not very likely. Saliba, 1994, A History of Arabic Astronomy, pp.177–178. AL-ṬŪSĪ gives an explanation of the reasons for the equation of time in his book *Memoir on Astronomy*, see Ragep, 1993, pp.286–290. See also Mercier, 1984, The Greek 'Persian Syntaxis' and the Zīj-I Īlkhānī, pp.43, 45.

g IBN AL-RAQQĀM (1250?–1315), an Andalusian astronomer. Samsó, 2014, Ibn al-Raqqam's al-Zīj al-Mustawfī and MS Rabat National Library; From Alexandria, Through Baghdad, Sidoli, van Brummelen (eds.), pp.301–302, 320. Values are probably based on AL-BATTĀNĪ. Two other zījes are also known: *Shāmil* and *Qawīm zīj*, which do not contain tables with the equation of time.

h AL-BAGHDĀDĪ (13th century) is the author of the *zīj*, which was apparently not widely used and contains contributions by other astronomers and is interesting from a historical point of view. The possible author of the equation of time from AL-BAGHDĀDĪ's *zīj* is ḤABASH AL-ḤĀSIB. Van Dalen, 1993, for the equation of time, see pp.141–152, for the *zīj* see chapter 4. Also Kennedy, 1957, p.48.

i IBN AL-ṢHĀTIR (c.1305–1375) was the most esteemed Islamic astronomer of the 14th century. He improved Ptolemaic models, and his work is interesting because his models are equal to COPERNICUS', which suggests the possibility that COPERNICUS knew his work.

j Also includes a table with values in arc minutes and seconds for each degree of true longitude. The table gives a correction of solar longitude for the change from mean to apparent solar time or for the difference between mean and apparent solar time. Kennedy, 1988, p.8.

Bibliography

- Akhmedov, A. 2000. Astronomy, Astrology, Observatories and Calendars. In: History of Civilizations of Central Asia, Bosworth, C. E. and Muhammad Seyfeydinovich Asimov (eds.), vol. IV, part 2. Paris: United Nations Educational, Scientific and Cultural Organization. ISBN 978-92-3-103654-5. URL: https://unesdoc.unesco.org/images/0012/001204/120455e.pdf (24 January 2024).
- Al-Bīrūnī, Abu-'r-Raihān. 1976. The Exhaustive Treatise on Shadows. Volume I, II. Translation and commentary by E. S. Kennedy. Aleppo: Institute for the History of Arabic Science, University of Aleppo.
- Bagheri, Mohammad. 1997. A Newly Found Letter of Al-Kāshī on Scientific Life in Samarkand. In: Historia Mathematica, 24, 3, 241–256. DOI 10.1006/hmat.1996.2145.
- Bagheri, Mohammad. 2006. Books I and IV of Kūshyār ibn Labbān's Jāmiʿ zīj: An Arabic Astronomical Handbook by an Eleventh Century Iranian Scholar, PhD dissertation. Utrecht: Utrecht University. ISBN 978-90-393-4388-3. URL: https://dspace.library.uu.nl/handle/1874/18830 (10 February 2024).
- Berggren, Lennart. 2011. Mathematik im mittelalterlichen Islam. Berlin: Springer. ISBN 978-3-540-76687-2.
- Bond, John David. 1921. The Development of Trigonometric Methods down to the Close of the XVth Century. In: Isis, 4, 2, 295–323. DOI 10.1086/358035. URL: https://www.jstor.org/stable/224250.
- Boyle, J. A. 1963. The Longer Introduction to the "Zij-i-ilkhani" of Nasir-ad-din Tusi. In: Journal of Semitic Studies, 8, 2, 244–254. DOI 10.1093/jss/8.2.244.
- Calvo, Emilia. 2007. Ibn Muʿādh. In: Biographical Encyclopedia of Astronomers. Hockey, Thomas, Virginia Trimble and Thomas R. Williams (eds.), 562–563. New York: Springer. ISBN 9780387310220. DOI 10.1007/978-0-387-30400-7_685.
- Cantor, Moritz. 1907. Vorlesungen über Geschichte der Mathematik /1: Von den ältesten Zeiten bis zum Jahre 1200 n. Chr. Leipzig: Teubner.
- De Vaux, Bernard Carra. 1892. L'Almageste d'Abū'l Wéfā Alboūzdjānī. In: Journal asiatique, 8th ser., 19, 408–471.
- Debarnot, Marie-Thérèse. 1978. Introduction du triangle polaire par Abu Nasr b. Iraq. In: Journal for the History of Arabic Sciences, 2, 1, 126–136.
- Debarnot, Marie-Thérèse. 1987. The Zij of Habash al-Hasib: A Survey of MS Istanbul Yeni Cami 784/2. In: From Deferent to Equant, King, David and George Saliba (eds.), Annals of the New York Academy of Sciences, Volume 500, 35–69. New York: New York Academy of Sciences. ISBN 0-89766-396-9.
- Delambre, Jean-Baptiste. 1819. Histoire de l'Astronomie du Moyen Age. Paris: Courcier. URL: https://gdz.sub.uni-goettingen.de/id/PPN618536701 (24 January 2024).
- Delambre, Jean-Baptiste. 1819. Histoire de l'Astronomie du Moyen Age. Reprint 2006. Paris: Éditions Jacques Gabay. ISBN 2-87647-251-1.
- Dold-Samplonius, Yvonne and Heinrich Hermelink. 1973. Al-Jayyānī. In: Dictionary of Scientific Biography, Gillispie, Charles Coulston (ed.), VII, 82–83. New York: Charles Scribner's Sons. SBN 684-10118-1.
- Gauss, Carl Friedrich. 1929a. Werke 12. Leipzig: Teubner.
- Gauss, Carl Friedrich. 1929b. Mittelpunktsgleichung nach Ulughbe in Zeitertien. In: Sezgin, Fuat (ed.), Islamic Mathematics and Astronomy, 55, 18–22. Reprint 1998. Frankfurt: Institut fur Geschichte der Arabisch-Islamischen Wissenschaften. ISBN 3-8298-4056-X.
- Goldstein, Bernard R. 1967. Ibn al-Muthannā's Commentary on the Astronomical Tables of al-Khwārizmī. New Haven, London: Yale University Press.
- Graßhoff, Gerd. 1990. The History of Ptolemy's Star Catalogue. New York: Springer. ISBN 0-387-97181-5.
- Hairetdinova, N. G. 1971. On the Oriental sources of Regiomontanus' trigonometrical treatise. In: Archives internationales d'histoire des sciences, 90–91, 61–66.
- Hairetdinova, N. G. 1986. On spherical trigonometry in the medieval near east and in Europe.
- In: Historia Mathematica, 13, 2, 136–146. DOI 10.1016/0315-0860(86)90026-1.
- Hall, Robert E. 1973. Al-Khāzinī, abu 'l-Fath. ʿAbd al-Rahmān. In: Dictionary of Scientific Biography, Gillispie, Charles Coulston (ed.), VII, 335–351. New York: Charles Scribner's Sons. SBN 684-10118-1.
- Hartner, Willy. 1970. Al-Bāttanī, abuʿ abd Allāh Muḥammad ibn Jābir ibn Sinān al-Raqqī al-Ḥarrānī al-Ṣābiʾ. In: Dictionary of Scientific Biography, Gillispie, Charles Coulston (ed.), I, 507–516. New York: Charles Scribner's Sons.
- Hartner, Willy. 1984. Oriens, Occidens, II. Hildesheim: Georg Olms Verlag. ISBN 3 487 07356 0.
- Henninger, Joseph. 1954. Über Sternkunde und Sternkult in Nord- und Zentralarabien. In: Zeitschrift für Ethnologie 79, 1, 82–117.

- Hyde, Thomas. 1665. Tabulæ longitudinis et latitudinis Stellarum fixarum, ex observatione Ulugh Beghi, Tamerlani magni nepotis, regionum ultra citra que Gjihun (Oxum) principis potentissimi. URL: https://archive.org/details/bub_gb_6EaIv_FppQkC/mode/2up (24 January 2024).
- Jensen, Claus. 1972. Abū Nasr Mansūr's Approach to Spherical Astronomy as Developed in His Treatise "The Table of Minutes". In: Centaurus, 16, 1, 1–19. DOI 10.1111/j.1600-0498.1972.tb00162.x.
- Juschkewitsch, A. P. 1964. Mathematik im Mittelalter. Leipzig: Teubner.
- Kari-Niazov, T. N. 1976. Ulugh Beg. In: Dictionary of Scientific Biography, Gillispie, Charles Coulston (ed.), XIII, 535–537. New York: Charles Scribner's Sons. ISBN 0-684-12925-6.
- Kennedy, Edward S. 1956. A Survey of Islamic Astronomical Tables. Transactions of the American Philosophical Society. New Series, 46, 2. Philadelphia: The American Philosophical Society.
- Kennedy, Edward S. 1960. The Planetary Equatorium of Jamshīd Ghiyāth al-Dīn al-Kāshī (d. 1429): an edition of the anonymous Persian ms. 75 [44b] in the Garrett Collection at Princeton University being a description of 2 computing instruments, the Plate of heavens and the Plate of conjunctions. Princeton Oriental Studies, Volume 18. Princeton, NJ: Princeton University Press.
- Kennedy, Edward S. 1988. Two Medieval Approaches to the Equation of Time. In: Centaurus, 31, 1, 1–8. DOI 10.1111/j.1600-0498.1988.tb00713.x.
- Krause, Max. 1936. Die Sphärik von Menelaos aus Alexandrien in der Verbesserung von Abū Nasr Mansūr b. Ali ibn Iraq. Dissertation. In: Abhandlungen der Gesselschaft der Wissenschaften zu Göttingen, Philologisch Historische Klasse, 3, 17. URL: https://www.deutsche-digitale-bibliothek.de/item/2TEP4KQXMSESGUAVX3LFZ2R2RTKJQGJ3 (24 January 2024).
- Krause, Max. 1936. Die Sphärik von Menelaos aus Alexandrien in der Verbesserung von Abū Nasr Mansūr b. Ali ibn Iraq. In: Sezgin, Fuat (ed.), Islamic Mathematics and Astronomy, 37. Reprint 1998. Frankfurt: Institut fur Geschichte der Arabisch-Islamischen Wissenschaften. ISBN 3-8298-4038-1.
- Kunitzsch, Paul. 1961. Untersuchungen zur Sternnomenklatur der Araber. Wiesbaden: Otto Harrassowitz.
- Lambert, Johann Heinrich. 1770. Beyträge zum Gebrauche der Mathematik und deren Anwendung, Zweyter Theil. Berlin: Buchladen der Realschule. URL: https://www.deutsche-digitale-bibliothek.de/item/CZHVDCRUOCGBHFICLHHCUSR2VPXTB3BG (9 July 2023).
- Lorch, R. P. 1975. The Astronomy of Jābir ibn Aflah. In: Centaurus, 19, 2, 85–107. DOI 10.1111/j.1600-0498.1975.tb00315.x.
- Luckey, Paul. 1951. Die Rechnenkunst bei Ǧamšīd B. Mascūd al-Kāšī. Deutsche Morgenländische Gesellschaft. Wiesbaden: Kommissionsverlag Franz Steiner.
- Luckey, Paul. 1953. Der Lehrbrief über den Kreisumfang (ar-Risāla al-Muḥītīya) von Gamšīd B. Mascūd al-Kāšī. In: Abhandlungen der Deutschen Akademie der Wissenschaften zu Berlin, 6. Berlin: Akademie Verlag.
- Mercier, Raymond. 1984. The Greek 'Persian Syntaxis' and the Zīj-I Īlkhānī. In: Archives Internationales d'Histoire des sciences, 34, 112, 35–60. Reprint 2004. Studies on the Transmission of Medieval Mathematical Astronomy. Aldershot: Ashgate. ISBN 0-86078-949-7.
- Mercier, Raymond. 2000. From Tantra to Zīj. In: Sic Itur Ad Astra: Studien zur Geschichte der Mathematik und Naturwissenschaften - Festschrift für den Arabisten Paul Kunitzsch zum 70. Geburtstag, Folkerts, Menso and Richard Lorch (eds.), 451–460. Wiesbaden: Harrassowitz Verlag. ISBN 3-447-04290-7.
- Millās Vendrell, Eduardo. 1963. El commentario de Ibn al-Mutannā's a las Tablas Astronomicas de al-Jwārizmī. Madrid: Associación para la Historia de la Ciencia Española.
- Nosonovsky, Michael. 2018. Abner of Burgos: The Missing Link between Nasir al-Din al-Tusi and Nicolaus Copernicus? In: Zutot 15, 25-30. DOI 10.1163/18750214-12151070.
- Neugebauer, Otto. 1962. The astronomical tables of al-Khwārizmī, Translation with Commentaries on the Latin Version, Suter, Heinrich (ed.), Supplemented by Corpus Christi College MS 283 by O. Neugebauer. København: Ejnar Munksgaard.
- Papadopoulos, Athanase. 2019. Nasir al-Din al-Tusi's treatise on the quadrilateral: The art of being exhaustive. Preprint, 17 February 2020. HAL id: hal-02463875. URL: https://hal.archives-ouvertes.fr/hal-02463875 (6 October 2021).
- Pedersen, Fritz S. 2002. The Toledan Tables. Copenhagen: The Royal Danish Academy of Sciences and Letters, C. A. Reitzeils Forlag. ISBN 87-7876-267-7.
- Pedersen, Olaf. 2011. A Survey of the Almagest. New York, Dordrecht, Heidelberg, London: Springer. ISBN 978-0-378-84825-9. DOI 10.1007/978-0-387-84826-6.
- Pingree, David. 1970. Abū Ma'shar. In: Dictionary of Scientific Biography, Gillispie, Charles Coulston (ed.), I, 32–39. New York: Charles Scribner's Sons.

- Pingree, David. 1985. The Astronomical Works of Gregory Chioniades, The Zīj al-'Alā ī. Part I: Text, Translation and Commentary. Amsterdam: Gieben. ISBN 9070265656.
- Pingree, David and E. S. Kennedy (trans., eds.). 1981. The Book of the Reasons behind the Astronomical Tables. Delmar, NY: Scholars' Facsimiles & Reprints. ISBN 0-8201-1298-4.
- Ragep, Jamil F. 1993. Nasīr al-Dīn al-Tūsī's Memoir on Astronomy (al-Tadhkira fī cilm al-hay'a), Volume I, Introduction, Edition and Translation. New York: Springer. e-ISBN 978-1-4757-2241-3. DOI 10.1007/978-1-4757-2241-3.
- Rius, Mònica. 2007. Hāshimī: ʿAlī ibn Sulaymān al-Hāshimī. In: Biographical Encyclopedia of Astronomers, I, 475, Hockey, Thomas, Virginia Trimble and Thomas R. Williams (eds.). Berlin: Springer. ISBN 978-0-387-31022. DOI 10.1007/978-0-387-30400-7_1433.
- Saisdan, A. S. 1973. Kūshyar ibn Labbān ibn Bāshahrī abu 'l-Ḥasan, al-Jīlī. In: Dictionary of Scientific Biography, Gillispie, Charles Coulston (ed.), VII, pp.531–533. New York: Charles Scribner's Sons. SBN 684-10118-1.
- Salam, Hala and Edward S. Kennedy. 1967. Solar and Lunar Tables in Early Islamic Astronomy. In: Journal of the American Oriental Society, 87, 4, 492–497. DOI 10.2307/597590.
- Saliba, George. 1994. A History of Arabic Astronomy: Planetary Theories During the Golden Age of Islam. New York: New York University Press. ISBN 0-8147-7962-X.
- Samsó, Julio. 2014. Ibn al-Raqqām's al-Zīj al-Mustawfi in MS Rabat National Library 2461. In: From Alexandria Through Baghdad, Sidoli, Nathan and Glen van Brummelen (eds.), 297–328. Berlin: Springer. ISBN 978-3-642-36735-9. DOI 10.1007/978-3-642-36736-6.
- Schoy, Carl. 1927. Die trigonometrischen Lehren des persischen Astronomen Abu'l'Raiḥân Muḥ. Ibn Aḥmad al-Bîrûnî dargestellt nach Al-Qânûn al-Mas'ûdî, Ruska, Julius and Heinrich Wieleitner (eds.). Hannover: Orient-Buchhandlung Heinz Lafaire.
- Schultz, Johann. 1797. Kurzer Lehrbegriff der Mathematik. Zum Gebrauch der Vorlesungen und für Schulen, 1. Königsberg: Friedrich Ricolovius. (Google Books, 20 June 2023).
- Sédillot, Louis-Pierre-Eugene Amelie (ed., trans.). 1847. Prolégomènes des Tables astronomiques d'Oloug Beg, publiées avec Notes et Variantes, et précédés d'une Introduction. Paris. (Google Books, 25 April 2020).
- Sédillot, Louis-Pierre-Eugene Amelie (ed., trans.). 1853. Prolégomènes des Tables astronomiques d'Oloug Beg, traduction et commentaire. Paris. (Google Books, 25 April 2020).
- Sédillot, Louis-Pierre-Eugene Amelie (ed., trans.). 1847–1853. Prolégomènes des Tables Astronomiques d'Oloug-Beg I, II. In: Sezgin, Fuat (ed.), Islamic Mathematics and Astronomy, 52–53. Reprint 1998. Frankfurt: Institut fur Geschichte der Arabisch-Islamischen Wissenschaften. ISBN 3-8298-4054-3.
- Sezgin, Fuat. 1978. Geschichte des Arabischen Schrifttums, VI. Leiden: Brill. ISBN 9004058788.
- Suter, Heinrich (ed.). 1914. Die astronomischen Tafeln des Muhammed ibn Mūsā al-Khwārizmī in der Bearbeitung des Maslama ibn Ahmed al-Madjrītī und der lateinischen Übersetzung des Adelard von Bath. København: Kongelige Danske Videnskabernes Selskab. URL: https://menadoc.bibliothek.uni-halle.de/urn/urn:nbn:de:gbv:3:5-12969 (10 February 2024).
- Suter, Heinrich (ed.). 1914. Die astronomischen Tafeln des Muhammed ibn Mūsā al-Khwārizmī in der Bearbeitung des Maslama ibn Ahmed al-Madjrītī und der lateinischen Übersetzung des Adelard von Bath. In: Sezgin, Fuat (ed.), Islamic Mathematics and Astronomy, 7. Reprint 1997. Frankfurt: Institut fur Geschichte der Arabisch-Islamischen Wissenschaften. ISBN 3-8298-4008-X.
- Toomer, Gerald J. 1973. Al-Khwārizmī, abū Ja'far Muḥammad ibn Musā. In: Dictionary of Scientific Biography, Gillispie, Charles Coulston (ed.), VII, 358–365. New York: Charles Scribner's Sons. SBN 684-10118-1.
- Van Brummelen, Glen. 1998. Mathematical Methods in the Tables of Planetary Motion in Kūshyār ibn Labbān's Jāmiʿ zīj. In: Historia Mathematica 25, 3, 265–280. DOI 10.1006/hmat.1997.2184.
- Van Brummelen, Glen. 2009. The Mathematics of the Heavens and the Earth. Princeton, NJ: Princeton University Press. ISBN 978-0-691-12973-0.
- Van Dalen, Benno. 1993. Ancient and Mediaeval Astronomical Tables: mathematical structure and parameter values. PhD dissertation. Utrecht: Utrecht University. ISBN 90-393-0167-0.
- Van Dalen, Benno. 1996. Al-Khwārizmī's astronomical tables revisited: analysis of the equation of time.
- In: Casulleras, Josep and Julio Samsó (eds.), From Baghdad to Barcelona: Studies in the Islamic Exact Sciences in Honour of Prof. Juan Vernet. Volume 1, 195–252. Barcelona: Universitat de Barcelona.
- Van Dalen, Benno. 2004. A Second Manuscript of the Mumtaḥan Zīj. In: Suhayl, 4, 9–44. URL: https://www.raco.cat/index.php/Suhayl/article/view/199637 (10 February 2024).
- Van Dalen, Benno. 2014. Islamic Astronomical Tables: Mathematical Analysis and Historical Investigation (Variorum Collected Studies). London: Routledge. ISBN 978-1472422385.

- Van Dalen, Benno. 2021. Ptolemaic Tradition and Islamic Inovation: The Astronomical Tables of Kūshyār ibn Labbān. Ptolemaeus Arabus et Latinus, Texts, Volume 2. Turnhout: Brepols. ISBN 978-2-503-59341-8.
- Van Der Waerden, Bartel Leendert. 1987. The Heliocentric system in Greek, Persian and Hindu Astronomy. In: From Deferent to Equant, King, David and George Saliba (eds.), Annals of the New York Academy of Sciences, Volume 500, 525–545. New York: New York Academy of Sciences. ISBN 0-89766-396-9.
- Vernet, Juan. 1954. Un antiguo tratado sobre el calendario judio en las Tabulae Probatae. In: Sefarad 14, 59–78. Reprinted in Vernet 1979, 213–232.
- Vernet, Juan. 1956. Las Tabulae probatae. In: Homenaje a Millás-Vallicrosa, vol. II, 501–522. Barcelona: Consejo Superior de Investigaciones Cientificas. Reprinted in Vernet 1979, 191–212.
- Vernet, Juan. 1974. Al-Majrīṭī, abu 'L-Qāsim Maslama ibn Aḥmad al-Faraḍī. In: Dictionary of Scientific Biography, Gillispie, Charles Coulston (ed.), IX, 39–40. New York: Charles Scribner's Sons. ISBN 0-684-10120-3.
- Vernet, Juan. 1976. Yaḥyā ibn Abī Manṣūr. In: Dictionary of Scientific Biography, Gillispie, Charles Coulston (ed.), XIV, 537–538. New York: Charles Scribner's Sons. ISBN 684-12926-5.
- Vernet, Juan. 1979. Estudios sobre historia de la ciencia medieval. Barcelona: Bellaterra.
- Von Braunmühl, Anton. 1900. Vorlesungen über Geschichte der Trigonometrie, Teil 1. Leipzig: Teubner. URL: https://archive.org/details/vorlesungenber00brauuoft/mode/2up (10 February 2024).
- Yamamoto, Keiji. 2007. Abū Maʿshar. In: Biographical Encyclopedia of Astronomers, I, 11, Hockey, Thomas, Virginia Trimble and Thomas R. Williams (eds.). Berlin: Springer. ISBN 978-0-387-31022. DOI 10.1007/978-0-387-30400-7_1433. URL: https://islamsci.mcgill.ca/RASI/BEA/Abu_Mashar_BEA.htm (10 February 2024).
- Youschkevitsch, A. P. 1970. Abū'l-Wafā'. In: Dictionary of Scientific Biography, Gillispie, Charles Coulston (ed.), I, 39–43. New York: Charles Scribner's Sons.
- Youschkevitsch, A. P. and B. A. Rosenfeld. 1973. Al-Kāshī, Ghiyāth al-Dīn Jamshīd Mas'ūd. In: Dictionary of Scientific Biography, Gillispie, Charles Coulston (ed.), VII, 255–262. New York: Charles Scribner's Sons. SBN 684-10118-1.

4

Early Chinese astronomy

At the winter solstice, the movement of the sun is the fastest, then it becomes slower. At the spring equinox, the movement of the sun is the mean [value]. And then gradually becomes slower. At the summer solstice, the movement of the sun is the slowest and it gradually becomes faster. At the autumn equinox, it is once more equal to the mean [value] and then it becomes faster.

Yixing, Xin Tangshu
(Tang, Qu, trans., 2018, pp.481–482)

4.1 The origins of Chinese astronomy

From its very beginnings, Chinese astronomy has been closely linked to the calendar. Observations with the gnomon, which was used in China at least as early as 1500 BCE, were integral to ancient Chinese astronomy.[1] Planetary observations in China are thought to have begun systematically around the 4th century BCE, when astronomical observations began to be officially recorded.[2]

Timekeeping was also the preserve of top government officials. The 6th century BCE work *Zhou li* also describes the civil servants in charge of timekeeping and their activities.[3] The description includes various groups of junior and senior officials measuring the time in spring, summer and autumn, including monitoring various astronomical and astrological phenomena and controlling wake-up times.[4] Various forms of water clocks, or *clepsydrae*, were used in China from around 2000 BCE and flourished during the XIA and SHANG Dynasties.[5] Originally ceramic, *clepsydrae* were replaced by robust bronze vessels in the Bronze Age. Astronomy and timekeeping also made significant advances later in the HAN Dynasty, a dynasty that left a great mark on Chinese history, something like the GUPTA Dynasty in India.[6]

The first known Chinese astronomical tables, *Santong Li*, which included solar eclipses and a calendar, were compiled by astronomer LIU XIN between 1 and 5 CE.[7] The knowledge that the Sun's motion is not uniform dates back to the second half of the 6th century, and Chinese astronomers in the SUI and TANG Dynasties built on this knowledge. Chinese astronomers have described the Sun's irregular motion in terms of the number of days it takes the Sun to travel 15° around the ecliptic (Figure 4.1).[8]

To escape the disorder left by civil war, astronomer ZHANG ZIXIN withdrew to an island around 560, where he spent 30 years making astronomical observations.[9] ZHANG is a rather mysterious figure in historical terms, but the astrological treatise *Tianwen Zhi*, part of the official *History of the Sui Dynasty*, provides some information about his work.[10] His findings were revolutionary at the time because, among other things, he discovered that the Sun and the planets do not move in uniform motions.[11]

1 Dubs, 1958, A record of the use of the gnomon is found in the *Book of Songs, Shi jing*, see Li, 2015, p.2095.
2 For an overview of astronomical records and tables from the HAN and TANG Dynasties, see Yabuuti, 1963a, pp.452–460.
3 *Rituals of the Zhou Dynasty*, 6th century BCE.
4 Deng, 2021, p.97.
5 XIA (c.2070–1600 BCE), the first Chinese dynasty, and SHANG (c.1600–1045 BCE), largely coinciding with China's bronze age.
6 HAN Dynasty (202 BCE – 9 CE and 25 – 220 CE).
7 Yabuuti, 1963a, p.452. The full set of Chinese astronomical tables from later periods is also given in the article.
8 Yabuuti, 1963a, table pp.462–463.
9 Tang, Qu, 2018, p.473. For some historical data, see also Needham, 1959, p.394, note a.
10 See also Sivin, 2009, pp.297–298.
11 ZIXIN's student, perhaps also involved in this discovery, was ZHANG MENG PIN, Needham, 1959, p.394.

These discoveries influenced astronomers in the SUI (581–618) and TANG (618–907) Dynasties, who, until then, had only known about uniform motion or the mean Sun.[12] The Sun's motion, ZHANG says, is: "...*slower after the vernal equinox and faster after the autumnal equinox.*"[13] LIU XIAOSUN, ZHANG MENGBIN, and ZHAO DAOYAN were the first generation of Chinese astronomers to put ZHANG ZIXIN's knowledge and achievements into practice. Unfortunately, the history of their work is very scarce.

The first Chinese astronomical model or system to include the Sun's irregular motion or the equation of the centre (Figure 4.2) was the *Sovereign Pole System* (*Huang ji Li*).[14] Based on LIU XIAOSUN's work, it was designed by LIU ZHUO in 600 CE, during the SUI Dynasty, and its description of the irregular motion of the Sun is very similar to the Mesopotamian system B.[15] It also included a table with the equation of the centre for the movement of the mean Sun for 12 months of the year. This system was also the basis for the tables of the equation of the centre used in astronomy later on, and the equation of the centre has been continuously refined by astronomers since then. However, due to strong criticism, notably from ZHANG ZHOUXUAN (c.526–612?), the system was never in public use. The *Linde Li* astronomical tables, which are a modification of LIU ZHUO's *Huang ji Li* tables, were probably compiled by LI CHUNFENG (602–670) around 630.

LIU ZHUO is also known as a surveyor. He requested gnomon measurements at various points along the meridian from the SUI Emperor in 604, but none were taken.[16] He proposed that a geometer with mathematical knowledge should measure lengths with ropes on the plains north and south of the Yellow River, which would be used to study the Earth, the sky and the stars. The YANGDI emperor was supposed to order the measurement, but the idea was not carried out due to LIU ZHUO's death.[17] LIU ZHUO was very likely familiar with quadratic equations, and he is thought to have used second-order interpolation in preparing the 604 calendar.[18] The surveys, led by the astronomer NANGONG YUE and the Buddhist monk YIXING, were then carried out in the 8th century between 723 and 726 during the TANG Dynasty.[19]

In 607, ZHANG ZHOUXUAN compiled a table with the equation of the centre in the *Great Patrimony System* (*Ta yeh li*), a system that was in public use for 11 years during the TANG

12 Yabuuti, 1963a, p.462.

13 Tang, Qu, trans., 2018, pp.481–482

14 This system shows similarities with Babylonian astronomy, but it is not clear whether this is an influence of Babylonian astronomy or an independent contribution by Chinese astronomers. Tang, Qu, 2018, p.473.

15 LIU ZHUO (544–610), Chinese astronomer and mathematician, already used quadratic interpolation with uniform intervals. The Mesopotamian system B is characterised by a linear zig-zag function. See Figure 4.1. For more information on System B, see Neugebauer, 1969, The Exact Sciences in Antiquity, chapter 5. For some information on the connection between Babylonian and Chinese astronomy, see also Needham, 1959, pp.394–396.

16 Needham, 1962, p.44 and Beer et al., 1961, p.4.

17 Needham, 1959, p.292, text and note 12. It also could have been the WENDI Emperor.

18 Needham, 1959, p.123. Li, Du, 1987, Chinese Mathematics: A Concise History. Also Meijering, 2011, A Chronology of Interpolation: From Ancient Astronomy to Modern Signal and Image Processing, p.320.

19 Needham, 1962, pp.44–53.

Dynasty.[20] The equation of the centre, which still closely resembles the zig-zag function (Figure 4.1) in this system, is probably based on his observations. YIXING (683–727, also romanised I-HSING), with the secular name ZHANG SUI, was a Buddhist monk and astronomer.[21] He made astronomical observations in various places in China, a number of them together with the astronomer NANGONG YUE, and is also known for his calendar. Around 724, YIXING and LIANG LINGZAN made an armillary sphere with a movable ecliptic ring that could be moved according to the precession of the equinoxes.[22] The instrument also had a movable lunar orbit ring, allowing YIXING to check the results against the work of his predecessors. He had obviously met Indian astronomers as well, or at least was familiar with their work, as a knowledge of Indian astronomy is evident in his works. Among his contributions to astronomy are improvements to the water clock, and to mathematics, the development of interpolation.[23] Unfortunately, all of his written works have been lost, and now known only indirectly. YIXING also criticised the work of predecessors - including LIU ZHUO's model of the motion of the mean Sun.[24] In 727, he compiled a table with the equation of the centre in the *Great Expansion System* (*Dayen Li*) whose curve is approximately sinusoidal and closer to the actual motion of the Sun than in the astronomical tables in use before then. It was also the first system to include procedures for determining the longitude of the planets, and was in official use between 729 and 762, a period of about 34 years.[25] In addition to the *Dayen Li* tables, the *Xuanming Li* were also in official use during the TANG Dynasty. These were composed by XU ANG in the 9th century and used between 822 and 892.[26]

4.2 Spread of Buddhism

With the spread of Buddhism to China around the 1st century during the Eastern HAN Dynasty, Buddhist monks also began to disseminate ancient Indian knowledge, especially in the fields of medicine and astronomy. The impact of Indian astronomy was initially quite slow, only becoming more rapid after the 5th century, when Indian astronomy saw its first really important works, including those of ĀRYABHAṬA I and later VARĀHAMIHIRA and BRAHMAGUPTA.

20 Sivin, 2009, p.378.

21 For brief excerpts and legends from YIXING's life and work, see Needham, 1959, pp.38, 139, 202–203, 234, 283.

22 The armillary sphere is said to have already used steel as a material, Needham, 1959, p.350. LIANG LINGZAN was a military engineer and government official during the TANG Dynasty. He is best known for the water clock he made with YIXING. This invention contained an element that is the precursor of the modern escapement in mechanical clocks, Needham, 1959, p.319. Needham et al., 1986, p.19, footnote 1. Also Deng, 2021, pp.138–140.

23 Ōhashi, 2011, Astronomy and Mathematics of Yixing, Nakamura, et al. (eds.), Mapping the Oriental Sky, Proceedings of the Seventh International Conference on Oriental Astronomy, Tokyo, pp.171–178.

24 Tang, Qu, 2018, p.481.

25 Sivin, 2009, pp.104, 379.

26 Yabuuti, 1963a, p.456.

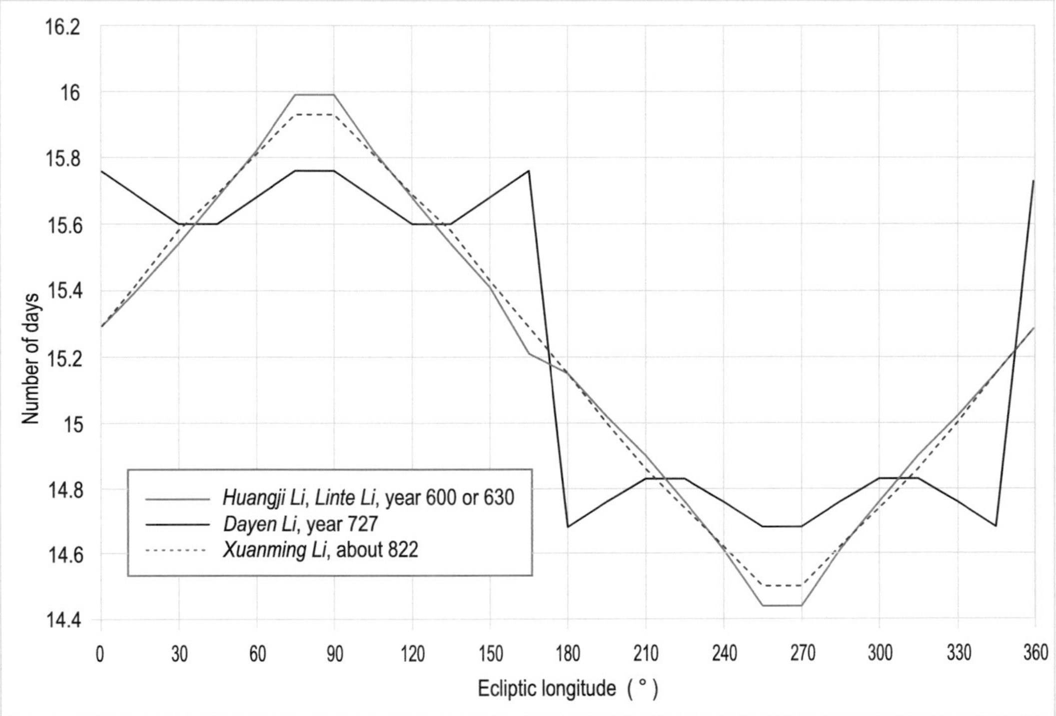

Figure 4.1 Number of days after 15° longitude path of the Sun as function of the longitude in different Chinese astronomical tables. Data source for the diagram: Yabuuti, 1963, pp.462–463.

We know that many Indian astronomers worked in China in several astronomical schools. Much of the credit for the spread of Indian astronomical knowledge goes to GAUTAMA SIDDHA (Chinese QUTAN XIDA), who served as court astronomer of the TANG Dynasty and translated or adapted the *Jiuzhi Li*, or in Sanskrit the *Navagraha*, a work based on seventh-century Indian astronomy; however, it is not clear whether this is a translation of several different texts or partly an original work.[27] It resembled the *Karaṇas* in form rather than the *Siddhāntas* and also contained inherited Greek astronomical knowledge.[28] In addition to astronomical content, it contained, for example, a table of sines, given in the same way as in the Indian literature, for radius $R = 3438$ and in steps of 3¾°.[29] In addition to GAUTAMA, the TANG Dynasty also saw the astronomy schools of KĀŚYAPA and KUMARA in China.[30]

27 The *Jiuzhi Li* was translated in 718 CE. Yabuuti, 1963b, p.493.
28 The work suggests some similarities with *Khaṇḍakhādyaka* of BRAHMAGUPTA.
29 Yabuuti, 1979, p.35.
30 Yabuuti, 1979, p.8.

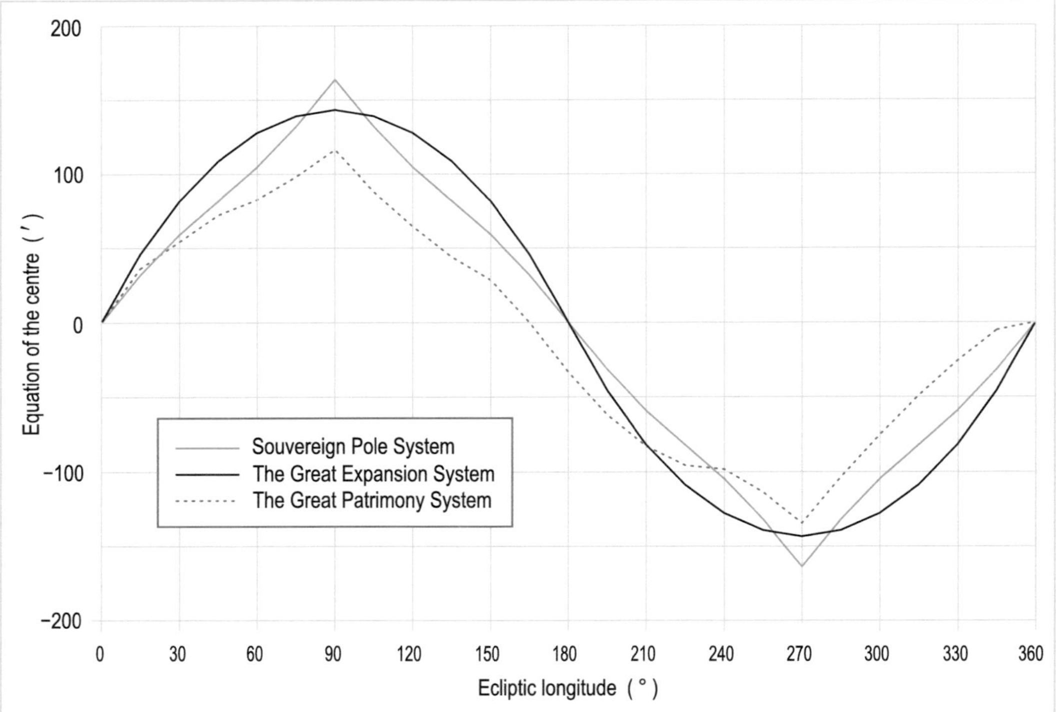

Figure 4.2 Equation of the centre in different Chinese astronomical tables. Data source for the diagram: Tang, Qu, 2018, Table 3, p.480.

Seventh- and eighth-century Chinese astronomy contains much of the simplified content that we find in a more complex form in the *Vāsiṣṭha Siddhānta*, *Pauliśa Siddhānta*, and *Romaka Siddhānta*, and which VARĀHAMIHIRA later compiled in the *Pañcasiddhāntikā*.[31]

Chinese astronomers also used separate degrees for calculations, with the ancient Chinese degrees based on the movement of the mean Sun, 1 du = the angle covered by the mean Sun in one day. This means that one ancient Chinese degree was slightly less than 1°. The connection between the angular degrees as we use them today and the ancient Chinese degrees du, taken from the work of *Xin Tangshu*, is:[32]

[4.1] $$360° = \left(365 + \frac{779.75}{3040}\right)\text{du} = 365.2565\ \text{du} \qquad \text{or} \qquad 1\ \text{du} = \frac{360°}{365.2565}$$

31 For more information see Yabuuti, 1979.

32 *Xin Tangshu*, History of the TANG Dynasty. Cullen, 1982, p.6, footnote 11. For additional explanation see also Yabuuti, 1979, p.12.

However, for the early period the following expression applies:[33]

$$1 \text{ du} = \frac{360°}{365.25} \quad [4.2]$$

Despite the adoption of Indian astronomical knowledge and later Islamic developments in astronomy, Chinese astronomy remained relatively immune to foreign influences until the arrival of the Jesuits. It was only after this that geometric models became more widely used in Chinese astronomy, and the use of sundials expanded.

A special place in Chinese history is occupied by the *Xinyixiang fayao* an illustrated monograph on astronomical instruments by SU SONG, written between 1088 and 1094 in the Northern SONG Dynasty. It is in three parts and richly illustrated, and is considered one of the best Chinese works of its kind.

4.3 Islamic zījes in China

GENGHIS KHAN's conquests were probably followed by a rather lively astronomical exchange between the Islamic world and China. The Chinese astronomer FU MENGCHI or FU MUZHAI worked at the observatory in Maragha, Iran, founded in 1259 by GENGHIS KHAN's grandson HULAGU KHAN. It was probably FU MENGCHI who provided the technical data on the lunisolar Sino-Uyghur calendar, which is described in several Islamic astronomical manuals from the Mongol period.[34] This calendar continued to be used in Iran for at least a century after the Mongol period and left a significant mark in Iranian astronomical almanacs until the 20th century. In 1263, the Mongol rulers set up an Islamic Astronomy and Astrology Bureau, and in 1271, they established an Islamic Observatory, headed by the astronomer ZHAMALUDING.[35] The earliest record of Arabic astronomical tables in China is in the official documents of the YUAN Dynasty, the *Mishujian zhi* (*Records of the Office of the Imperial Secretariat*). The Imperial Secretariat, established in 1272, was the YUAN Dynasty body in charge of collecting the books of the previous dynasties, as well as forbidden books. According to the *Jichi zhujia li*[36] (*Mathematical Astronomy of the Various Schools of Jichi*), the collection of *zījes* of the Islamic Observatory of the YUAN Dynasty comprised at least 48 volumes.[37]

33 Cullen, 1996, p.28.
34 Van Dalen, 2002, p.19.
35 ZHAMALUDING was an astronomer originally from Persia. Li, 2016, p.22, footnote 4.
36 *Jichi* is the Chinese translation of *zīj*.
37 Li, 2016, p.22.

However, there was no more systematic translation of Islamic literature into Chinese during the YUAN Dynasty.[38] The only extant translation or adaptation of the Islamic *zīj* in Chinese is the *Huihui Li*. After the conquest of Beijing, the *Imperial Secretariat* was moved to Nanjing by the HONGWU emperor, who established the MING Dynasty, where the translation or adaptation of the *Huihui Li* began in 1382. *Huihui Li* is clearly at least partly the work of Chinese astronomers, but it also has several common themes with the tables in *Sanjufini zīj*, by an unknown author.[39] In both works, we also find a table containing the equation of time.[40] Perhaps the two works have a common ancestor and are based on ZHAMALUDING's observations.

Among the few translations of Islamic works into Chinese, *Tianwen shu*, a 1382 translation of KŪSHYĀR[41] IBN LABBĀN's *al-Madkhal fī Sinā at Ahkām al-Nujūm* (*Introduction to Astrology*) is worth mentioning.

Bibliography

- BEER, Arthur, Ho PING-YÜ, Lu GWEI-DJEN, Joseph NEEDHAM and Edwing G. PULLEYBLANK. 1961. An 8th-Century Meridian Line: I-Hsing's Chain of Gnomons and the Pre-history of the Metric System. In: Vistas in Astronomy, 4, 3–28. DOI 10.1016/0083-6656(61)90001-0.
- BHAT, Ramakrishna. 2021. Varahamihira's Brhat Samhita I-II. New Delhi: Motilal Banarsidas Publishers. ISBN 9788120810600.
- CULLEN, Christopher. 1982. An Eighth Century Chinese Table of Tangents. In: Chinese Science, 5, 1–33. URL: https://www.jstor.org/stable/43290348.

38 Much of the "credit" for this probably goes to KUBLAI KHAN, who is said to have been reluctant to transfer knowledge. Li, 2016, p.23, footnote 6.

39 The author is said to have written the text in 1366 and dedicated it to the Mongol Prince RADNA, Governor of Tibet. See Kennedy, Hogendijk, 1988.

40 Van Dalen, 2002, p.25.

41 In China known under the Chinese name KUOSHIYAER.

- CULLEN, Christopher. 1996. Astronomy and mathematics in ancient China: the Zhou bi suan jing. Cambridge: Cambridge University Press. ISBN 0 521 55089 0.
- DENG, Kehui. 2021. The Ancient Chinese Timekeeping Instruments. In: The Studies of Heaven and Earth in Ancient China, History of Science and Technology in China, Jiang, X. (ed.), 95–151. Singapore: Springer Nature Singapore. DOI 10.1007/978-981-15-7841-0_4.
- DUBS, Homer H. 1958. The Beginnings of Chinese Astronomy. In: Journal of the American Oriental Society, 78, 4, 295–300. DOI 10.2307/595793. URL: https://www.jstor.org/stable/595793.
- KENNEDY, Edward S. and Jan HOGENDIJK. 1988. Two Tables from an Arabic Astronomical Handbook for the Mongol Viceroy of Tibet. In: A Scientific Humanist. Studies in Memory of Abraham Sachs, Leichty, Erie et al. (eds.). Occasional Publications of the Samuel Noah Kramer Fund 9. Philadelphia: The University Museum. Reprinted in: Kennedy, 1998, Astronomy and astrology in the medieval Islamic world, 233–242. Aldershot; Brookfield, VT: Ashgate. ISBN 9780860786825.
- LI, Geng. 2015. Gnomons in Ancient China. In: Handbook of Archaeoastronomy and Ethnoastronomy, Ruggles, C. L. N. (ed.), 2095–2104. New York: Springer. ISBN 978-1-4614-6140-1.
- LI, Liang. 2016. Arabic Astronomical Tables in China: Tabular Layout and its Implications for the Transmission and Use of the Huihui lifa. In: EASTM, 44, 21–68.
- LI, Yan and Shírán DU. 1987. Chinese Mathematics: A Concise History. Oxford: Clarendon Press. ISBN 0-19-858181-5.
- MEIJERING, E. 2011. A Chronology of Interpolation: From Ancient Astronomy to Modern Signal and Image Processing. In: Proceedings of the IEEE, 90, 3, 319–342. DOI 10.1109/5.993400.
- MORGAN, Daniel Patrick and Damien CHAUSSENDE (eds.). 2019. Monographs in Tang Official Historiography. Perspectives from the Technical Treatises of the History of Sui (Sui shu). Cham: Springer. DOI 10.1007/978-3-030-18038-6. e-ISBN 978-3-030-18038-6.
- NEEDHAM, Joseph. 1959. Science and Civilization in China: Mathematics and the Sciences of the Heavens and the Earth, volume 3. Cambridge: Cambridge University Press.
- NEEDHAM, Joseph. 1962. Science and civilisation in China: Physics and Physical Technology, volume 4, part 1. Cambridge: Cambridge University Press.
- NEUGEBAUER, Otto. 1969. The exact sciences in antiquity. New York: Dover Publications. ISBN 0-486-22332-9.
- ÔHASHI, Yukio. 2011. Astronomy and Mathematics of Yixing. In: Mapping the Oriental Sky. Proceedings of the Seventh International Conference on Oriental Astronomy, Nakamura, T. (ed.), 171–178. Tokyo: National Astronomical Observatory of Japan.
- SIVIN, Nathan. 2009. Granting the Seasons. The Chinese Astronomical Reform of 1280, With a Study of Its Many Dimensions and a Translation of its Records. New York: Springer. DOI 10.1007/978-0-387-78956-9. e-ISBN 978-0-387-78956-9.
- TANG, Quan and Anjing QU. 2018. A Comparison of Chinese and Indian Solar Theories in the Sui and Early Tang Periods (Seventh–Eighth Centuries C.E.). In: Journal for the History of Astronomy, 49, 4, 472–495. DOI 10.1177/0021828618808876
- VAN DALEN, Benno. 2002. Islamic Astronomical Tables in China. The Sources for the Huihui li. In: History of Oriental Astronomy, 19–30, Ansari, S. M. Razaullah (ed.). Proceedings of the Joint Discussion-17 at the 23rd General Assembly of the International Astronomical Union, Kyoto. Dordrecht: Springer. DOI 10.1007/978-94-015-9862-0. ISBN 978-1-4020-0657-9.
- VAN DALEN, Benno and Michio YANO. 1998. Islamic Astronomy in China: Two New Sources for the Huihui li ("Islamic Calendar"). In: Highlights of Astronomy, 11, 2, 697–700. DOI 10.1017/S1539299600018499.
- YABUUTI, Kiyoshi. 1963a. Astronomical Tables in China, from the Han to the T'ang Dynasties. In: Chūgoku chūsei Kagaku gijutsushi no kenkyū: Studies in the history of medieval Chinese science and technology, 445–492. Tokyo: Kadokawa Shoten.
- YABUUTI, Kiyoshi. 1963b. The Chiuchih-li - an Indian astronomical book in the Tang Dynasty. In: Chūgoku chūsei Kagaku gijutsushi no kenkyū: Studies in the history of medieval Chinese science and technology, 493–538. Tokyo: Kadokawa Shoten.
- YABUUTI, Kiyoshi. 1979. Researches on the Chiu-chih li – Indian Astronomy under the T'ang Dynasty. Astronomical Tables in China, from the Han to the T'ang Dynasties. In: Acta asiatica, 36, 7–48.
- YABUUTI, Kiyoshi. 1987. The Influence of Islamic Astronomy in China. In: From Deferent to Equant, King, David and George Saliba (eds.), Annals of the New York Academy of Sciences, Volume 500, 547–559. New York: New York Academy of Sciences. ISBN 0-89766-396-9.

This page is intentionally left blank.

5

Toledan and Alfonsine Tables

In hac autem tabula est quaedam linea quae intitulatur "Aequatio dierum", cuius utilitas est ut sciamus vertere dies naturales mediocres sive aequales in dies naturales differentes, et differentes in mediocres.

Roger of Hereford, Diversi astrologi, 12th century
(Pedersen, ed., 2009, Diversi astrologi)

5.1 From late antiquity to the early Middle Ages

Manuals, written by Roman authors from the 3rd century onwards, contributed significantly contribution to the transmission of Roman knowledge to the CAROLINGIANS. Among the most important authors of this period are SOLINUS, CALCIDIUS, MACROBIUS, MARTIANUS CAPELLA, BOETHIUS and ISIDORE OF SEVILLE. SOLINUS is credited with transmitting geographical knowledge and BOETHIUS with mathematical knowledge. MACROBIUS, CAPELLA and, above all, CALCIDIUS are credited with passing on astronomical knowledge.

The most important works of this period are the 7th century encyclopaedic *Etymologiae (Etymologies)* of ISIDORE OF SEVILLE and BEDE's treatise on time and calendars, *De temporum ratione*, completed in 725.[1] ISIDORE's *Encyclopaedia* was one of the main sources of ancient knowledge in the early Middle Ages and it was completed after his death by his pupil BRAULIO.[2] The work provides, among other things, information on the division of the day into time units, and BEDE's treatise includes variable day length and shadow length for different places and time zones, known as PTOLEMY's climata.[3] Neither of these works mention the so-called *natural days* or how the length varies the variable length of time between two consecutive sunrises or middays, nor do they mention the equation of time.[4] However, the authors of this period distinguished between the seasonal and equinoctial hours, as can be seen in BEDE's *Reckoning of Time*.[5]

THE VENERABE BEDE, Lat. BEDA VENERABILIS, was probably born around 673 or perhaps a year earlier. The sources of information about his life are mainly the writings in his works.[6] As a child, perhaps because his parents could not support him, he entered St Peter's Monastery near Wearmouth in North Cumbria around 680 and later moved to its sister monastery of Jarrow, founded by BENEDICT BISCOP, who was also BEDE's first teacher. Later, CEOLFRIST also taught BEDE for a long time. BEDE lived at the Jarrow Monastery for the rest of his life, except for a few short trips. He was only later titled Venerable[7], though sources from his time call him SAINT

1 ISIDORE OF SEVILLE (c.560–636), Archbishop of Seville, Lat. ISIDORUS HISPALENSIS. The word *etymology*, Gr. *etumologiã*, means the origin, formation and development (of words), the part of grammar that deals with forms. ISIDORE drew on the works of Roman authors, including PLINY's *Naturalis historia*.

2 BRAULIO, Lat. BRAULIUS CAESARAUGUSTANUS (c.585–651), later bishop of Zaragoza. For English and German translations of the encyclopaedia, see Barney et al., 2014, the Etymologies of Isidore of Seville and Möller, 2008, Die Enzyklopädie des Isidor von Sevilla. For a modern Latin edition, see Lindsay, 1911. After BRAULIO's death, Seville slowly began to decline as a cultural centre and Toledo slowly began to take over.

3 Wallis (trans.), 2012, pp.89–96.

4 ISIDORE's treatise *On the Nature of Things* does not mention this. CASSIODORUS and MACROBIUS also do not mention it in their works. FLAVIUS MAGNUS AURELIUS CASSIODORUS (490–around 585), historian, statesman and monk, is credited with preserving classical Roman knowledge during the period of increasing barbarism in the early Middle Ages. AMBROSIUS THEODOSIUS MACROBIUS (early 5th century), a government official active in North Africa, also belongs to the small circle of people who helped to transmit Roman knowledge from antiquity to the Middle Ages. For his collected works in Latin with commentaries, see von Jan, 1848.

5 Wallis (trans.), 2012, pp.14–15. BEDE's work *De temporum ratione*, or *The Reckoning of Time*, was important for the calculation of calendars in the early Middle Ages.

6 Browne, 1919, The Venerable Bede: His Life and Writings.

7 Venerable, Lat. *venerabilis*.

Bede and the like, well before any formal ecclesiastical appellation took place. His monastic life was divided between prayer, study and teaching. His most important work is an account of English church history, and he left a large body of other works covering historical, cosmological, religious, calendar and naturalistic content, which he also drew from the works of Isidore of Seville.[8] Bede's *On the Nature of Things* is thus based on Isidore's work of the same title.[9] Bede is credited with popularising Pliny's *Naturalis historia*, which also contains considerable astronomical concepts or data and became required reading for Carolingian astronomers. *Naturalis historia* was also appreciated by Alcuin of York[10] and Charlemagne. Charlemagne was a patron of education; he increased the number of schools so that Latin became a widely spoken language, and written works, and thus education, became more accessible.[11] Astronomy, which was still closely linked to astrology, was more or less limited to observations and was probably only accessible by the highest ecclesiastical circles. Records of eclipses and comets are known from this period, but astronomy was mainly used for calendaring and determining the time of evening prayer. In addition to Pliny's work, the Carolingian apparently had at their disposal Isidore's *Etymologiae*, Martianus Capella's[12] *On the Marriage of Philology and Mercury*[13], Macrobius's *Commentary on the Dream of Scipio*, and the 9th-century *De universo* of Maurus Hrabanus[14]. Calcidius'[15] 4th-century translation and commentary on Plato's *Timaeus* and Aratus' *Phenomena* were also among the works available. The available works offered drawings of the planetary movements, but this was not based on Ptolemy's epicycles or the *Almagest*. Calcidius thus included diagrams showing the eccentric and epicyclic movements of the Sun, which also show the different lengths of the seasons.[16] Figure 5.1 shows an approximate reconstruction of the diagrams of the eccentric motion of the Sun, where the different manuscripts show errors or offsets of the signs of the zodiacal circle.

The manuscripts of Martianus Capella's *On the Marriage of Philology and Mercury* also contain eccentric diagrams illustrating the irregular paths of the Sun at different times of the year (Figure 5.2).[17] Martianus' procedure for calculating the size of the Moon is also interesting.[18]

8 For an English translation of Isidore's work, see Embach (trans.), 1969, On the Nature of Things (De Natura Rerum), by Isidore of Seville (ca. 560–636).

9 Kendall and Wallis (trans.). 2010. Bede: On the Nature of Things and On Times. Liverpool University Press.

10 Alcuin of York (8th century, born around 732?), poet, teacher and cleric, the first representative of the so-called Carolingian Renaissance.

11 Dutton, 2004, Carolingian Civilization, p.92.

12 Martianus Minneus Felix Capella was active in Carthage at the time of the decline of the Roman Empire in the 5th century. His writing style was clearly influenced by Apuleius' novel *The Golden Ass*. See also Eastwood, 2000, Astronomical Images and Planetary Theory in Carolingian Studies of Martianus Capella.

13 *De nuptiis philologiae et mercurii* is an allegorical encyclopaedia in which the author brings together several fields or disciplines, in the modern sense grammar, dialectics, rhetoric, arithmetic, geometry, astronomy and music. His work was widely used in the Middle Ages and is also a kind of foundation for the educational content we know today.

14 Benedictine monk, poet, encyclopaedist and military writer.

15 Calcidius, a 4th century philosopher, but nothing is known about him.

16 Eastwood, Graßhoff, 2003, pp.199–200.

17 Eastwood, 2000, pp.20–21.

18 Carman, 2017, Martianus Capella's calculation of the size of the Moon. For the works of Martianus Capella, see also Stahl, 1959, Dominant Traditions in Early Medieval Latin Science.

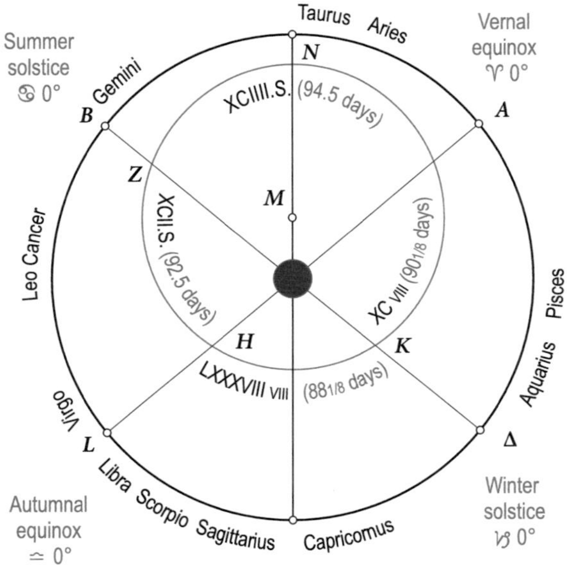

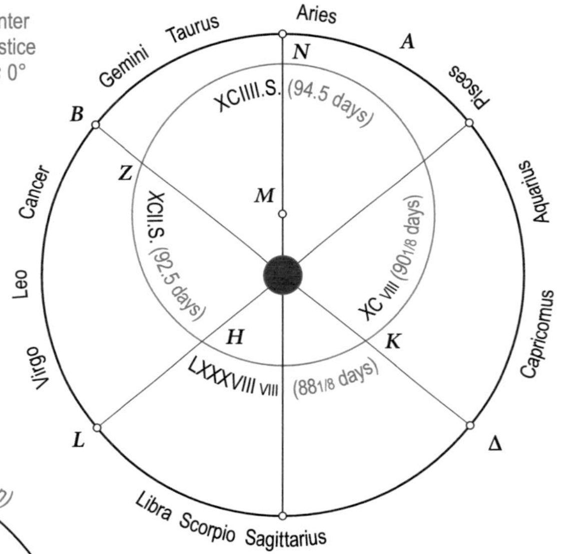

Figure 5.1 An approximate reconstruction of CALCIDIUS' 4th century eccentricity diagrams from manuscripts written in the 10th century (left image) and the 12th century (below). The Roman numerals appear to indicate the number of days between the solstices and the equinoxes and are consistent with the values known to HIPPARCHUS, see Figure 0.2. The grey text is the author's commentary. Taken from Eastwood, Grasshof, 2004, p.199. Original source: Bamberg Staatsbibliothek, MS Class. 18, f. 35v (left) and London, British Library MS Add. 15293, f. 22r (below).

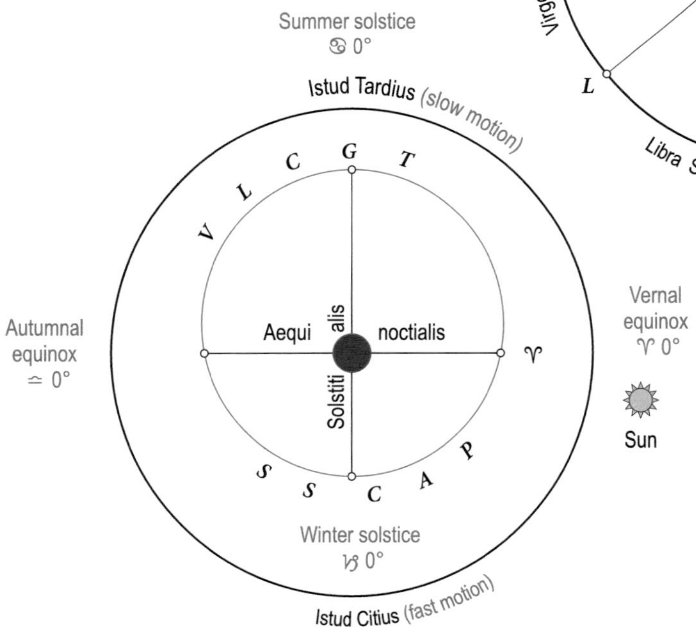

Figure 5.2 An approximate reconstruction of the diagrams of MARTIANUS CAPELLA from a 9th century manuscript. The letters next to the inner circle symbolise the signs of the zodiac and the grey text is the author's commentary. Summarised from Eastwood, 2000, p.20, Figure 11, and p.21, Figure 12, respectively. Original source: Leiden, University Library, MS Voss. Lat. F.48, f.81r.

5.2 Toledan Tables

5.2.1 The cultural centres of Córdoba and Toledo

The earliest known written sources relating to the development of astronomy or astrology in Andalusia record astrologers working in Córdoba at the court of the emirs AL-ḤAKAM I (reigned 796–822) and his son ʿABD AL-RAḤMĀN II (reigned 822–852).[19] According to the historian IBN ḤAYYĀN,[20] the astrologer ʿABBĀS BEN NĀṢIḤ[21] (flourished c.800–850) brought several works from Baghdad to Córdoba that were of great importance for the later development of astronomy in Andalusia. The works are said to have included an unknown *zīj*, PTOLEMY's *Handy Tables*, AL-KHWĀRIZMĪ's *Zīj al-Sindhind*, and an Indian astronomical table known by its Arabic title *Al-Arkand*. Astronomy in Córdoba reached its peak with MASLAMA AL-MAJRĪṬĪ, Latin METHILEM, who systematically studied ancient astronomical works at the beginning of the 11th century.

Libraries played a similar role in the Arab world as the Library played in Alexandria. AL-ḤAKAM's Library in Córdoba has a special place among them. It was founded by the second UMAYYAD Caliph of Córdoba, AL-ḤAKAM II also known as AL-MUSTANṢIR (915–976, reigned 961–976), following the example of similar libraries in the Arab world. The library enjoyed a high reputation as one of the largest in the Arab world. In Córdoba, as in Alexandria, the library's wealth was due to the systematic translations of ancient works encouraged by the bibliophilic caliph, which also made the library an important state institution. Shortly after the death of AL-ḤAKAM II, due to a misunderstanding of its importance, it was systematically destroyed on the orders of AL-ḤAKAM's successor. Virtually nothing remains of it.[22] The reign of ʿABD AL-RAḤMĀN III and his son AL-ḤAKAM II is considered the peak of UMAYYAD art in Spain.

Much of the credit for Toledo's rise as a centre of science goes to YAḤYĀ AL-MA'MŪN (reigned 1044–1075), a member of the Berber dynasty that ruled Toledo in the 11th century. For example, a water clock was built in Toledo during his time, which followed the phases of the moon.[23] Another important figure in the development of science in Toledo was the qadi or judge ṢĀʿID, also known as ṢĀʿID AL-ANDALUSI.[24] He was also a historian and the author of *Kitāb ṭabaqāt al-umam* (*Categories of Nations*), a history of science that includes a detailed account of the development of science in Andalusia from the 9th to the 11th centuries. ṢĀʿID was familiar with Greek works, especially PTOLEMY's *Almagest*, *Geography* and *Tetrabiblos*, as well as THEON's

19 Samso, 2020.

20 Samsó, 2020. Quoted in IBN ḤAYYĀN, *al-Muqtabis fī tārīkh al-Andalus*. IBN ḤAYYĀN (d. 1076), Andalusian historian.

21 Lévi-Provençal, Abbās b. Nāṣiḥ, Encyclopédie de l'Islam.

22 Wasserstein, 1990, The library of al-Ḥakam II al-Mustanṣir and the culture of Islamic Spain; Manuscripts of the Middle East, 5, pp.99–105.

23 AHMAD or MUHAMMAD IBN KHALAF AL-MURĀDĪ described the workings of 18 clepsydrae in *Kitāb al-asrār fī natā'ij al-afkār* (*Secrets of the Results of Ideas*). For translation and commentary, see Lisa et al. (eds.), 2008, The Book of Secrets in the Results of Ideas: Incredible Machines from 1000 Years Ago, Ibn Khalaf al-Murādī.

24 ṢĀʿID AL-ANDALUSĪ (1029–1070) also worked on mathematics and astronomy. Only fragments of his works are known.

Commentaries on the *Almagest* and the *Handy Tables*. We know that he was also well acquainted with the works of the early Islamic period of the 9th and 10th centuries, especially those of AL-KHWĀRIZMĪ, THĀBIT IBN QURRA, ABŪ MA'SHĀR, IBN AL-ĀDAMĪ[25] and AL-HAMDĀNĪ[26]. However, he does not appear to have been familiar with works written after 950, which indicates the isolation of Andalusian culture after the fall of the Córdoban Caliphate.

5.2.2 First translated Arabic tables

Ancient mathematical and astronomical knowledge was transmitted to Europe via Islamic Spain, which saw the arrival of Arabic translations of ancient manuscripts and other works of Arab scholars, and, in the late Middle Ages, from Constantinople via Italy. The Arabic tables, including AL-KHWĀRIZMĪ's, were translated into Latin by PETRUS ALFONSI in 1116, but it is not known which tables from the original Arabic edition of *Zīj al-Sindhind* were included in the Latin translation because the original Arabic edition survives only in fragments.[27] PETRUS faced several problems in preparing the translation, since AL-KHWĀRIZMĪ used elements of both Indian and Ptolemaic astronomy, and there was still a need to develop appropriate astronomical terminology in Latin. The translation, though not very good, is the first known written work in Latin describing the movement of the planets, but it is not known whether astronomers also used it in practice. In as little as ten years, it was followed by a better translation by ADELARD OF BATH[28], based on AL-MAJRĪṬĪ's review of AL-KHWĀRIZMĪ's tables, perhaps also commented on by AL-MAJRĪṬĪ's disciple IBN AL-ṢAFFĀR.[29] The Latin translation of AL-MAJRĪṬĪ's partial review of AL-KHWĀRIZMĪ's *zījes* is preserved. ADELARD also faced many problems with terminology. It was ABRAHAM IBN EZRA who finally untangled the mess.[30]

25 IBN AL-ĀDAMĪ (10th century), astronomer. His otherwise lost *zīj Naẓm al-'iqd*, completed after his death by his disciple AL-QĀSIM IBN MUḤAMMAD IBN HISHĀM AL-MADA'INI and published in 949/950, was based on Indian sources. IBN YUNUS and ṢĀ'ID AL-ANDALUSI also knew about the *zīj*. This *zīj* could also be the source of the theory of trepidation. IBN AL-ĀDAMĪ is also said to be the source of a visit by Indian astronomers to the Caliph AL-MANṢŪR in Baghdad in 770.

26 AL-HAMDĀNĪ (10th century), a Yemeni-born man best known as a geographer, was also an astronomer and politician.

27 PETER OF TOLEDO, Lat. PETRUS ALFONSI (11/12th century) arranged a series of Arabic astronomical tables, the *Tabulae astronomicae*. See Tolan, 1993, Petrus Alfonsi and his Medieval Readers, pp.55–61. For the equation of time from AL-MAJRĪṬĪ's comments on AL-KHWĀRIZMĪ's work, see Suter, 1914, pp.181–182, Table 67, 68, *pagina examinationis dierum*.

28 Born in Bath, England, ADELARD OF BATH (c.1080–1152) was educated in Tours, France, where he later taught for a time in Laon. He was one of the first translators to translate Arabic-Greek works comprehensively into Latin. He also pioneered the introduction of the teaching of Arabic science into the curriculum. He also translated EUCLID's *Elements* into Latin. We know that in around 1110, perhaps a little later, his path led him to Salerno in Sicily, where he probably came into contact with Arabic. He also travelled to Cilicia in Turkey, Syria and possibly Palestine. According to the translation of the annotated edition of AL-KHWĀRIZMĪ's Tables by AL-MAJRĪṬĪ, it seems likely that he also spent some time in Spain.

29 IBN AL-ṢAFFĀR, AL-MAJRĪṬĪ's student, worked in Andalusia where he taught arithmetic, geometry and astronomy. His brother MUḤAMMAD was a famous maker of astronomic instruments, and two of his astrolabes have been preserved.

30 Steinschneider, 1880, Abraham ibn Esra, Abhandlungen zur Geschichte der mathematischen Wissenschaften mit Einschluß ihrer Anwendungen, 3, pp.57–128. Lipton, 1978, The Rational Evaluation of Astrology in the Period of Arabo-Latin Translation ca. 1126–1187 AD, p.158. Ibn Ezra, 1947, El Libro de los Fundamentos da las tablas astronómicas, Millás Vallicrosa (ed.). Millás Vallicrosa, 1960, Nuevos estudios sobre historia de la ciencia española, pp.110–112. Sela, 2003, Abraham Ibn Ezra and the Rise of Medieval Hebrew Science, pp.22–27. Sela, 2007, Abraham Ibn Ezra The Book of Reasons. See also chapter 6.

AL-KHWĀRIZMĪ's tables were later translated and edited by ROBERT OF CHESTER and HERMANN OF CARINTHIA.[31]

5.2.3 Toledan Tables

Of all the Arabic astronomical tables translated into Latin, the *Toledan Tables* were the most widely used and were also the basis for other similar works. They were most likely created in the second half of the 11th century and were compiled by a group of astronomers in Toledo, possibly under the direction of qadi ṢĀ'ID.[32] AL-ZARQĀLĪ (Lat. AZARQUIEL) is the author of at least part of the tables. He worked in Toledo and Córdoba after the conquest of Toledo (1085), producing precise astronomical instruments and working in the ṢĀ'ID's group. He is also supposed to have conducted observations of the Sun and the Moon. He observed the Sun for about 25 years and recorded the results in the form of tables in the period between 1075 and 1080. The contents of his works are known from citations in both Latin and Arabic sources. He discovered that the Sun's apogee does not stand still, but varies by about 1° over 279 years. On the basis of this finding, he designed a system to model the motion of the Sun with variable eccentricity. His work also included theoretical astronomy and thaumaturgy.

The Latin edition of the *Toledan Tables*, which exist in many versions, was edited by GERARD OF CREMONA. In addition to GERARD, the most important translators in Toledo included JOHN OF SEVILLE, followed by ALVREDUS ANGLICUS, MICHAEL SCOTUS and HERMANNUS ALEMANNUS, among others. The Toledo school of translation has much to do with the translation of Arabic works and without the Mozarabs and Jews, or the multilingual local population, translation would certainly not have been so intense. The local people were invaluable as interpreters between Arabic and the spoken language, and the translators then transcribed the interpreted texts into literary

♈	0	2;10
	10	3; 3
	20	3;54 ∗
♉	30	4;38
	40	5;10
	50	5;29
♊	60	5;33
	70	5;20
	80	4;53
♋	90	4;20
	100	3;48
	110	3;21
♌	120	3; 5
	130	3; 5
	140	3;17
♍	150	3;45
	160	4;26
	170	5;16
♎	180	6; 4
	190	6;49
	200	7;31
♏	210	7;48
	220	7;54
	230	7;41 ∗
♐	240	7; 7 ∗
	250	6;12
	260	5; 4
♑	270	3;49
	280	2;37
	290	1;31
♒	300	0;41
	310	0;10
	320	0; 1
♓	330	0;10
	340	0;38
	350	1;21
	360	2;10

Table 5.1 The values of AL-BATTĀNĪ's equation of time in time-degrees from the *Toledan Tables* for every 10° of longitude. Temporal units are obtained if the values are multiplied by 4. Source: Nallino, 1907, II, pp.61–64. ∗ Corrections, see Pedersen, 2002b, p.971.

31 HERMANN OF CARINTHIA also HERMANN OF DALMATIA (c.1100–1160) is one of the most important translators from Arabic in the 12th century, with works including EUCLID's *Elements* and PTOLEMY's *Planispherium*. He probably at least edited, if not translated, AL-KHWĀRIZMĪ's astronomical tables. We know that he was also quite familiar with THEODOSIUS' *Spherics*. Very little is known about his life. The only period for which we have any reliable information is the time of his activity in Spain between 1138 and 1143. For more information, see Paić-Vukić, 2019.

32 Pedersen, 2002b, The Toledan Tables, III, p.757, 07k. Glick, 2005, Sa'id al-Andalusi, Medieval Science, Technology, and Medicine: An Encyclopedia, pp.34–35.

Latin.[33] The *Toledan Tables* mostly contain the data or tables of other astronomers, based mainly on the works of AL-KHWĀRIZMĪ and AL-BATTĀNĪ (Table 5.1).[34] Data on translations into Latin are scarce; the extant translations were probably created at the end of the 12th century, but it is possible that a Latin translation could have been made earlier.

The *Toledan Tables* are based on the Ptolemaic model (see Chapter 1) with the following parameters:[35]

▪ Longitude of apogee	$\lambda_\Gamma = 82°15'$
▪ Maximum value of the equation of the centre	$c_{max} = 1°59'10''$
▪ Obliquity of the ecliptic	$\varepsilon = 23°33'30''$
▪ Eccentricity	$e = 2;4,45$
▪ Radius	$R = 60$ or $R = 150$
▪ Mean daily motion of the Sun	0;59, 8,11,28,27,30°
▪ Epoch	Era of Hijra 622, 15 June at noon
▪ True longitude for Toledo at the epoch	$\lambda_0 = 114;5,29°$

The true longitude for Toledo at the epoch is based on AL-BATTTĀNĪ's value 113;58,4, corrected to the meridian through Toledo. The radius value of 150, which is given in various places in the *Toledan Tables*, was already used by AL-KHWĀRIZMĪ and is taken from Indian sources. The motion of the Sun, or mean anomaly, is determined in a similar way to PTOLEMY's *Handy Tables* - by taking the partial contributions for each time interval, the sum of which is the period between the reference moment and the moment of observation (see Example 1.1).[36] When determining the time period, it should be borne in mind that the *Toledan Tables* are based on the Islamic lunar calendar.[37] The right ascension, which is also listed in the tables, is calculated using the following equation:[38]

[5.1] $$\mathrm{Sin}\,\alpha_\odot = \frac{\mathrm{Sin}(90° - \varepsilon)}{\mathrm{Sin}\,\varepsilon} \cdot \frac{\mathrm{Sin}\,\delta_\odot}{\mathrm{Sin}(90° - \delta_\odot)} \cdot 150$$

The equation of time can then be calculated in an identical way as in chapter 1.

33 Burnett, 1977, A Group of Arabic-Latin Translators Working in Northern Spain in the Mid-12th Century. The Journal of the Royal Asiatic Society of Great Britain and Ireland, 1, pp.62-108.

34 For AL-BATTTĀNĪ's work see Nallino, 1903/1907, Opus Astronomicum, I, pp.48–49 (text) and II, pp.61–64 (tables). For more information, see Pedersen, 2002b, The Toledan Tables: A review of the manuscripts and textual versions with an edition, Toomer, 1968, A Survey of the Toledan Tables, Osiris, 15, pp.5–174, and Zinner, 1935, Die Tafeln von Toledo, Osiris, 1, pp.747–774. The tables were compiled in Islamic Spain, probably around 1070. The original *Toledan Tables* are lost and we know them only from Latin translations.

35 Nallino, 1903, I, p.44:10–15. Pedersen, 2002b, pp.411, 970, 1140–1141. Zinner, 1935, pp.762–764. For the radius 150, see McCarthy, Byrne, 2003, Al-Khwarizmi's Sine Tables and a Western Table with the Hindu Norm of $R = 150$.

36 For mean anomaly tables, see Pedersen, 2002b, p.1147.

37 For the conversion of the dates of the Islamic lunar calendar into Julian or Gregorian dates, see Neugebauer, 1937, Hilfstafeln zur technischen Chronologie, III. Birashk, 1993, A Comparative Calendar of the Iranian, Muslim Lunar and Christian Eras for Three Thousand Years. Spuler, Mayr (eds.), 1961, Wüstenfeld-Mahler'sche Vergleichungs-Tabellen.

38 Pedersen, 2002b, p.411. The calculation is very similar to PTOLEMY's calculation of right ascension, except that the *Toledan Tables* use the sine function and a radius of 150. See also Equation 1.13.

5.3 Dissemination of astronomical knowledge

5.3.1 Raymond of Marseilles and Roger of Hereford

RAYMOND OF MARSEILLES, who worked in the middle of the 12th century, was one of the first astronomers we know to have used the *Toledan Tables*, as evidenced by his text from 1141, although we do not know which edition or translation was available to him. RAYMOND, like PETRUS ALFONSI, was apparently self-taught. He was well acquainted with the Toledo translations of Arabic works. Although not a translator, he was probably familiar with the translations of Arabic texts from astronomy, astrology and probably alchemy. The work of both men was particularly influential in the further development of astronomy in England.

The equation of time, or *aequatio dierum* in Latin, is mentioned and tabulated in the treatise *Diversi astrologi*[39] attributed to ROGER OF HEREFORD, who worked between 1175 and 1198.[40] All we know about him is that did astronomy in Hereford; there are no detailed data on his origins and life. He adopted astronomical tables from the work of RAYMOND OF MARSEILLES, which were fundamentally based on the *Toledan Tables*. He was familiar with the works of Arabic and Greek authors, with much of this knowledge apparently coming from the translations of JOHN OF SEVILLE and HERMANN OF CARINTHIA.[41] He dedicated a work on calculating the church calendar, *The Computus*, to GILBERT FOLIOT, who was Bishop of Hereford until 1163 and later Bishop of London (and passed away in 1187). ROGER is also the author of the astronomical treatises *De ortu et occasione signorum* and *Theorica planetarum*, which is one of the first works of its kind in Latin, and apparently also contains elements of PTOLEMY's *Almagest* and the *Toledan Tables*. ROGER also worked on an extensive astrological work called *Liber de quatuor partibus judiciorum astronomie*, which he never finished. He wrote several other astronomical and astrological treatises, and is probably the author of *On the Rising and Setting of the Signs of the Heavens*, these being the signs of the zodiac.[42]

In addition to ROGER, other translators were probably associated with Hereford in the 12th and 13th centuries and made important contributions to the transmission of Arabic knowledge to Europe. They included DANIEL OF MORLAY and ALEXANDER NECKAM, and perhaps ALVREDUS ANGLICUS.[43]

39 Pedersen (ed.), 2009, (Dj241) *In hac autem tabula est quaedam linea quae intitulatur "Aequatio dierum", cuius utilitas est ut sciamus vertere dies naturales mediocres sive aequales in dies naturales differentes, et differentes in mediocres.* Table {BB11} is mentioned in the source but is not part of the online text, see Pedersen (ed.), 2009, (Dj249).

40 Burnett, 2007. See also Haskins, 1927, pp.123–128.

41 JOHN OF SEVILLE (12th century, active after 1133), Latin JOHANNES HISPALENSIS or JOHANNES HISPANIENSIS. For his life and work, see Thorndike, 1959.

42 Pedersen, 2004, The Treatise on the Rising and Setting of the Signs, Ascribed to Roger of Hereford.

43 Russell, 1932, Hereford and Arabic Science in England about 1175–1200. Thorndike, 1923, A History of Magic and Experimental Science II, pp.181–187 and 260.

5.3.2 Almagesti minor

The *Almagesti minor* is a Latin edition or summary of PTOLEMY's *Almagest* with some additional content based on the work of Arabic authors. The work originated in northern France, with the earliest manuscripts probably dating from around 1240. The author is unknown, but it may have been WALTER OF LILLE, whose identity is not clear.[44] The causes and contents necessary for the calculation of the equation of time are described in Book III, paragraphs 19–25, in more detail than the *Almagest* itself gives.[45] Paragraph 25 describes the conversion of average days into what the author calls *"different days"* (Lat. *dies differentes*).[46] *Almagesti minor* was used and referred to in the works of many later astronomers, such as JOHANNES ANDREE SCHINDEL, JOHN OF GMUNDEN, ALBERTUS DE BRUDZEWO, PEUERBACH and REGIOMONTANUS, to name but a few.

> *Causa inequalitatis dierum ex diverso motu Solis proveniens ab alterutra longitudine media incipit et ad oppositam desinit, et differentia diei mediocris ad dies differentes maior, cum ex hoc collecta fuerit, ex duplo differentie maxime motus medii et motus diversi perficitur.*
>
> *Almagesti minor, III.19, Zepeda, 2019, p.264.*

5.3.3 Campanus of Novara

Information about CAMPANUS' life is scarce, and the only thing we know for certain is that he died in Viterbo in 1296. Despite his faint trail, CAMPANUS was a well-known mathematician who was also highly regarded abroad. He was associated with France, Spain and England, but there is no evidence that he actually travelled much or even that he ever left Italy.[47] The earliest evidence of his activity is a translation or adaptation of EUCLID's *Elements*, apparently written between 1255 and 1261, based on translations by ADELARD OF BATH and ROBERT OF CHESTER.[48] At the time of the work's composition, JACQUES PANTALÉON, who was later elected Pope URBAN IV, was Patriarch of Jerusalem,[49] and CAMPANUS dedicated his most important work, the *Theorica Planetarum*, to him.

> *Sol habet unum circulum super cuius circumferentiam corpus eius semper equaliter mouetur omni die naturali 59 minuta et 8 secunda; ...*
>
> *Campanus of Novara, Theorica planetarum, III.3, Benjamin, Toomer (trans.), 1971, p.142.*

44 Perhaps this is the same person as WALTER OF CHÂTILLON. Zepeda, 2019, pp.11–19.

45 Zepeda, 2019, pp.265–275.

46 In the literature, it is also commonly called *natural day*, Lat. *dierum naturalium*.

47 For comments on his acquaintance with ROGER BACON, see Benjamin, Toomer, 1971, pp.7–8.

48 Clagett, 1953, The Medieval Latin Translations from the Arabic of the Elements of Euclid, with a Special Emphasis on the Versions of Adelard of Bath. Busard, Folkerts, 1992a, 1992b, Robert of Chester's Redaction of Euclid's Elements, the so-called Adelard II Version. Volume I-II.

49 JACQUES PANTALÉON, (c.1200–1264), Pope URBAN IV from 1261 to 1264.

Although it was not an original work, being based on PTOLEMY's *Almagest*, it had a profound influence on the further development of astronomy in Europe.[50] This influence is particularly evident in the works of astronomers from the German-speaking area, especially JOHN OF GMUNDEN, and also later of PEUERBACH and REGIOMONTANUS.[51]

CAMPANUS' second important astronomical work, probably written in 1268 or a little later, is the *Computus maior*, which describes the cycles of the Sun and Moon, or the determination of time using them. The work also provides a procedure for calculating the date of Easter, and it describes the seasonal hours in Chapter 6. CAMPANUS is also known for his own design of the equatorium, but he was also an astrologer, which may have been decisive in him having served three popes.[52] He was certainly well acquainted with the translations of works from the Toledo school of translation.[53] His knowledge of the *De figura* by THĀBIT IBN QURRA, *De elementis arithmeticae* by JORDANUS NEMORARIUS[54] and the *Treatise on Ratio and Proportion*[55] by AḤMAD IBN YŪSUF is evident from his works.[56] For *Theorica planetarum*, he also apparently had the help of AL-FARGHĀNĪ's work, via the Latin translation by JOHN OF SEVILLE.[57] CAMPANUS also wrote astronomical tables based on the *Toledan Tables*, which contain values for the meridian through Novara, but these were not printed and survive only as manuscripts.[58]

50 Apart from the *Almagest*, the main sources on which CAMPANUS' work was based were translations of the astronomical works of AL-FARGHĀNĪ, GEBER and the *Toledan Tables*. AL-FARGHĀNĪ, who was active in Baghdad in the 9th century, wrote a summary of the *Almagest*, a work that, especially in its Latin translation, had a profound influence on the development of astronomy. He is also the author of a treatise on the astrolabe. Sezgin, 1978, pp.149–151. JĀBIR IBN AFLAḤ, Lat. GEBER (12th century), is known for his commentaries or corrections on the *Almagest*, where, for example, he abandoned MENELAUS' theorem and replaced it with right-angled spherical triangles.

51 Traces of knowledge of CAMPANUS' work can also be found in the legacies of BERNARD OF VERDUN (Lat. BERNARDUS DE VIRDUNO), ANDALÒ DI NEGRO and, apparently, JOHN OF LIGNÈRES. Benjamin, Toomer, 1971, pp.36–37. Very little is known about BERNARD OF VERDUN (13th century). His work was based on the *Almagest* and the work of AL-BATTĀNĪ and ARISTOTLE, and he was also familiar with the work of AL-BIṬRŪJĪ, and is known for his treatise on the torquetum. GEBER or IBN AFLAḤ may have influenced the development of the torquetum, but the real inventor, at least according to the available sources, was probably a European astronomer who knew Arabic. Lorch, 1977, p.33. The earliest known descriptions of the torquetum are by BERNARD OF VERDUN and FRANCO DE POLONIA, both dating from the 13th century. Poulle, 1964, Bernard de Verdun et le Turquet, Isis, 55, pp.200–208 in Thorndike, 1945, Franco de Polonia and the Turquet, Isis, 36, pp.6–7. See also Gunther, 1923, Early Sciences in Oxford, II, pp.35–36 and 370–375. PETER APIAN discusses the tourquet in the last five folios of his *Astronomicum caesareum* of 1540.

52 As the equatorium had been described 200 years earlier by IBN AL-SAMḤ and AL-ZARQĀLĪ, it seems likely that his description of the equatorium was based on these sources. Toomer, Campanus of Novara, Encyclopedia.com.

53 For comments on his works, see Benjamin, Toomer, 1971, Campanus of Novara and Medieval Planetary Theory, pp.12–24.

54 JORDANUS NEMORARIUS (13th century), also JORDANUS DE NEMORE, left behind treatises on arithmetic, algebra, geometry and astronomy. Høyrup, 1988, Jordanus de Nemore, 13th Century Mathematical Innovator.

55 These are commentaries on Book V of the *Elements*, translated into Latin by GERARD OF CREMONA. Carmody, 1955, pp.130–131.

56 Sezgin, 1974, Geschichte des arabischen Schrifttums, Band V, pp.288–290. AḤMAD IBN YŪSUF (8th/9th century) is the author of several treatises and commentaries that influenced the subsequent development of mathematics in Europe. He influenced the work of LEONARDO OF PISA or FIBONACCI, JORDANUS NEMORARIUS and apparently also GERARD OF CREMONA. See also Steinschneider, 1888.

57 Carmody, 1943, Alfragani differentie in quibusdam collectis scientie astrorum.

58 Benjamin, Toomer, 1971, p.15.

5.3.4 Astronomers in Paris

In the late 13th and early 14th centuries, there was a group of skilled astronomers in Paris, among whom were Peter of Dacia, William of St. Cloud, Peter of St. Omer and John of Sicily. They all used the *Toledan Tables* in their work.

Peter of Dacia

Peter of Dacia, Latin Petrus de Dacia, a Danish astronomer, presumably the same figure as Peter Nightingale or Peter Philomena, probably studied in Bologna from 1280, leaving in 1292 for Paris. Three of his works are known: the mathematical *Expositio super Algorismum* or comments on Johannes de Sacrobosco's *Algorismus*, the common textbook of calculation with Hindu-Arabic numerals, *Kalendarium*, and the astronomical treatise about the equatorium *Tractatus instrumenti eclipsium*.[59] In preparing the calendar, he may have collaborated with William of St. Cloud, who, at the end of the 13th century, was leading Parisian astronomer.[60] The equation of time in *Tractatus instrumenti eclipsium* is given for every few degrees for each sign of the zodiac. Values are positive, given in whole minutes, with a maximum of 32 minutes at the ecliptic longitude of 6° Scorpio.[61] The *Tractatus instrumenti eclipsium* describes an innovative astronomical tool, specifically an improved equatorium, for calculating solar and lunar eclipses using the *Toledan Tables*, the oldest known such instrument.[62] Later, makers produced similar instruments in different shapes with different functions. Peter may also have worked with William of St. Cloud to construct the equatorium. Peter spent the last years of his life in Denmark, and was a canon of Roskilde Cathedral from 1303.

> *Quia igitur propter inaequalitatem ascensionum buth solis superaddditarum revolutionibus caeli diurnis efficiuntur dies naturales inaequales, excrementum inaequalitatis, quod vocatur aequatio dierum, cum gradu solis accipitur.*
>
> *Peter of Dacia, Tractatus instrumenti eclipsium, Pedersen (ed.), 1978*

William of Saint-Cloud

We know that William lived in Saint-Cloud, near Paris, France, around 1290, and that is more or less all that history has to offer about his life. In 1285, he probably observed a solar eclipse, whence his subsequen complaint of eye problems due to too much time spent observing. This event is said to be the origin of his recommendation to use a camera obscura for eclipse

59 Pedersen, 2012.
60 Pedersen, 2002b, p.985. Chabás, Goldstein, 2003, The Alfonsine Tables of Toledo, pp.244–248.
61 Pedersen, 1978, p.49, also Pedersen, 1983, p.505.
62 North 1976, II, p.268, note 2.

observation. Three of his works are known: his *Directorium* about compasses, a calendar called *Kalendarium regine*, and an astronomical almanac, *Almanach planetarum*.[63] In the *Almanach planetarum*, the maximum value of the equation of time is 31 minutes, and the individual values are given in whole minutes for several days of the month.[64]

Peter of St. Omer

Peter of St. Omer, Latin Petrus de Sancto Audomaro, worked between 1298 and 1308 as a theologian at the Paris University, however little is known about his life and work. He is associated with three extant written works.[65] He is the author of the *Tractatus de semissis*, in which he also describes the length of the day and the equation of time.[66] Among his more important writings is a commentary on Jacob ben Machir's[67] innovative combined instrument, which combined the quadrant and the astrolabe, *Novus quadrans correctus a Petro Dane de Sancto Audomaro*.[68] These two treatises were written in 1293/94. His third extant work, *The ars et operatio novi quadrantis...*, which also deals with the quadrant, was probably written after 1300.[69] He is also known for his treatise on eclipses.

John of Sicily

We know little about John of Sicily's life beyond that he was active in the astronomical circle in Paris at the end of the 13th century. In his only extant work, *Commentaries on the Canons of the Toledan Tables, "Quoniam cujusque actionis quantitatem..."*, the equation of time is also mentioned.[70] These Canons were written by al-Zarqālī and translated by Gerard of Cremona. The commentary was written in 1290 or 1291.

> *Semper enim debet haec aequatio superaddi: cuius rei causa est quia aequatio dierum in tabulis constituta sumit initium a 19'o gradu aquarii, ubi est principium additionis ad diem mediocrem ut ex eo fiat differens.*
>
> *Scriptum Johannis de Sicilia super canones Azarchelis de tabulis Toletanis, Pedersen (ed.), 1986, II, p.199*

63 For more on the publication *Kalendarium regine*, see Harper, Richard Irwin, 1966, The Kalendarium Regine of Guillaume de St-Cloud. PhD Dissertation. Atlanta, GA: Emory University.

64 Pedersen, 2014, p.30.

65 Pedersen, 1983, pp.42–44. Peter of St. Omer may be the same person as Peter of Dacia.

66 Pedersen, 1979, pp.33–35 and 95–96. For the values of the equation of time of which he is the putative author, see Pedersen, 2002b, pp.984–985. Chabás, Goldstein, 2003, The Alfonsine Tables of Toledo, pp.244–248. John of Lignères is also considered to be a possible author of the value of the equation of time.

67 Jacob ben Machir ibn Tibbon, also called Don Profeit Tibbon, Profiatus, (c.1236–1312), French Jewish physician, translator, and astronomer, he served as regent of the faculty of medicine at the University of Montpellier. Ibn Tibbon translated into Hebrew a large number of works including Arabic versions of Euclid's *Elements*, the *Almagest* and Arabic works by al-Ghazālī and ibn Rushd (Averroës).

68 Pedersen, 1983, p.42.

69 Here, the authorship of the treatise, which is attributed to Peter of St. Omer, is not entirely clear.

70 Pedersen, 1986, pp.117–118.

The text also seems to make predictions for eclipses in 1295, a few years in the future. It is not clear whether JOHN OF SICILY predicted the eclipses himself or just relayed information from elsewhere.[71]

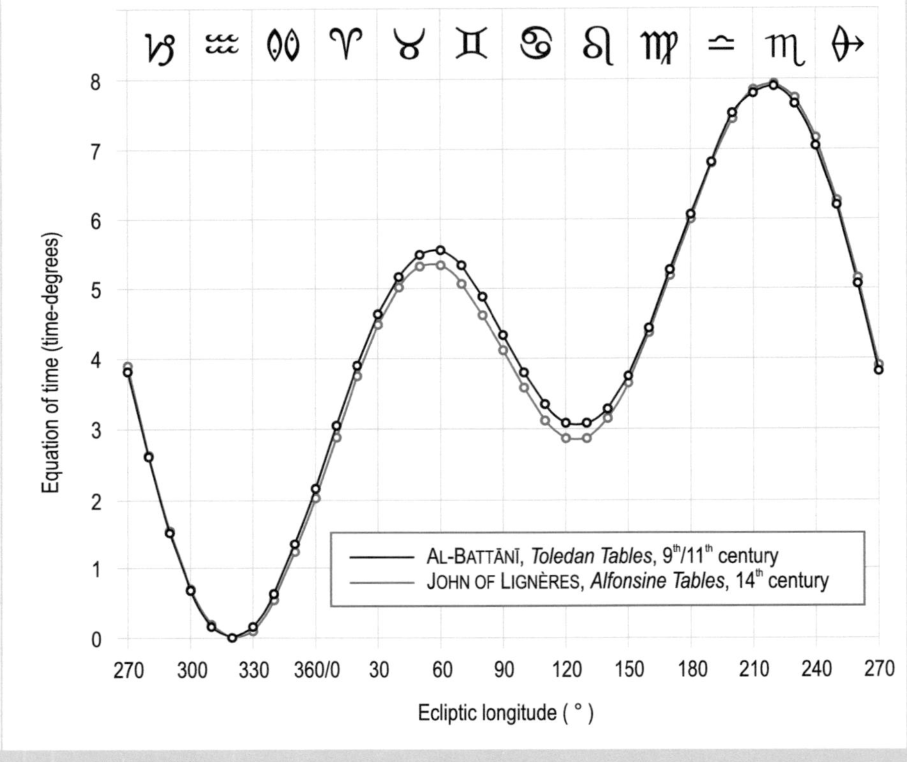

Figure 5.3 The equation of time from the *Toledan* and *Alfonsine Tables*. The equation of time in the *Toledan Tables* is identical to AL-BATTĀNĪ's values, and the authorship of the value of the equation of time presented in the *Alfonsine Tables* is attributed to PETER OF ST. OMER or JOHN OF LIGNÈRES. Sources of data: Pedersen, 2002b, The Toledan Tables, pp.972–975 and 985.

Tables by an unknown author and tables from Toulouse

The equation of time is mentioned in the *Astronomical Tables* of an unknown author in 1290.[72] The value of the equation of time in these tables is given in time-degrees, with a maximum value of 7°54′ or 31.6 minutes.[73] The equation of time is also mentioned in the *Tables of Toulouse*. The tables were in use at the end of the 13th century and in the first half of the 14th century and appear to be based on the *Toledan Tables*.[74] The equation of time was supposed to be published in the *Tabula de equacionibus dierum in circulo directo*.[75]

71 Pedersen, 1986, pp.8–9.
72 This work may be one of the sources of comments of JOHN OF SICILY. Pedersen, 1984b, p.163.
73 Pedersen, 2002a, p.86 (Ap351). The maximum value is the same as in the *Toledan Tables*. Chabás, Goldstein, 2003, p.186.
74 Pedersen, 1998, p.3.
75 Pedersen, 1998, p.10.

When FERDINAND III, in the first half of the 13th century, occupied a large part of Andalusia, including Córdoba and Seville, many more cultural treasures of the Islamic world, including astronomical tables, became accessible to Christianity. Jewish astronomers also played an important role in transferring knowledge to the Christian world and translating it into Latin.

5.4 Alfonsine Tables

5.4.1 First edition of the Alfonsine Tables

The main authors of the *Alfonsine Tables*, which originated around 1272 and followed the *Toledan Tables*, were ISAAC BEN SID and YEHUDA BEN MOSHE HA-KOHEN.[76] Jewish astronomers played a very important role in preparing astronomical works in the Castilian language, as they not only possessed fairly extensive literature but had the linguistic expertise to translate the Arabic works. The tables were created under the patronage of FERDINAND's son ALFONSO X, after whom they were named.[77] Vigorous diplomatic contacts between the Castille and Paris courts most likely helped to bring the tables to Paris. Additional proof of the transfer of astronomical knowledge from Castille to Paris is provided by JOHN VIMOND's astronomical chart from 1320.[78] The *Alfonsine Tables* were also known in Italy - we know that, for example, ANDALÒ DI NEGRO from Genoa knew of them.[79] Astronomical tables known as the *Parisian Alfonsine Tables*, most likely based on the *Alfonsine Tables*, were published in Paris in the 14th century, around 1320 or a little later. The original edition of the Castilian *Alfonsine Tables* is quite different from the first printed edition.[80] The tables themselves are of uncertain nature because they have not survived; only the text is preserved. Other works were translated or written at ALFONSO's request, notably a textbook on astronomical instruments in 1276 and 1277. This comprised a total of 19 books, based on translations of scholarly treatises from Arabic, which were then compiled into a final book or textbook.[81] Much of the credit for the introduction of the *Alfonsine Tables* in England goes to WILLIAM REDE (c.1320–1385?), who adapted the *Alfonsine Tables* to the meridian

76 Wegener, 1905a, Die astronomischen Werke Alfons X, p.138. ISAAC BEN SID, also known as RABIÇAG, a Jewish astronomer; almost nothing is known about his life. YEHUDA BEN MOSHE HA-KOHEN, Jewish astronomer and translator and rabbi of the synagogue in Toledo, and at the same time one of the most influential members of the Jewish community in Toledo. He was also the physician to ALFONSO X.

77 The *Toledan Tables* were followed by the *Alfonsine Tables*, a collection of astronomical data that was also based on Arabic sources. ALFONSO X (1221–1284) was the patron of various professional works, mainly translated from Arabic. In addition to the two Jewish astronomers, there were at least fifteen scholars working under his patronage. For historical background and context, see Chabás, Goldstein, 2003, pp.225–241.

78 Chabás, Goldstein, 2014, Essays on Medieval Computational Astronomy, pp.227–307.

79 ANDALÒ DI NEGRO (1260–1334), an Italian astronomer and geographer. Chabás, Goldstein, 2003, p.248. North, 1977, The Alfonsine Tables in England, p.270. ANDALÒ DI NEGRO (also DEL NEGRO) also wrote an important treatise on the astrolabe, *De compositione astrolabii*. Raynaud et al., 2019, Andalò di Negro's De compositione astrolabii: a critical edition with English translation and notes.

80 See section 5.4.3, JOHN OF SAXONY.

81 Wegener, 1905a, pp.140–164.

through Oxford and the epoch of 1340.[82] REDE, aka READE, of the diocese of Exeter, was a bishop and astronomer. He was active in public life and known for his private library, which was very large for the time. His Tables were widely used in England in the 14th and 15th centuries.

The *Alfonsine Tables* were used for more than 300 years and have been published in numerous editions and reprints. They continued to be used long after the publication of the *Prutenic Tables* (1551), and the last Spanish edition was issued 14 years after the publication of KEPLER's *Rudolphine Tables* in 1641.

5.4.2 Equation of time in the Alfonsine Tables

The equation of time is always positive in the *Alfonsine Tables*, as in the *Toledan Tables*. The example of the calculation of the true longitude below is summarised from the literature and is based on the latest Paris edition of 1553.[83]

Also, in this case Equation 1.6 applies, and we additionally note that the mean longitudes λ_M are tabulated as a function of time. The equation of the centre c is tabulated as a function of the mean anomaly and the longitude of the apogee is:[84]

[5.2] $$\lambda_\Gamma = \lambda_{\Gamma 0} + p$$

λ_Γ longitude of the apogee
$\lambda_{\Gamma 0}$ longitude of the apogee at the epoch = 71°.423
p total precession, consisting of secular precession and trepidation from the epoch (day 0 year 1)

Example 5.1 Calculate the true longitude of the Sun on 20 September 1477 at 6 h 1 min 36 s for Toledo.[85]

From Table II (Wegener, 1905b, p.46) it follows that:

1477.0 + 263 days + 6 hours + 1 minute + 36 seconds = 1477.72.

The value of the precession and trepidation p is determined by interpolation and Table I (Wegener, 1905b, p.46).

Value for the year	1470.0 =19°.473
For 7.72 calculate the interpolated value	0°.073

p = 19°.546

82 Ker, Piper (eds.), 1968–2002, Medieval manuscripts in British libraries, I, pp.193–196. Harper, 1975, The Astronomical Tables of William Rede, Isis, 66, 3, 369–378. North, 1992, pp.65–102, 103–174.
83 Wegener, 1905b, pp.6, 18–20. For ERASMUS REINHOLD's comments on the equation of time from the *Alfonsine Tables* see Wegener, 1905b, pp.12–13.
84 Wegener, 1905b, p.19.
85 Summarised by Wegener, 1905b, example on p.19.

The value of the apogee for the initial moment or the epoch 0 January, year 1, is given in Table III (Wegener, 1905b, p.46), $\lambda_{\Gamma 0} = 71^{\circ}.423$.

Longitude of the apogee $= \lambda_{\Gamma 0} + p = 90^{\circ}.969$.

The mean longitude (medius motus) can be calculated by using Tables IV and V (Wegener, 1905b, pp.47–48):

1470.0	$288^{\circ}.896$
7 years	$0^{\circ}.298$
200 days	$197^{\circ}.129$
60 days	$59^{\circ}.139$
3 days	$2^{\circ}.957$
6 hours	$0^{\circ}.246$
1.6 minutes	$0^{\circ}.001$

Mean longitude $\lambda_M = 188^{\circ}.666$.

Mean anomaly (Equation 1.5) $\kappa_M = \lambda_M - \lambda_\Gamma = 97^{\circ}.697$.

The equation of the centre is given in Table VI (Wegener, 1905b, p.49), argument is mean longitude and interpolated value is: $c = -2^{\circ}.163$.

True longitude (Equation 1.6) $\lambda_{\odot} = \lambda_M + c = 186^{\circ}.503$.[86]

The values of the equation of time in the *Alfonsine Tables for Toledo* and in the later edition of *Tabulae Resolutae Salamanca* are in time-degrees, and in the later editions of *Alfonsine Tables* also in minutes and seconds.[87]

5.4.3 Parisian Alfonsine Tables

The preparation and popularisation of the *Parisian Alfonsine Tables* is attributed to the circle of astronomers including John of Murs, John Vimond, John of Lignères and John of Saxony.

John of Murs

John of Murs (born 1290/1300 – died after 1344), Lat. Johannes de Muris, was born in the diocese of Lisieux in Normandy. He studied at the Sorbonne between 1317 and 1321, and published treatises on arithmetic, geometry, astronomy and music.[88] His twelve-chapter handbook *De arte mensurandi*, written around 1343, collects and organises mathematical methods relevant to astronomy.[89] In addition to astronomy, he was also involved in the reform of

86 For the sign of the equation of the centre see Equation 1.5 or 1.6 and explanation in Chapter 1.
87 For example, *Alfonsine Tables* of 1524, issued in Venice.
88 For more information on his work, see Chabás, Goldstein, 2002, pp.277–281.
89 Busard, 1998, Johannes de Muris, De Arte Mensurandi.

the calendar, though this did not succeed. Nevertheless, his work was an important contribution to the later Gregorian reform of the Julian calendar. We relate him closely to the *Parisian Alfonsine Tables*, for which he certainly contributed some of the content. An important contribution to astronomy is also his equation of time, included in the tables from the year 1321 (Figure 5.4), which is given in minutes and seconds and was unprecedented in that period.[90]

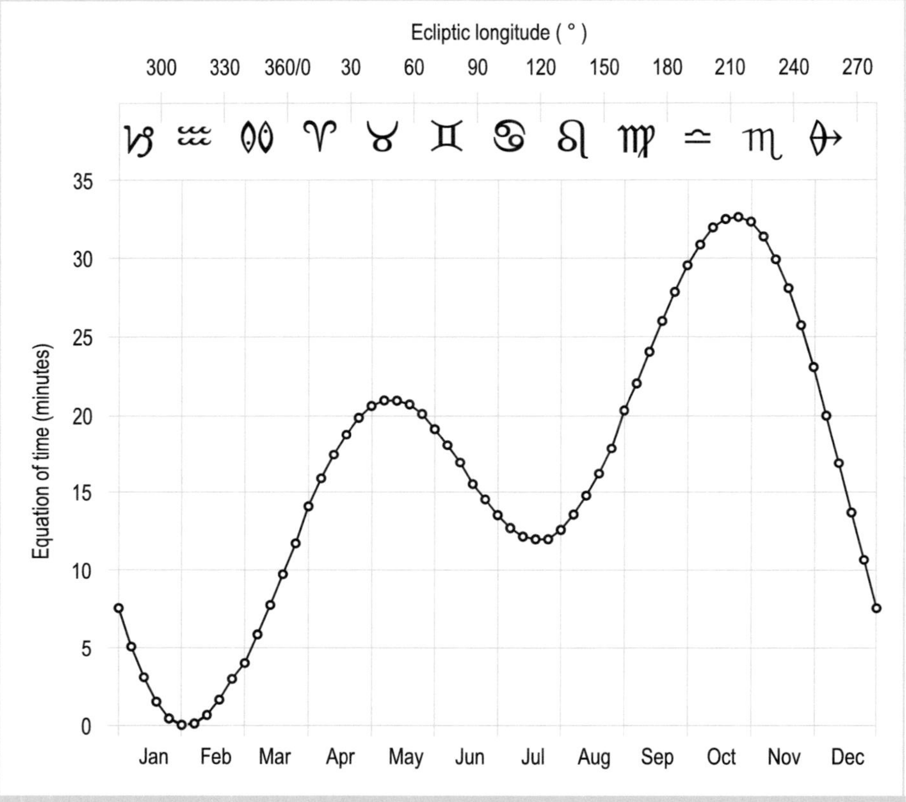

Figure 5.4 The equation of time of John of Murs from 1321. Data source for the diagram: Chabás, Goldstein, 2012, p.41.

John of Lignères

John of Lignères, in the diocese of Amiens, lived in Paris from about 1320 to 1335. He contributed a great deal to the dissemination of astronomical knowledge from the *Alfonsine Tables*. His astronomical works include astronomical tables, canons to the tables, the theory of planetary motion and treatises on astronomical instruments. In his works on astronomical

90 At that time, his tables were among a few that contained the equation of time for certain days of the year in minutes and seconds of time. The table manuscripts are kept by the library of Ajuba in Lisbon 52-XII-35, folio 59v, Chabás, Goldstein, 2012, p.41.

instruments, he described the saphea, the equatorium and the directorium, a special astrolabe-like instrument for use in astrology. The saphea is a special version of the astrolabe, attributed to AL-ZARQĀLĪ. WILLIAM THE ENGLISHMAN[91] attempted to describe such an astrolabe in 1231, although he probably had not seen it, and a detailed description was available in 1263, when IBN TIBBON translated AL-ZARQĀLĪ's treatise. JOHN OF LIGNÈRES wrote two treatises on the equatorium, one describing an equatorium similar to CAMPANUS'. The equation of time attributed to him is published in the manuscript Pal. Lat. 1412, folio 101v, Biblioteca Apostolica Vaticana (see Figure 5.3).[92]

John of Saxony

JOHN OF SAXONY, Latin JOHANNES DE DANECOWE, possibly born in Magdeburg, probably a pupil of JOHN OF LIGNÈRES, lived in Paris between 1327 and 1335 and his involvement in astronomy probably dates back to the end of the 13th century.[93] He left several written works, the Ephemerides, the *Canons to the Alfonsine Tables*, and a commentary on AL-QABISI's[94] astrological treatise. His *Canons to the Alfonsine Tables* were a great success, being included in the first printed edition of the *Alfonsine Tables*, printed by ERHARD RATDOLT in Venice in 1483.[95] The *Canons* also mention the equation of time.[96]

91 WILLIAM THE ENGLISHMAN (13th century), astronomer and astrologer, active in France. His works date from between 1220 and 1231.

92 For more information about the equation of time attributed to JOHN OF LIGNÈRES see Chabás, Saby, 2022, pp.72–75.

93 Also known as JOHANNES DE SAXONIA and DANECOWE OF SAXONY.

94 Lat. ALCABITIUS (10th century), so named because he was originally from Qabiza, Iraq, astrologer, astronomer and mathematician. He was based in Aleppo, today's Syria.

95 In this edition, there is a reference suggesting that JOHN OF SAXONY was a disciple of JOHN OF LIGNÈRES. Wegener, 1905a, p.168. ERHARD RATDOLT (1442–1528), German printer, active in Venice from 1476 to 1486 and later in Augsburg. He also printed EUCLID's *Elements* and SACROBOSCO's *De sphaera*. He was the first to print text on one page in more than two colours. RATDOLT's typesetting is superb. This is of relevance when considering the cultural context of dissemination in print, because the Venetian printers' typographic excellence contrasts starkly against the northern European typography of the Reformation era.

96 Poulle, 1984, Les Tables Alphonsines avec les Canons de Jean de Saxe, table with the equation of time, pp.218–219.

Bibliography

- ANONYMOUS. St. Isidore of Seville. In: Encyclopedia.com [online source]. URL: https://www.encyclopedia.com/history/encyclopedias-almanacs-transcripts-and-maps/st-isidore-seville (23 April 2023).
- BARNEY, Stephen A., W. J. LEWIS, J. A. BEACH and Oliver BERGHOF. 2014. Etymologies of Isidore of Seville. Cambridge: Cambridge University Press. ISBN 978-0-521-14591-6.
- BENJAMIN, Francis S. Jr. and Gerald J. TOOMER (trans., eds.). 1971. Campanus of Novara and Medieval Planetary Theory: Text in Latin and English. Madison, WI: The University of Wisconsin Press. ISBN 9780299059606.
- BIRASHK, Ahmad. 1993. A Comparative Calendar of the Iranian, Muslim Lunar, and Christian Eras for Three Thousand Years (1260 B.H.–2000 A.H./639 B.C.–2621 A.D.). Costa Mesa, CA, New York: Mazda Publishers. ISBN 0-939214-95-4.
- BREHAUT, Ernest. 1912. An Encyclopedist of the Dark Ages: Isidore of Seville. New York: Columbia University, Longmans, Green & Co. London: P. S. King & Son. URL: https://archive.org/details/anencyclopedist00brehgoog (24 April 2023).
- BROWNE, George Forest. 1919. The Venerable Bede His Life and Writings. Society for Promoting Christian Knowledge. New York: MacMillan Comp. URL: https://archive.org/details/venerablebedehis00browuoft (4 December 2022).
- BURNETT, Charles. 1977. A Group of Arabic-Latin Translators Working in Northern Spain in the Mid-12th Century. In: The Journal of the Royal Asiatic Society of Great Britain and Ireland, 1, 62–108. URL: https://www.jstor.org/stable/25210854.
- BURNETT, Charles. 1978. Arabic into Latin in Twelfth Century Spain: the Works of Hermann of Carinthia. In: Mittellateinisches Jahrbuch 13, 100–134.
- BURNETT, Charles. 2007. Roger of Hereford. In: Biographical Encyclopedia of Astronomers, Hockey, Thomas, Virginia Trimble and Thomas R. Williams (eds.), 982. New York: Springer. ISBN 9780387310220.
- BUSARD, Hubert L. L. 1998, Johannes de Muris, De Arte Mensurandi. Stuttgart: Franz Steiner Verlag. ISBN 3-515-07410-4.
- BUSARD, Hubert L. L. 1998. Über eine Vorlage des Traktats Algorismus de minutiis von Johannes de Lineriis. In: Sudhoffs Archiv, 82, 1, 74–91. URL: https://www.jstor.org/stable/20777653.
- BUSARD, Hubert L. L. and Menso FOLKERTS. 1992a. Robert of Chester's Redaction of Euclid's Elements, the so-called Adelard II Version. Volume I. Basel: Birkhäuser. DOI 10.1007/978-3-0348-8636-9. ISBN 978-3-7643-2658-6.
- BUSARD, Hubert L. L. and Menso FOLKERTS. 1992b. Robert of Chester's Redaction of Euclid's Elements, the so-called Adelard II Version. Volume II. Basel: Birkhäuser. DOI 10.1007/978-3-0348-8604-8. ISBN 978-3-7643-2727-9.
- CARMAN, Christián C. 2017. Martianus Capella's calculation of the size of the moon. In: Archive for History of Exact Sciences, 71, 1, 193–210. DOI 10.1007/s00407-016-00185-0.
- CARMODY, Francis J. 1943. Alfragani differentie in quibusdam collectis scientie astrorum. Berkeley, CA.
- CARMODY, Francis J. 1956. Arabic Astronomical and Astrological Sciences in Latin Translation A Critical Bibliography. Berkeley, CA: University of California Press. URL: https://archive.org/details/Carmody1956AAAS (28 January 2024).
- CASULLERAS, Josep. 1996. Las "Tablas astronómicas" de Pedro Alfonso In: Estudios sobre Pedro Alfonso de Huesca, Ducay, María Jesús Lacarra (ed.), 349–366. Huesca: Instituto de Estudios Altoaragoneses. ISBN 84-8127-047-4.
- CHABÁS, José and Bernard R. GOLDSTEIN. 2003. The Alfonsine Tables of Toledo. Dordrecht: Kluwer Academic Publisher. ISBN 1-4020-1572-0.
- CHABÁS, José and Bernard R. GOLDSTEIN. 2009. John of Murs's Tables of 1321. In: Journal for the History of Astronomy, 40, 3, 297–320. DOI 10.1177/002182860904000303.
- CHABÁS, José and Bernard R. GOLDSTEIN. 2012. A Survey of European Astronomical Tables in the Late Middle Ages. Leiden: Brill Academic Pub. ISBN 978-9004230583.
- CHABÁS, José and Bernard R. GOLDSTEIN. 2014. Essays on Medieval Computational Astronomy. Leiden: Brill. ISBN 9789004281745.
- CHABÁS, José and Bernard R. GOLDSTEIN. 2019. The Master and the Disciple: The Almanac of John of Lignères and the Ephemerides of John of Saxony. In: Journal for the History of Astronomy, 50, 1, 82–96. DOI 10.1177/0021828618820215.
- CHABÁS, José and Marie-Madeleine SABY. 2022. The Tables of 1322 by John of Lignères, An Edition with Commentary. Turnhout: Brepols. e-ISBN 978-2-503-59610-5.

- Clagett, Marshall. Adelard of Bath. In: Encyclopedia.com, updated May 18 2018 [online source]. url: https://www.encyclopedia.com/people/philosophy-and-religion/philosophy-biographies/adelard-bath (30 October 2022).
- Clagett, Marshall. 1953. The Medieval Latin Translations from the Arabic of the Elements of Euclid, with Special Emphasis on the Versions of Adelard of Bath. In: Isis, 44, 1–2, 16–42. doi 10.1086/348186. url: https://www.jstor.org/stable/227639.
- Clagett, Marshall. 1970. Adelard of Bath. In: Dictionary of Scientific Biography, Gillispie, Charles Coulston (ed.), I, 61–64. New York: Charles Scribner's Sons.
- Dadić, Žarko. 1995. Herman the Dalmatian as Astronomer. In: Hvar Observatory Bulletin, 19, 1, 39–53. url: https://ui.adsabs.harvard.edu/abs/1995HvaOB..19...39D/abstract (18 September 2022).
- Dutton, Paul Edward. 2004. Carolingian Civilization. Toronto: University of Toronto Press. isbn 9781551114927.
- Eastwood, Bruce S. 2000. Astronomical Images and Planetary Theory in Carolingian Studies of Martianus Capella. In: Journal for the History of Astronomy, 31, 1–28. doi 10.1177/002182860003100101.
- Eastwood, Bruce S. 2007. Ordering the Heavens, Roman Astronomy and Cosmology in the Carolingian Renaissance. Medieval and Early Modern Philosophy and Science, Volume 8. Lleiden: Brill. doi 10.1163/ej.9789004161863.i-453.
- Eastwood, Bruce and Gerd Graβhoff. 2003. Planetary Diagrams - Descriptions, Models, Theories. From Carolingian Deployments to Copernican Debates. In: The Power of Images in Early Modern Science, Lefèvre, W., J. Renn in U. Schoepflin (eds.), 197–226. Basel: Birkhäuser. doi 10.1007/978-3-0348-8099-2_11. isbn 978-3-7643-2434-6.
- Eastwood, Bruce and Gerd Graβhoff. 2004. Planetary Diagrams for Roman Astronomy in Medieval Europe, ca. 800–1500. In: Transactions of the American Philosophical Society, 94, 3, 1–158.
- Embach, Carolyn (trans.). 1969. On the Nature of Things (De Natura Rerum), by Isidore of Seville (ca. 560–636 AD). url: https://www.researchgate.net/publication/315663902_On_the_Nature_of_Things_De_Natura_Rerum_by_Isidore_of_Seville_ca_560-636_AD_Translated_by_Carolyn_Embach_1969 (22 April 2023).
- Evans, Gillian R. 1976. The "Sub-Euclidean" Geometry of the Earlier Middle Ages, up to the Mid-Twelfth Century. In: Archive for the History of Exact Sciences, 16, 2, 105–118. doi 10.1007/BF00349633. url: https://www.jstor.org/stable/41133464.
- Glick, Thomas F. 2005. Sa'id al-Andalusi. In: Medieval Science, Technology, and Medicine: An Encyclopedia, Thomas F. Glick, Steven J. Livesey and Faith Wallis (eds.). New York, London: Routledge. isbn 978-0-415-96930-7.
- Grant, Edward. Jordanus de Nemore. In: Encyclopedia.com [online source]. url: https://www.encyclopedia.com/science/dictionaries-thesauruses-pictures-and-press-releases/jordanus-de-nemore (30 October 2022).
- Gunther, Robert Theodore 1923. Early Sciences in Oxford, Volume II. London: Dawsons of Pall Mall.
- Hamerly, Don. 2017. The Córdovan Library of Caliph al-Hakam II. In: Libraries - Traditions and Innovations, Papers from the Library History Seminar XIII, Kimball, Melanie A. and Katherine M. Wisser (eds.), 4–15. Berlin: De Gruyter. e-isbn 9783110450842. doi 10.1515/9783110450842.
- Harper, Richard. 1975. The Astronomical Tables of William Rede. In: Isis, 66, 3, 369–378. doi 10.1086/351475. url: https://www.jstor.org/stable/228843.
- Haskins, Charles Homer. 1927. Studies in the History of Mediaeval Science. Cambridge, MA: Harvard University Press. url: https://archive.org/details/studiesinhistory00hask/mode/2up (28 January 2024).
- Herz, Norbert. 1893. Über die Alphonsinischen Tafeln und die im Besitze der k. k. Hofbibliothek in Wien befindlichen Handschriften derselben. In: Sitzungsberichte der Akademie der Wissenschaften mathematisch-naturwissenschaftliche Klasse, Abtheilung 2a, 102, 99–117. url: https://www.zobodat.at/pdf/SBAWW_102_2a_0099-0117.pdf (28 January 2024).
- Høyrup, Jens. 1988. Jordanus de Nemore, 13th Century Mathematical Innovator: an Essay on Intellectual Context, Achievement and Failure. In: Archive for History of Exact Sciences, 38, 4, 307–363. doi 10.1007/BF00357064.
- Ibn 'Ezra, Abraham R. 1947. El Libro de los Fundamentos da las tablas astronómicas de R. Abraham Ibn 'Ezra. Edición Crítica, con Introducción y Notas por Millás Vallicrosa, Jose Maria (ed.). Madrid-Barcelona: Catedrático de la Universidad de Barcelona.
- Ker, Neil Ripley and Alan J. Piper (eds.). 1968–2002. Medieval manuscripts in British libraries, I. Oxford: Clarendon Press.
- Kren, Claudia. Bernard of Verdun. In: Encyclopedia.com [online source]. url: https://www.encyclopedia.com/science/dictionaries-thesauruses-pictures-and-press-releases/bernard-verdun (24 November 2022).
- Lévi-Provençal, E. Abbās b. Nāṣiḥ. In: Encyclopédie de l'Islam [online source]. doi 10.1163/9789004206106_eifo_SIM_0026 (12 November 2022).
- Lindsay, Wallace Martin. 1911. Isidori Hispalensis episcopi Etymologiarum sive Originvm. Oxford: Clarednon Press. London: Henry Prowde. url: https://archive.org/details/isidori01isiduoft/mode/2up (23 April 2023).

- Lipton, Joshua David. 1978. The Rational Evaluation of Astrology in the Period of Arabo-Latin Translation ca. 1126–1187 AD. Dissertation. Los Angeles, CA: University of California.
- Lisa, Massimiliano, Mario Taddei and Edoardo Zanon (eds.). 2008. The Book of Secrets in the Results of Ideas: Incredible Machines from 1000 Years Ago, Ibn Khalaf al-Murādī. Milano: Leonardo3. isbn 978-88-6048-013-2.
- Lorch, Richard P. Jābir Ibn Aflah Al-Ishb. In: Encyclopedia.com [online source]. url: https://www.encyclopedia.com/science/dictionaries-thesauruses-pictures-and-press-releases/jabir-ibn-aflah-al-ishb (24 November 2022).
- Lorch, Richard P. 1975. The Astronomy of Jabir ibn Aflah. In: Centaurus, 19, 2, 85–107. doi 10.1111/j.1600-0498.1975.tb00315.x.
- Lorch, Richard P. 1977. The Astronomical Instruments of Jabir ibn Aflah and the Torquetum. In: Centaurus, 20, 1, 11–35. doi 10.1111/j.1600-0498.1976.tb00214.x.
- McCarthy, Daniel P. and John G. Byrne. 2003. Al-Khwarizmi's Sine Tables and a Western Table with the Hindu Norm of R = 150. In: Archive for History of Exact Sciences, 57, 3, 243–266. doi 10.1007/s00407-002-0058-6.
- McVaugh, Michael. 1974. The Translation of Greek and Arabic Science into Latin. A Source Book in Medieval Science, Grant, Edward (ed.). Cambridge, MA: Harvard University Press. isbn 674-82360-5.
- Mercier, Raymond. 1988. Astronomical Tables in the Twelfth Century. In: Adelard of Bath an English Scientist and Arabist of the Early Twelfth Century, Burnett, Charles (ed.). London: The Warburg Institute - University of London. isbn 978-0854810703.
- Millás Vallicrosa, José María. 1960. Nuevos estudios sobre historia de la ciencia española. Barcelona: Instituto Luis Vives de Filosofía.
- Möller, Lenelotte. 2008. Die Enzyklopädie des Isidor von Sevilla. Wiesbaden: marixverlag. isbn 978-3-86539-177-3.
- Moreton, Jennifer. 1995. Before Grosseteste: Roger of Hereford and Calendar Reform in Eleventh- and Twelfth-Century England. In: Isis, 86, 4, 562–586. doi 10.1086/357318. url: https://www.jstor.org/stable/235377.
- Nallino, Carlo Alfonso (ed., trans.). 1903. Al-Battānī sive Albatenii, Opus Astronomicum. Pubblicazioni del reale osservatorio di Brera in Milano, XL, I. In: Sezgin, Fuat (ed.), Islamic Mathematics and Astronomy, 11. Reprint 1997. Frankfurt: Institut fur Geschichte der Arabisch-Islamischen Wissenschaften. isbn 3-8298-4011-X.
- Nallino, Carlo Alfonso (ed., trans.). 1907. Al-Battānī sive Albatenii, Opus Astronomicum. Pubblicazioni del reale osservatorio di Brera in Milano, XL, II. In: Sezgin, Fuat (ed.), Islamic Mathematics and Astronomy, 12. Reprint 1997. Frankfurt: Institut fur Geschichte der Arabisch-Islamischen Wissenschaften. isbn 3-8298-4011-X.
- Nallino, Carlo Alfonso. 1969. Al-Battānī sive Albatenii, Opus Astronomicum. Pubblicazioni del reale osservatorio di Brera in Milano, XL, I-II, reprint of 1903-1907 Edition. Frankfurt: Minerva.
- Neugebauer, Paul Viktor. 1937. Hilfstafeln zur technischen Chronologie, Teil III. In: Astronomische Nachrichten, 261, 6264, 461–524.
- North, J. D. 1976. Richard of Wallingford: An Edition of His Writings with Introduction, English Translation and Commentary. Volume 2. Oxford: Oxford University Press. isbn 978-0199227457.
- North, J. D. 1977. The Alfonsine Tables in England. In: Prismata: Festschrift für Willy Hartner, Maeyama, Y. and W. G. Salzer (eds.), 269–301. Wiesbaden: Franz Steiner Verlag. isbn 3-515-02683-5.
- North, J. D. 1989. Stars, Minds and Fate: Essays in Ancient and Medieval Cosmology. London: Hambledon Press. isbn 0-907628-94-X.
- North, John D. 1992. Natural Philosophy in Late Medieval Oxford and Astronomy and Mathematics. In: The History of the University of Oxford, II, Late Medieval Oxford, Catto, J. I. and Ralph Evans (eds.). Oxford: Clarendon Press. isbn 978-0-199510-122.
- Norton, Stephen D. Robert of Chester. In: Encyclopedia.com [online source]. url: https://www.encyclopedia.com/science/encyclopedias-almanacs-transcripts-and-maps/robert-chester (30 October 2022).
- O'Connor J. J. and E. F. Robertson. 2009. Campanus of Novara. In: MacTutor, School of Mathematics and Statistics, University of St Andrews, Scotland [online source]. url: https://mathshistory.st-andrews.ac.uk/Biographies/Campanus/ (30 October 2022).
- Paić-Vukić, Tatjana. 2019. The Biography of Hermann the Dalmatian: Separating Conjectures from Verifiable Claims. In: Exploring the Commonalities of the Mediterranean Region: Proceedings of the TUBA-EMAN Symposium, Nuri, Ahmet (ed.), 6 – 7 November 2017, Yurdusev. Ankara: Turkish Academy of Sciences. doi 10.53478/TUBA.2019.010.
- Pedersen, Fritz Saaby (ed.). 1978. Petrus de Dacia: Tractatus Instrumenti Eclipsium. In: Cahiers de l'Institut du Moyen-Âge Grec et Latin, 25, 1–102. url: https://cimagl.saxo.ku.dk/download/25/25pedersen1-102.pdf (30 October 2022).
- Pedersen, Fritz Saaby (ed.). 1979. Petrus de St. Audomaro: Tractatus de Semissis. In: Cahiers de l'Institut du Moyen-Âge Grec et Latin, 33, 1–113. url: http://cimagl.saxo.ku.dk/download/33/33Pedersen1-116.pdf/ (30 October 2022).

- Pedersen, Fritz Saaby. 1983. Petri Philomenae de Dacia et Petri de S. Audomaro opera quadrivialia, Part I. Copenhagen: Societas Linguae & Litterarum Danicarum. ISBN 87-12-15896-8.
- Pedersen, Fritz Saaby. 1984a. Petri Philomenae de Dacia et Petri de S. Audomaro opera quadrivialia, Part II. Copenhagen: Societas Linguae & Litterarum Danicarum. ISBN 87-12-15897-6.
- Pedersen, Fritz Saaby. 1984b. A Paris Astronomer of 1290. In: Cahiers de l'Institut du Moyen-Âge Grec et Latin, 48, 163–188. URL: https://cimagl.saxo.ku.dk/download/48/48Pedersen163-188.pdf (30 October 2022).
- Pedersen, Fritz Saaby. 1986. Johannis de Sicilia super canones Azarchelis de tabulis Toletanis. In: Cahiers de l'Institut du Moyen-Âge Grec et Latin, 51-52, 1–396. URL: https://cimagl.saxo.ku.dk/download/51/51-52Pedersen.pdf (30 October 2022).
- Pedersen, Fritz Saaby. 1998. The Toulouse Tables: A List of Manuscripts. In: Cahiers de l'Institut du Moyen-Âge Grec et Latin, 68, 3–12. URL: https://cimagl.saxo.ku.dk/download/68/68Pedersen3-12.pdf (30 October 2022).
- Pedersen, Fritz Saaby. 2001. Anonymous Parisian Astronomer of 1290, I. In: Cahiers de l'Institut du Moyen-Âge Grec et Latin, 72, 169–269. URL: https://cimagl.saxo.ku.dk/download/72/72Pedersen169-269.pdf (30 October 2022).
- Pedersen, Fritz Saaby. 2002a. Anonymous Parisian Astronomer of 1290, II. In: Cahiers de l'Institut du Moyen-Âge Grec et Latin, 73, 61–166. URL: https://cimagl.saxo.ku.dk/download/73/73Pedersen61-166.pdf (30 October 2022).
- Pedersen, Fritz Saaby. 2002b. The Toledan Tables: A review of the manuscripts and the textual versions with an edition. Copenhagen: The Royal Danish Academy of Sciences and Letters, C. A. Reitzeils Forlag. ISBN 87-7876-267-7.
- Pedersen, Fritz Saaby. 2004. The Treatise on the Rising and Setting of Signs Ascribed to Roger of Hereford. In: Cahiers de l'Institut du Moyen-Âge Grec et Latin, 75, 3–6. URL: https://cimagl.saxo.ku.dk/download/75/75Pedersen3-6.pdf (18 September 2022).
- Pedersen, Fritz Saaby. 2009. The tract "Diversi astrologi" ascribed to Roger of Hereford. URL: https://aigis.igl.ku.dk/fsp/texts2009/diversi.html (18 September 2022).
- Pedersen, Fritz Saaby. 2012. Petrus Philomena de Dacia. Bergen: University of Bergen [online source]. URL: https://wiki.uib.no/medieval/index.php/Petrus_Philomena_de_Dacia (11 September 2022).
- Pedersen, Fritz Saaby (ed.). 2013. William of Saint-Cloud (Guillelmus de Sancto Clodoaldo): Canons to Almanach Planetarum. URL: https://aigis.igl.ku.dk/fsp/texts2009/almanach-plain.html (11 September 2022).
- Pedersen, Fritz Saaby (ed.). 2014. William of Saint-Cloud: Almanach Planetarum, An edition of the canons, a few samples from the tables, and a foray into the numbers. Copenhagen: Centre for the Aristotelian Tradition, Saxo Institute. URL: https://cimagl.saxo.ku.dk/download/83/83Pedersen1-135.pdf (30 October 2022).
- Pedersen, Olaf. 1967. The life and work of Peter Nightingale. In: Vistas in Astronomy, 9, 3–10. DOI 10.1016/0083-6656(67)90006-2.
- Pedersen, Olaf. 1974. Peter Philomena of Dacia. In: Dictionary of Scientific Biography, Gillispie, Charles Coulston (ed.), X, 540–542. New York: Charles Scribner's Sons. ISBN 0-684-10121-1.
- Pedersen, Olaf. Peter Philomena of Dacia, also known as Petrus Dacus, Petrus Danus, Peter Nightingale. In: Encyclopedia.com [online source]. URL: https://www.encyclopedia.com/science/dictionaries-thesauruses-pictures-and-press-releases/peter-philomena-daciaalso-known-petrus-dacus-petrus-danus-peter-nightingale (30 October 2022).
- Poulle, Emmanuel. 1964. Bernard de Verdun et le Turquet. In: Isis, 55, 2, 200-208. DOI 10.1086/349829. URL: https://www.jstor.org/stable/228186.
- Poulle, Emmanuel. 1973. John of Lignères. In: Dictionary of Scientific Biography, Gillispie, Charles Coulston (ed.), VII, 122–128. New York: Charles Scribner's Sons. SBN 684-10118-1.
- Poulle, Emmanuel. 1973. John of Murs. In: Dictionary of Scientific Biography, Gillispie, Charles Coulston (ed.), VII, 128–133. New York: Charles Scribner's Sons. SBN 684-10118-1.
- Poulle, Emmanuel. 1973. John of Saxony. In: Dictionary of Scientific Biography, Gillispie, Charles Coulston (ed.), VII, 139–141. New York: Charles Scribner's Sons. SBN 684-10118-1.
- Poulle, Emmanuel. 1973. John of Sicily. In: Dictionary of Scientific Biography, Gillispie, Charles Coulston (ed.), VII, 141–142. New York: Charles Scribner's Sons. SBN 684-10118-1.
- Poulle, Emmanuel. 1976. William of Saint-Cloud. In: Dictionary of Scientific Biography, Gillispie, Charles Coulston (ed.), XIV, 389–391. New York: Charles Scribner's Sons. ISBN 0-684-12926-4.
- Poulle, Emmanuel. 1984. Les Tables Alphonsines avec les Canons de Jean de Saxe. Paris: Centre national de la recherche scientifique. ISBN 2-22203430-2.
- Poulle, Emmanuel. 1988. The Alfonsine Tables and Alfonso-X of Castille. In: Journal for the history of astronomy, 19, 2, 97–113. DOI 10.1177/002182868801900202.
- Poulle, Emanuel. John of Lignères, or Johannes de Lineriis. In: Encyclopedia.com [online source]. URL: https://www.encyclopedia.com/science/dictionaries-thesauruses-pictures-and-press-releases/john-ligneres-or-johannes-de-lineriis (30 October 2022).

- Poulle, Emanuel. John of Saxony. In: Encyclopedia.com [online source]. URL: https://www.encyclopedia.com/science/dictionaries-thesauruses-pictures-and-press-releases/john-saxony (30 October 2022).
- Poulle, Emmanuel. John of Sicily. In: Encyclopedia.com [online source]. URL: https://www.encyclopedia.com/science/dictionaries-thesauruses-pictures-and-press-releases/john-sicily (30 October 2022).
- Poulle, Emanuel. Raymond of Marseilles. In: Encyclopedia.com [online source]. URL: https://www.encyclopedia.com/science/dictionaries-thesauruses-pictures-and-press-releases/raymond-marseilles (19 November 2022).
- Poulle, Emmanuel. William of Saint-Cloud. In: Encyclopedia.com [online source]. URL: https://www.encyclopedia.com/science/dictionaries-thesauruses-pictures-and-press-releases/william-saint-cloud (30 October 2022).
- Poulle, Emanuel. William the Englishman. In: Encyclopedia.com, updated June 27 2018 [online source]. URL: https://www.encyclopedia.com/literature-and-arts/art-and-architecture/architecture/william-englishman (30 October 2022).
- Ragep, Jamil and Marvin Bolt. 2007. Ādamī: Abū ʿAlī al-Ḥusayn ibn Muḥammad al-Ādamī. In: Biographical Encyclopedia of Astronomers, Hockey, Thomas, Virginia Trimble and Thomas R. Williams (eds.), 12. New York: Springer. ISBN 9780387310220. URL: https://islamsci.mcgill.ca/RASI/BEA/Adami_BEA.htm (12 November 2022).
- Raynaud, Dominique, Samuel Gessner and Bernardo Mota. 2019. Andalò di Negro's De compositione astrolabii: a critical edition with English translation and notes. In: Archive for History of Exact Sciences, 73, 1, 551–617. DOI 10.1007/s00407-019-00235-x.
- Richter-Bernburg, Lutz. 1987. Sāʿid, the Toledan Tables, and Andalusī Science. In: From Deferent to Equant, King, David and George Saliba (eds.), Annals of the New York Academy of Sciences, Volume 500, 373–401. New York: New York Academy of Sciences. ISBN 0-89766-396-9.
- Rico y Sinobas, Manuel. 1863–1867. Libros del saber de astronomía del rey D. Alfonso X de Castilla (1863–67) - Alfonso X , Rey de Castilla, 1221–1284. Madrid: Eusebio Aguado. URL: https://bibliotecadigital.jcyl.es/es/consulta/registro.cmd?id=13284 (28 January 2024).
- Rusell, Jossiah C. 1932. Hereford and Arabic Science in England about 1175–1200. In: Isis, 18, 1, 14–25. DOI 10.1086/346685. URL: https://www.jstor.org/stable/224477.
- Samsó, Julio. 2015. Al-Andalus, a Bridge between Arabic and European Science. In: Alhadra, 1, 101–125.
- Samsó, Julio. 2020. Ibn al-Zarqālluh, Andalusian Astronomy in the Eleventh Century. In: Inference, 5, 3. URL: https://inference-review.com/article/andalusian-astronomy-in-the-eleventh-century (12 November 2022).
- Sela, Shlomo. 2003. Abraham Ibn Ezra and the Rise of Medieval Hebrew Science. Leiden: Brill. ISBN 978-9004129733.
- Sela, Shlomo. 2007. Abraham Ibn Ezra The Book of Reasons. Leiden: Brill. ISBN 978-9004157644.
- Sezgin, Fuat. 1974. Geschichte des arabischen Schrifttums, Band V. Lleiden: Brill.
- Sezgin, Fuat. 1978. Geschichte des arabischen Schrifttums, Band VI. Lleiden: Brill. ISBN 90 04 05878 8.
- Spuler, Bertold and Joachim Mayr (eds.). 1961. Wüstenfeld-Mahler'sche Vergleichungs-Tabellen. Deutsche Morgenländische Gesellschaft. Wiesbaden: Franz Steiner Verlag.
- Stahl, William. Martianus Capella. In: Encyclopedia.com, updated 14 May 2018 [online source]. URL: https://www.encyclopedia.com/people/literature-and-arts/classical-literature-biographies/martianus-capella (22 April 2023).
- Stahl, William H. 1959. Dominant Traditions in Early Medieval Latin Science. In: Isis, 50, 2, 95–124. DOI 10.1086/348742. URL: https://www.jstor.org/stable/226732.
- Stahl, William H. 1965. To a Better Understanding of Martianus Capella. In: Speculum, 40, 1, 102–115. DOI 10.2307/2856467.
- Steinschneider, Moritz. 1880. Abraham ibn Esra, Abhandlungen zur Geschichte der mathematischen Wissenschaften mit Einschluß ihrer Anwendungen. In: Zeitschrift für Mathematik und Physik. Supplement zur historisch-literarischen Abtheilung des XXV. Jahrgangs, 57–128. URL: https://sammlungen.ub.uni-frankfurt.de/freimann/content/titleinfo/7401598?lang=en (1 May 2023).
- Steinschneider, Moritz. 1888. Iusuf ben Ibrahim und Ahmed ben Iusuf. In: Biblioteca Mathematica, Zeitschrift für Geschichte der Mathematik, Neue Folge 2, 2, 49–52. URL: https://archive.org/details/sim_bibliotheca-mathematica_1888_2_2/mode/2up (30 October 2022).
- Suter, Heinrich. 1914. Die astronomischen Tafeln des Muhammed ibn Mūsā al-Khwārizmī in der Bearbeitung des Maslama ibn Ahmed al-Madjrītī und der lateinischen Übersetzung des Adelard von Bath. København: Kongelige Danske Videnskabernes Selskab. URL: https://menadoc.bibliothek.uni-halle.de/urn/urn:nbn:de:gbv:3:5-12969 (10 February 2024).
- Suter, Heinrich (ed.). 1914. Die astronomischen Tafeln des Muhammed ibn Mūsā al-Khwārizmī in der Bearbeitung des Maslama ibn Ahmed al-Madjrītī und der lateinischen Übersetzung des Adelard von Bath. In: Sezgin, Fuat (ed.), Islamic Mathematics and Astronomy, 7. Reprint 1997. Frankfurt: Institut fur Geschichte der Arabisch-Islamischen Wissenschaften. ISBN 3-8298-4008-X.

- Thorndike, Lynn. 1923. A History of Magic and Experimental Science II. New York: The Macmillan Company. URL: https://archive.org/details/in.ernet.dli.2015.201717/mode/2up (28 January 2024).
- Thorndike, Lynn. 1945. Franco de Polonia and the Turquet. In: Isis, 36, 1, 6–7. DOI 10.1086/347897. URL: https://www.jstor.org/stable/225669.
- Thorndike, Lynn. 1959. John of Seville. In: Speculum, 34, 1, 20–38. DOI 10.2307/2847976. URL: https://www.jstor.org/stable/2847976.
- Tolan, John. 1993. Petrus Alfonsi and his Medieval Readers. Gainesville, FL: University Press of Florida. ISBN 0-8130-1238-4.
- Toll, Christopher. Al-Hamdānī, Abū Muḥammad Al-Ḥasan Ibn Aḥmad Ibn Ya'qūb. In: Encyclopedia.com [online source]. URL: https://www.encyclopedia.com/science/dictionaries-thesauruses-pictures-and-press-releases/al-hamdan (12 November 2022).
- Toomer, Gerald J. 1968. A Survey of the Toledan Tables. In: Osiris, 15, 5–174. DOI 10.1086/368631.
- Toomer, Gerald J. 1971. Campanus of Novara. In: Dictionary of Scientific Biography, Gillispie, Charles Coulston (ed.), III, 23–29. New York: Charles Scribner's Sons.
- Toomer, Gerald J. Campanus of Novara. In: Encyclopedia.com, updated May 18 2018 [online source]. URL: https://www.encyclopedia.com/people/science-and-technology/mathematics-biographies/campanus-novara#2830900765 (30 October 2022).
- Vernet, Juan. 1976. Al-Zarqālī. In: Dictionary of Scientific Biography, Gillispie, Charles Coulston (ed.), XIV, 592–595. New York: Charles Scribner's Sons. ISBN 684-12926-5.
- Vogel, Kurt. 1973. John of Gmunden. In: Dictionary of Scientific Biography, Gillispie, Charles Coulston (ed.), VII, 117–122. New York: Charles Scribner's Sons. SBN 684-10118-1.
- Von Jan, Ludwig. 1848. Macrobii Ambrosii Theodosii Opera quae supersunt. Quedlinburg, Leipzig: Gottfried Bass. URL: https://archive.org/details/macrobiiambrosi01jangoog/mode/2up?q=dies (23 April 2023).
- Wallis, Faith (trans.). 2012. Bede The Reckoning of Time. Translated, with introduction, notes and commentary. Liverpool: Liverpool University Press. ISBN 978-0-85323-693-1.
- Wasserstein, David John. 1990. The library of al-Hakam II al-Mustansir and the culture of Islamic Spain. In: Manuscripts of the Middle East, 5, 99–105. URL: https://www.academia.edu/40309741/The_library_of_al_Hakam_II_al_Mustansir_and_the_culture_of_Islamic_Spain (3 March 2024).
- Wegener, Alfred. 1905a. Die astronomischen Werke Alfons X. In: Bibliotheca Mathematica, Zeitschrift für Geschichte der mathematischen Wissenschaften, III. Folge, 6, 129–185. Leipzig: Teubner.
- Wegener, Alfred. 1905b. Die Alfonsinischen Tafeln für den Gebrauch eines modernen Rechners. Dissertation. Berlin: Friedrich-Wilhelms-Universität zu Berlin. URL: https://www.digi-hub.de/viewer/!fulltext/BV041981724/45/ (28 January 2024).
- Wielen, Roland and Ute Wielen. 2017a. Alfred Wegener und das Astronomische Rechen-Institut. Mit einer Anwendung seiner umgerechneten Alfonsinischen Tafeln auf den „Astronomischen Kalender" für 1448. Heidelberg: Astronomisches Rechen-Institut Zentrum für Astronomie Universität Heidelberg. DOI 10.11588/heidok.00024001.
- Wielen, Roland and Ute Wielen. 2017b. Supplement zu „Alfred Wegener und das Astronomische Rechen-Institut". Heidelberg: Astronomisches Rechen-Institut Zentrum für Astronomie Universität Heidelberg. DOI 10.11588/heidok.00024002.
- Wreglesworth, J. Braulio, St. In: Encyclopedia.com [online source]. URL: https://www.encyclopedia.com/religion/encyclopedias-almanacs-transcripts-and-maps/braulio-st (23 April 2023).
- Zepeda, Henry. 2019. The First Latin Treatise on Ptolemy's Astronomy: The Almagesti minor (c. 1200). Ptolemaeus Arabus et Latinus, Volume I. Turnhout: Brepols Publishers. ISBN 978-2-503-58137-8.
- Zinner, Ernst. 1935. Die Tafeln von Toledo. In: Osiris, 1, 75–114. DOI 10.1086/368453.
- Zinner, Ernst. 1936. Petrus de Dacia, ein mittelalterlicher dänischer Astronom. In: Archeion, 18, 318–329.

Internet links

- Cahiers de l'Institut du Moyen-Âge Grec et Latin, University of Copenhagen. URL: https://cimagl.saxo.ku.dk/access/.
- Swetz, Frank J. (Pennsylvania State University), "Mathematical Treasure: The Alfonsine Tables", Convergence (July 2018). URL: https://www.maa.org/press/periodicals/convergence/mathematical-treasure-the-alfonsine-tables.
- Islamic Scientific Manuscripts Initiative. URL: https://ismi.mpiwg-berlin.mpg.de/biographies-list.

This page is intentionally left blank.

6 Astronomical methods before Copernicus

This book then is called a Ζῆζι (zīj) according to the Persians because, like a loom, the stars are woven on it and arranged in the weaving, and thanks to such weavings all sorts of forms are shown, according to the metaphor of the looms ...

Gregor Chioniades, Zīj al-'Alā'ī
(Mercier, 2000, From Tantra to Zīj, p.454
or Pingree, 1985, The Astronomical Works of Gregory Chioniades, I, p.37).

6.1 Byzantine astronomy

6.1.1 Transfer of knowledge to Constantinople

Despite the decline of Alexandrian mathematics that coincided with HYPATIA's death, PTOLEMY's knowledge of mathematics and astronomy was preserved until the Arabian conquest of Alexandria in 641. Alexandria was then under Byzantine rule which, in addition to the Neoplatonic school in Athens, also contributed to the spread of knowledge to Constantinople.[1] STEPHANUS OF ALEXANDRIA, who wrote the commentary on the *Handy Tables*, which is very likely based on THEON's work, spent many years in Byzantium at the court of HERACLIUS.[2] The knowledge of the Byzantine astronomers in the early period was based on Alexandrian mathematics and astronomy, especially that written in the *Almagest*. Later, Byzantine astronomers gained knowledge from abroad, particularly from Persia. A summary of important Byzantine contributions is given in Table 6.1.

Arabic astronomy has influenced Byzantine works since at least the 11th century.[3] Byzantine astronomy reached its peak in the 13th and 14th centuries; from this period, several astronomical works have been preserved, which also contain the information needed to calculate the equation of time. GREGOR CHIONIADES (around 1240–1320), who worked in the second half of the 13th century, studied astronomy and astrology with SHAMS AL-DĪN AL-BUKHĀRĪ at the Mongol court in Tabriz. He is the author of *Zīj al-'Alā'ī*,[4] and probably also the author of the translations of KHĀZINĪ's *al-Sanjarī zīj* and AL-ṬŪSĪ's *Īlkhānī zīj*. His translations into Greek later contributed to the spread of ancient knowledge to Italy.[5] AL-KHĀZINĪ, recorded as a slave of Greek or Byzantine origin, received an astronomical education after conversion and, in the first half of the 12th century worked at the court of the Seljuq ruler, SANJAR IBN MALIK-SHĀH (reigned 1118–1157), to whom he dedicated the *al-Sanjarī zīj*. AL-KHĀZINĪ was known for his work in mechanics, astronomy and instrumentation. He undoubtedly knew the works of his predecessors AL-BĪRŪNĪ, THĀBIT IBN QURRA and AL-BATTĀNĪ. It seems that their works also affected the *al-Sanjarī zīj*. He is also very likely to have performed or at least participated in astronomical observations, and apparently had good instruments available.

1 Lempire, 2014, Byzantine astronomy under the reign of Heraclius (610–641), Late Antique Astronomy in Syria: Heritage, Contacts and Particularities. TOPOI – Humboldt-Universität zu Berlin. Berlin, 11 April 2014. For a brief overview of Byzantine mathematics, see Wußing, 2009, 6000 Jahre Mathematik, pp.211–217.

2 Lempire, 2016, Le commentaire astronomique aux Tables Faciles de Ptolémée attribué a Stéphanos d'Alexandrie. As with many other ancient names, we know very little about STEPHANUS. HERACLIUS (about 575–641), Lat. FLAVIUS HERACLIUS AUGUSTUS, Gr. Φλάβιος Ἡράκλειος, FLAVIOS IRAKLIOS, introduced Greek as the official language in Byzantium.

3 Jones, 1987, An eleventh century manual of Arabo-Byzantine astronomy.

4 *Ala'i*, astronomy manual with tables. The tables in the second half are only available in Greek.

5 LEO THE MATHEMATICIAN, who worked at the beginning of the 9th century, has similar merits. He systematically copied the Ancient Greek works of PLATO, APOLLONIUS, EUCLID, ARHIMEDES and PTOLEMY, some of whose Greek texts may have been preserved thanks to him. Instead of Greek letters, LEO already used numerical symbols as numbers, created various machines, and also designed an optical telegraph system for the needs of the Empire.

Author	Era	Source or book
Codex Vaticanus Graecus[a]	1058	Folio 307v contains a table with the equation of time, obviously based on both Persian and Greek astronomic traditions
Gregor Chioniades[b]	second half of the 13th century	*Zīj al-'Alā'ī*
Theodore Metochites[c]	13/14th century	*Introduction to Astronomy*
George Lapithes(?)[d]	14th century	Translation of *Toledan Tables*
Georgius Chrysococces[e]	c.1347	*Persian Syntaxis*
Theodore Meliteniotes[f]	14th century	*Tribiblos Astronomique*
Georgius Gemistus, Plethon[g]	15th century	Astronomical handbook

Table 6.1 Notable Byzantine astronomers and their works.

Notes

a Neugebauer, 1960, Studies in Byzantine Astronomical Terminology, p.34.
b Pingree, 1985, The Astronomical Works of Gregory Chioniades, I, p.37, the tables in the second part are in Greek.
c It follows from this text that Ptolemy's and Theon's work was very well known and appreciated in Byzantium. Paschos, Simelidis, 2017, Introduction to Astronomy by Theodore Metochites.
d Pingree, 1976, The Byzantine Version of the "Toledan Tables": The Work of George Lapithes? In: Dumbarton Oaks Papers, Vol. 30, pp.85,87–132.
e *Persian Syntaxis* is based on a similar work of Gregory Chioniades or on *Zīj-i Īlkhānī* of al-Ṭūsī. It contains tables with the equation of time. Mercier, 1984, The Greek 'Persian Syntaxis' and the Zīj-I Īlkhānī, pp.43,45.
f Manuscript preserved, *Codex Vaticanus Graecus 792.* The procedure for calculating the true longitude is given in *Tribiblos* II.4. For comments, see Leurquin, 1993a, Théodore Méliténiote, Tribiblos Astronomique, Livre II Commentaire, pp.628–632. Time corrections are given in Tribiblos II.5. For comments, see Leurquin, 1993a, pp.633–647. Also, Leurquin, 1993b, Théodore Méliténiote, Tribiblos Astronomique, Livre II Édition Critique et Traduction, pp.37–48, 49–77.
g Tihon, Mercier, 1998, pp.188–191.

6.1.2 Rebirth of Greek astronomy

Among the greatest names of Byzantine astronomy and science is Theodore Metochites, who has substantial credit for the rebirth of Greek astronomy.[6] In addition to astronomy, he performed the highest state duties.[7] State duties were also performed by Theodore Meliteniotes, who operated several decades later.[8] His main work the *Tribiblos*, is based on Ptolemy's work and the work of his predecessors, and was written before 1352. In the second

6 Theodore Metochites (1270–1332), Gr. Θεόδωρος Μετοχίτης.
7 Bazzani, 2006, Theodore Metochites, a Byzantine Humanist. In: Byzantion, vol. 76, pp.32–52.
8 Theodore Meliteniotes (around 1320 – 1393), Gr. Θεόδωρος Μελιτηνιώτης.

book, similar to THEON's *Commentary*, he summarises PTOLEMY's work. Among the last great names in Byzantine astronomy was GEORGIUS GEMISTUS, later called PLETHON, who was also the author of an *Astronomical Manual*.[9] GEMISTUS was a teacher to BESSARION, who was born in Trapezund and later attained the position of Cardinal in Italy. When BESSARION worked in Italy, he had contact with intellectuals in Europe and Byzantium; he probably had considerable credit for transferring knowledge from Byzantium to Europe. However, the written treasures from the Constantinople libraries ultimately became known in Europe after the Turkish siege of Byzantium, which fell in 1453 with many scholars fleeing to Italy and taking with them invaluable antique works, some of which were previously unfamiliar in Europe.[10]

6.2 Jewish astronomers

6.2.1 Baraitha diShmuel

The *Baraitha diShmuel* is an ancient Hebrew astronomical work, thought to date from the Talmudic period, and the first four chapters were probably written in Babylon.[11] Its mathematical framework is clearly based on Babylonian arithmetic methods, and it also shows clear influence from Indian astronomy. The first four chapters are loosely attributed to the astronomer SHEMUEL (d. 254 CE), who was active during the reign of King SHAPUR I, and who is said to have known him personally. SHAPUR I supported the spreading of Greek and Indian astronomical knowledge in the SASANID kingdom, continuing the work already laid down by his father, King ARDASHIR I.[12] Chapter 3 of *Baraitha diShmuel* describes astronomical concepts such as oblique ascension, length of daylight, ortive amplitude, length of daylight as a function of seasons, shadow lengths, eclipses and lunar phases. The oblique ascension is the same as in the Babylonian system A (Figure 6.1).[13]

The shadow length references are based on the same equation as found, for example, in Indian literature, which assumes a gnomon 12 units long:[14]

9 GEORGIUS GEMISTUS (1355/1360–1452/1454), Gr. Γεώργιος Γεμιστός; PLETHON, Gr. Πλήθων.

10 Wilson, 1967, The Libraries of the Byzantine World.

11 Beller, 1988, p.61. Babylonian Talmudic period (around 200–500 CE).

12 Pingree, 1963, Astronomy and astrology in early India and Iran.

13 Beller, 1988, p.58. See also Neugebauer, 1969, The Exact Sciences in Antiquity, p.159.

14 Beller, 1988, pp.59–60. The gnomon of 12 aṅgulas was in common use in ancient India, where the term probably originates. Tables of shadow lengths corresponding to the equation are found in various ancient Indian sources. Beller, 1988, pp.59–60. *Arthashāstra* (*Arthaśāstra*) describes the annual variation in shadow length in II 20.41–42 and the diurnal variation in II 20.39–40. Kangle (trans.), 1960, The Kauṭilya Artha-śāstra, p.140. The *Arthashāstra* is a treatise on public administration, economic policy and warfare; it can be compared with the works of MACHIAVELLI in the Middle Ages, and is considered to be authored or part-authored by KAUTILYA, aka CHANAKYA, adviser to the ruler CHANDRAGUPTA (crowned in 321 BCE) and one of the founders of political science and economics in the modern sense. A very similar formula is also found in a work from the 3rd century, *Yavanajātaka*, by SPHUJIDHVAJA. Pingree, (ed.), 1978, The Yavanajātaka of Sphujidhvaja, vol. 2, p.189. Later also in VARĀHAMIHIRA, *Pañcasiddhāntikā* IV.48–49, written by VARĀHAMIHIRA in the 6th century. See Neugebauer, Pingree, 1971, II, p.46, Equation 5.

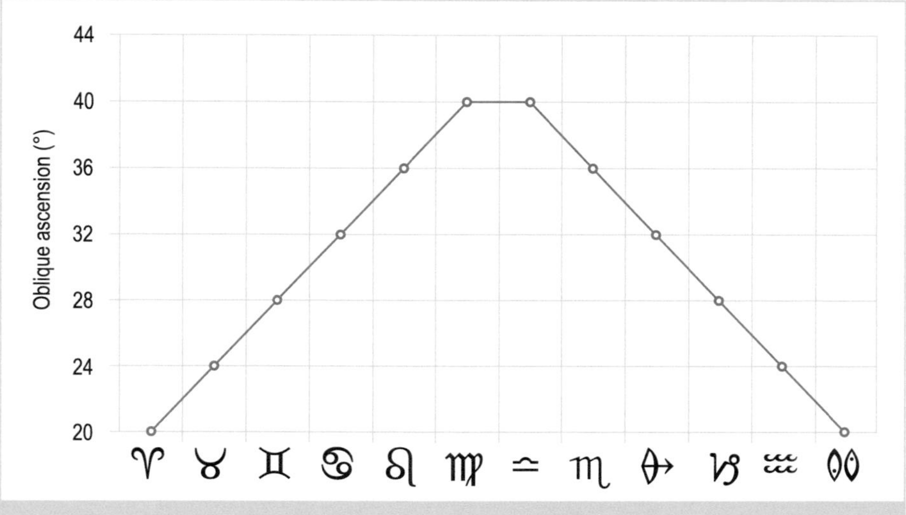

Figure 6.1 Oblique ascension from *Baraitha diShmuel* or the Babylonian system A. Taken from Beller, 1988, p.58, see also Neugebauer, 1969, p.159.

$$t = \frac{72}{s + 12 - s_0}$$

t	seasonal hours of the day from sunrise
s	shadow length of the gnomon
s_0	shadow length at midday

The *Baraitha diShmuel* supplied data for Jewish astronomers working much later, being found, for instance, in *Sefer ha-`ibbur* by ABRAHAM BAR ḤIYYA.[15] BAR ḤIYYA used the moment of creation in the Jewish calendar as the starting point or epoch. The initial moment or epoch of creation apparently dates from the original edition of the *Baraitha diShmuel* and was later replaced, probably in the 8th century, by the Jewish calendar year 4536 or 776 CE.[16] IBN EZRA, in writing the procedure for calculating the beginning of the new moon, or molad in Hebrew, also cited the moment of creation as the starting point, which probably also comes from *Baraitha diShmuel.*[17]

6.2.2 Abraham bar Ḥiyya

ABRAHAM BAR ḤIYYA (1065?–1135?) or SAVASORDA, also known as ABRAHAM JUDAEUS, was the first to systematically compile astronomical tables in Hebrew.[18] The tables' epoch is 21 September 1104 at midday.[19] The tables were based on AL-BATTĀNĪ's work and on the meridian for Raqqa, which BAR ḤIYYA, perhaps unaware that it was Raqqa, mistook as the meridian

15 Filipowski (ed.), 1851, Sefer ha-Ibbur (Book of Intercalation), R. Abraham b. Hiyya ha-Nasi.

16 The starting point of the Jewish calendar, 1 Tishrei Anno Mundi 1, corresponds to Monday 7 October 3761 BCE according to the proleptic Julian calendar. Dershowitz, Reingold, 2008, Calendrical Calculations, p.15, Table 1.2.

17 Beller, 1988, pp.55–56. The process of determining the new Moon is very important because of the Jewish lunar calendar.

18 Millás Vallicrosa, 1959, La obra Séfer Hešbón mahlekot ha-kokabim de R. Abraham Bar Hiyya ha-Bargeloni.

19 Epoch 21 September 1104 = 29 Elul 4865 (Hebrew) = 29 Dhū'l-Ḥijja 497 (Arabic) or *JDN* 2124548. Mercier, 2014, Astronomical Tables of Abraham Bar Hiyya, p.165. Millás Vallicrosa, 1959, p.54.

through Jerusalem.[20] He also began to use a new starting point, replacing the Hijra epoch used by AL-BATTĀNĪ with the moment of creation according to Jewish tradition, 6 October 3761 BCE (0 Tishrei, Anno Mundi 1), and adapting the Tables to the Jewish calendar.[21] In his *La obra Séfer Hešbón mahlekot ha-kokabim*, BAR ḤIYYA also gives a table of the difference between the lengths of days and nights, or the equation of time.[22] His most important work is a work on practical geometry, *Ḥibbur ha-Meshiḥah we ha-Tishboret*. PLATO OF TIVOLI translated the work into Latin as the *Liber embadorum* in 1145.[23] It is the first European work to provide a complete solution of the quadratic equation $x^2 - ax + b = 0$. In the same year, ROBERT OF CHESTER translated the *Algebra* of AL-KHWĀRIZMĪ, so this year can be seen as the arrival of algebra in medieval Europe. SAVASORDA knew the works of EUCLID, THEODOSIUS OF BITHYNIA, MENELAUS, AUTOLYCUS, APOLLONIUS OF PERGA, EUDEMUS OF RHODES[24] and HERON OF ALEXANDRIA. He was also well-acquainted with the works of AL-KHWĀRIZMĪ and AL-KARAJĪ[25]. He has also helped translate several works into Latin. He may have collaborated with RUDOLF OF BRUGES on *De astrolabia*. His work had a considerable influence on LEONARDO FIBONACCI.

6.2.3 Abraham ibn Ezra

ABRAHAM IBN EZRA (c.1089–1167), was a Jewish philosopher, astronomer, astrologer and physician, and wrote on mathematics and astronomy. He was the first to succeed in distinguishing clearly between the Hindu and Ptolemaic astronomic terms. He was also a translator from Arabic into Hebrew, a poet and a Hebrew grammarian. He travelled constantly from 1140 to 1160, and did not write his most important works until late in life. His work shows influence from Neoplatonism. Three of his papers are on numbers, number theory and basic arithmetic operations. He used zero in his works, which shows the strong influence of Indian knowledge. He translated AL-BĪRŪNĪ's commentaries on AL-KHWĀRIZMĪ's tables (the Arabic original is lost), and also treated permutations and combinations. His treatises on the calendar, the *Book on Intercalation* and the astrolabe are also well known. He also wrote many astrological treatises. IBN EZRA thus contributed greatly to the transmission of knowledge, and his few modern translations remain of great philological or historical importance.

6.2.4 Levi ben Gerson

In addition to the astronomers mentioned above, Jewish astronomers have also contributed significantly to the transfer of Arabic astronomical knowledge to Europe. Among the most important Jewish astronomers are LEVI BEN GERSON and ABRAHAM ZACUT[26], who benefitted

20 Goldstein, Chabas, 2017, p.73.

21 Goldstein, Chabas, 2017, p.73. Hijra epoch, July 15, 622 used by astronomers, July 16, 622 used by historians. Neugebauer, 1975, p.1066, note 4. See also Neugebauer, 1962, The Astronomical Tables of Al-Khwarizmi, pp.10–11, Fig. 1.

22 Millás Vallicrosa, 1959, p.111.

23 PLATO OF TIVOLI (12th century), reputedly Italian by birth and known for his works. He wrote several works in Barcelona between 1132 and 1146.

24 EUDEMUS OF RHODES, a peripatetic philosopher, 4th century BCE.

25 AL-KARAJĪ (953–c.1029), an Arab mathematician based in Baghdad, one of the founders of modern algebra.

26 ABRAHAM ZACUTO, romanised ZACUT.

from a wealth of astronomical literature in Hebrew, the work of their predecessors.[27] These resources were more important to their work than Arab sources, to which they may not have had easy access.[28] Their work also includes tables containing the equation of time. LEVI BEN GERSON (1288–1344), Lat. GERSONIDES, also known by his Hebrew acronym RALBAG, was born in Provence, where he probably spent most of his life. Today he is better known more for his philosophical writings than his work as an astronomer.[29] Few details are known about his life, but his achievements rank him among the most important astronomers of the 14th century, although he was almost unknown outside the Jewish community in his time. He criticised PTOLEMY and contributed to the improvement of the lunar movement model, and is considered to be the inventor of Jacob's Staff; he also worked with astrolabes and camera obscura. His works are written in Hebrew, and it is not apparent that he knew any other languages; he most likely drew on the extensive astronomical literature that was available in Hebrew during his life. He based his astronomy on the achievements of Islamic and Jewish astronomers[30]; also important is his contribution to the development of mathematics,[31] where he was ahead of his time. He contributed pioneering treatises on combinatorics, worked on trigonometry and commented on EUCLID. His greatest mathematical work, *Maaseh hoshev*, "*Art of Calculation*", has been preserved in editions from 1321 and 1322. The trigonometric treatise *De sinibus, chordis et arcubus*, "*On Sines, Chords, and Arcs*", was completed in 1342, and the theoretical musical work *De numeris harmonicis*, of which the Latin translation has been preserved, was completed in 1343, shortly before his death.

LEVI BEN GERSON did not use the same epoch as PTOLEMY when calculating the equation of time, but chose a longitude of ♒ 20° for the smallest value of the equation, which was basically very similar to the contemporary definition of the equation:[32]

[6.1] $$E = \alpha_0 - \alpha_\odot - (\lambda_{M0} - \lambda_M) = \alpha_0 - \lambda_{M0} - (\alpha_\odot - \lambda_M) = E_0 - (\alpha_\odot - \lambda_M)$$

$\alpha_\odot$	right ascension
α_0	right ascension for reference moment
λ_M	mean longitude
λ_{M0}	mean longitude for reference moment
E_0	equation of time for reference moment ♒ 20°, $E_0 = 4;2°$.

27 ABRAHAM IBN EZRA (12th century), MOSES MAIMONIDES (12th century), JACOB BEN TIBBON (13th century), ISAAC BEN SID (13th century), ISAAC ISRAELI (14th century), JACOB BEN DAVID BONJORN (14th century), IMMANUEL BEN JACOB BONFILS (14th century), ISAAC AL-ḤADIB (14th century). ZACUT also cites other astronomers: JUDAH BEN ASHER II (14th century), ḤAYYIM OF BRIVIESCA (14th century) and JUDAH BEN VERGA (15th century).

28 For the Hebrew edition of IBN AL-MUTHANNĀ's *Commentary* on the Astronomical Tables of AL-KHWĀRIZMĪ see Goldstein, 1967.

29 For more information about his work, see Goldstein, 1985, The Astronomy of Levi ben Gerson (1288–1344): A Critical Edition of Chapters 1–20 with Translation and Commentary. Also Kellner, 1979, R. Levi Ben Gerson: A Bibliographical Essay. Studies in Bibliography and Booklore, Vol. 12, (1979), pp.13–23.

30 AL-BATTĀNĪ, BAR ḤIYYA, IBN EZRA, AVERROËS or IBN RUSHD (1126–1198).

31 For more information about his work on philosophy, astronomy and mathematics, see Freudenthal, 1992, Studies on Gersonides, A Fourteenth-Century Jewish Philosopher-Scientist. For a brief summary of his contributions to mathematics, see also Carlebach, 1910, Lewi ben Gerson als Mathematiker.

32 Goldstein, 1974, The Astronomical Tables of Levi ben Gerson, pp.102–103.

The argument of LEVI BEN GERSON' equation of time of is the true longitude, and the results are in time-degrees. The parameters for calculating the value of the equation of time are: λ_{Γ} = ♋ 3°, $e = 2;14$, $\varepsilon = 23;33°$, and the value of the equation of time at ♒ 20° is $E_0 = 4;2°$.

6.2.5 Immanuel ben Jacob Bonfils

BONFILS, about whose life very little is known, was active in Tarascon, southern France, in the 14th century.[33] The astronomical tradition of the Jewish community in southern France was based on translations of Arabic works into Hebrew, and BONFILS' astronomical tables are based partly on the Julian calendar and partly on the Jewish calendar. BONFILS' work is clearly based on the work of ABRAHAM BAR ḤIYYA, or at least partly on that of AL-BATTĀNĪ, but the values of the equation of time, which is preserved in several manuscripts, differ from those of AL-BATTĀNĪ.

Three of his astronomical works are known:

- *Six Wings* comprises six tables that have the name *Wing*. The *Wing 3* table contains data on the daylight hours and solar eclipses.[34]
- Astronomical Tables from 1340, gives values partly for the Jewish calendar and partly for the Julian calendar. Several manuscripts have been preserved that also contain the equation of time, partly in minutes and seconds, partly in time-degrees. The values are based on the work of ABRAHAM BAR ḤIYYA.[35]
- *A treatise*, an astronomical work, in which the equation of time is discussed at the beginning.[36]

The BONFILS' tables of 1340 also contain the lengths of the seasonal hours for latitude 44°, arguing the longitude of the Sun for the individual signs of the zodiacal cycle. According to the values in the seasonal hours tables, and according to the author, the values in these tables were obviously derived from LEVI BEN GERSON's tables.[37]

6.2.6 Jacob Bonjorn

JACOB BONJORN's tables were based on Ptolemaic models, and the values were calculated using LEVI BEN GERSON's astronomical parameters, making BONJORN one of BEN GERSON's most important "disciples".[38] His father, DAVID BONJORN DEL BARRI, was an astronomer in Perpignan in southern France. BONJORN was part of the Jewish astronomical school in southern France and also contributed to the astronomical activities associated with the ruler PERE EL CERIMONIÓS

33 For more information on his life and work, see Steinschneider, 1896, Mathematik bei den Juden, pp.38–42, and Sarton, 1927–48, Introduction to the History of Science, III, pp.1517–1520.
34 Solon, 1971, The Six Wings of Immanuel Bonfils and Michael Chrysokokkes.
35 Goldstein, Chabas, 2017, p.77.
36 Solon, 1971, p.10, and Steinschneider, 1964, p.160.
37 Goldstein, Chabas, 2017, pp.97–98.
38 BONJORN is also known as PO'EL, BONA DIEI, EN BONET, YOMTOB... Chabás, 1991, p.279.

– according to sources, he entered royal service in 1356. BONJORN compiled the *Tables*, which were widely used, in around 1361. The tables were also used by ABRAHAM ZACUT, as can be seen in the canons of the *Almanac perpetuum*.[39]

6.2.7 Abracam Zacut

ABRAHAM ZACUT (1452–1515?), a Jewish astronomer, was probably born in Salamanca, where he was a student of ISAAC ABOAB.[40] Initially, he worked in Salamanca under the patronage of Bishop GONZALEZ DE VIVERA, but it is not clear whether he lectured at Salamanca University. In 1493 he was in Lisbon, working for King JUAN II. Some sources claim that he advised navigators at sea, but these claims cannot be verified and naval navigation is not mentioned in any of ZACUT's extant works. However, his *Almanach perpetuum* did have a significant impact on the development of Spanish and Portuguese naval navigation at the end of the 15th century.[41] Due to the persecution of the Jews, he went to Tunisia where he worked for some time in Carthage, and later in the Ottoman Empire. His works evidence only limited access to Arabic astronomical sources, but knew the works of Jewish astronomers well. ZACUT published a table with the equation of time in the *Ḥibbur* tables, which is his earliest work, in *Tabule verificate for Salamanca*, and in *Almanach perpetuum*.[42]

6.2.8 Raphael Levi of Hannover

In the 18th century, LEIBNIZ's pupil and secretary, the mathematician RAPHAEL LEVI OF HANNOVER (1685–1779), made a major contribution to the understanding of Jewish astronomical literature. LEVI was the first to approach MAIMONIDES' treatises by means of mathematics, in particular by means of spherical trigonometry. He is the author of two books, the first in 1756 and the second in 1757, and of several Hebrew manuscripts that have survived but were never published. He, too, has probably been unjustly forgotten for this very reason.[43] As far as we know, he was the first proponent of the Copernican doctrine among Jewish astronomers. was largely self-taught, and later studied under LEIBNIZ, as Jews were not allowed to study at German universities in the 17th century.[44] Most of the Jewish mathematical and astronomical literature of this period is concerned with solving practical problems, such as practical examples of arithmetics, calendaring and the like.

39 Cantera Burgos, 1935, Abraham Zacut, p.105.

40 For more information about his work see Chabás, Goldstein, 2000, Astronomy and the Iberian Peninsula: Abraham Zacut and the Transition from Manuscript to Print. Transactions of the American Philosophical Society. For information on the life of ZACUT, see Cantera, Burgos, 1931, Notas para la historia de la astronomía en la España medieval: El judio salmantino Abraham Zacut. Revista de la Academia de Ciencias Exactas, Físicas y Naturales de Madrid 27: 63–98 and 1935 (reprint 1960), Abraham Zacut. Biblioteca de la cultura española, Siglo XV. M. Aguilar, Madrid.

41 For ZACUT's contributions and the importance of his work for the further development of navigation, see Wagner, 1918, Die Entwicklung der wissenschaftlichen Nautik im Beginn des Zeitalters der Entdeckungen nach neuern Anschauungen, Annalen der Hydrographie und maritimen Meteorologie, 46.

42 Chabás, Goldstein, 2000, pp.29, 56, 108. Zacuto, 1496, Almanach Perpetuum, p.42. Latin edition of *Almanach Perpetuum* was first published in 1496, it was based on a Hebrew text compiled in 1478. Goldstein, Chabás, 2018.

43 For more information on MAIMONIDES' astronomy, see Neugebauer, 1949 and Baneth, 1898–1903.

44 The study was allowed at Italian universities.

6.3 Astronomy in Vienna in the 14th and 15th centuries

6.3.1 John of Gmunden

John of Gmunden (around 1380/1384–1442), Lat. Johannes de Gamundia, German Johannes von Gmunden, from Gmunden am Traunsee, was an Austrian mathematician, astronomer and instrument maker. He studied in Vienna, where he also lectured from 1406. He is considered one of the founders of the Vienna School of Astronomy alongside Heinrich von Langenstein[45], Georg Müstinger[46], the head of the Klosterneuburg monastery, and the Czech astronomer Johannes Andree Schindel[47], who also lectured for a time in Vienna. John of Gmunden initially lectured on non-mathematical subjects, including the works of Petrus Hispanus[48], and later on mathematical works.[49] He was known for making models of instruments for lectures, and students also learned the basics of using the astrolabe and the equatorium from him. His works, mostly preserved as unedited manuscripts, show a good knowledge of other authors' works. He wrote the treatise *Algorismus de minutiis*, which is clearly based on the work of John of Lignères, and a treatise on sines, arcs and chords, *De sinibus, chordis et arcubus*,[50] which is also based on the work of predecessors including Levi ben Gerson, John of Murs, John of Lignères and Dominic de Clavasio[51]. His works also show his knowledge of Campanus' equatorium, ibn Tibbon's quadrant, Richard of Wallingford's equatorium (which Wallingford called the *albion*)[52], and Franco de Polonio's torquetum.[53] John's work on the astrolabe is based on a treatise on the astrolabe by Christian of Prachatice, the

45 Lat. Henricus de Hassia Germanus. He studied at the Sorbonne, where he later lectured and served as Vice-Rector. He was forced to leave Paris because of religious differences and from 1384, he taught theology and astronomy at the University of Vienna. He was also the Rector of the University of Vienna for one year.

46 Klug, 1943, pp.23–24.

47 Also Jan Ondřejův or Jan Šindel (c.1370–1455/1457?), Lat. Johannes Pragensis, Czech astronomer and theologian. He also worked as a doctor. We know that he was also in contact with Bianchini. Mikuláš of Kadaně (1350–1419), an imperial clockmaker, constructed the famous Prague Astronomical Clock based on his suggestions and calculations. The Prague Astronomical Clock, the first of its kind, is still in operation. For his work in Nuremberg see Pilz, 1977, 600 Jahre Astronomie in Nürnberg, pp.47-48.

48 The identity of Petrus Hispanus (13th century) is not entirely clear. His *Summule logicales magistri Petri Hispani* (*Collection of logic matters of master Peter of Spain*) made an important contribution to the development of logic. His work was still taught in European universities in the 17th century. For more data, see Spruyt, 2005, and Spruyt, 2019.

49 Klug, 1943, p.18.

50 Busard, 1971, For a summary, see also Folkerts, 2006, Die Beiträge von Johannes von Gmunden zur Trigonometrie, Simek, Chlench (eds.), pp.71–89.

51 Dominicus de Clavasio (14th century), mathematician, physician and astrologer. Born near Turin, he later worked and taught in Paris.

52 For some comments on Wallingford's *albion* see North, 1976, II, pp.127–248 or pp.249–284.

53 The torquetum was a sophisticated astronomical instrument, an analogue calculator for converting horizontal, equatorial and ecliptic coordinates. Franco de Polonia (13th century, probably died in 1285) studied in Paris and was a canon of Wroclaw Cathedral. He also wrote a first known treatise on the torquetum. We know that he produced a reworked version of the *Toledan Tables*, but it is lost. Ożóg, 2009, p.64. For more information, see Birkenmajer, 1972, pp.444–448 (in French).

first printed work of its kind in medieval Europe.[54] Based on his lectures, CHRISTIAN wrote the *Composition*, Lat. incipit *Quamvis de astrolabii composicione tam modernorum quam veterum dicta habentur pulcherrima*, and *Use of the Astrolabe*, Lat. incipit *Quia plurimi ob nimiam quandoque accurtacionem* in 1407.[55] JOHN apparently brought the cylindrical sundial to Vienna. His legacy also includes calculated shadow curves for Vienna, Venice, Rome, Nuremberg, Prague and Klosterneuburg. His tables are quite similar to those of CAMPANUS OF NOVARA.[56] The equation of time is published in the manuscripts of his tables, the argument of the equation of time is the ecliptic longitude of the Sun, and the values are in time-degrees.[57] The equation of time is almost identical to that of the *Alfonsine Tables*.[58] The only difference is in a few values, which are probably transcription errors.[59]

More guidance, relevant to practical astronomy, is given in Part 2 of the *Sphera solida* treatise. Determining the altitude of the Sun, or the zenith distance at noon, is described in chapter 7, and determining the ortive amplitude in chapter 8.[60] Chapter 10 describes the length of the day and night, and chapter 11 the determination of the time according to the position of the Sun.[61] His important contributions to the development of astronomy were later lost in the shadows of the works of PEUERBACH and REGIOMONTANUS.[62]

6.3.2 Georg Peuerbach

We know nothing about the early years of PEUERBACH's life, but we do know that he was admitted to the University of Vienna in 1446. It is not known with whom he studied, as JOHN OF GMUNDEN, the last known astronomer of importance in Vienna before PEUERBACH, died in 1442. PEUERBACH (1423–1461) may, however, have had access to JOHN's written works and instruments. After his studies, he lectured at universities in Germany, France and Italy. We know

54 Hadravová, Hadrava, 2006, Johannes von Gmunden und seine Version des Astrolabtraktats des Christian von Prachatitz, Simek, Chlench (eds.), pp.151–159. CHRISTIAN OF PRACHATICE also KŘIŠT'AN Z PRACHATIC worked at the University of Prague. As a young man, he was friends with JAN HUS, later a religious reformer. He left written works on medicine and medicinal herbs and mathematics. His mathematical treatises are based, among other things, on the work of JOHANNES DE SACROBOSCO.

55 Hadravová, Hadrava, Prachatiz, Christannus de. Encyclopedia.com.

56 Benjamin, 1954.

57 One of the manuscripts is held by the Nuremberg City Library, MS Cent. VI, 18, ff.46v–47r. Neske, 1997, Die Handschriften der Stadtbibliothek Nürnberg, Band 4, p.173.

58 Pedersen, 2002b, The Toledan Tables, p.985, Table BB. The authorship of the table is attributed to PETER OF ST. OMER and JOHN OF LIGNÈRES respectively.

59 The differences are in the longitudes Aquarius 8°, Aries 8°, Virgo 12°, and Sagittarius 3°. However, for Leo longitudes of 11°–30°, the values in the table are clearly shifted.

60 Chlench, 2007, Johannes von Gmunden deutsch, Der Wiener Codex 3055, pp.123 and 163–164 and pp.123–125 and 164–165 respectively.

61 Chlench, 2007, pp.125–129 and 166–169.

62 For his works see also Chlench, 2007.

that he knew NICOLAS OF CUSA[63], and towards the end of his life, he had a fruitful collaboration with REGIOMONTANUS, who was also his student. At the suggestion of BESSARION, who visited Vienna in 1460, he began to translate the *Almagest*, but only managed to translate six books before his death. The work was completed by REGIOMONTANUS and dedicated to BESSARION.[64] PEUERBACH's most important work is the *Tabulae eclipsium* of 1459, which includes a table with the equation of time.[65] The work is based on the *Alfonsine Tables*. The parameters for calculating PEUERBACH's values are:[66]

$\varepsilon = 23;33;44°$
$e = 2;16;42$
$\lambda_\Gamma = 91;2;4°$
$E_0 = 4;5;36°$

6.3.3 Regiomontanus

REGIOMONTANUS refers to the equation of time in book III, chapter XXX of *Epytoma Joannis De monte regio In almagestum ptolemei.*[67] Despite his short life, JOHANNES MÜLLER or REGIOMONTANUS (1436–1476) was the best-known mathematician and astronomer of his time. At the age of 12, he began studying at the University of Leipzig and later in Vienna, where he completed his studies at the age of 16. In Vienna, he met PEUERBACH, and his work was influenced by BIANCHINI, as REGIOMONTANUS studied his works. Also preserved is his rich correspondence with BIANCHINI.[68] At the Cardinal's request, he accompanied BESSARION to Italy where he studied Greek manuscripts, especially the *Almagest*, a copy of which BESSARION owned. In 1464 he lectured in Padua. He was not satisfied with the accuracy of the trigonometric tables in foreign sources, so he greatly improved the sine tables.[69] He later worked in Hungary, Nuremberg and Rome.[70] He died in Italy, most likely of the plague.

63 NICHOLAS OF CUSA (1401–1464), Ger. NICOLAUS CUSANUS, NIKOLAUS CRYFFTZ or KREBS, was one of the most important German thinkers of the 15th century. He was also a church reformer, administrator, canon law expert and cardinal. Among other things, he was a legal adviser at the Council of Basel in 1431, Bishop of Brixen and adviser to the papal curia.

64 The work is also based on the Latin adaptation of the *Almagest*, *Almagesti minor*, see Zepeda, 2019, pp.109–114.

65 Peuerbach, 1514, Tabulae eclypsium magistri Georgij Peurbachij, Table *Tabula equationis dierum*. See also Pietroni, 2022, p.153, Figure 57.

66 Pietroni, 2022, p.152.

67 Reprint of Edition from Venice, 1496. Regiomontani, 1972, Opera Collectanea, pp.114–115. For further information, see Zinner, 1990, Regiomontanus: His Life and Work, pp.41–42.

68 Curtze, 1902a, Urkunden zur Geschichte der Mathematik im Mittelalter und der Renaissance, 1, 2.

69 Glowatzki, Göttsche, 1990, Die Tafeln des Regiomontanus ein Jahrhundertwerk.

70 For his work in Nuremberg see Pilz, 1977, pp.58–93, 109–110.

6.4 Giovanni Bianchini and his tables

Very little is known about BIANCHINI's (1410–after 1469) life. He was known for his highly reliable astronomical calculations. In 1427, he started working in the Ferrara accounting chamber, heading it 1430 onwards. His important works are *Tabulae astronomicae* and *Tabulae magistrales*, a collection of tables for solving problems in spherical trigonometry. The *Tabulae primi mobilis*, completed by BIANCHINI in around 1452, are preserved in several manuscripts.[71] They contain 100 pages of tables and 50 pages of canons. The auxiliary tables, which BIANCHINI apparently compiled as a computational aid for use with the main tables, are known as *Tabulae magistrales* and are the oldest known tool tables for solving computational problems in astronomy.[72] The tables were circulated chiefly as manuscripts, limiting their reach and hence also BIANCHINI's historical prominence. The tables are also the oldest known to be in decimal, rather than sexagesimal, form. They are given for two radii: $R = 60{,}000$ and $R = 10{,}000$.[73] Individual manuscripts of the *Tabulae magistrales* contain, depending on the manuscript, six or eight tables of trigonometric functions. From PEUERBACH's letter to JOHANN NIHIL in 1456, we know that the tables were used by both PEUERBACH and REGIOMONTANUS.[74] In 1460, REGIOMONTANUS published a copy of BIANCHINI's *Tables* and supplemented them with additional content under the title *Canones breviati in tabulas Ioannis de Blanchinis*.

BIANCHINI knew and built on the knowledge of his predecessors. For example, he took the obliquity of the ecliptic as $\varepsilon = 23;30;33°$, the same as in the *Toledan Tables*, though other values differ from the *Toledan Tables*. He also did not use only one fixed epoch.[75] BIANCHINI used the names sinus primus and sinus secundus for the trigonometric functions sine and cosine, respectively. The tables tabulate, for a radius $R = 10{,}000$, the values of various mathematical expressions involving the trigonometric functions sine and cosine. For most of the parameters in the tables, values are given for the argument ζ from 0;10° to 90;0° in intervals of 0;10°.[76] The tabulated values (*Tabula magistralis* 1–8) correspond to the modern mathematical expressions as given in Equations 6.2 to 6.8.[77]

71 Rosińska, 1987, Tables of Decimal Trigonometric Functions from ca. 1450 to ca. 1550, p.420. BIANCHINI's written works were written between 1440 and 1464.

72 We know that BIANCHINI asked JOHANNES SCHINDEL, REGIOMONTANUS and PAOLO TOSCANELLI for their opinion on the tables. Rosińska, 1987, p.421.

73 Chabás, 2016, An analysis of the Tabulae magistrales by Giovanni Bianchini. Rosińska, 1981, Tables trigonometriques de Giovanni Bianchini. Radii of 60,000 and 10,000 correspond to $R = 60$ and $R = 1$ respectively.

74 Chabás, 2016, An analysis of the Tabulae magistrales by Giovanni Bianchini, p.544, note 2. JOHANN NIHIL was court astrologer to FREDERICK III OF HABSBURG, and it was probably this knowledge, in addition to his mathematical and astronomical skills, that opened the door for PEUERBACH to the court of LADISLAUS V, FREDERICK's nephew. PEUERBACH later also served FREDERICK III.

75 Chabás, 2016, p.545. Chabás, Goldstein, 2009, p.8.

76 Decimal tables, most of which did not live up to the reputation and quality of BIANCHINI's, were later also compiled by APIAN, COPERNICUS and RHETICUS. See Rosińska, 1987.

77 Chabás, 2016, pp.545–550.

[6.2] $$T_1(\zeta) = R\frac{\cos\varepsilon}{\sin\zeta} \quad \text{or} \quad T_1(\zeta) = R\cos\varepsilon\,\mathrm{cosec}\,\zeta$$ *Tabula magistralis prima*

$R = 10{,}000$, ε obliquity of ecliptic, Bianchini's value 23;33,30°

The second table gives the sine of ascensional difference θ, see Equation 0.15. BIANCHINI did the calculations for Ferarra, taking latitude $\varphi = 45°$, hence $\tan\varphi = 1$. In this case:

$$\sin\theta = \tan\delta$$

When $\zeta = 90°$, $\sin\zeta = 1$ and $\delta = \varepsilon$ or $\sin\delta = \tan\varepsilon$. It follows that, for T_2 as a function of longitude λ, where $R = 60{,}000$:

[6.3] $$T_2(\lambda) = R\tan\delta$$ *Tabula magistralis secunda*

[6.4] $$T_3(\zeta) = R\frac{\sin\varepsilon}{\cos\varepsilon}\cos\alpha \quad \text{or} \quad T_3(\lambda) = R\tan\varepsilon\cos\alpha$$ *Tabula magistralis tercia*

Right ascension α can be calculated using Equation D.37, and $\tan\delta$ or the tangent of declination is given by Equation 6.3 (*Tabula magistralis secunda*).

Tabula magistralis quarta gives the tangent function for radius $R = 10{,}000$

[6.5] $$T_4(\zeta) = R\frac{\sin\zeta}{\cos\zeta} = R\tan\zeta$$

Tabula magistralis quinta gives the cosecant function for radius $R = 10{,}000$

[6.6] $$T_5(\zeta) = \frac{R}{\sin\zeta} = R\,\mathrm{cosec}\,\zeta$$

Tabula magistralis sexta contains values from the *Tabula magistralis quarta* multiplied by constant $\tan\varepsilon$, where $\varepsilon = 23;33,30$.

[6.7] $$T_6(\zeta) = R\tan\zeta\tan\varepsilon$$

For the argument of right ascension *Tabula magistralis septima* contains oblique ascension (Equation 0.14) and the semidiurnal arc. The semidiurnal arc is the arc described by a celestial body between the rising and passing through the meridian, or between the passing through the meridian and setting. *Tabula magistralis octava* gives the inverse values of the table T_1, it applies $T_1 \times T_8 = 10^8$.

[6.8] $$T_8(\zeta) = R\frac{\sin\zeta}{\cos\varepsilon} \quad \text{or} \quad T_8(\zeta) = R\sin\zeta\sec\varepsilon$$

Bianchini's tables, which clearly have roots in the *Toledan Tables*, give the equation of time in time units – minutes and seconds – which was unusual for this period; the usual practice was to use time-degrees.[78] Bianchini's equation of time from the 1526 edition, *Tabula aequationis dierum* is given for the ecliptic longitude of the Sun in 1° increments; the maximum converted to time-degrees is 7;54°, which is also the maximum in the *Toledan Tables*.[79] Similar is the equation of time from the Vatican Library manuscript, MS Pal. Lat. 1375 folio 258v, *Tabula horarum meridiei et equationis dierum ad meridianum ferarrie et bononie*. Another, different, equation of time by Bianchini, can also be found in the manuscript Pal. Lat. 1375, folio 90r, *Tabula equanoms dierum*. Again, the argument is the Sun's ecliptic longitude in 1° increments.

78 Chabás, Goldstein, 2009, The Astronomical Tables of Giovanni Bianchini.
79 Pietroni, 2022, pp.136–137, see also Chabás, Goldstein, 2003, p.186.

6.5 German astronomers and ephemeris makers

6.5.1 Bernhard Walther

BERNHARD WALTHER (1430–1504), unjustly almost forgotten, was a pupil of REGIOMONTANUS and a leading German astronomer.[80] He was known for his precise observations and his data were later used by COPERNICUS, BRAHE and KEPLER, among other astronomers.[81] WALTHER was the first astronomer known to have used a clock for indicating the time of astronomical observations, and his work also marks the beginning of systematic astronomical observations in medieval Europe.[82] BERNHARD WALTHER's clock based measurements began as early as 1484, eight years after REGIOMONTANUS' death. On 16 January 1484, he wrote:[83]

> *16 Ianuarij, obſeruaui Mercurium horologio bene correcto quod uidelicet de meridie ad meridiem diei præcedentis ad unguem redijt. Vidi uero ☿ de mane in contactu horizontis et in eodem inſtanti appedi pondus horologio habenti 56, den ticulos in rota horaria, quæ ſemel reuertebatur et inſuper 35 de ticuli quo, cetru Solis in horizote apparuit, unde ſequitur Mercuriu eo die una hora et 37 mi. ante Solem ortum ſuiſſe, quod ſere concordat calculo.*

English translation:[84]

> *... I observed Mercury with a well regulated clock, which gave the time precisely from noon the day before to noon. I saw Mercury in the morning in contact with the horizon, and simultaneously I put a weight on the clock, which had 56 teeth in the hour wheel. It made one complete turn and 35 teeth more, by the time the centre of the Sun appeared on the horizon, whence it follows that Mercury on that day rose one hour 37 minutes before the Sun, which almost agrees with the calculation.*

The clock was thus weight driven and of somewhat unusual construction, apparently designed to measure intervals of time rather absolute than time – its large 56-toothed gear turned a full revolution in one hour (Lat. *rota horaria*). Thus, for example, one revolution plus the 35 teeth more (93 teeth in total) corresponds to a time interval of 1 hour and 37 minutes. WALTHER must have started the clock at the beginning of the observation by hanging a weight.[85]

80 For more information on his life and work, see Pilz, 1977, pp.93–100 or Kremer, 2010, War Bernhard Walther, Nürnberger astronomischer Beobachter des 15. Jahrhunderts, auch ein Theoretiker?

81 Kremer, 1981.

82 Beaver, 1971, Bernard Walther: Innovator in Astronomical Observation, p.40.

83 Schönner, 1544, Scripta clarissimi mathematici, 50r-v.

84 Beaver, 1970, p.41.

85 Repsold, 1908, pp.15–16. For the original Latin text, see Snell, 1618, Coeli et siderum ..., p.32.

Doppelmayr notes, for example, that Peter Hele and Andreas Henlein, who were active around 1500, were competent clockmakers, so it is possible that Nuremberg was already familiar with conventional time-keeping clocks at that time.[86] When he was based in Nuremberg, Regiomontanus also published a calendar, which was well circulated and included tables with the length of the day.[87] The length of the day was particularly important in Nuremberg because of the famous Nuremberg Astronomical Clock, which had been running since at least 1374 and was in use for 400 years.[88] The clock was adjusted several times a year to keep synchronise it with the the Sun. In 1488, a new rule for setting the clock was introduced, probably mainly because of Walther.[89] Interestingly, Walther's pupil Conrad Heinfogel (d. 1517) used Stöffler's and Pflaum's *Almanac for the years 1499 to 1531* as a kind of diary, where he apparently also wrote down a table with the equation of time, which suggests that Walther was obviously well acquainted with it. The *Almanac* also states that Walther recalculated and corrected the Nuremberg Clock on 2 August 1488, which certainly required the values of the equation of time, in other words the local clock time at solar noon.

6.5.2 Johannes Stöffler and Jakob Pflaum

Johannes Stöffler (1452–1531) was a German mathematician, astronomer, astrologer, geographer, astronomical instrument maker and professor at the University of Tübingen.[90] Among others he taught Philipp Melanchthon and Sebastian Münster.[91] Stöffler used Ptolemy's system of epicycles, and his ephemerides were based on the *Alfonsine Tables*. In addition to the ephemerides, he worked on calendars and left a work on the construction and use of the astrolabe, *Elucidatio fabricae ususque astrolabii*, which was published in several editions in Latin, German and French, and eventually in English translation in 2007.[92] The equation of time appears in his astronomical tables of 1514 *Justingensis germani tabulae astronomicae*, in the *Tabula aequationis dierum cum noctibus moderna*. The equation of time, given in minutes and seconds, takes only positive values; the minimum of 0;0 min is at 21°–26° Aquarius and the maximum of 32;52 min at 7°–8° Scorpio. In addition to Stöffler, a very famous ephemeris maker of the same period was the German astronomer, mathematician and astrologer Jakob Pflaum (c.1450–early 16th century), Latin Jacob Prunus. Ephemerides by both authors were published in numerous editions, which shows the high demand for this type of work at the time, and Stöffler and Pflaum collaborated on some of the works.[93]

86 Doppelmayr, 1730, pp.286–287. Also Repsold, 1908, p.16.
87 Gaab, 2006, pp.68, 70.
88 Gaab, 2006, p.60.
89 Gaab, 2006, p.71, also footnote 26.
90 Hartfelder, 1893.
91 Philipp Melanchton (1497–1560), German theologian and Protestant reformer. Sebastian Münster (1488–1552), German cartographer and cosmographer.
92 Stoeffler's Elucidatio - The Construction and Use of the Astrolabe, Alessandro Gunella and John Lamprey (trans.), Classic Science Press.
93 Pflaum Jakob, Deutsche Digitale Bibliothek, Kultur und Wissen online.

6.5.3 Johannes Virdung

Johannes Virdung (1463–1538/1539), one of the most important astrologers of his time, is less well known, despite his extensive bibliographical legacy of at least 80 items. He began his studies in Leipzig in 1481 and continued his studies in Kraków, where he listened to lectures by Albertus de Brudzewo and Johann von Glogau, and obtained his Master's degree in 1491.[94] From 1491, he lectured in Heidelberg. He produced calendars and published astrological writings, among other things. The equation of time is published in the *Tabulae resolutae de supputandis siderum motibus* of 1542 in the *Tabula aequationis dierum cum noctibus*. The equation of time is identical to Stöffler's, with the same minimum and maximum values, and the two authors' tables apparently differ only because of a few handwriting or typesetting errors (Table 6.2).

6.5.4 Johannes Schöner

Johannes Schöner, Latin Johannes Schonerus (1477–1547), studied in Erfurt and was ordained priest in 1500.[95] He taught mathematics at the Gymnasium in Nuremberg from 1526, where he remained until his retirement in 1547. He is best known for his globes – with the globe he made in 1515 being the first to depict the continent of America (which had by then been depicted on maps by Matthias Ringmann and Martin Waldseemüller).[96] Schöner was active not only as a mathematician, but also as a geographer, cartographer, astronomer, astrologer and printer or publisher of scientific works.

He was also familiar with the work of Copernicus, and we know that, in 1540, he received a letter from Rheticus quoting Copernicus' doctrine. Schöner's contribution to the printing of Regiomontanus' legacy is also very important. Based on Regiomontanus' and Walther's work, Schöner in 1544 published the fundamental work of practical astronomy *Scripta clarissimi mathematici M. Ioannis Regiomontani*. During his stay in Nuremberg, Schöner worked not only as a printer, but also in astronomy and astrology. This period saw the production of his work *Opusculum geographicum*, which also contains a chapter on the daily rotation of the Earth, which he categorically rejected in the text. He also published ephemerides and astronomical

94 Hammel, 2007. Albert of Brudzewo, Latin: Albertus de Brudzewo (1446–1495), mathematician and astronomer. Johann von Glogau, also Johannes Schelling von Glogau, Latin Johannes Glogoviensis (c.1445–1507), philosopher and mathematician. After 1491, Copernicus may have been among his students. For the information or planetary models of Islamic astronomers that may have been available to Copernicus in Kraków, see Rosinska, 1974.

95 For a comprehensive overview of his life and work, see Maruska, 2008. Andreas Schöner, one of his sons, was apparently also a mathematician, and in 1552 he published a work on gnomonics, Gnomonice Andreae Schoneri Noribergensis. Maruska, 2008, p.73.

96 Matthias Ringmann (1481/82–1510/11), also known by his Latin name Philesius Vosigena, was an Alsatian humanist and cosmographer. Together with the cartographer Martin Waldseemüller (c.1470–1518?), Lat. Hylacomylus, they were the first to use the name America for the continent, on a map published in the 1507 *Universalis cosmographia*. They later worked together on a new edition of Ptolemy's *Geography*.

tables.[97] His table of the equation of time, published in *Tabulae resolvtae astronomicae* from 1588, is identical to those of Stöffler and Virdung, respectively.[98]

Luculentissima quaedam terrae totius descriptio
Inscription on the Schöner's globe of 1515

6.5.5 Peter Apian

Peter Apian (1495? or 1501?–1552), Lat. Petrus Apianus, stated the equation of time or, as he called it, the *equation of days*, Lat. *equatio dierum*, in the fifth chapter of *Astronomicum caesareum*. He was born in Leisnig, Saxony and was a professor of mathematics in Ingolstadt.[99] Little is known about his life. His legacy comprises a large collection of works on astronomy, calendars and astronomical instruments, which are characterised by their great didactic value. Among his most famous works are *Cosmographia* from 1524 and *Astronomicum caesareum* from 1540, both masterpieces, both mathematically and in terms of printing. He produced sundials and his works contain a tabled equation of time (Figure 6.2). Apian's equation of time takes only positive values, and zero on January 31st. The dates are valid for the Julian calendar.[100] His equation of time is a masterwork, and he represents it in the book using a multi-part volvelle in which the external circle carries the date, the inner circle carries the equation of time, and a silk string provides a straight line for conducting the calculation, similar to the other devices in Apian's book.[101] The calculation assume circular orbits, Apian being a convinced advocate of Ptolemy's astronomical principles, and hence deviate slightly from the true values. Apian's works are among the last astronomical works based on the *Almagest*. A few years after *Astronomicum caesareum*, Copernicus' work was published. After Apian's death, his son Philipp Apian succeeded him in Ingolstadt as professor of mathematics.

97 For comments about his treatises on equatorium see North, 1976, volume II, pp.276–277.

98 *Tabulæ æquationum dierum com noctibus suis.*

99 For more information on his life and work, see Röttel et al., 1997, Peter Apian - Astronomie, Kosmographie und Mathematik am Beginn der Neuzeit. For more information on the life and work of Peter and Philipp Apian, see Schöner, 1994, Mathematik und Astronomie an der Universität Ingolstadt im 15. und 16. Jahrhundert, pp.358–429.

100 For tabulated values, see Ionides, 1934, Medieval Astronomical Constants, pp.329–330.

101 Apian, 1540, Astronomicum Caesareum, chapter 5. For an explanation, see also Apian, Ionides, 1936, Caesars' Astronomy, p.368. The equation of time in Apian's book is presented in chapter 5 in the form of a circular nomogram.

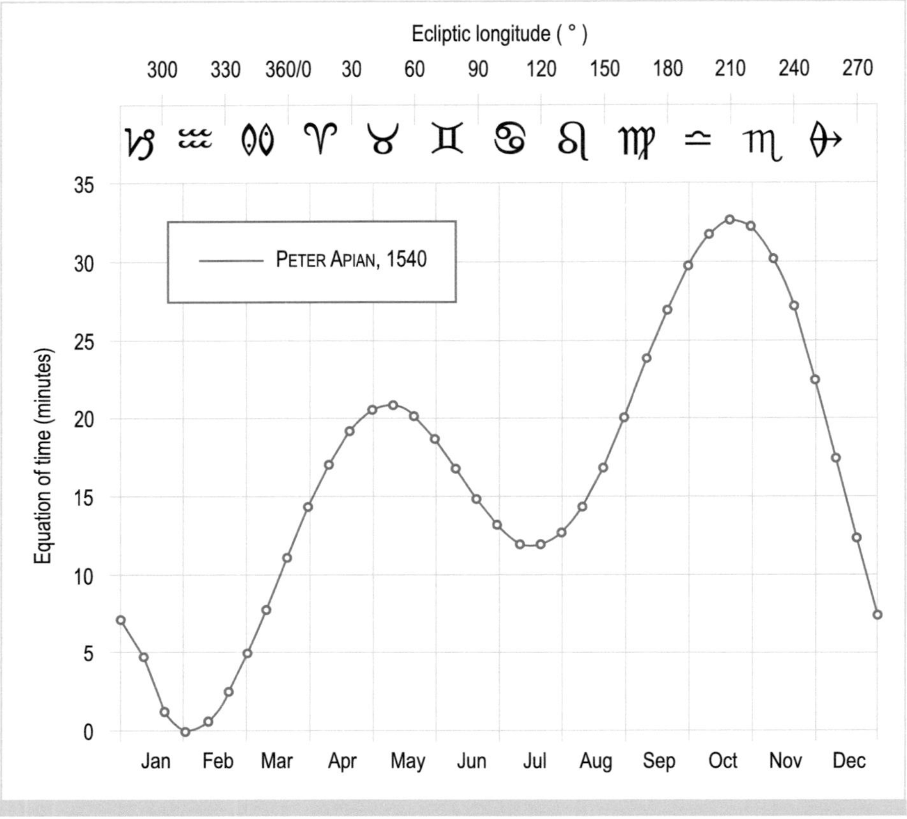

Figure 6.2 The equation of time of PETER APIAN from *Astronomicum caesareum*, published in 1540. Data source for the diagram: Ionides, 1934, pp.330–331.

6.6 Holbrooke, James of Zalesie and others

6.6.1 John Holbrooke

JOHN HOLBROOKE, Lat. JOANNES HOLIBROCUS, also JOANNES DE SACRO FONTE, was an English mathematician active in the first half of the 15th century; he received his doctorate in 1418.[102] For the first two decades of the 15th century, he was at Peterhouse College, Cambridge. He was Chancellor of Cambridge University in 1427 and 1428.[103] He is also credited with the development of astrology. The manuscript *Codex Holbrookensis* comprises several works on astronomical subjects, apparently produced separately, and is a good indication of HOLBROOKE's output.[104] HOLBROOKE's equation of time is mentioned in FABRICIUS' *Bibliotheca latina mediae et infimae aetatis, IV* where he also mentions HOLBROOKE's *Novae tabulae midiorum motuum, et aequationes dierum, et continuatio libelli de reductione Tabularum Alphonsiper annos Christi, menses, dies et horas.*[105]

A table with HOLBROOKE's equation of time, *Tabula equationis dierum cum noctibus suis*, is available in the British Library manuscript, Egerton 889, folios 150v–151r. The argument is the true longitude of the Sun in degrees, and the values are in time-degrees.[106]

6.6.2 James of Zalesie

JAMES OF ZALESIE, Latin JAKUB DE ZALESIE, of Silesian descent, was a professor of astronomy and astrology and a doctor of medicine. He worked as a physician at the court of CASIMIR IV JAGIELLON.[107] In the service of the Hungarian Bishop MIKLÓS STOLZ of present-day Oradea in Romania, he lectured at the University of Bologna from 1469 to 1470.[108] Apparently because of disagreements with the Spanish astronomer BARTHOLOMEI DE VALENCIA, he wrote the treatise *Rationes magistri Jacobi Zalesse Poloni contra magistri Bartholomei de Valencia de dierum aequationibus et earum causis.*[109] The associated dispute over the length of the sidereal hours and

102 Robson, Röhrkasten (ed.), 2010, p.298. Toomer, 1998, p.99. Pedersen,

103 Ball Rouse, 1889, A History of the Study of Mathematics at Cambridge, p.9.

104 Riehl Leader et al., 1988, p.151. British Library, Egerton manuscript 889.

105 Fabricius, 1735, Bibliotheca latina mediae et infimae aetatis, IV, pp.241–242. JOHANN ALBERT FABRICIUS (1668–1736), German classicist and bibliographer.

106 Pietroni, 2022, p.124, Figure 49.

107 CASIMIR JAGIELLON (1427–1492), Pol. KAZIMIERZ JAGIELLOŃCZYK, Grand Prince of Latvia and King of Poland.

108 Láng, 2010, Unlocked Books: Manuscripts of Learned Magic in the Medieval Libraries of Central Europe (Magic in History), p.249. Pol. JAKUB Z ZALESIA. For more information on his life and work, see Birkenmajer, 1972, pp.541–544 (in French). He was the third representative of the Polish astronomical-mathematical school at the University of Bologna in the 15th century, after MARCIN KRÓL, who was in Bologna in 1448/1449, and WOJCIECH D'OPATÓWI, who was in Bologna in 1454–1456.

109 Birkenmajer, 1972, p.543, note. The text is preserved in manuscript form in Biblioteca Medicea Laurenziana, Florence, cod. *Ashburnham 214*, folios 13r–20r. Ożóg, 2009, p.139. For more information on the treatise, see Birkenmajer, 1912, "De diebus naturalibus" de Barthélémy Berp de Valentia. Bulletin International de l'Académie des sciences de Cracovie, Classe des Sciences mathématiques et naturelles, Série A, pp.339–379.

the normal or average hours is said to have taken place between the astronomers within the walls of the University of Bologna in 1471.[110] JAMES' *De dierum aequationibus contra propositiones Bartholomaei de Valentia* is mentioned in *De Sphere of Marcus Manilius Made an English Poem* of 1675.[111]

6.6.3 Other notable astronomers

LUCA GAURICO (1475–1558), Latin LUCAS GAURICUS, was an Italian astrologer, astronomer and mathematician. When he taught mathematics at Ferrara, JOSEPH JUSTUS SCALIGER was among his pupils.[112] Commissioned by Pope NICHOLAS V, he edited GEORGE OF TREBIZOND's[113] translation of the *Almagest*, and he described the equation of time in *Super tabulis directionum Ioannis Monteregiensis quod dam supplementum*. In the *Tabula aequationis dierum* on pages 31 and 32, he tabulated the values that apparently correspond to the mean Sun's right ascension, which are given in hours and minutes.[114] He was involved in preparing the 1521 and 1524 editions of the *Alfonsine Tables*.[115]

HENRICUS BAERSIUS, mathematician and printer at the University of Louvain, was active in the first half of the 16th century. His identity, also known as VEKENSTYL, is known from a letter to JOHANNES DANTISCUS, the Polish envoy to Benelux.[116] As he described himself, he was the treasurer of St Peter's Church in Louvain and a mathematician. He was also clearly interested in astronomy, as evidenced by the written works that he printed himself in 1530 and 1537. A table with an equation of time identical to STÖFFLER's or VIRDUNG's, with some minor variations or individual errors by scribes or typesetters, is found in the astronomical tables of 1528.[117] The only remaining known book from his printing house is a booklet from 1530 containing meditations and seven gospel psalms by the Benedictine monk GUILIELMUS LAPIDANUS, Dutch WILLEM VAN DEN STEEN.[118] The colophon shows that he printed the books himself.

110 Ożóg, 2009, p.139. For more information on the dispute, see Birkenmajer, 1972, pp.545–547 (in French).

111 Sherburne, 1675, Appendix, p.77.

112 SCALIGER is considered the father of modern chronology, see Appendix A.

113 GEORGE OF TREBIZOND, Gr. Γεώργιος Τραπεζούντιος (1395–1486), Byzantine philosopher and humanist. He probably came to Italy in around 1430, perhaps to Venice. He later worked in Rome. He also died in Italy in dire poverty.

114 The text is in Latin, and a detailed translation with a commentary would be needed to understand the meaning of the table with the so-called equation of time. The actual equation of time should correspond to the difference between the solar time at midday and the mean time given in the table.

115 Wegener, 1905a, pp.168–169.

116 Vervliet, 1978, pp.188–189.

117 Meeus, Savoie, 1995, p.188, also Baers ou Vekenstyl, 1976. All these equation of time tables are obviously based on the *Alfonsine Tables*.

118 Vervliet, 1978, pp.190–191.

Longitude		STÖFFLER	VIRDUNG	BAERSIUS
♈	7°	10; 8	10; 8	10;12
	16°	13; 8	13; 8	13;12
♉	6°	18;28	18;28	18;26
♊	21°	18; 4	*17; 4	18; 4
♌	9°	*11;25	*11;25	11;52
♍	5°	16;40	16;40	16; 4
	24°	22;48	*23;48	22;48
♎	1°	25;28	*25;24	25;28
	2°	*45;44	25;44	*25; 4
	8°	27;40	27;40	*27;44
♏	1°	32;28	32;28	*32;24
	2°	32;32	32;32	*32;24
	3°	32;36	32;36	*32;29
	4°	*22;40	32;40	*32;26
	13,	32;34	32;34	32;36
	21°	31;46	31;46	31;40
♑	6°	13; 8	13; 8	*1; 8
♒	1°	*2; 0	*2; 0	3; 0
♓	24°	5;36	5;36	5;32
	25°	5;58	5;58	5;48
	26°	6;23	6;23	6; 8
	27°	6;47	6;47	6;36

Table 6.2 Comparison of the equation of time between tha tables of STÖFFLER (1514), VIRDUNG (1542) and BAERSIUS or VEKENSTYL (1528). The asterisk indicate obvious scribal or typesetting errors, even where there are traces of copying of tables. The errors in SCHÖNER's table with the equation of time are also very similar.

Bibliography

- Albertus de Brudzewo, Indexentry. In: Deutsche Biographie, [online version]. URL: https://www.deutsche-biographie.de/pnd102419035.html (17 September 2022).
- Alfonso X., Alfonsus de Cordoba and Luca Gaurico. 1524. Alfonsi Hispaniarum Regis Tabule [et] L. Gaurici Artium doctoris egregij Theoremata quorum hic est index ... In calce huius libri seorsum annexe sunt tabul[a] e Elisabeth Regin[a]e nuper castigat[a]e & in ordinem redact[a]e per. L. Gauricum cum additionibus & Nouis problematibus eiusdem Gaurici. Venice: Junta: URL: https://nbn-resolving.org/urn:nbn:de:urmel-0fb92c27-9ad0-490e-82d0-274b93fea45f4-00004523-00154 (31 October 2022).
- Alfonso X. 1545. Astronomicae Tabulae. (GoogleBooks, 9 November 2022).
- Apianus, Peter. 1540. Astronomicum Caesareum. Ingolstadt: Arnt von Aych. URL: https://archive.org/details/astronomicumcsar00apia/mode/2up (29 January 2024).
- Apianus, Peter and S. A. Ionides. 1936. Caesars' Astronomy: (Astronomicum Caesareum). Osiris, 1, 356–389. DOI 10.1086/368431.
- Baers ou Vekenstyl, Henri. 1976. Les Tables Astronomiques de Louvain de 1528, Poullet, Emanuell and Alfred de Smet (eds.). Bruxelles: Edition Cultur et Civilisation.
- Ball, Walter William Rouse. 1889. A History of the Study of Mathematics at Cambridge. Cambridge: Cambridge University Press. (GoogleBooks, 17 September 2022).
- Baneth, E. 1898. Maimuni's Neumondsberechnung. Berichte uber die Lehranstalt fur die Wissenschaft des Judenthums in Berlin, Teil 1, 16, 1–30. URL: https://sammlungen.ub.uni-frankfurt.de/cm/periodical/titleinfo/2273546 (29 January 2024).
- Baneth, E. 1899. Maimuni's Neumondsberechnung. Berichte uber die Lehranstalt fur die Wissenschaft des Judenthums in Berlin, Teil 2, 17, 31–62. URL: https://sammlungen.ub.uni-frankfurt.de/cm/periodical/titleinfo/2273546 (29 January 2024).
- Baneth, E. 1902. Maimuni's Neumondsberechnung. Berichte uber die Lehranstalt fur die Wissenschaft des Judenthums in Berlin, Teil 3, 20, 63–116. URL: https://sammlungen.ub.uni-frankfurt.de/cm/periodical/titleinfo/2273546 (29 January 2024).
- Baneth, E. 1903. Maimuni's Neumondsberechnung. Berichte uber die Lehranstalt fur die Wissenschaft des Judenthums in Berlin, 21, 117–195. URL: https://sammlungen.ub.uni-frankfurt.de/cm/periodical/titleinfo/2273546 (29 January 2024).
- Bazzani, Marina. 2006. Theodore Metochites, a Byzantine Humanist. In: Byzantion, 76, 32–52. URL: https://www.jstor.org/stable/44173322.
- Beaver, D. 1970. Bernard Walther: Innovator in Astronomical Observation. In: Journal for the History of Astronomy, 1, 1, 39–43. DOI 10.1177/002182867000100105. URL: https://www.researchgate.net/publication/234359433_Bernard_Walther_Innovator_in_Astronomical_Observation (29 January 2024).
- Beller, Eliyahu. 1988. Ancient Jewish mathematical astronomy. In: Archive for History of Exact Sciences, 38, 1, 51–66. DOI 10.1007/BF00329980.
- Benjamin, Francis S. Jr. 1954. John of Gmunden and Campanus of Novara. In: Osiris, 11, 221–246. DOI 10.1086/368580. URL: https://www.jstor.org/stable/301672.
- Biobibliographisches Handbuch der Kalendermacher von 1550 bis 1750 [online source]. URL: https://www.presseforschung.uni-bremen.de/dokuwiki/doku.php?id=cuno_d._j._jacob (30 October 2022).
- Birkenmajer, Aleksander. 1912. "De diebus naturalibus" de Barthélémy Berp de Valentia. In: Bulletin International de l'Académie des sciences de Cracovie, Classe des Sciences mathématiques et naturelles, Série A, 339–379. Reprint 2023.
- Birkenmajer, Aleksander. 1972. Etudes d'histoire des sciences en Pologne. Studia Copernicana, IV. Wroclaw: Polska Akademia Nauk.
- British Library. Detailed record for Egerton 889, Catalogue of Illuminated Manuscripts. URL: https://www.bl.uk/catalogues/illuminatedmanuscripts/record.asp?MSID=524 (17 September 2022).
- Busard, Hubert L. L. 1971. Der Traktat De sinibus, chordis et arcubus von Johannes von Gmunden. In: Denkschriften der Akademie der Wissenschaften, 116, 3, 73–113. Vienna: Springer. DOI 10.1007/978-3-7091-4013-0_1.
- Cantera Burgos, Francisco. 1931. El judio salmantino Abraham Zacut: Notas para la historia de la astronomía en la España medieval. In: Revista de la Academia de Ciencias 28. Madrid: C. Bermejo.
- Cantera Burgos, Francisco. 1935. Abraham Zacut. Madrid: Biblioteca de la cultura española, Siglo XV. Madrid: M. Aguilar.

- Carlebach, Joseph. 1910. Lewi ben Gerson als Mathematiker. Berlin: Lamm.
- Carman, Christián C. 2017. Martianus Capella's calculation of the size of the moon. In: Archive for History of Exact Sciences, 71, 1, 193–210. doi 10.1007/s00407-016-00185-0.
- Chabás, José. 1991. The Astronomical Tables of Jacob ben David Bonjorn. In: Archive for History of Exact Sciences, 42, 4, 279–314. doi 10.1007/BF00375156. url: https://www.jstor.org/stable/41133906.
- Chabás, José. 1998. Astronomy in Salamanca in the Mid-fifteenth Century: The Tabulae resolutae. In: Journal for the History of Astronomy, 29, 95, 167–175. doi 10.1177/002182869802900209.
- Chabás, José. 2016. An analysis of the Tabulae magistrales by Giovanni Bianchini. In: Archive for History of Exact Sciences, 70, 5, 543–552. doi 10.1007/s00407-016-0178-z. url: https://www.jstor.org/stable/24913252.
- Chabás, José and Bernard R. Goldstein. 2000. Astronomy in the Iberian Peninsula: Abraham Zacut and the Transition from Manuscript to Print. In: Transactions of the American Philosophical Society, 90, Part 2. American Philosophical Society. isbn 0-87169-902-8.
- Chabás, José and Bernard R. Goldstein. 2003. The Alfonsine Tables of Toledo. Dordrecht: Kluwer Academic Publisher. isbn 1-4020-1572-0.
- Chabás, José and Bernard Goldstein. 2009. The Astronomical Tables of Giovanni Bianchini. Medieval and Early Modern Philosophy and Science, Volume 10. Lleiden: Brill. isbn 978-90-47-42959-3.
- Chabás, José and Bernard R. Goldstein. 2018. New evidence on Abraham Zacut's astronomical tables. In: Archive for History of Exact Sciences, 72, 1, 21–62. doi 10.1007/s00407-017-0199-2.
- Chlench, Kathrin. 2007. Johannes von Gmunden deutsch, Der Wiener Codex 3055. Wien: Fassbaender. isbn 978-3-900538-99-6.
- Curtze, Maximilian. 1900. Urkunden zur Geschichte der Trigonometrie im christlichen Mittelalter. In: Bibliotheca Mathematica, 3, 1, 321–416.
- Curtze, Maximilian. 1902a. Urkunden zur Geschichte der Mathematik im Mittelalter und der Renaissance, 1. In: Abhandlungen zur Geschichte der mathematischen Wisenschaften mit Einschluß ihrer Anwendungen, 12. Leipzig: Teubner.
- Curtze, Maximilian. 1902b. Urkunden zur Geschichte der Mathematik im Mittelalter und der Renaissance, 2. In: Abhandlungen zur Geschichte der mathematischen Wisenschaften mit Einschluß ihrer Anwendungen, 13. Leipzig: Teubner.
- Dannenfeldt, Karl H. 1976. Stephanus of Alexandria. In: Dictionary of Scientific Biography, Gillispie, Charles Coulston (ed.), XIII, 37–38. New York: Charles Scribner's Sons. isbn 0-684-12925-6.
- De Albuquerque, Luis. 1976. Zacuto, Abraham bar Samuel bar Abraham. In: Dictionary of Scientific Biography, Gillispie, Charles Coulston (ed.), XIV, 583–584. New York: Charles Scribner's Sons. isbn 684-12926-5.
- Dershowitz, Nachum and Edward M. Reingold. 2008. Calendrical Calculations. Cambridge: Cambridge University Press. isbn 978-0-521-70238-6.
- Doppelmayr, Johann Gabriel. 1730. Historische Nachricht von den nürnbergischen Mathematicis und Künstlern, welche fast von dreyen Seculis her Durch ihre Schrifften und Kunst-Bemühungen die Mathematic und mehreste Künste in Nürnberg. Nurnberg: Peter Conrad Monath. url: https://archive.org/details/b30455820 (29 January 2024).
- Fabricius, Johann Albert. 1735. Bibliotheca latina mediae et infimae aetatis, IV, Books IX–XI. Hamburg: sumtu viduae felgineriae. (GoogleBooks, 17 September 2022).
- Fletcher, Richard. 2006. The Moorish Spain. Oakland, CA: University of California Press. isbn 978-0520248403.
- Folkerts, Menso. 2009. Die Beiträge von Johannes von Gmunden zur Trigonometrie. In: Johannes von Gmunden (ca. 1384–1442), Astronom und Mathematiker Simek, Rudolf and Kathrin Chlench (eds.). Studia Medievalia Septentrionalia 12, 71–89. Wien: Fassbaender. isbn 9783900538958.
- Freudenthal, Gad (ed.). 1992. Studies on Gersonides A Fourteenth-Century Jewish Philosopher-Scientist. Leiden: Brill. isbn 9789004096417.
- Gaab, Hans. 2006. Die große Nürnbergische Uhr. In: Acta Historica Astronomiae, 28, Dick, Wolfgang R. and Jürgen Hammel (eds.), 43–90. Frankfurt am Main: Harri Deutsch. isbn 978-3-8171-1771-0.
- Glowatzki, Ernst and Helmut Göttsche. 1990. Die Tafeln des Regiomontanus ein Jahrhundertwerk. München: Institut für Geschichte der Naturwissenschaften. isbn 3-89241-001-1.
- Goldstein, Bernard R. 1967. Ibn al-Muthanna's Commentary on the Astronomical Tables of al-Khwarizmi. New Haven, CT: Yale University Press. isbn 0-300-00498-2.
- Goldstein, Bernard R. 1974. The Astronomical Tables of Levi Ben Gerson. Transactions published by The Connecticut Academy of Arts and Sciences, Volume 45. New Haven, CT: The Connecticut Academy of Arts and Sciences. isbn 0-208-01425-X.

- Goldstein, Bernard R. 1985. The Astronomy of Levi ben Gerson (1288–1344): A Critical Edition of Chapters 1–20 with Translation and Commentary (Studies in the History of Mathematics and Physical Sciences). New York, Berlin: Springer. ISBN 978-0387961323.
- Goldstein, Bernard R. and José Chabás. 2001. The Maximum Solar Equation in the Alfonsine Tables. In: Journal for the History of Astronomy, 32, 109, 345–348. DOI 10.1177/002182860103200404.
- Goldstein, Bernard R. and José Chabás. 2017. Analysis of the astronomical tables for 1340 compiled by Immanuel ben Jacob Bonfils. In: Archive for History of Exact Sciences, 72, 1, 71–108. DOI 10.1007/s00407-016-0181-4.
- Goldstein, Bernard R. and José Chabás. 2018. New evidence on Abraham Zacut's astronomical tables. In: Archive for History of Exact Sciences, 72, 1, 21–62. DOI 10.1007/s00407-017-0199-2. URL: https://www.jstor.org/stable/45211938.
- Green, Daniel W. E. and Thomas R. Williams. 2007. Walther Bernard. In: Biographical Encyclopedia of Astronomers, Hockey, Thomas, Virginia Trimble and Thomas R. Williams (eds.), 1193–1194. New York: Springer. ISBN 9780387310220.
- Grössing, Helmuth. 1985. Johannes von Gmunden in seiner Zeit. In: Mitteilungen der Österreichischen Gesellschaft für Geschichte der Naturwissenschaften, 3–4, 66–72.
- Grössing, Helmuth. 1997. Muestinger, Georg. In: Neue Deutsche Biographie 18, 562 [online version]; URL: https://www.deutsche-biographie.de/pnd137901003.html#ndbcontent (30 October 2022).
- Günther, Siegmund. 1891. Schöner, Johann. In: Allgemeine Deutsche Biographie, 32, 295–297, [online version]. URL: https://www.deutsche-biographie.de/pnd118795341.html#adbcontent (28 October 2022).
- Günther, Siegmund. 1896. Virdung, Johann. In: Allgemeine Deutsche Biographie 40, 9–10 [online version]. URL: https://www.deutsche-biographie.de/pnd100963897.html#adbcontent (30 October 2022).
- Hadravová, Alena and Petr Hadrava. Prachatiz, Christannus de. In: Encyclopedia.com [online source]. URL: https://www.encyclopedia.com/science/dictionaries-thesauruses-pictures-and-press-releases/prachaticz-cristannus-de (30 October 2022).
- Hadravová, Alena and Petr Hadrava. 2006. Johannes von Gmunden und seine Version des Astrolabtraktats des Christian von Prachatitz. In: Johannes von Gmunden (ca. 1384–1442), Astronom und Mathematiker, Simek, Rudolf and Kathrin Chlench (eds.). Studia Medievalia Septentrionalia 12, 151–159. Wien: Fassbaender. ISBN 9783900538958.
- Hairetdinova, N. G. 1970. On the oriental sources of the Regiomontanus' trigonometrical treatise. In: Archives internationales d'histoire des sciences, 23, 61–66.
- Hammel, Jürgen. 2007. Johannes Virdung. In: Biographical Encyclopedia of Astronomers, Hockey, Thomas, Virginia Trimble and Thomas R. Williams (eds.), 1180. New York: Springer. ISBN 9780387310220.
- Hartfelder, Karl. 1893. Stoeffler, Johannes. In: Allgemeine Deutsche Biographie 36, 317–318 [online version]. URL: https://www.deutsche-biographie.de/pnd118798979.html#adbcontent (30 October 2022).
- Heiberg, Johan Ludvig. 1899. Byzantinische Analekten. In: Abhandlungen zur Geschichte der Mathematik, 9, 161–174.
- Hellman, C. Doris and Noel M. Swerdlow. Peurbach (or Peuerbach), Georg. In: Encyclopedia.com [online source]. URL: https://www.encyclopedia.com/science/dictionaries-thesauruses-pictures-and-press-releases/peurbach-or-peuerbach-georg (30 October 2022).
- Hoffman, J. E. 1971. Cusa, Nicholas. In: Dictionary of Scientific Biography, Gillispie, Charles Coulston (ed.), III, 512–516. New York: Charles Scribner's Sons.
- Hoffman, J. E. Cusa, Nicholas. In: Encyclopedia.com [online source]. URL: https://www.encyclopedia.com/science/dictionaries-thesauruses-pictures-and-press-releases/cusa-nicholas (30 October 2022).
- Ionides, S. A. 1934. Medieval Astronomical Constants. In: Publications of the Astronomical Society of the Pacific, 46, 274, 325–338. DOI 10.1086/124511.
- Ivor Bulmer, Thomas. 1971. Eudemus of Rhodes. In: Dictionary of Scientific Biography, Gillispie, Charles Coulston (ed.), IV, 460–465. New York: Charles Scribner's Sons. SBN 684-10115-7.
- Jones, Alexander. 1987. An eleventh century manual of Arabo-Byzantine astronomy. Amsterdam: Gieben. ISBN 9050630146.
- Kangle, R. P. (trans.). 1960. The Kauṭilya Artha-śāstra. Part II, An English Translation with Critical and Explanatory Notes. Delhi: Motilal Banarsidass. URL: https://archive.org/details/kangle1960/Artha%20Shastra%20Part%202%20-%20R.P.%20Kangle_text.pdf (29 January 2024).
- Kellner, Menachem Marc. 1979. R. Levi Ben Gerson: A Bibliographical Essay. In: Studies in Bibliography and Booklore, 12, 13–23. URL: https://www.jstor.org/stable/27943474.
- Kern, Ralf. 2010a. Wissenschaftliche Instrumente in ihrer Zeit. Band 1: Vom Astrolab zum mathematischen Besteck. Köln: Verlag der Buchhandlung Walther König. ISBN 978-3-86560-865-9.

- Kern, Ralf. 2010b. Wissenschaftliche Instrumente in ihrer Zeit. Band 2: Vom 15. – 19. Jahrhundert. Köln: Verlag der Buchhandlung Walther König. isbn 978-3-86560-772-0.
- Kish, George. 1970. Apian, Peter. In: Dictionary of Scientific Biography, Gillispie, Charles Coulston (ed.), I, 178–179. New York: Charles Scribner's Sons.
- Klug, Rudolf. 1943. Johannes von Gmunden, der Begründer der Himmelskunde auf deutschem Boden. In: Sitzungsberichte der Akademie der Wissenschaften in Wien. Philosophisch-Historische Klasse, 222, 1–93. Wien, Leipzig: Hölder-Pichler-Tempsky. url: https://digi.landesbibliothek.at/viewer/image/AC02949962/1/LOG_0000/ (30 October 2022).
- Kremer, Richard L. 2010. War Bernhard Walther, Nürnberger astronomischer Beobachter des 15. Jahrhunderts, auch ein Theoretiker? In: Astronomie in Nürnberg, Wolfschmidt, Gudrun (ed.), 157–183. Nuncius Hamburgensis, Beiträge zur Geschichte der Naturwissenschaften, Band 3. Hamburg: tredition science. isbn 978-3-86850-609-9.
- Kremer, Richard, Mathieu Husson and José Chábas (eds.). 2022. Alfonsine Astronomy: The Written Record. Turnhout: Brepols. isbn 978-2-503-59520-7.
- Kren, Claudia. 1971. Dominicus De Clavasio. In: Dictionary of Scientific Biography, Gillispie, Charles Coulston (ed.), IV, 157–158. New York: Charles Scribner's Sons. sbn 684-10115-7.
- Kren, Claudia. Dominicus De Clavasio. In: Encyclopedia.com [online source]. url: https://www.encyclopedia.com/science/dictionaries-thesauruses-pictures-and-press-releases/dominicus-de-clavasio (30 October 2022).
- Láng, Benedek. 2010. Unlocked Books: Manuscripts of Learned Magic in the Medieval Libraries of Central Europe (Magic in History). University Park, PA: Pennsylvania State University Press. isbn 978-0271033785.
- Leichter, Joseph Gerard. 2004. The Zij as-Sanjari of Gregory Chioniades. Providence, RI: Brown University. url: https://archive.org/details/TheZijAs-sanjariOfGregoryChioniades/mode/2up (10 February 2024).
- Lempire, Jean. 2014. Byzantine astronomy under the reign of Heraclius (610–641). In: Late Antique Astronomy in Syria: Heritage, Contacts and Particularities, Workshop. TOPOI – Humboldt-Universität zu Berlin. Berlin, 11 April 2014. url: https://hdl.handle.net/2078.1/180957 (10 February 2024).
- Lempire, Jean. 2016. Le commentaire astronomique aux Tables Faciles de Ptolémée attribué a Stéphanos d'Alexandrie. Louvain-la-Neuve: Université Catholique de Louvain, Institut Orientaliste & Peteers Press. isbn 978-90-429-3074-2.
- Leurquin, Régine. 1990. Théodore Méliténiote, Tribiblos Astronomique. Livre I. Amsterdam: Gieben. isbn 9050630456.
- Leurquin, Régine. 1993a. Théodore Méliténiote, Tribiblos Astronomique. Livre II, Commentaire. Amsterdam: Adolf M. Hakkert. isbn 9025610250.
- Leurquin, Régine. 1993b. Théodore Méliténiote, Tribiblos Astronomique. Livre II, Édition Critique et Traduction. Amsterdam: Adolf M. Hakkert. isbn 9025610242.
- Levey, Martin. 1970. Abraham bar Hiyya Ha-Nasi. In: Dictionary of Scientific Biography, Gillispie, Charles Coulston (ed.), I, 22–23. New York: Charles Scribner's Sons.
- Levey, Martin. 1971. Ibn Ezra, Abraham ben Meir. In: Dictionary of Scientific Biography, Gillispie, Charles Coulston (ed.), IV, 502–503. New York: Charles Scribner's Sons. sbn 684-10115-7.
- Maruska, Monika. 2008. Johannes Schöner – 'Homo est nescio qualis'. Leben und Werk eines fränkischen Wissenschafters an der Wende vom 15. zum 16. Jahrhundert. Dissertation. Wien: Universität Wien. url: https://utheses.univie.ac.at/detail/1061# (30 October 2022).
- Meeus, Jean and Denis Savoie. 1995. L'Équation du temps. In: l'Astronomie, 109, 188–193.
- Mercier, Raymond. 1984. The Greek 'Persian Syntaxis' and the Zīj-I Īlkhānī. In: Archives Internationales d'Histoire des sciences, 34, 112, 35–60. Reprint 2004. Studies on the Transmission of Medieval Mathematical Astronomy. Aldershot: Ashgate. isbn 0-86078-949-7.
- Mercier, Raymond. 2000. From Tantra to Zīj. In: Sic Itur Ad Astra: Studien zur Geschichte der Mathematik und Naturwissenschaften - Festschrift für den Arabisten Paul Kunitzsch zum 70. Geburtstag, Folkerts, Menso and Richard Lorch (eds.), 451–460. Wiesbaden: Harrassowitz Verlag. isbn 3-447-04290-7.
- Mercier, Raymond. 2014. Astronomical Tables of Abraham Bar Hiyya. In: Time, Astronomy, and C
- alendars in the Jewish tradition, Stern, Sacha and Charles Burnett (eds.), 155–207. Lleiden, Boston: Brill. e-isbn 978-90-04-25966-9.
- Millás Vallicrosa, José Maria. 1959. La obra Séfer Hešbón mahlekot ha-kokabim de R. Abraham Bar Hiyya ha-Bargeloni. Barcelona: Consejo superior de investigationes científicas, Instituto Arias Montano.
- Miller, Clyde Lee. Cusanus, Nicolaus [Nicolas of Cusa]. In: The Stanford Encyclopedia of Philosophy (Winter 2021 Edition), Edward N. Zalta (ed.) [online source]. url: https://plato.stanford.edu/archives/win2021/entries/cusanus/ (30 October 2022).

- Mundy, John. 1943. John of Gmunden. In: Isis, 34, 3, 196–205. doi 10.1086/347790. url: https://www.jstor.org/stable/225838.
- Neska, Ingeborg. 1997. Die Handschriften der Stadtbibliothek Nürnberg, Band 4: Die lateinischen mittelalterlichen Handschriften. Wiesbaden: Harrassowitz Verlag. isbn 9783447037747.
- Neugebauer, Otto. 1949. The Astronomy of Maimonides and Its Sources. In: Hebrew Union College Annual, 22, 322–363. Reprint 1983, Astronomy and History, Selected Essays, 381–423. Springer. isbn 0-387-90844-7.
- Neugebauer, Otto. 1960. Studies in Byzantine Astronomical Terminology. In: Transactions of the American Philosophical Society, 50, 2, 1–45. doi 10.2307/3231944.
- Neugebauer, Otto. 1969. The Exact Sciences in Antiquity. New York Dover. sbn 486-22332-9.
- Neugebauer, Otto and David Pingree. 1971. The Pancasiddhāntikā of Varāhamihira, Volume II.
- Historisk-Filosofiske Skrifter 6, 1. Kopenhagen: Det Kongelige Danske Videnskabernes Selskab.
- North, John. 1976. Richard of Wallingford, An edition of his writings with introductions, English translation and commentary. Volume II. Oxford: Clarendon Press. isbn 0 19 858139 4.
- Oestmann, Günther, Elly Dekker and Peter Schiller. 1993. Schicksalsdeutung und Astronomie, Der Himmelsglobus des Johannes Stoeffler von 1493. Katalog zur gleichnamigen Ausstellung im Württembergischen Landesmuseum, Stuttgart, 2.12.1993–6.3.1994. Stuttgart: Württembergisches Landesmuseum. isbn 3-929055-28-7.
- Ożóg, Krzysztof. 2009. The Role of Poland in the Intellectual Development of Europe in the Middle Ages. Kraków: Societas Vistulana. isbn 978-83-61033-36-3.
- Paschos, Emmanuel A. and Christos Simelidis. 2017. Introduction to Astronomy by Theodore Metochites. Singapore: World Scientific. isbn 9789813207486. doi 10.1142/10378.
- Peuerbach, Georg. 1514. Tabulae eclypsium magistri Georgij Peurbachij. Vienna. (GoogleBooks, 9 October 2022).
- Pflaum, Jakob, Indexentry. In: Deutsche Biographie, [online version]. url: https://www.deutsche-biographie.de/pnd100232639.html (16 October 2022).
- Pflaum, Jakob, Indexentry. In: Deutsche Digitale Bibliothek, Kultur und Wissen online, [online source]. url: https://www.deutsche-digitale-bibliothek.de/person/gnd/100232639 (30 October 2022).
- Pietroni, Silvia. 2022. A Survey on the Evolution of the Equation of Time during the Centuries. In: Gerbertus, 17, 1–207. url: http://www.icra.it/gerbertus/ (30 October 2022).
- Pilz, Kurt. 1977. 600 Jahre Astronomie in Nürnberg. Nürnberg: Hans Carl. isbn 9783418004471.
- Pingree, David. 1963. Astronomy and astrology in early India and Iran. In: Isis, 54, 2, 229–246. doi 10.1086/349703.
- Pingree, David. 1976. The Byzantine Version of the "Toledan Tables": The Work of George Lapithes? In: Dumbarton Oaks Papers, 30, 85+87–132. doi 10.2307/1291391.
- Pingree, David, (ed.). 1978. The Yavanajātaka of Sphujidhvaja. Sphujidhvaja, Volume I-II. Cambridge, MA: Harvard University Press. isbn 0-674-96373-3. url: https://archive.org/details/PingreeYJ/1978%20 Yavanajataka%201/mode/2up (10 February 2024).
- Pingree, David. 1985. The Astronomical Works of Gregory Chioniades, The Zīj al-'Alā ī. Part I: Text, Translation and Commentary. Amsterdam: Gieben. isbn 9070265656.
- Pingree, David. 1986. The Astronomical Works of Gregory Chioniades, The Zīj al-'Alā ī. Part II: Tables. Amsterdam: Gieben. isbn 9070265508.
- Regiomontani, Joannis. 1972. Opera collectanea, Schmeidler, Felix (ed.). Osnasbrück: Otto Zeller Verlag. isbn 3-535-00816-6.
- Repsold, Johann Adolf. 1908. Zur Geschichte der Astronomischen Messwerkzeuge von Purbach bis Reichenbach, 1450 bis 1830. Leipzig: Verlag von Wilhelm Engelmann.
- Riehl Leader, Damian, Victor Morgan and Peter Searby. 1988. A History of the University of Cambridge: Volume 1, The University to 1546. Cambridge: Cambridge University Press. isbn 978-0521328821.
- Ries, Rotraud. 2008. Die Residenzstadt Hannover als Kommunikationsraum für Juden und Christen um 1700. In: Leibniz und das Judentum, Cook, Daniel J., H. Rudolph and Christoph Schulte (eds.). Ergebnisse einer internationalen Konferenz in Potsdam, 49–77. Stuttgart: Steiner Verlag. isbn 978-3-515-09251-7.
- Robson, Michael and Jens Röhrkasten (eds.). 2010. Franciscan organisation in the mendicant context. Formal and informal structures of the friars' lives and ministry in the Middle Ages. Münster: LIT Verlag. isbn 978-3-643-10820-3.
- Rosen, Edward. Schöner, Johannes. In: Encyclopedia.com [online version]. url: https://www.encyclopedia.com/science/dictionaries-thesauruses-pictures-and-press-releases/schoner-johannes (30 October 2022).
- Rosińska, Grażyna. 1974. Naṣīr al-Dīn al-Ṭūsī and Ibn al-Shāṭir in Cracow? In: Isis, 65, 2, 239–243. doi 10.1086/351260.

- Rosińska, Grażyna. 1981. Tables trigonometriques de Giovanni Bianchini. In: Historia Mathematica, 8, 1, 46–55. doi 10.1016/0315-0860(81)90004-5.
- Rosińska, Grażyna. 1987. Tables of Decimal Trigonometric Functions from ca. 1450 to ca. 1550. In: From Deferent to Equant, King, David and George Saliba (eds.), Annals of the New York Academy of Sciences, Volume 500, 419–426. New York: New York Academy of Sciences. isbn 0-89766-396-9.
- Röttel, Karl, W. Kaunzner, R. Finsterwalder and Menso Folkerts. 1997. Peter Apian - Astronomie, Kosmographie und Mathematik am Beginn der Neuzeit. Eichstätt: Polygon-Verlag. isbn 978-3928671125.
- Sabra, A. I. 1971. Al-Farghānī. In: Dictionary of Scientific Biography, Gillispie, Charles Coulston (ed.), IV, 541–545. New York: Charles Scribner's Sons. sbn 684-10115-7.
- Sarton, George. 1948. Introduction to the History of Science, Volume III/2. Washington, DC: Carnegie Institution of Washington.
- Schmeidler, Felix. 2007. Schöner, Johannes. In: Neue Deutsche Biographie, 23, 405–406, [online version]. url: https://www.deutsche-biographie.de/pnd118795341.html#ndbcontent (30 October 2022).
- Schöner, Christoph. 1994. Mathematik und Astronomie an der Universität Ingolstadt im 15. und 16. Jahrhundert. Berlin: Duncker & Humblot. isbn 3-428-08118-8.
- Schöner, Johannes. 1544. Scripta clarissimi mathematici M. Ioannis Regiomontani, de Torqueto, astrolabio armillari, regula magna Ptolemaica, baculoque astronomico, & obseruationibus cometarum. Norimbergae: Montanum, Neuber. url: https://archive.org/details/ARes35402/mode/2up (10 February 2024).
- Schöner, Johannes. 1588. Tabulae Resolvtae Astronomicae Iohannis Schoneri Mathematici Clarissimi. url: https://archive.org/details/bub_gb_TipeTa5YIKQC/mode/2up (10 February 2024).
- Schulte, Christoph. 2004. Die Mathematik in der jüdischen Aufklärung. In: 10+5=Gott. Die Macht der Zeichen, Tyradellis, Daniel and Michal Friedlander (eds.), 47–60. Berlin: DuMont Literatur und Kunst Verlag. isbn 978-3-8321-7874-1.
- Schulte, Christoph. 2008. Leibniz und sein "Schüler" Raphael Levi. In: Leibniz und das Judentum, Cook, Daniel J., H. Rudolph and Christoph Schulte (eds.). Ergebnisse einer internationalen Konferenz in Potsdam, 35–47. Stuttgart: Steiner Verlag. isbn 978-3-515-09251-7.
- Schulte, Christoph. 2017. Raphael Levi. Biographie (Version II, 2017). In: haskala.net. Das online-Lexikon zur jüdischen Aufklärung, Schulte, Christoph (ed.). url: https://www.uni-potsdam.de/en/haskala/haskala-in-biographien/levi (10 February 2024).
- Schulte, Christoph. 2022. Raphael Levi, Mathematiker und Astronom des 18. Jahrhunderts. [online source]. url: https://www.math.uni-potsdam.de/fileadmin/user_upload/Institutskolloquium/Schulte_22_04_15/Schulte_Kolloquium22apr15.pdf (10 February 2024).
- Sherburne, Edward (ed.). 1675. The Sphere of Marcus Manilius Made an English Poem: With Annotations and an Astronomical Appendix. London: Nathanael Brooke. (GoogleBooks, 17 September 2022).
- Snell, Willebrord. 1618. Coeli & siderum in eo errantium observationes Hassicae, illustrissimi principis Wilhelmi Hassiae Lantgravii auspicijs quondam institutae. Et specilegium biennale ex observationibus Bohemicis V. N. Tychonis Brahe. Nunc primum publicante Willebrordo Snellio. R. F. Quibus accesserunt, Ioannis Regiomontani & Bernardi Walteri Observationes Noribergicae. Lugdunum Batavorum: Justum Colsterum. url: https://archive.org/details/bub_gb_ss35ABDJP3gC/mode/2up (10 February 2024).
- Solon, Peter. 1971. The Six Wings of Immanuel Bonfils and Michael Chrysokokkes. In: Centaurus, 15, 1, 1–20. doi 10.1111/j.1600-0498.1971.tb00147.x.
- Spruyt, Joke. Peter of Spain. In: Stanford Encyclopedia of Philosophy (Winter 2019 Edition), Edward N. Zalta (ed.). url: https://plato.stanford.edu/entries/peter-spain/ (30 October 2022).
- Spruyt, Joke. 2005. Peter of Spain (13th century). In: Encyclopedia.com [online source]. url: https://www.encyclopedia.com/humanities/encyclopedias-almanacs-transcripts-and-maps/peter-spain-13th-century (30 October 2022).
- Stehliková, Dana. 2017. Christian von Prachatitz und sein lateinischer Herbarius: Zum Forschungsstand des ältesten Kräuterbuchs aus dem mittelalterlichen Böhmen / Cristannus of Prachatice and his Latin Herbarium. The state of research on the oldest herbarium of mediaeval Bohemia. In: Sudhoffs Archiv, 101, 2, 139–159. url: https://www.jstor.org/stable/26385713.
- Steinmetz, Max. 1986. Johann Virdung von Hassfurt, sein Leben und seine astrologischen Flugschriften. In: Astrologi hallucinati, stars at the end of the world in Luther's time, Zambelli, Paola (ed.), 195–214. Berlin: De Gruyter. doi 10.1515/9783110852134.195.
- Steinschneider, Moritz. 1867. Abraham Judäus - Savasorda und Ibn Esra. Zur Geschichte der Mathematischen Wissenschaften im 12. Jahrhundert. Zeitschrift für Mathematik und Physik, 12, 1–44.

- Steinschneider, Moritz. 1896. Mathematik bei den Juden. In: Bibliotheca Mathematica, Neue Folge, 10, 38–42. url: https://sammlungen.ub.uni-frankfurt.de/freimann/content/titleinfo/375269?lang=de (10 February 2024).
- Stöffler, Johannes. 1514. Joannis Stöfler Justingensis germani Tabulae Astronomicae. Tübingen: Anselmus. url: http://nbn-resolving.de/urn:nbn:de:bvb:12-bsb00001937-6 (30 October 2022).
- Tihon, Anne and Raymond Mercier. 1998. Georges Gémiste Pléthon, Manuel d'Astronomie. Louvain-la-Neuve: Bruylant-Academia. isbn 2872095055.
- Thorndike, Lynn. 1936. Johann Virdung of Hassfurt again. In: Isis, 25, 2, 363–371. doi 10.1086/347086. url: https://www.jstor.org/stable/225374.
- Thorndike, Lynn. 1937. Faust and Johann Virdung of Hassfurt. In: Isis, 26, 2, 321. doi 10.1086/347179. url: https://www.jstor.org/stable/224916.
- Thorndike, Lynn. 1943. Another Virdung Manuscript. In: Isis, 34, 4, 291–293. doi 10.1086/347820. url: https://www.jstor.org/stable/225630.
- Thorndike, Lynn. 1945. Franco de Polonia and the Turquet. In: Isis, 36, 1, 6–7. doi 10.1086/347897. url: https://www.jstor.org/stable/225669.
- Thorndike, Lynn. 1947. Notes and Correspondence, Johann Virdung of Hassfurt: dates of birth and death. In: Isis, 37, 1/2, 74. doi 10.1086/347973.
- Thorndike, Lynn. 1950. Pre-Copernican Astronomical Activity. In: Proceedings of the American Philosophical Society, 94, 4, 321–326. url: https://www.jstor.org/stable/3143515.
- Tihon, Anne. Chioniades, George (or Gregory). In: Encyclopedia.com [online source]. url: https://www.encyclopedia.com/science/dictionaries-thesauruses-pictures-and-press-releases/chioniades-george-or-gregory (23 March 2024).
- Uiblein, Paul. 1974. Johannes von Gmunden, der Begründer der Wiener Mathematikerschule, Kanonikus von St. Stephan in Wien und Pfarrer von Laa a. d. Thaya (+ 1442). In: Beiträge zur Wiener Diözesangeschichte 15, 3, 17–19. url: https://www.edw-presseservice.at/Wiener-Dioezesangeschichte1960-96/661/index.html (30 October 2022).
- Van Brummelen, Glen. 2018. The end of an error: Bianchini, Regiomontanus, and the tabulation of stellar coordinates. In: Archive for History of Exact Sciences, 72, 5, 547–563. url: https://www.jstor.org/stable/45211959.
- Van Helden, Albert. David (1564–1617) and Johannes (1587–1616) Fabricius. In: The Galileo Project [online source]. url: http://galileo.rice.edu/sci/fabricius.html (10 February 2024).
- Vervliet, H. D. L. 1978. Henricus Baers, Libellus Novus De Compositione Ac Usu Quadrantis Perquam Utile Leuven, Henricus Baers, 1537 [Nk 2III-Nat I, 4]. In: Post-Incunabula en Hun Uitgevers in de Lage Landen/Post-Incunabula and Their Publishers in the Low Countries, Vervliet, H. D. L. (ed.). Dordrecht: Springer. doi 10.1007/978-94-009-8828-6_89.
- Virdung, Johannes. 1542. Tabulae resolutae de supputandis siderum motibus. Nurnberg: Johannes Petreium. (GoogleBooks, 17 September 2022).
- Vogel, Kurt. 1973. John of Gmunden. In: Dictionary of Scientific Biography, Gillispie, Charles Coulston (ed.), VII, 117–122. New York: Charles Scribner's Sons. sbn 684-10118-1.
- Vogel, Kurt. John of Gmunden. In: Encyclopedia.com [online source]. url: https://www.encyclopedia.com/science/dictionaries-thesauruses-pictures-and-press-releases/john-gmunden (30 October 2022).
- Wagner, Herrmann. 1918. Die Entwicklung der wissenschaftlichen Nautik im Beginn des Zeitalters der Entdeckungen nach neuern Anschauungen. In: Annalen der Hydrographie und maritimen Meteorologie, 46, März/April, 105–118.
- Wagner, Herrmann. 1918. Die Entwicklung der wissenschaftlichen Nautik im Beginn des Zeitalters der Entdeckungen nach neuern Anschauungen. In: Annalen der Hydrographie und maritimen Meteorologie, 46, Mai/Juni, 153–173.
- Wagner, Herrmann. 1918. Die Entwicklung der wissenschaftlichen Nautik im Beginn des Zeitalters der Entdeckungen nach neuern Anschauungen. In: Annalen der Hydrographie und maritimen Meteorologie, 46, Juli/August, 215–233.
- Wagner, Herrmann. 1918. Die Entwicklung der wissenschaftlichen Nautik im Beginn des Zeitalters der Entdeckungen nach neuern Anschauungen. In: Annalen der Hydrographie und maritimen Meteorologie, 46, September/Oktober, 276–283.
- Wegener, Alfred. 1905a. Die astronomischen Werke Alfons X. In: Bibliotheca Mathematica, Zeitschrift für Geschichte der mathematischen Wissenschaften, III. Folge, 6, 129–185. Leipzig: Teubner.
- Wilson, Nigel G. 1967. The Libraries of the Byzantine World. In: Greek, Roman and Byzantine Studies, 8, 1, 33–80.
- Wußing, Hans. 2009. 6000 Jahre Mathematik, Volume 1. Berlin, Heidelberg: Springer. doi 10.1007/978-3-540-77192-0. isbn 978-3-540-77189-0.
- Zacuto, Abraham. 1496. Almanach Perpeetum Celestium Motuum. Facsimile Reprint, 1915. Munich: J. B. Obernetter.

Manuscripts

- Stadtbibliothek Nürnberg, Manuscript Cent. VI, 18, ff.46v–47r, Johannes de Gamundia: Tabulae, Tabula aequationis dierum com noctibus suis.
- Biblioteca Apostolica Vaticana, Manuscript Pal. Lat. 1375. URL: https://digi.vatlib.it/view/MSS_Pal.lat.1375 (30 October 2022).
- Biblioteca Medicea Laurenziana, Florence. Ashburnham 214, Tractatus Astrologicus. Lectiones MAG. JACOBI GALESSE in propositiones falsas. MAG. BARTH. DE VALENTIA de dierum Equationibus. Judicium anni 1470. Cod. cart. del XV secolo (1467), in quarto.

This page is intentionally left blank.

7

From Copernicus to Kepler

Cuius duæ ſunt cauſæ, cũ inæqualitas apparentiæ Solaris, tum etiam obliquitatis ſigniferi diſpari aſcenſio, prima quæ propter inæqualem Solis apparentem'q; motum exiſtit.

Nicolaus Copernicus, 1543, On Revolutionibus, III, 26

7.1 Nicolaus Copernicus

NICOLAUS COPERNICUS is considered the father of the modern heliocentric system, though possible influences of antique astronomy, predominantly Greek and Arabic, on his mindset are certainly not excluded.[1] While he did not tabulate the equation of time in his works, he mentions the equation of time in Chapter 26 of book III of *De revolutionibus*.[2] His equation of time is the difference between the right ascensions of the true and the mean Sun, which moves along the equator, and not along the ecliptic as in antiquity, and thus corresponds to the contemporary definition of the equation.[3] COPERNICUS appears to have used similar assumptions as PTOLEMY in the *Almagest*. He takes the longitude of the apogee to be $\lambda_\Gamma = 96;40°$ and the maximum value of the solar equation $c_{max} = 1;51°$, which distinguishes him from PTOLEMY whose accounts are based on $\lambda_\Gamma = 65;30°$ and $c_{max} = 2;23°$.[4] COPERNICUS' right ascension value table covers only one quadrant, from which we can then calculate the values of other quadrants. The values differ slightly from PTOLEMY's. Similar to PTOLEMY's equation of time, both contributions to COPERNICUS' equation of time can be expressed as the difference in the value at two different moments:[5]

[7.1]

$$\Delta E_1 = (\lambda_{\odot 2} - \lambda_{\odot 1}) - (\bar{\lambda}_2 - \bar{\lambda}_1) = c_2 - c_1$$
$$\Delta E_2 = (\alpha_{\odot 2} - \alpha_{\odot 1}) - (\lambda_{\odot 2} - \lambda_{\odot 1})$$

ΔE_1	contribution to the equation of time due to non-uniform motion
ΔE_2	contribution to the equation of time due to obliquity of the ecliptic

$\bar{\lambda}_{1,2}$	mean longitude	for both temporal moments
$\lambda_{\odot 1,2}$	true longitude	
$\alpha_{\odot 1,2}$	right ascension	
$c_{1,2}$	equation of centre	

from which the known expression follows:

[7.2]

$$\Delta E = \Delta E_1 + \Delta E_2 = (\alpha_{\odot 2} - \alpha_{\odot 1}) - (\bar{\lambda}_2 - \bar{\lambda}_1)$$

1 Nosonovsky, 2018, Abner of Burgos: The Missing Link between Nasir al-Din al-Tusi and Nicolaus Copernicus? In: Zutot 15, 25-30.
2 Copernicus, 1970, De Revolutionibus Orbium Coelestium, Rosen (trans.), Complete Works, II, pp.170–172.
3 Neugebauer, Swerdlow, 1984, Mathematical Astronomy and Copernicus's Revolutionibus, Part I, p.172.
4 Neugebauer Swerdlow, 1984, Part I, pp.172–179.
5 Neugebauer Swerdlow, 1984, Part I, p.173, equation 52, p.174, equation 53.

or for a particular epoch:

$$\Delta E = (\alpha_{\odot} - \bar{\lambda}) - (\alpha_0 - \bar{\lambda}_0)$$

$\bar{\lambda}$	mean longitude
$\bar{\lambda}_0$	mean longitude, for the epoch
$\alpha_{\odot}$	right ascension
α_0	right ascension, for the epoch

Among the astronomers who did not accept COPERNICUS' science was GIOVANNI BATTISTA RICCIOLI, who published the book the *New Almagest*, Lat. *Almagestum novum*, in 1651, which deals with the equation of time in Chapters 25 and 26 in the first book.

7.2 Rheticus and his successors

7.2.1 From Rheticus to Pitiscus

GEORG JOACHIM VON RHETICUS (1514–1574), real name GEORG VON LAUCHEN, Austrian mathematician, astronomer, cartographer and instrument maker, was the author, together with ACRONIUS, of the *Tabulae astronomicae in gratiam studiosae iuventutis seorsim aeditae*, published in 1542.[6] The first part of *De aequatione dierum* contains, obviously, the equation of time, the second part *De ascensionibus signorum in sphaera recta, & obliqua ad latitudinem. 52. gradum Vitebergae* has a right ascension and oblique ascension for 52° N and Wittemberg.[7] The values are obviously based on the *Alfonsine Tables*, as they are the same as in the 1524 edition of the *Alfonsine Tables*. In 1536, RHETICUS was awarded a master's degree and soon afterwards became Professor of Mathematics at the University of Wittemberg. He is best known for his work on trigonometry and trigonometric tables.[8] His fundamental works on trigonometry are the *Canon doctrinae triangulorum* of 1551, which contains the first tables of all six trigonometric functions, and a much more extensive work, published posthumously in 1599, the *Opus palatinum de triangulis*. RHETICUS studied with COPERNICUS for a time, was one of the first astronomers to accept COPERNICUS' teachings, and is also credited with the publication of COPERNICUS' work. In 1540, he published a review of COPERNICAN astronomy, *Narratio prima*, in which Earth is a planet that rotates on its axis and revolves around the Sun. He later brought COPERNICUS' manuscript to Nuremberg, where *De revolutionibus* was published in 1543.

6 JOHANNES ACRONIUS FRISIUS (1520–1564), Dutch mathematician, named after his hometown Akkrum in Friesland.

7 This rare work exists in several different editions. Some editions only contain Part II with right ascension and oblique ascension. For example, one of the editions of 1542, Bamberg City Library, VD16 J 280, urn:nbn:de:bvb:12-bsb00034792-9, and the edition of 1545, Bavarian State Library, Munchen, VD16 J 282, urn:nbn:de:bvb:12-bsb10889110-9.

8 For more information on trigonometric tables in this period and on RHETICUS' work, see Roegel, 2021a and 2021b. For a good historical introduction, see also Busard, 1971, Der Traktat "de Sinibus, Chordis at Arcubus" von Johannes von Gmunden, pp.73–81.

ACRONIUS, who worked with RHETICUS on the astronomical tables, was admitted to the University of Basel in 1542. In 1547, he became professor of mathematics, and in 1549 of logic. He was also a doctor. His mathematical and astronomical works include *De motu terrae*, *De sphaera*, *De astrolabio et annuli astronomici confectione* and an unpublished manuscript or chronicle *Cronicon* and *Prognosticon astronomica*.[9]

VALENTHINUS OTHO (around 1545/1546–1603) was a mathematician and astronomer who visited RHETICUS in Košice in 1574, where the latter died shortly afterwards. OTHO, with the help of patrons, struggled to continue RHETICUS' work, which he only completed in 1596 with the publication of the *Opus palatinum de triangulis*. The basis of the work is a table of sines for the first quadrant or 0 to 90° in 45″ intervals and for a radius of 1015. In addition, for radius 1010, the sine and cosine values are given in intervals of 10″, and the manuscript is six pages long for each degree, giving a measure of the work's mathematical complexity. In addition to COPERNICUS' manuscript of *De revolutionibus*, which RHETICUS owned, OTHO inherited other manuscripts from COPERNICUS, which BARTHOLOMEO PITISCUS, German, BARTHOLOMÄUS PITISCUS (1561–1613) later included in a *Thesaurus mathematicus*, published in Frankfurt am Main in 1613.

PITISCUS, who succeeded OTHO as professor of mathematics at Heidelberg, also improved RHETICUS' trigonometric tables.[10] The expression trigonometry is a contraction of his first appearance in *Trigonometria: sive de solutione triangulorum tractatus brevis et perspicuus*, published in Heidelberg in 1595 as the final part of a SCULTETUS' treatise *Sphaericorum libri tres methodicé conscripti et utilibus scholiis expositi*.[11] A revised edition of PITISCUS' work was published in Augsburg in 1600 under the title *Trigonometriae sive de dimensione triangulorum libri quinque*. PITISCUS' contribution to mathematics is very important because, in addition to the improved RHETICUS' tables, he also wrote works on trigonometry, addressing both plane and spherical trigonometry, in addition to the use of all six trigonometric functions.[12]

7.2.2 Erasmus Reinhold

ERASMUS REINHOLD was appointed Professor of Mathematics and Astronomy at the University of Wittenberg in 1536. Together with RHETICUS, who was also mathematician at the university, he was among the first to realise the importance of COPERNICUS' work. He was not satisfied with COPERNICUS' tables and tried to improve them. In 1551, after seven years of work, he

9 Folkerts, 2001, Acronius and Atrocianus are not the same person, as some biographical sources suggest. For the works of ATROCIANUS, see for example Guerra, Harich-Schwarzbauer and Hindermann (eds.), 2018.

10 Hilfstein, 1986/1987, Was Valentinus Otho a mathematics professor at the University of Heidelberg?

11 ABRAHAM SCULTETUS (1566–1625), German professor of theology.

12 For a comprehensive overview of PITISCUS' bibliography, see Archibald, 1949a, Bartholomäus Pitiscus (1561–1613), Mathematical Tables and Other Aids to Computation, 3, pp.390–397, and Pitiscus Revision of the Opus Palatinum Canon or Archibald, 1949b, Rheticus, with special reference to his Opus Palatinum, pp.556–561. Also Von Braunmühl, 1900, pp.220–226.

published the *Prutenic Tables* under the patronage of Duke ALFRED OF PRUSSIA; these were later reprinted several times. The *Prutenic Tables* were based on COPERNICUS' circular eccentric orbit model and proved that such a model could be used in practice.[13] The University of Wittenberg did not have its own observatory, and while REINHOLD used a wooden quadrant of his own to make astronomical observations, these were not the main focus of his activities. His work was mainly theoretical. For several decades, until the publication of KEPLER's *Rudolphine Tables*, the *Prutenic Tables* were among the most important astronomical handbooks of the time. They formed the basis for the ephemerides calculated by various astronomers in the period up to KEPLER, and in this connection it is worth mentioning the work of MICHAEL MÄSTLIN, GIOVANNI A. MAGINI and DAVID ORIGANUS.[14] In the *Prutenic Tables*, the equation of time is given in two tables (Figure 7.1).[15]

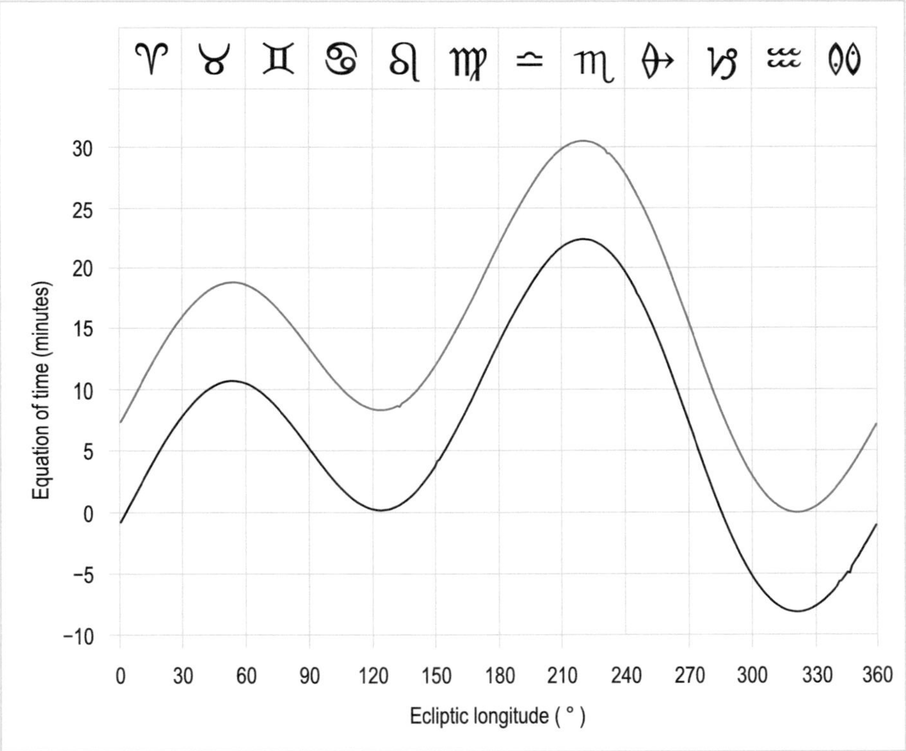

Figure 7.1 Equation of time in the *Prutenic Tables*. Sources for the diagram: Reinhold, 1585, Tabulae Prutenicae, ff.5–6.

13 For the equation of time for a circular eccentric orbit see Chapter 14. See also Morando, Savoie, 1996.

14 Oestmann, 2003, Ephemeridenwerke des 16. Jahrhunderts, eine wichtige Arbeitsgrundlage für Astronomen und Astrologen, p.155.

15 Reinhold, 1551 or 1585, Table, ff.5-6. See also Chapter 14, Table 14.2, note p.

One table, based on REGIOMONTANUS' data, roughly corresponds to the *Alfonsine Tables*. In the second table, the values are partly negative, calculated, according to the author, for the epoch of Christ's birth. The maximum (positive) value of this equation of time is about 22 minutes.[16] The values of the two equations of time differ by about 8 minutes and 9 seconds.

7.2.3 Ciprianus Leovitius

CIPRIANUS LEOVITIUS (1514?–1574),[17] now almost forgotten, born in Hradec Kralove in Bohemia, was the author of several works. Of great importance are his ephemerides, which were also used by TYCHO BRAHE.[18] He studied mathematics abroad, first in Wroclaw in 1540, then in Leipzig, and in Wittemberg in 1544, where he studied Latin alongside mathematics and astronomy with MELANCHTHON. We know that he was in Nuremberg in 1547, where he was in contact with JOHANNES SCHÖNER, and later also in Augsburg. He knew TYCHO BRAHE personally. The two men apparently had a friendly relationship, although TYCHO, at least according to the sources, was not very enthusiastic about LEOVITIUS' involvement in astrology. LEOVITIUS was, in the true sense of the word, what we now call an astronomical reckoner. His calculations were based on REGIOMONTANUS' tables, which were identical to SCHÖNER's, for an epoch in 1349.[19]

LEOVITIUS' works, which were outstanding in terms of their calculation, have been translated into other languages, and some have been published in several editions. In 1552 he compiled a table of the oblique ascension for 48° N latitude or Nuremberg, *Tabulae ascensionum obliquarum; et positionem particularis, ad latitudinem grad 48. minut 8.*[20] Unfortunately, his surviving legacy today consists only of more or less unedited manuscripts.[21] It is the manuscripts that best testify to the breadth of his work. Specifically, a group of manuscripts with the oblique ascension for each degree of the signs of the zodiacal circle and latitudes from 1° to 62°, preserved in the Vatican Apostolic Library, Pal. Lat. 1429, 1431 and 1432 Part A, comprising 748 folios or almost 1500 pages of full-page tables.[22]

7.2.4 Johannes Stadius

JOHANNES STADIUS (1527–1579), a Flemish astronomer, was the next very famous ephemeris maker after STÖFFLER. His teacher was most probably GEMMA FRISIUS[23], and it was GEMMA

16 This equation of time is interesting - it would require a thorough study of the *Prutenic Tables* to understand it well.

17 The year of birth 1524 also appears in the sources. For an explanation, see Mayer, 1903, pp.134–135.

18 A shorter biography and description of the works is available in Latin in Wydra, 1778, pp.20–25. For some comments see also Pilz, 1977, pp.224–226. For his works see Zinner, 1941.

19 Zinner, 1941, p.227.

20 Manuscript Pal. Lat. 1429, ff. 88r-100r.

21 For an overview of the extant manuscripts, see Oestmann, 2002, pp.9–12.

22 Biblioteca Apostolica Vaticana, *Tabulae ascensionum obliquarum cum latitudine meridionali, ad eleuationem poli grad.*

23 GEMMA FRISIUS (1508–1555), physician and lecturer in medicine at the University of Louvain. He is remembered for his achievements in mathematics and geography, and for the many globes and astronomical instruments he produced. His pupil was GERHARD MERCATOR, Latin GERARDUS MERCATOR (1512–1594), who attended FRISIUS' private lectures.

who, in a letter of 1555, encouraged him not to be afraid to publish the *Ephemerides Joannis Stadii*, a very important work based on the *Prutenic Tables* and a great influence on TYCHO BRAHE and NOSTRADAMUS. After observing a solar eclipse in Copenhagen in 1560, TYCHO turned to astronomical observations, and found STADIUS' *Ephemerides* of great help. In addition to taking the heliocentric system as a basis, STADIUS also abandoned the use of the *Alfonsine Tables*. FRISIUS' letter was published in several editions of *Ephemerides*.[24]

STADIUS also describes the equation of time, giving the values for 1° of ecliptic longitude in the table *Canon equationis dierum naturalium ex Copernici doctrina* on pages 120 and 121.[25] The equation of time is apparently based on the COPERNICAN model or a circular (eccentric) orbit. The values in the 1585 edition are evidently from the *Prutenic Tables*, since they are the same as those of ERASMUS REINHOLD's, *Prutenicae tabulæ coelestium motuum*, 1551, *Canon prior aequationis Ptolomæi & Copernici.*

7.3 Tycho Brahe

TYCHO BRAHE was an advocate of a geo-heliocentric model (Figure 7.4). The beginnings of a geo-heliocentric planetary model, as conceived by TYCHO, can be found in MARTIANUS CAPELLA, MACROBIUS and JOHN SCOTUS.[26] NĪLAKAṆṬHA's work also suggests a familiarity with such a model.[27] BRAHE knew the equation of time, but we do not know the sources from which it would be possible to conclude that he used it.[28] Obviously, he did not think that its commonly used form was correct.[29] TYCHO's values of the equation of time are about 7 to 8 minutes smaller than the actual values (Figure 7.3).[30]

He published the equation of time in two tables in *Astronomiae instauratae progymnasmata*. The *Progymnasmata* publication was produced gradually over a long period of time, initially printed in TYCHO's printing house in Uraniborg. It was still incomplete at the time of TYCHO's

24 For some of FRISIUS' other letters, see Curtze, 1874.

25 Stadius, 1585, pp.120–121.

26 Eastwood, 2001, Johannes Scottus Eriugena, Sun-Centred Planets, and Carolingian Astronomy. JOHN SCOTUS ERIUGENA (9th century), one of the most important philosophers of the CAROLINGIAN period, was also a translator.

27 In his commentaries on *Āryabhaṭīya*, *Āryabhaṭīyabhāṣya*, NĪLAKAṆṬHA presented a partially heliocentric model where Mercury, Venus, Mars, Jupiter and Saturn orbit the Sun and the Sun orbits Earth. Ramasubramanian et al., 1994, p.788.

28 SAMUEL VINCE in his work A Complete System of Astronomy, 1797, Chapter 4, Book 1, p.49 (138), states that BRAHE knew part of the equation of time due to non-uniform circular motion. This actually coincides with the equation of time from LONGOMONTAUS' book *Astronomia danica*.

29 Thoren, Christianson, 1990, The Lord of Uraniborg, p.491. A table with TYCHO's equation of time is included in DELAMBRE's book Historie de l'Astronomie Moderne, Paris 1821, vol. I, p.160, explanation on page 159. TYCHO is said to have deducted 7′36″ or ⅛ hours from each value. We do not know the reasons for such thinking. Thoren, Christianson, 1990, pp.491–494. For a table with BRAHE's equation of time, see also Ricciolli, 1665, Almagestum Novum, Table 12.

30 For a comparison, see Delambre, 1821, vol. 1, reprint 2006, pp.159–160. DELAMBRE used the following expression:
$E = -8'13'' \sin(\lambda_\odot - \lambda_\Gamma) - 9'56.''23 \sin 2\lambda_\odot + 12.''928 \sin 4\lambda_\odot - 0.''3726 \sin 6\lambda_\odot$.

death, but KEPLER later completed the work.[31] In the first table of the *Tabella aequationis dierum naturalium*, in the section describing the Sun's motion, the values are in whole minutes.[32] In the section on the Moon, the second table, *Tabula aequationis temporis*, only shows the contribution to the equation of time due to the obliquity of the ecliptic. In this case, since TYCHO used the equation of time as a mathematical tool to determine the position of the Moon, or more precisely to determine the annual inequality or annual equation of the Moon, the equation of time does not contain the equation of the centre. The equation of time, which contains only the part due to the obliquity of the ecliptic, is called the *aequatio Tychonica*, and the attempt to determine the exact motion of the Moon proved to be a tough nut to crack, both for TYCHO and for KEPLER. Both TYCHO and KEPLER tried to find an explanation for the irregular motion of the Moon, a long-known phenomenon, based on some kind of equation of time.[33] While several phenomena influence the irregular motion of the Moon, variation and evection have the greatest impact, and the annual inequality is also of historical interest.

The term evection was first used by ISMAIL BOULLIAU.[34] Evection, sometimes called a second lunar anomaly, is a disturbance in the Moon's motion with a period of 31⅘ days. The discovery is attributed to CLAUDIUS PTOLEMY, who described the phenomenon in Book IV of the *Almagest*. It is defined as:[35]

$$\text{evection} = 4586'' \cdot \sin(2D - M_{☽})$$

D	mean elongation of the Moon
$M_{☽}$	mean anomaly of the Moon (see Chapter 17)

Variation is the second largest disturbance of the Moon's motion, after evection, due to the Sun's influence. It is defined by the expression:[36]

$$\text{variation} = 2370'' \cdot \sin 2D$$

BRAHE discovered the variation of the Moon from observations in 1590, apparently independently of the Persian mathematician ABŪ AL-WAFĀʾ, but his discovery of the Moon's annual inequality, which is also attributed to him by some sources, is questionable.[37]

31 For more information, see Anschütz, 1886, p.163, note 1. TYCHO's observations were collected and edited by J. L. E. Dreyer and published in *Tychonis Brahe Opera Omnia*, Parts X–XIII, 1913–1927.

32 Brahe, 1610, p.113.

33 See Kepler, 1618, Epitome Astronomiae Copernicanae, ch.6, pp.711–712. Also Kepler, 1992, Epitome Astronomiae Copernicanae, pp.601–602.

34 Neugebauer, 1975, III, p.1109.

35 Brouwer, Clemence, 1961, Methods of celestial mechanics, pp.326–328. Smart, 1960, Celestial Mechanics, pp.289, 308–309. Danjon, 1980, Astronomie Générale, pp.287–289. Montenbruck, Pfleger, 2004, p.153. Moulton, 1964, Celestial Mechanics, pp.359–360.

36 Brouwer, Clemence, 1961, pp.324–326. Danjon, 1980, pp.277–279. Smart, 1960, p.289. Montenbruck, Pfleger, 2004, p.153. For historical comment see Neugebauer, 1975, III, pp.1109–1110.

37 Anschütz, 1886, pp.167–171 for Kepler's contribution see pp.201–219. See also Kepler, 1992, p.602.

The Moon's annual inequality, also known as the Moon's annual equation, is a minor contribution to the Moon's true longitude, which has a period of one anomalistic year, determined by two successive passages of Earth through perihelion. Because the Sun is closer in January, the Sun's influence is greater then than in early June. In addition to the Sun, the annual inequality is also influenced by the eccentricity of the orbit along which the Earth-Moon barycentre, or the centre of mass of these two bodies, orbits the Sun.

The annual inequality in the longitude of the Moon reaches its maximum value of about $11'9''$ halfway along Earth's path, or more precisely, halfway along the Earth-Moon barycentre[38], between the perihelion and aphelion, i.e. at the values of the Sun's mean anomaly $M = 90°$ and $M = 270°$. It is zero when passing through the perihelion and aphelion, or at $M = 0$ and $M = 180°$. The annual inequality is given by:[39]

$$\text{annual inequality} = -668'' \cdot \sin M$$

M mean anomaly of the Sun (see Chapter 17)

Christian Longomontanus references it in *Astronomia danica* (Table 7.1), in which case the maximum is approximately 9 minutes, which equals the contribution to the equation of time due to the obliquity of the ecliptic.[40] *Astronomia danica* otherwise describes Tycho's planetary system and is the best written source of its kind. Tycho's equation of time is also referenced by Kepler in the *Rudolphine Tables.*[41]

♈	0	0; 0
	10	3;16
	20	6;11
♉	30	8;25
	40	9;42
	50	9;51
♊	60	8;47
	70	6;36
	80	3;33
♋	90	0; 0
	100	3;33
	110	6;36
♌	120	8;47
	130	9;51
	140	9;42
♍	150	8;25
	160	6;11
	170	3;16
♎	180	0; 0
	190	3;16
	200	6;11
♏	210	8;25
	220	9;42
	230	9;51
♐	240	8;47
	250	6;36
	260	3;33
♑	270	0; 0
	280	3;33
	290	6;36
♒	300	8;47
	310	9;51
	320	9;42
♓	330	8;25
	340	6;11
	350	3;16
	360	0; 0

Table 7.1 Excerpt from Longomontanus' table with the equation of time or *aequatio Tychonica*. Source: *Astronomia danica*, Pars altera, Theoria planetarum, Book I, Chapter 3, p.59.

38 Barycentre = centre of mass.

39 Montenbruck, Pfleger, 2004, p.153. Brouwer, Clemence, 1961, Methods of celestial mechanics, pp.328–329. Smart, 1960, p.289. Moulton, 1964, Celestial Mechanics, p.348. For historical comment see Neugebauer, 1975, III, pp.1110–1111.

40 Longomontanus, 1640, Astronomia Danica, Book 1, Chapter 3, pp.198–199. See also Thoren, Christianson, 2007, The Lord of Uraniborg: A Biography of Tycho Brahe, p.494, note 25. A later advocate of such an equation of time was Ismael Boulliau, Lat. Ismael Bullialdus. Taton, Wilson (eds.), 2003, Planetary Astronomy from the Renaissance to the Rise of Astrophysics, Part A, p.197.

41 Reichert (ed.), 2014, Die Rudolphinischen Tafeln, p.178. Unlike *Astronomia danica* by Longomontaus with the values expressed in temporal minutes and seconds, the values are expressed in time degrees.

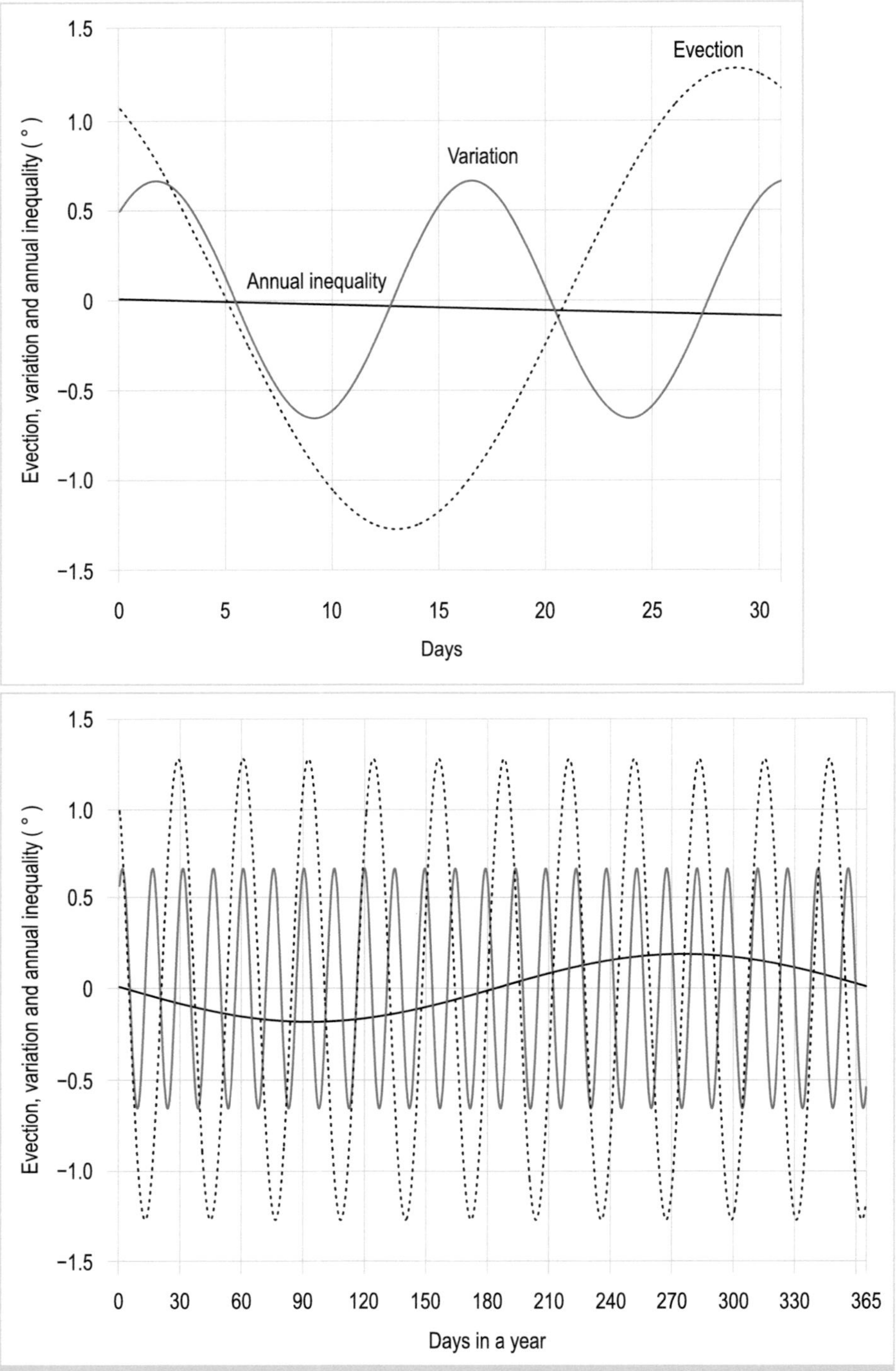

Figure 7.2 Evection, variation and annual inequality calculated for the year 2025. $M_{☽}$, M and D are calculated by expressions 17.40, 17.41 and 17.43.

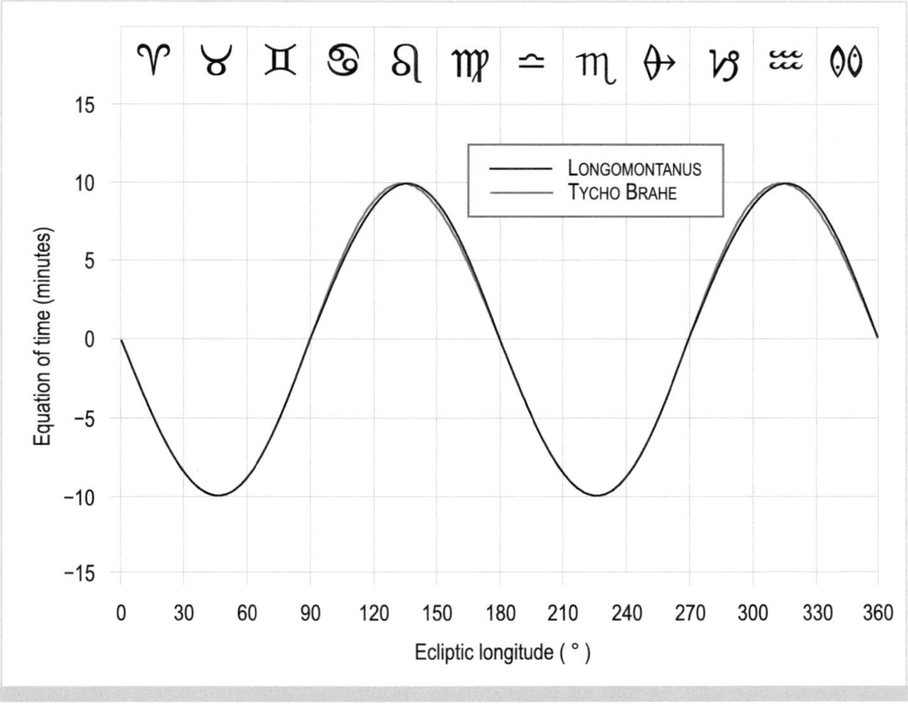

Figure 7.3 TYCHO BRAHE's equation of time or *aequatio Tychonica*, which LONGOMONTANUS also adopted. Sources for the diagram: Longomontanus, *Astronomia danica*, I.3, as well as Reichert (ed.), 2014, p.178.

7.4 Some other works before Kepler

7.4.1 David Origanus

DAVID ORIGANUS[42] was a professor of mathematics in Frankfurt/Oder and a maker of ephemerides and almanacs. The ephemerides *Ephemerides novae* of 1599 are based on a multi-year work and the COPERNICAN model or *Prutenic Tables.*[43] ORIGANUS discusses the equation of time in Chapter 5, where a table *Tabula ostendens quantitatem dierum & noctium in latitudine 52 grad. 20 min; cujusmodi est Francofurti Marchionum* is given, covering also a day and night lengths for the latitude of Frankfurt/Oder.[44] His second ephemerides of 1609, *Novae motuum coelestium ephemerides Brandenburgicae*, however, like those of LONGOMONTANUS, indicate a strong influence from TYCHO BRAHE. ORIGANUS advocated a geo-heliocentric system based on BRAHE's model (Figure 7.4), but he was also a proponent of the rotation of Earth on its own axis. In the 1609 work, he gives several tables with the equation of time - for the *Prutenic Tables*, or the COPERNICAN model, and for BRAHE's equation of time, which is the same as LONGOMONTANUS' equation of time (Table 7.1).

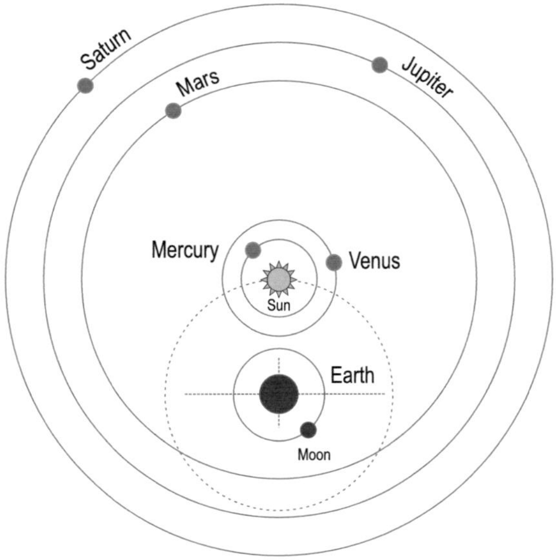

Figure 7.4 TYCHO BRAHE's geo-heliocentric planetary model. After Origanus, 1609, p.122.

42 Also known as DAVID TOST in German. For more information on his life and work, see Omodeo, 2011.
43 Omodeo, 2011, p.439.
44 Origanus, 1599, p.48.

7.4.2 Heinrich Rantzau

HEINRICH RANTZAU, also RANZOW (1526–1598), Latin RANZOVIUS, pseudonym CHRISTIANUS CILICIUS CIMBER, was a German humanist, writer, statesman, astrologer and collaborator of TYCHO BRAHE. From 1556 to 1598, he was administrator (Latin *produx cimbricus*), administering the territories of the principalities of Schleswig and Holstein, which were under Danish administration. He was a Renaissance scholar, bibliophile and humanist, and performed other duties as a statesman. His vision of general peace in Europe based on freedom of religion and thought was far ahead of its time. The table with the equation of time is published on pages 210 and 211 of his 1593 work *Diarium sive calendarium romanum*. This table is also apparently based on the COPERNICAN model or the *Prutenic Tables* and the values are the same as those of STADIUS.

7.4.3 Some other works

JACOB CUNO the Elder (c.1526–1583/84), also known as CONON, observed the weather in Berlin from September 1555, and in the same year was made Prince's Mathematician and Astronomer to the Court. His son JACOB CUNO the Younger (c.1558–1595) was admitted to the University of Frankfurt am Oder at an early age (1569) and then at the University of Leipzig in 1573.[45] JACOB CUNO the Younger is the author of the 1581 treatise on the different lengths of the days of the year, *Themata de diervm natvralivm differentia et aeqvatione*. He also wrote the 1583 *Theses de calendario Juliano, quo ecclesia nunc utitur, viciato ejusdem restitutione: Ut Pascha reliquaque festa mobilia mutationem suam inde habentia, debito tempore, secundum decreta Patrum & S. Conciliorum in posterum celebrente*r, published by J. Eichhorn in Frankfurt am Oder.

GIOVANNI PAOLO GALLUCCI (1538–1621) was a teacher, writer, translator and cartographer. He studied in Padua before moving to Venice, where he worked in science and the humanities for the rest of his life. His most important work is *Theatrum mundi et temporis* (*Theatre of the World and Time*), published in Venice in 1588. Here, GALLUCCI presents celestial phenomena and describes them in both astronomical and astrological terms. A special feature of the book is the presentation of the 48 constellations of PTOLEMY in a trapezoidal projection, showing the brightest stars of each constellation together with the corresponding mythological figure. GALLUCCI mentions the equation of time in his *Speculum uranicum* of 1593.

ILLARIO ALTOBELLI (1560–1637), a contemporary of GALILEO who taught mathematics at the University of Verona between 1599 and 1605, left several written works and is also known for his observation of the 1604 supernova. In *Tabulae regiae diuisionum duodecim partium coeli, et siderum obuiationum*, TYCHO's equation of time appears on page 100 in the table *Aequationes dierum Thiconica*.[46]

45 Friedländer, 1887, p.215, Jacobus Cuno, magistri Jacobi Cunonis, insignis astronomi electoris Brandenburgici, filius, Francophordianus. Erler, 1909, Band 1, p.71, Cuno Iac. Francoforden.

46 In the edition available on archive.org this table appears on page 52.

ELIAS MOLERI, aka ELIE DE MOLERI, Swiss theologian and astronomer, studied mathematics in Lausanne under GERVASIUS CURIANUS.[47] The equation of time is mentioned in his astronomical-religious treatise *Astronomicvs tēs psēfoforias ecleipsium solis et lunae epilogismos* of 1607.[48]

MOLERI also describes the lengths of the day in his *Opus novum astronomicum, ex quo doctrina de supputandis multis annis deliquiis iuxta tabulas prutenicas perspicue demonstratur, ex typis explicatur.*

7.5 Johannes Kepler and Rudolphine Tables

In Chapter 15 of the *Rudolphine Tables*, KEPLER describes the equation of time, which he calls a temporal (daily) compensation, as a difference in right ascensions of the mean and true Sun.[49]

KEPLER lists three types of the equation of time, or three contributions to the equation of time, which he tabulates directly or indirectly in the *Rudolphine Tables*, rounded to the nearest minute:[50]

- Tychonic equation (*aequatio Tychonica*) is the contribution due to the obliquity of the ecliptic, $\alpha_\odot - \lambda_\odot$ and is known as reduction of the ecliptic to the equator (Equation 14.18). By applying NAPIER rules on the triangle formed by the Sun, the vernal equinox, and the projection of the Sun on celestial equator (see Appendix D or Figure D.7) we can write:[51]

[7.3] $$T = \alpha_\odot - \lambda_\odot = \arctan(\cos\varepsilon \tan\lambda_\odot) - \lambda_\odot$$

The values of this term are tabulated in the *Rudolphine Tables*.[52] There is also a reference to TYCHO's equation of time in the correspondence between KEPLER and DAVID FABRICIUS.[53]

47 Graf, 1888, pp.72–75, also Verdun, 2005, p.17. CURIANUS aka GERVAIS DE LA COURT taught mathematics and astronomy at the Academy of Lausanne from 1594 to 1597. His lectures included the standard works of EUCLID, SACROBOSCO and one of the last classical philosophers of the Neoplatonic school, PROCLUS DIADOCHUS (5th century). Verdun, 2005, p.23, note 33.

48 Moleri, 1687, pp.143–145.

49 Reichert (ed.), 2014, p.53. For the first use of the *Rudolphine Tables* for calendar making see Herbst, 2007, Die erstmalige Benutzung von Kepler Rudolphinischen Tafeln für die Herstellung eines Schreibkalenders, Acta Historica Astronomiae, 40, 160–169.

50 Bialas, 1969, Die Rudolphinischen Tafeln von Johannes Kepler, Mathematische und astronomische Grundlagen, p.19. Also Reichert (ed.), 2014, pp.A.63–A.64.

51 Bialas, 1969, p.19, Equation 1.10. See also Appendix D and Chapter 15.

52 Reichert (ed.), 2014, p.178.

53 Anschütz, 1886, p.166, f.1. DAVID FABRICIUS (1564–1617), Lutheran pastor and astronomer, corresponded with KEPLER. He worked in Osteel in eastern Friesland, observing the Sun and sunspots.

- Astronomical equation (*aequatio temporis astronomica*) is the difference between the mean and the true anomaly, $M - \nu$ or the equation of the centre (Equation 15.31):[54]

$$A = M - \nu \approx 2e \sin E$$

In the *Rudolphine Tables*, the values are tabulated indirectly using the expression:[55]

[7.4] $$E = T - A = -(\lambda_\odot - \alpha_\odot) + (\nu - M)$$

e	eccentricity
E	eccentric anomaly
$\alpha_\odot$	right ascension
$\lambda_\odot$	true longitude
ν	true anomaly
M	mean anomaly

- The physical equation (*aequatio temporis physica*)[56] is based on incorrect assumptions and therefore has no practical value. It is interesting for historical reasons, however. KEPLER is said to have assumed, based on his lunar theory and the observations of TYCHO BRAHE, that the sunlight indirectly influences the movement of the Moon. He also posed the question of whether this might impact the speed of rotation of Earth.[57]

KEPLER arrived at the third part of the equation of time from an erroneous model of Earth's rotation, which he thought was faster in the winter and slower in the summer.[58] According to KEPLER, the total contribution of rotation due to solar radiation should be 5¼ days, which means $5\frac{1}{4} \cdot 360° = 1860°$ per year with steady rotation.[59] This rotation should be added to 360 days in a year, so the whole number of days in a year should be 365¼ days, which corresponds to the actual number of days in a year.[60] KEPLER's temporal compensation can be expressed through a geometric construction in the same way as the mean anomaly.[61]

The mean anomaly can be determined from Figure 7.5, where the mean anomaly is proportional to the surface of a sector ΓZS or consequently also to the surface ΓBS.

54 Bialas, 1969, p.19.
55 Bialas, 1969, pp.19–20. For tabulated values, see Reichert (ed.), 2014, p.178.
56 This is also indicated by SAMUEL VINCE in his work *A Complete System of Astronomy*, 1797, Chapter 4, Book 1, p.49 (138).
57 The first person who realised that these two phenomena are mutually exclusive was JEREMIAH HORROCKS. Horrocks to G. Crabrius, August 12, 1637, cited in *Opera Omnia*, Frisch (ed.), 6: 562. Also Stephenson, 1987, p.199, note 78. For a detailed understanding of this part of KEPLER's work, the reader should study the *Mysterium cosmographicum*.
58 Anschütz, 1887, p.6, also Anschütz, 1886, p.210.
59 Reichert (ed.), 2014, p.52. Kepler, 1992, pp.408–409, see also pp.316, 317 and 526. For an explanation, see Stephenson, 1987, p.166, Figure 39 and text on p.200.
60 360 days should follow from 12 synodic months, 30 days each. Linton, 2004, From Eudoxus to Einstein: A History of Mathematical Astronomy, p.219.
61 Stephenson, 1987, pp.166–167, Figure 39, and pp.198–201. Linton, 2004, p.219.

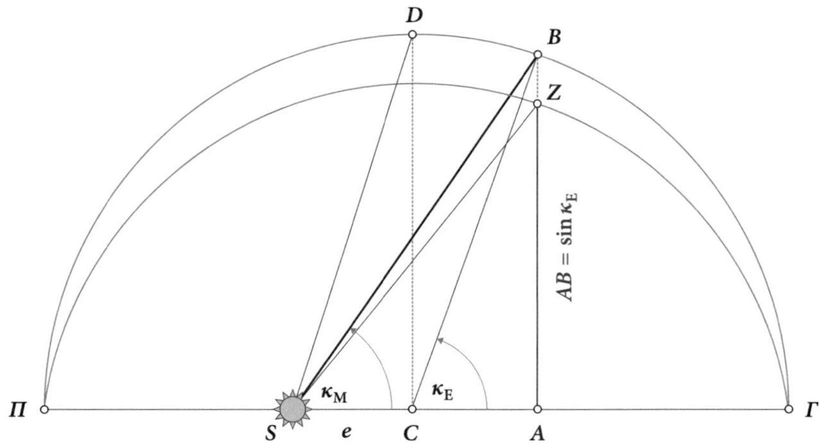

Figure 7.5 Geometrical model from which we can derive KEPLER's equation. Adopted from Smart, 1931, p.111, Figure 48 and Stephenson, 1987, p.166, Figure 39.

[7.5] $$\Gamma BS = \Gamma BC + \triangle BCS = \frac{\kappa_E}{2\pi}\pi + \frac{1}{2}CS \cdot AB$$

If we take into consideration that the radius $R = 1$, the circumference equals 2π and the area π, from which it is derived that the angle κ_M is proportional to double the surface of the geometric form ΓBS. From the geometric model, we thus obtain KEPLER's equation where the anomalies or pseudo-angles κ_M and κ_E are the angles measured from the apogee:[62]

[7.6] $$\Gamma BS = \frac{\kappa_E}{2\pi}\pi + \frac{1}{2}AC \cdot BS$$
$$= \frac{\kappa_E}{2} + \frac{e}{2} \cdot \sin \kappa_E = \frac{1}{2}(\kappa_E + e \cdot \sin \kappa_E)$$

In the same way, we also evaluate the temporal compensation that is supposed to contribute 5¼ days. The path of Earth's movement is proportional to the surface of the shape ΓBS, whereas the entire circle is 5¼ days, thus giving us the expression for the highest value of the imaginary temporal compensation under the consideration of radius $R = 1$:

[7.7] $$P = 5\,¼\ \text{day} \cdot \frac{\text{area } \triangle CSD}{\text{area of a circle}} = 7560\ \text{min} \cdot \frac{\triangle CSD}{\pi}$$

In adition: $$\triangle CSD = CS \cdot \frac{DC}{2} = \frac{e}{2}$$

62 Smart, 1931, Textbook on Spherical Astronomy, pp.111–114 or Stephenson, 1987, pp.200–201.

By taking into consideration an eccentricity of 0.018 from the expression above, we obtain the highest value of the physical equation, P_{max}, which is 21 min and 40 s or 5°25′.[63]

Values of the physical equation or *aequatio physica*, *P*, are given in the *Rudolphine Tables* indirectly, by the so-called *excessus physicus*, E_P for which:[64]

$$E_P = P + (\nu - M)$$

$$P = P_{max} \sin \nu = 5°25' \sin \nu$$

Taking the approximation for the equation of the centre $\nu - M \approx -2e \sin\nu$:[65]

$$E_P = (5°25' - 2e) \sin \nu$$

$$E_P = 3°21' \sin \nu$$

The values are tabulated to the nearest minute and are rounded using the following equation:[66]

[7.8] $$E = T + E_P = -(\lambda_\odot - \alpha_\odot) + 3°21' \sin \nu = -(\lambda_\odot - \alpha_\odot) + 3°21' \sin\left(\lambda_\odot - \underbrace{6°♋}_{\text{apogee}}\right)$$

In addition to the contribution due to the obliquity of the ecliptic and non-uniform circular motion, KEPLER also mentions the contribution of precession to the equation of time.[67]

In Chapter 6 of the *Epitome*, KEPLER described two monthly inequalities - the first, as a result of evection (*inaequalitates menstruas*), which is time-dependent (*temporanea*), and the second, invariable, as a result of variation, which is constant or unchanging (*perpetua*).[68]

63 Reichert (ed.), 2014, p.52, also Anschütz, 1887, p.6 or Stephenson, 1987, p.201.

64 Bialas, 1969, p.20.

65 Bialas, 1969, p.20.

66 Bialas, 1969, p.20, for tabulated values see Reichert (ed.), 2014, p.178.

67 Reichert (ed.), 2014, p.53, Equation I.15. Stephenson, 1987, Kepler's Physical Astronomy, pp.197–201. Kepler, 1953, Epitome Astronomicae Copernicae, Gesammelte Werke Band 7, p.184, lines 3–5, 31–36, pp.408–409, lines 39–17 and p.409, lines 20–24.

68 Kepler, 1992, p.449, De inaequalitatibus menstruis.

7.6 Jost Bürgi

7.6.1 Bürgi's logarithms and sine function

JOST BÜRGI (1552–1632), a Swiss instrument maker and mathematician, is another professionally remarkable but almost forgotten figure.[69] In 1572, as a clock and instrument maker, he entered the service of Prince WILHELM IV in Kassel. WILHELM IV showed considerable interest in astronomy and had an observatory built, which certainly contributed to the flourishing of astronomy in Kassel during this period.[70] In Kassel, during the reign of WILHELM IV, the height of the Sun was systematically observed and recorded using a four-foot quadrant, in addition to other observations. Solar altitude measurement from 1559 and 1561 are preserved in tabular form.[71]

BÜRGI went to Prague in 1605, where he remained a clockmaker until 1631, before returning to Kassel, where he died a year later. He obviously had no formal classical education, wrote in German, and had no knowledge of Greek or Latin, which must have limited his work, but he was clearly a good mathematician and clock or instrument maker. The legacy of clocks and globes commissioned by the Princes of Kassel is the pinnacle of precision mechanics at the time. In Kassel, he came into contact with CHRISTOPH ROTHMANN, PAUL WITTICH (1555?–1587) and NICOLAUS REIMER URSUS (1551–1600). In Prague, he worked with KEPLER. WITTICH and URSUS had previously worked with TYCHO when he was still on the island of Hven in Denmark.[72] URSUS clearly held BÜRGI in high esteem, as is evident from the introduction to his 1588 *Fundamentum astronomicum*:[73]

> *Hæc autem maximè profunda atq; perplexa anteà doctrina, ad quãtam facilitatem mei charifsimi Præceptoris Iufti Byrgi Heluetij indefeꝛo ftudio, fedulisque meditationibus, atque diuturnis cogitationibus, meoq; hoc opufculo ac ftylo, meaque quam vobis hic exhibeo Methodo, redacta, correcta, emendataque fit, non eft meum dicere.*
>
> *Ursus, Fundamentum astronomicum, 1588*

69 For information on his life and work see Wolf, 1858, Biographien zur Kulturgeschichte der Schweiz, pp.57–80 or Staudacher, 2014, Jost Bürgi, Kepler und der Kaiser.

70 Von Mackensen, 1988, Die erste Sternwarte Europas mit ihren Instrumenten und Uhren; 400 Jahre Jost Bürgi in Kassel. Leopold, 1986, Astronomen, Sterne, Geräte - Landgraf Wilhelm IV. und seine sich selbst bewegenden Globen, pp.15–34.

71 Von Mackensen, 1988, Die erste Sternwarte Europas mit ihren Instrumenten und Uhren: 400 Jahre Jost Bürgi in Kassel, p.11.

72 ROTHMANN was also with BRAHE for a short time. ROTHMANN, whose birth and death dates are unknown, left no written work, but we know that he collaborated with BÜRGI on mathematical work. Lenke, Roudet, 2014, pp.223–242.

73 For the German translation see Launert (ed., trans.), 2012, p.27. URSUS's work contains, among other things, the basics of sexagesimal calculus, chapters on plane and spherical trigonometry, and a description of BÜRGI's procedure for calculating the sine function. Launert (ed., trans.), 2012, pp.64–66.

WITTICH also brought to Kassel the trigonometric identity that apparently originated in the work of JOHANNES WERNER and was later used by ROTHMANN and BÜRGI:[74]

$$\sin\alpha\sin\beta = \frac{1}{2}(\sin(90° - \alpha + \beta) - \sin(90° - \alpha - \beta))$$

The above expression, in its modern form, is written as:

$$\sin\alpha\sin\beta = \frac{1}{2}(\cos(\alpha + \beta) + \cos(\alpha - \beta))$$

BÜRGI is known for his own calculation of the sine function, but he is not known to have recorded anywhere how he developed his algorithm. It may have been the above identity that sparked BÜRGI's interest in the sine function, as the calculation required sine tables. BÜRGI's calculation process begins in a familiar way. He drew regular polygons around the circle and calculated the sines of the angles 18°, 30°, 36°, 45° and 60° that follow from the circumscribed decagon, hexagon, pentagon, square and triangle.[75] Using the division of the angle 1½° or the geometric construction in Figure 7.6, he calculated the value of sin 1°.[76]

The angle 1½° is divided into six parts, or angles of 15′. In the figure, ZA is the radius and U is the projection of point E onto it. M, R, T, P and Q are the projections of points A, P, S, ... onto the UE. It follows from geometric relationships that $UT = \sin 45'$, but it is also true that $\frac{1}{6}\,UE \approx \sin 15'$, or that $UT + \frac{1}{6}\,UE$ is a good approximation to sin 1°. Using this approximation of sin 1° and trigonometry, BÜRGI was able to continue the calculation for the whole sine table.

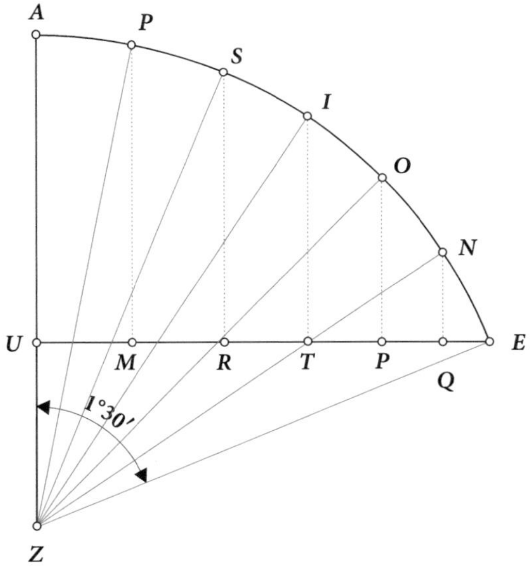

Figure 7.6 Division of the angle 1½° into six parts, geometric construction for calculating sin 1°. Summarised by Folkerts et al., 2016, Figure 2.

74 List, Bialas, 1973, Die Coss von Jost Bürgi, p.78. Van Brummelen, 2009, pp.264–265, also p.265 note 128. JOHANNES WERNER (1468–1522), a German astronomer from Nuremberg, studied the works of REGIOMONTANUS. His work *De triangulis sphaericis* was at least as high in quality as REGIOMONTANUS' *De triangulis omnimodis* but remained unpublished until 1907. Björnbo, 1907, contains the Latin text with some commentaries in German. When it came to knowledge transfer, WITTICH's contribution was obviously quite important. See, Roegel, 2010, p.3 note 4

75 See also Figure 0.6 and the explanation in the Introduction.

76 Folkerts et al., 2016, Jost Bürgi's method for calculating sines.

BÜRGI described a tabular method for calculating the value of the sine function in his treatise *Fundamentum astronomiae*, which is preserved as a manuscript. BÜRGI's notation is unusual, as the table of calculations results is written from right to left in what looks like Arabic script, so the procedure may be based on older ancient sources.[77]

For ease of understanding, BÜRGI's values are tabulated below in the usual left-to-right order, opposite to how they appear in BÜRGI's text, and given in decimal rather than the BÜRGI's sexagesimal.[78]

A	A_1	B	B_1	...	E
0		0			0
	$63 = 2 + 61$		2064		
2		$63 = 63 + 0$			2235060
	$61 = 4 + 57$		2001		
4		$124 = 61 + 63$			4402208
	$57 = 6 + 51$		1877		
6		$181 = 57 + 124$			6435596
	$51 = 7 + 44$		1696		
7		$232 = 51 + 181$			8273441
	$44 = 8 + 36$		1464		
8		$276 = 44 + 232$			9859902
	$36 = 9 + 27$		1188		
9		$312 = 36 + 276$			11146776
	$27 = 10 + 17$		$876 = 339 + 537$		
10		$339 = 27 + 312$			12094962
	$17 = 11 + 6$		$537 = 356 + 181$		
11		$356 = 17 + 339$			12675649
	$6 = {}^{12}\!/_{2}$		$181 = {}^{362}\!/_{2}$		
12		$362 = 6 + 356$			12871192

In column A, the approximations of the individual sine values (0, 10°, 20°, ... 90°) are rounded to the integer values. They range up to 12 because BÜRGI calculated them for a radius $R = 12$. Reading up from the bottom of column A_1, we can follow the underlying rule: first, halve the bottom value of column A and write it in the bottom row of column A_1. Next, add the penultimate value of column A to the result and write the sum one row higher in column A_1, and so on. The values obtained correspond approximately to the approximations of the sines of the intermediate angles of 5°, 15°, 25°,... rounded to integers.

Next, column B is computed by entering 0 ($= \sin 0°$) in the top row and working downwards in a similar way to the previous column.

The more repetitions, the more accurate the result. After four such iterations, BÜRGI reached the values in column E (but as mentioned in sexagesimal notation).

77 For a possible interpretation of Indian sources, see Wagner, Hunziker, 2019.

78 Folkerts et al., 2016, p.140. While BÜRGI's values in the first column are not exact integer approximations for the radius $R = 12$, the order of magnitude is appropriate.

The sines 10°, 20°,... are calculated from the values in the last column as the ratio of the values in column E for the sine of the given angle and sin 90°. For example, for the values of sin 10° and sin 20°:

sin 10° = 2,235,060 : 12,871,192 = 0.173 648 **25** (exact value 0.173 648 18...)
sin 20° = 4,402,208 : 12,871,192 = 0.342 020 **22** (exact value 0.342 020 14...)
where bold digits indicates deviations from the true values.

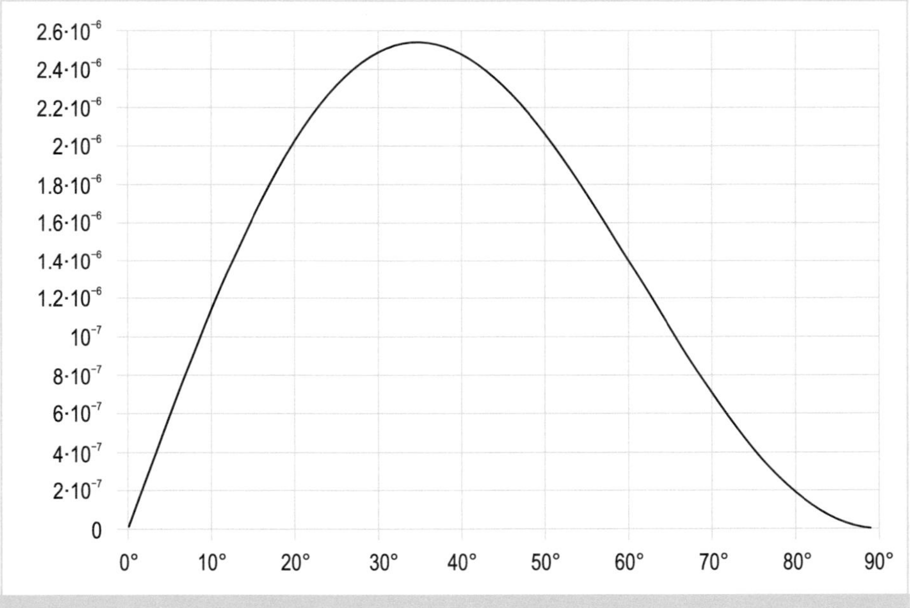

Figure 7.7 Error in the calculated values after four iterations for radius $R = 180$, taking the integer approximations of the sine function, from sin 0° to sin 90°, as the first column. See also Roegel, 2016, p.4.

BÜRGI is also the author of a treatise titled *Coss*, which describes the elements of algebra, the division of angles, and the calculation of the chord and sine functions.[79] He is also the author of the table of sequences that calculate logarithms and is considered one of the fundamental works in this field.[80] BÜRGI prepared the tables independently of JOHN NAPIER, so both can be considered fathers of logarithms.[81] Shortly after NAPIER, HENRY BRIGGS, who was presumably familiar with BÜRGI's work on the calculation of the sine function, calculated decimal logarithms.[82]

79 List, Bialas, 1973.

80 Clark, 2019, Jost Bürgi's Aritmetische und Geometrische Progreß Tabulen (1620).

81 Folkerts et al., 2016, p.137 note 17. See also Waldvogel, 2012, Jost Bürgi and the discovery of the logarithms. For a good commentary on the significance of BÜRGI and NAPIER's contributions, see Roegel, 2010, pp.29–31.

82 Nicollier, 2018, p.19.

KEPLER, in the *Rudolphine Tables*, states the following about the BÜRGI logarithms:[83]

> *Diese logistischen Hochzahlen haben auch Justus Bürgi viele Jahre vor der neperschen Ausgabe den Weg zu genau diese Logarithmen gewiesen. Er war jedoch ein zögerlicher Mensch und hütete seine Geheimnisse, so hat er sein geistiges Erzeugnis gleichsam bei der Geburt im Stich gelassen und nicht zur öffentlichen Verwendung aufgezogen.*
>
> *Kepler, 1627, Tabulae Rudolphinae (Reichert (ed.), 2014, p.29)*

KEPLER left Prague in 1612, so BÜRGI must have prepared the Tables well before they were printed in Prague in 1620 under the title *Progreß Tabulen*. The values in the tables mathematically correspond to the following sequence:[84]

$$a_0 = 1 \qquad a_{n+1} = a_n + \frac{a_n}{10000} \qquad n = 0, 1, \ldots, 23027$$

7.6.2 Clocks and globes

The earliest known example of an irregularly shaped gear, apparently used to compensate or correct the time in a clock, was incorporated in an astronomical clock designed by RICHARD OF WALLINGFORD at St Albans Abbey. WALLINGFORD described the clock in 1327 in his *Tractatus horologii astronomici* (*Treatise on the Astronomical Clock*), which survives in fragments.[85]

The masterwork of the Italian clockmaker GIOVANNI DONDI, also from the 14th century, is also well documented;[86] an astronomical clock, also with irregularly shaped gears, was made by DONDI in 1348–1364 and described in his treatise *Tractatus astrarii.*

Clocks and globes were made for the princes of Kassel in the 16th century by the watchmakers and precision instrument makers EBERHARD BALDEWEIN of Marburg, HANS BUCHER, CHRISTIAN HEIDEN and GERHARD EMMOSER before BÜRGI.

BÜRGI probably acquired his mathematical knowledge from CONRAD DASYPODIUS in Strasbourg and his watchmaking skills from ISAAC and JOHANN HABRECHT of Diessenhofen in Switzerland,

83 Tabulæ Rudolphinæ, 1627, Præcepta, Caput III, p.21. German translation from Reichert (ed.), 2014, Die Rudolphinischen Tafeln.

84 Waldvogel, 2012, p.15.

85 North, 1976, Richard of Wallingford, I Translation, pp.444–523, II Commentary, pp.309–370, III Appendices, pp.61–74. For the commentary on possible interpretation of non-circular wheels see II. 4.6, pp.342–348 and II, p.70. For the equation of time see North, 1976, III, Alfonsine Tables for the Albion, p.100.

86 Addomine et al., 2018, A landmark in the history of non-circular gears design: The mechanical masterpiece of Dondi's astrarium.

but there is no definite evidence for this.[87] Between 1571 and 1574, these two clockmakers built the astronomical clock in the cathedral of Strasbourg according to the designs of Dasypodius. Bürgi's most important works include instruments, specifically clocks and five surviving globes. His 1591 table clock has a gear movement that allows the two corrections to the uniform movement that are part of the equation of time.[88] The clock is kept by the Kassel Museum and its movement is basically a kind of analogue computer, not only showing the variable lengths of the days, but also the variable lengths of the lunar months.[89] Bürgi was also well-versed in sundial planning. An example of a sundial with gnomons on different planes is Bürgi's sundial from around 1590. It combines one horizontal sundial, two vertical sundials and one on a tilted face.[90] Some further surviving examples of clocks and globes with corrections to the equation of time from this period are:[91]

- Dresden planet clock from 1568 by Eberhard Baldewein, now in the Mathematisch-Physikalischer Salon, Zwinger in Dresden.
- The Great Globe of 1575, made by Baldewein and Hans Bucher, a watchmaker.
- Gerhard Emmoser's Globe of 1579.
- Bürgi's Globe of 1582?
- A Bürgi's clock with an armillary sphere, probably made in 1585?
- Bürgi's Moon anomaly clock, made in 1590–91.
- Bürgi's small globe of 1594, which has a special place among Bürgi's globes.[92]

87 The study of mathematics in Strasbourg was apparently at a fairly high level at that time. Dasypodius (c.1530–1600) became professor of mathematics at Strasbourg in 1558 and is known for the didactic material he compiled for his students. See also Waldvogel, 2012, pp.3–4.

88 Karlsen, 1983, p.66. Also Wenzel, 1983b, p.156.

89 Clock catalogue number 23, see for more information: Wenzel, 1983, Alte Uhren, Heft 1 and Heft 2 in Karlsen 1983.

90 Zinner, 1957, Deutsche und Niederlandische astronomische Instrumente des 11. Bis 18. Jahrhunderts, p.79, Table 9. Astronomisch-Physikalische Kabinett der Staatlichen Kunstsammlungen Kassel, inv. no. 1979-31. Leopold, 1986, pp.48–53.

91 Leopold, 1986, pp.69, 98–100, 109, 143–144, 155, 170, 183.

92 For an explanation of the mechanism of the equation of time on the globe, see Leopold, Pechstein, 1977, pp.53–58.

Bibliography

- Addomine, M, G. Figliolini and E. Pennestri. 2018. A landmark in the history of non-circular gears design: The mechanical masterpiece of Dondi's astrarium. In: Mechanism and Machine Theory Journal, 122, 219–232. doi 10.1016/j.mechmachtheory.2017.12.027.
- Alfonso X. 1492. Tabulae Astronomice Alfonsi Regis. (Google Books, 9 November 2022).
- Alfonso X., Alfonsus de Cordoba and Luca Gaurico. 1524. Alfonsi Hispaniarum Regis Tabule [et] L. Gaurici Artium doctoris egregij Theoremata quorum hic est index ... In calce huius libri seorsum annexe sunt tabul[a] e Elisabeth Regin[a]e nuper castigat[a]e & in ordinem redact[a]e per. L. Gauricum cum additionibus & Nouis problematibus eiusdem Gaurici. Venice: Junta. url: https://nbn-resolving.org/urn:nbn:de:urmel-0fb92c27-9ad0-490e-82d0-274b93fea45f4-00004523-00154 (31 October 2022).
- Altobelli, Ilario. 1628. Tabulae regiae diuisionum duodecim partium coeli, et siderum obuiationum, ad ... Macerata: Baptistae Bonomi. (Google Books, 5 February 2023). url: https://archive.org/details/tabulaeregiaediu00alto/mode/2up? (5 August 2024).
- Anschütz, C. S. J. 1886. Ueber die Entdeckung der Variation und der jährlichen Gleichung des Mondes. In: Zeitschrift fur Mathematik und Physik, Schlömilch, O., E. Kahl and M. Cantor (eds.), Historisch-literarische Abtheilung, 31, 161–171, 201–219. Leipzig: Teubner.
- Anschütz, C. S. J. 1887. Ueber die Entdeckung der Variation und der jährlichen Gleichung des Mondes. In: Zeitschrift fur Mathematik und Physik, Schlömilch, O., E. Kahl and M. Cantor (eds.), Historisch-literarische Abtheilung, 32, 1–15. Leipzig: Teubner. (Google Books, 26 August 2023).
- Archibald, R. C. 1949a. Bartholomäus Pitiscus (1561–1613). In: Mathematics of Computational Tables and Other Aids to Computation, 3, 25, 390–397. url: https://www.jstor.org/stable/2002363.
- Archibald, R. C. 1949b. Rheticus, with special reference to his Opus Palatinum. In: Mathematics of Computational Tables and Other Aids to Computation, 3, 28, 552–561. url: https://www.jstor.org/stable/2002258.
- Bialas, Volker. 1969. Die Rudolphinischen Tafeln von Johannes Kepler Mathematische und astronomische Grundlagen. Bayerische Akademie der Wissenschaften, Mathematisch-Naturwissenschaftliche Klasse, Abhandlungen, Neue Folge, Heft 139. Munchen: München: Verlag der Bayerischen Akademie der Wissenschaften in Kommission bei der C.H.Beck'schen Verlagsbuchhandlung München. url: https://gateway-bayern.de/BV003028967 (3 March 2024).
- Bialas, Volker. 1978. Ephemerides of the early seventeenth century. In: Vistas in Astronomy, 22, part 1, 21–26. doi 10.1016/0083-6656(78)90004-1.
- Björnbo, Axel Anthon (ed.). 1907. De triangulis sphaericis libri quatuor; De meteoroscopiis libri sex. Leipzig: Teubner: url: https://archive.org/details/ioannisvernerid00rhgoog (10 February 2024).
- Brahe, Tycho. 1610. Astronomiae instauratae progymnasmata. Frankfurt: Gottfried Tambach. url: https://archive.org/details/A021051/mode/2up (26 November 2022).
- Brouwer, Dirk and Gerald M. Clemence. 1961. Methods of Celestial Mechanics. New York, London: Academic Press.
- Busard, Hubert L. L. 1971. Der Traktat De sinibus, chordis et arcubus von Johannes von Gmunden. In: Denkschriften der Akademie der Wissenschaften, 116, 3, 73–113. Vienna: Springer. doi 10.1007/978-3-7091-4013-0_1.
- Busard, Hubert L. L. 1975. Pitiscus, Bartholomeo. In: Dictionary of Scientific Biography, Gillispie, Charles Coulston (ed.), XI, 3–4. New York: Charles Scribner's Sons. isbn 0-684-10122-X.
- Busard, Hubert L. L. Pitiscus. In: Encyclopedia.com [online source]. url: https://www.encyclopedia.com/science/dictionaries-thesauruses-pictures-and-press-releases/pitiscus-bartholomeo (30 October 2022).
- Clark, Kathleen. 2019. Jost Bürgi's Aritmetische und Geometrische Progreß Tabulen (1620), Edition and Commentary. New York: Springer. isbn 978-1-4939-3160-6. doi 10.1007/978-1-4939-3161-3.
- Copernicus, Nicolaus. 1543. De Revolutionibus. Nürnberg: Johannes Petreius. url: https://archive.org/details/nicolaicopernici00cope_1/mode/2up (10 February 2024).
- Copernicus, Nicholaus. 1978. Complete Works, II. Rosen, E. (transl.) and J. Dobrzycki (ed.). Warsaw: Polish Scientific Publishers.
- Cuno, Jacob. 1581. Themata De Diervm Natvralivm Differentia Et Aeqvatione. In: Pvblicam Collationem Ad XII. Avgusti proposita A M. Iacobo Cvnone Filio. Frankfurt an der Oder: Eichhorn. url: https://katalog.ub.uni-leipzig.de/Record/0-1109495129 (30 October 2022).
- Curtze, Maximilian. 1874. Fünf ungedruckte Briefe von Gemma Frisius. In: Archiv der Mathematik und Physik, 56, 313–325.
- Danjon, André. 1980. Astronomie Générale. Second edition, Reprint 1994. Paris: Albert Blanchard. isbn 2-85367-068-6.

- Delambre, Jean-Baptiste. 1821. Historie de l'Astronomie Moderne, Volume I, reprint 2006. Paris: Éditions Jacques Gabay. ISBN 2-87647-252-X.
- Dreyer, John Louis Emil. 1913–1927. Tychonis Brahe Opera Omnia, Volumes X-XIII.
- Dutka, Jacques. 1997. A note on Kepler's equation. In: Archive for History of Exact Sciences, 15, 1, 59–65. DOI 10.1007/BF00376451.
- Eastwood, Bruce Stansfield. 2001. Johannes Scottus Eriugena, Sun-Centred Planets, and Carolingian Astronomy. In: Journal for the History of Astronomy, 32, 4, 281–324. DOI 10.1177/002182860103200.
- Eis, Gerhard. 1953. Acronius, Johannes. In: Neue Deutsche Biographie 1, 39 [online source]. URL: https://www.deutsche-biographie.de/pnd130005304.html#ndbcontent (30 October 2022).
- Erler, Georg. 1909. Die iüngere Matrikel der Universität Leipzig, 1559 – 1809, Band 1, Die Immatrikulationen vom Wintersemester 1559 bis zum Sommersemester 1634. Leipig: Giesecke & Devrient. URL: https://digital.slub-dresden.de/werkansicht/dlf/451750/1 (30 October 2022).
- Folkerts, Menso. 2001. Eine Verwechslung mit Folgen: Die Humanisten Acronius und Atrocianus. In: Sudhoffs Archiv, 85, 1, 55–63. URL:https://www.jstor.org/stable/20777786.
- Folkerts, Menso. 2014. Eine bisher unbekannte Schrift von Jost Bürgi zur Trigonometrie. In: Arithmetic, Geometrie und Algebra in der frühen Neuzeit, Gebhardt, Rainer (ed.), 107–114. Annaberg-Bucholz: Adam-Ries-Bund. ISBN 978-3-944217-06-2.
- Folkerts, Menso, Dieter Launert and Andreas Thom. 2016. Jost Bürgi's method for calculating sines. In: Historia Mathematica, 43, 2, 133–147. DOI 10.1016/j.hm.2016.03.001.
- Friedländer. Ernst. 1887. Aeltere Universitäts-Matrikeln. I. Universität Frankfurt a.O. Leipzig: T. Hirzel. URL: https://archive.org/details/bub_gb_6rVBAAAAYAAJ/mode/2up (30 October 2022).
- Gaab, Hans and Pierre Leich (eds.). 2019. Simon Marius and His Research. Historical & Cultural Astronomy. Cham: Springer. DOI 10.1007/978-3-319-92621-6. ISBN 978-3-319-92620-9.
- Gallucci, Giovanni Paolo. 1593. Speculum Uranicum. URL: https://www.e-rara.ch/zut/doi/10.3931/e-rara-49833?lang=it, https://archive.org/details/ARes01204 (30 October 2022).
- Gaurici, Lucae. 1560. Super tabulis directionum Ioannis Monteregiensis quod dam supplementum. Rome: Vincentius Luchinus. (Google Books, 18 September 2022).
- Gingerich, Owen. 1973. From Copernicus to Kepler: Heliocentrism as Model and as Reality. In: Proceedings of the American Philosophical Society, 117, 6, 513–522. URL: https://www.jstor.org/stable/986462.
- Gingerich, Owen. 1975. Crises verses aesthetic in the Copernican revolution. In: Vistas in Astronomy, 17, 85–94. DOI 10.1016/0083-6656(75)90050-1.
- Graf, Johann Heinrich. 1888. Geschichte der Mathematik und der Naturwissenschaften in bernischen Landen, Band I, das XVI Jahrhundert. Bern: J. K. Wyss. (Google Books, 18 September 2022).
- Guerra, Christian, Henriette HAarich-Schwarzbauer and Judith Hindermann (eds.). 2018. Johannes Atrocianus: Text, Übersetzung, Kommentar. Hildesheim: Georg Olms Verlag. ISBN 978-3-487-42255-8.
- Günther, Siegmund. 1887. Origanus, David. In: Allgemeine Deutsche Biographie, 24, 422, [online source]. URL: https://www.deutsche-biographie.de/sfz73754.html (16 June 2023).
- Herbst, Klaus-Dieter. 2007. Die erstmalige Benutzung von Kepler Rudolphinischen Tafeln für die Herstellung eines Schreibkalenders. In: Kepler, Galilei, das Fernrohr und die Folgen; Acta Historica Astronomiae, 40, Gaulke, Karsten and Jürgen Hammel (eds.), 160–169. Frankfurt am Main: Harri Deutsch. ISBN 978-3-8171-1867-0.
- Handelmann, Gottfried Heinrich. 1888. Rantzau, Heinrich. In: Allgemeine Deutsche Biographie 27, 278–279 [online source]. URL: https://www.deutsche-biographie.de/pnd11898716X.html#adbcontent (30 October 2022).
- Hilfstein, Erna. 1986/87. Was Valentinus Otho a mathematics professor at the University of Heidelberg? In: Organon, 22–23, 221–225.
- Karlsen, Helge B. J. 1983. A 16th Century Astronomical Tableclock by Jost Bürgi (Justus Borgen) with non-uniform motions of the Sun and Moon. In: Antiquarian Horology - Proceedings of the Antiquarian Horological Society, 14, 1, 63–75.
- Kepler, Johannes. 1618. Epitome astronomiae copernicanae. Linz: J. Plancus. URL: https://archive.org/details/epitomeastronomi02kepl/mode/2up (10 February 2024).
- Kepler, Johannes. 1627. Tabulæ Rudolphinæ. (Google Books, 9 July 2023).
- Kepler, Johannes. 1866. Opera Omnia, volume 6, Frisch, Christian (ed.). Frankfurt am Main, Erlangae: Verlag Heyder & Zimmer.
- Kepler, Johannes. 1953. Epitome Astronomicae Copernicae, Gesammelte Werke Band 7. Von Dyck, Walther, Max Caspar and Franz Hammer (eds.). München: Kepler-Kommission der Bayerischen Akademie der Wissenschaften.

- Kepler, Johannes. 1991. Epitome Astronomicae Copernicae, Gesammelte Werke Band 7. Von Dyck, Walther, Max Caspar and Franz Hammer (eds.). München: C. H. Beck'sche Verlagsbuchhandlung. ISBN 3406016510.
- Kirchvogel, Paul A. 1967. Wilhelm IV, Tycho Brahe, and Eberhard Baldewein — the missing instruments of the Kassel Observatory. In: Vistas in Astronomy, 9, 1, 109–121. DOI 10.1016/0083-6656(67)90019-0.
- Kish, George. 1972. Gemma, Frisius Reiner. In: Dictionary of Scientific Biography, Gillispie, Charles Coulston (ed.), V, 349. New York: Charles Scribner's Sons. SBN 684-10116-5.
- Launert, Dieter (ed., trans.). 2012. Astronomischer Grund, Fundamentum Astronomicum (1588). Acta Historica Astronomiae, 47. Frankfurt am Main: Harri Deutsch. ISBN 978-3-8171-1896-0.
- Launert, Dieter. 2015. Nova Kepleriana: Bürgis Kunstweg im Fundamentum Astronomiae, Entschlüsselung seines Rätsels. München: Verlag der Bayerischen Akademie der Wissenschaften. ISBN 9783769601305.
- Lenke, Nils and Nicolas Roudet. 2014. Johannes, Christoph und Bartholomaeus Rothmann. Einige biographische Ergänzungen zu den Gebrüdern Rothmann. In: Acta Historica Astronomiae, 52, 223–242. Leipzig: Akademische Verlagsanstalt. ISBN 978-3-944913-42-1.
- Leopold, J. H. and K. Pechstein. 1977. Der kleine Himmelsglobus 1594 von Jost Bürgi. Luzern: Edition Joseph Fremersdorf.
- Leopold, J. H. 1986. Astronomen, Sterne, Geräte - Landgraf Wilhelm IV. und seine sich selbst bewegenden Globen. Luzern: Edition Joseph Fremersdorf. ISBN 3857660236.
- Linton, C. M. 2004. From Eudoxus to Einstein: A History of Mathematical Astronomy. Cambridge: Cambridge University Press. ISBN 978-0521827508.
- List, Martha and Volker Bialas. 1973. Die Coss von Jost Bürgi in der Redaktion von Johannes Kepler. Ein Beitrag zur frühen Algebra. München: Verlag der Bayerischen Akademie der Wissenschaften. ISBN 3 7696 2544 7. URL: https://publikationen.badw.de/de/003090688 (10 February 2024).
- Longomontani, Christiaanii S. 1622. Astronomia Danica. Amsterdam: Ioh. & Cornelivm Blaev. URL: https://reader.digitale-sammlungen.de/en/fs1/object/display/bsb10049403_00001.html (10 February 2024).
- Malpangotto, Michela. 2016. The original motivation for Copernicus's research: Albert of Brudzewo's Commentariolum super Theoricas novas Georgii Purbachii. In: Archive for History of Exact Sciences, 70, 1, 361–411. DOI 10.1007/s00407-015-0171-y.
- Mayer, Joseph. 1903. Der Astronom Cyprianus Leovitius (1514–1574) und seine Schriften. In: Bibliotheca Mathematica, III, 4, 134–159. Leipzig, Berlin: Teubner.
- Moleri, Elias. 1607. Astronomicvs tēs psēfoforias ecleipsium Solis et Lunae epilogismos. Geneva: Jacobus Stoer. DOI 10.3931/e-rara-427. URL: https://archive.org/details/bub_gb_rXD1oYkVZYgC/mode/2up (10 February 2024).
- Moleri, Elias. 1687. Opus novum astronomicum, ex quo doctrina de supputandis multis annis deliquiis iuxta tabulas prutenicas perspicue demonstratur, ex typis explicatur. Lyon: Petrum Dipalduium. (Google Books, 4 February 2024).
- Montenbruck, Oliver and Thomas Pfleger. 2004. Astronomie mit dem Personal Computer, 4. Auflage. Berlin: Springer. ISBN 3-540-21204-3.
- Morando Bruno and Denis M. Savoie. 1996. Etude de la théorie du Soleil des Tables pruténiques/A study of the theory of the Sun according to the Prutenic Tables. In: Revue d'histoire des sciences, 49, 4, 543–567. DOI 10.3406/rhs.1996.1268.
- Moulton, Forest Ray. 1964. Celestial Mechanics. New York: Macmillan.
- Multhauf, Lettie S. Rothmann, Christoph. In: Encyclopedia.com [online source]. URL: https://www.encyclopedia.com/science/dictionaries-thesauruses-pictures-and-press-releases/rothmann-christoph (8 July 2023).
- Murana, Marco. 2007. Rheticus. In: Biographical Encyclopedia of Astronomers. Hockey, Thomas, Virginia Trimble and Thomas R. Williams (eds.), 966–967. New York: Springer. ISBN 9780387310220.
- Neri, Davide. 2007. Gallucci, Giovanni Paolo. In: Biographical Encyclopedia of Astronomers. Hockey, Thomas, Virginia Trimble and Thomas R. Williams (eds.), 402–403. New York: Springer. ISBN 9780387310220.
- Neugebauer, Otto. 1975. A History of Ancient Mathematical Astronomy. Vol. I-III. Berlin, New York: Springer. ISBN 978-0387069951
- Neugebauer, Otto and Noel M. Swerdlow. 1984. Mathematical Astronomy in Copernicus's de Revolutionibus, Part I-II. New York, Berlin: Springer. ISBN 3-540-90939-7.
- Nicollier, Grégoire. 2018. How Bürgi computed the sines of all integer angles simultaneously in 1586. In: Mathematische Semesterberichte, 65, 1, 15–34. DOI 10.1007/s00591-017-0209-0.
- North, John. 1976. Richard of Wallingford, An edition of his writings with introductions, English translation and commentary, I-III. Oxford: Clarendon Press. ISBN 0 19 858139 4.
- Nosonovsky, Michael. 2018. Abner of Burgos: The Missing Link between Nasir al-Din al-Tusi and Nicolaus Copernicus? In: Zutot 15, 25-30. DOI 10.1163/18750214-12151070.

- Nový, Luboš. 1970. Bürgi, Joost. In: Dictionary of Scientific Biography, Gillispie, Charles Coulston (ed.), II, 602–603. New York: Charles Scribner's Sons.
- Oestmann, Günther. 2002. Cyprianus Leovitius, der Astronom und Astrologe Ottheinrichs. In: Tagungsband des Symposiums Pfalzgraf Ottheinrich: Politik, Kunst und Wissenschaft im 16. Jahrhundert, 348–359. Regensburg: Stadt Neuburg.
- Oestmann, Günther. 2003. Ephemeridenwerke des 16. Jahrhunderts, eine wichtige Arbeitsgrundlage für Astronomen und Astrologenkiefer. In: Gemeinnützige Mathematik: Adam Ries und seine Folgen, Kiefer, Jürgen and Karin Reich (eds.). Akademie gemeinnütziger Wissenschaften zu Erfurt, Acta Academiae Scientiarum, 8, 149–164. Erfurt: Erfurter Akademie Verlag. ISBN 978-3-932295-56-0.
- Oestmann, Günther. 2004. Heinrich Rantzau und die Astrologie: ein Beitrag zur Kulturgeschichte des 16. Jahrhunderts. Disquisitiones historiae scientiarum 2. Braunschweig: Braunschweigisches Landesmuseum. ISBN 9783927939653.
- Oestmann, Günther. 2019. "Die 'Planetenuhr' und 'Bergkristaluhr': Zwei Hauptwerke Jost Bürgis im Kunsthistorisches Museum zu Wien. In: Acta Historica Astronomiae, 66, Dick, Wolfgang and Jürgen Hammel (eds.), 37–58. Leipzig: Akademische Verlagsanstalt. ISBN 978-3-8171-1867-0.
- Omodeo, Pietro Daniel. 2011. David Origanus's Planetary System (1599 and 1609). In: Journal for the History of Astronomy, 42, 4, 439–454. DOI 10.1177/002182861104200402.
- Origanus, David. In: Biobibliographisches Handbuch der Kalendermacher von 1550 bis 1750 [online source]. URL: https://www.presseforschung.uni-bremen.de/dokuwiki/doku.php?id=origanus_david&s[]=origanus (16 June 2023).
- Origanus, David. 1599. Ephemerides novae. Frankfurt/Oder: Andreas Eichorn. (Google Books, 16 June 2023).
- Origanus, David. 1609. Nouae motuum coelestium ephemerides Brandenburgicae, annorum 60, incipientes … Frankfurt/Oder: Johannes Eichorn and David Reichard. (Google Books, 16 June 2023).
- Pogo, A. 1935. Gemma Frisius, His Method of Determining Differences of Longitude by Transporting Timepieces (1530), and His Treatise on Triangulation (1533). In: Isis, 22, 2, 469–506. DOI 10.1086/346920. URL: https://www.jstor.org/stable/225130.
- Ramasubramanian, K., M. D. Srinivas and M. S. Sriram. 1994. Modification of the earlier Indian planetary theory by the Kerala astronomers (c. 1500 AD) and the implied heliocentric picture of planetary motion. In: Current Science, 66, 10, 784–790. URL: https://www.jstor.org/stable/24098820.
- Rantzau, Heinrich. 1593. Diarium sive Calendarium Romanum. Wittemberg: Christoph Axin. URL: https://opacplus.bsb-muenchen.de/title/BV001540833 (30 October 2022).
- Reichert, Jürgen (ed.). 2014. Die Rudolphinischen Tafeln. Würzburg: Königshausen & Neumann. ISBN 978-3-8260-5352-8.
- Reimer Ursus, Nicolaus. 1588. Fundamentum Astronomicum. Strasbourg: Bernhard Jobin. URL: https://archive.org/details/den-kbd-all-130017588436-001/mode/2up (10 February 2024).
- Reinhold, Erasmus. 1551. Prutenicae Tabulæ Coelestium Motuum. Tübingen. URL: https://www.e-rara.ch/zut/wihibe/content/pageview/143071 (5 March 2024).
- Reinhold, Erasmus. 1585. Prutenicae Tabulæ Coelestium Motuum. Wittenberg: Matthaeus Velack. URL: https://archive.org/details/ARes34403/page/n6/mode/2up (10 February 2024).
- Rheticus, Georg Joachim and Johannes Acronius. 1542. Tabulae Astronomicae In Gratiam Studiosae Iuventutis Seorsim Aeditae: I De aequatione dierum. II De ascensionibus Signorum in sphaera recta, & obliqua ad latitudinem. 52. gradum. URL: https://collections.thulb.uni-jena.de/rsc/viewer/HisBest_derivate_00004523/BE_1100_0449.tif (31 October 2022).
- Rheticus, Georg Joachim and Johannes Acronius. 1542. Tabulae Astronomicae In Gratiam Studiosae Iuventutis Seorsim Aeditae: II De ascensionibus Signorum in sphaera recta, & obliqua ad latitudinem. 52. gradum. Vitebergae: Klug. URL: https://www.digitale-sammlungen.de/en/view/bsb00034792?page=,1 (30 October 2022).
- Riccioli, Giovanni Battista. 1665. Almagestum Novum. (Google Books, 25 February 2020).
- Roegel, Denis. 2010. Bürgi's "Progress Tabulen" (1620): logarithmic tables without logarithms. Update 10 January 2013. URL: https://locomat.loria.fr/buergi1620/buergi1620doc.pdf (9 July 2023).
- Roegel, Denis. 2015. Jost Bürgi's skilful computation of sines. [Research Report] LORIA - Université de Lorraine, hal-01220160. URL: https://inria.hal.science/hal-01220160 (8 July 2023).
- Roegel, Denis. 2016. A note on the complexity of Bürgi's algorithm for the computation of sines. [Research Report] LORIA - Université de Lorraine, hal-01284671. URL: https://inria.hal.science/hal-01284671 (12 August 2023).
- Roegel, Denis. 2021a. A reconstruction of the tables of Rheticus' Opus Palatinum (1596). Research Report, inria-00543932, version 2, 2 September 2021. URL: https://hal.inria.fr/inria-00543932v2 (30 October 2022).

- Roegel, Denis. 2021b. A reconstruction of the tables of Rheticus' Canon doctrinæ triangulorum (1551). Research Report, inria-00543931v2. LORIA (Université de Lorraine, CNRS, INRIA). URL: https://hal.inria.fr/inria-00543931v2 (30 October 2022).
- Rosen, Edward. 1975. Rheticus, Georg Joachim. In: Dictionary of Scientific Biography, Gillispie, Charles Coulston (ed.), IX, 395–398. New York: Charles Scribner's Sons. ISBN 0-684-10122-X.
- Rosen, Edward. Rheticus. In: Encyclopedia.com [online source]. URL: https://www.encyclopedia.com/science/dictionaries-thesauruses-pictures-and-press-releases/rheticus-george-joachim (30 October 2022).
- Schenkel, Michael (ed.). 1992. Johannes Kepler Gesammelte Werke, Register zu Band VII, Epitome Astronomiae Copernicae. Berichte der Kepler-Kommission, Heft 3. München: Kepler-Kommission der Bayerischen Akademie der Wissenschaften.
- Scultetius, Abraham. 1595. Abrahami Sculteti Grünbergensis Silesii Sphaericorum libri tres methodicé conscripti & utilibus scholiis expositi. Accessit De solutione triangulorum tractatus breuis & perspicuus Bartholomaei Pitisci Grünbergensis. Heidelberg: Matthaei Harnisch. URL: https://archive.org/details/bub_gb_CZ8jx2-edGYC/mode/2up (3 February 2024).
- Smart, William Marshall. 1931. Textbook on Spherical Astronomy, Reprint 1999. Cambridge: Cambridge University Press. ISBN 0 521 29180 1.
- Smart, William Marshall. 1953. Celestial Mechanics. Reprint 1960. London: Longmans, Green and Co.
- Stadius, Johannes. 1585. Ephemerides Joannis Stadii ... secundum Antverpiae longitudinem ab anno 1583–1606. London. URL: https://www.digitale-sammlungen.de/en/view/bsb10158955?q=%28Ephemerides+Joannis+Stadii%29&page=,1 (30 October 2022).
- Staudacher, Fritz. 2014. Jost Bürgi, Kepler und der Kaiser, 4., überarbeitete und erweiterte Auflage. Zürich: NZZ Libro. ISBN 978-3-03810-345-5.
- Stephenson, Bruce. 1987. Kepler's Physical Astronomy. New York, Berlin: Springer. ISBN 3-540-96541-6.
- Taton, R. and C. Wilson (eds.). 2003. Planetary Astronomy from the Renaissance to the Rise of Astrophysics, Part A. Cambridge: Cambridge University Press. ISBN 9780521542050.
- Thoren, Victor E. 1967. Tycho and Kepler on the Lunar Theory. In: Publications of the Astronomical Society of the Pacific, 79, 470, 482–489. DOI 10.1086/128534.
- Thoren, Victor E. 1988. Prosthaphaeresis Revisited. In: Historia Mathematica 15, 1, 32–39. DOI 10.1016/0315-0860(88)90047-X.
- Thoren, Victor E. and John Robert Christianson. 1990. The Lord of Uraniborg: A Biography of Tycho Brahe. Cambridge: Cambridge University Press. ISBN 0521351588.
- Van Bummelen, Glen. 2009. The Mathematics of the Heavens and the Earth. Princeton, NJ: Princeton University Press. ISBN 978-0-691-12973-0.
- Verdun, Andreas. 2005. Astronomie und Mathematik in "bernischen Landen" im 16. und 17. Jahrhundert. In: Orion, 63, 330, 15–27. DOI 10.5169/seals-897780.
- Vincent, Clare and Bruce Chandler. 1989/1990. A Case Study of Dependency: The Mathematical Competence of a Renaissance Clockmaker (Jost Bürgi). In: Jahrbuch der Kunsthistorischen Sammlung in Wien, 85/86, Sonderheft Nr. 335.
- Vince, Samuel. 1797. A Complete System of Astronomy. Cambridge: J. Burges. URL: https://archive.org/details/bim_eighteenth-century_a-complete-system-of-ast_vince-samuel_1797_1, https://archive.org/details/bim_eighteenth-century_a-complete-system-of-ast_vince-samuel_1799_2 (4 February 2024).
- Vips in Freiburg. Martin Waldseemüller Hylacomylus, [online source]. URL: http://www.virtuelles-freiburg.com/vip/wald.html (30 October 2022).
- Von Braunmühl, Anton. 1900. Vorlesungen über Geschichte der Trigonometrie, I. Leipzig: Teubner. URL: https://archive.org/details/vorlesungenber00brauuoft (30 October 2022).
- Von Mackensen, Ludolf. 1988. Die erste Sternwarte Europas mit ihren Instrumenten und Uhren; 400 Jahre Jost Bürgi in Kassel. München. Callwey Verlag. ISBN 3 7667 0875 9.
- Wagner, Roy and Samuel Hunziker. 2019. Jost Bürgi's methods of calculating sines, and possible transmission from India. In: Archive for History of Exact Sciences, 73, 3, 243–260. DOI 10.1007/s00407-019-00225-z.
- Waldvogel, Jörg. 2012. Jost Bürgi and the discovery of the logarithms. Seminar für Angewandte Mathematik, Eidgenössische Technische Hochschule, Zürich. URL: https://www.sam.math.ethz.ch/sam_reports/reports_final/reports2012/2012-43_rev1.pdf (10 February 2024).
- Waldvogel, Jörg. 2016. Jost Bürgi's Artificium of 1586 in modern view, an ingenious algorithm for calculating tables of the sine function and the discovery of the logarithms. In: Elemente der Mathematik, 71, 89–99. URL: http://www.sam.math.ethz.ch/~waldvoge/Projects/lichtenpaper.pdf (10 February 2024).

- Wenzel, Johann. 1983a. Das Getriebe im Deckel einer Tischuhr von Jost Bürgi. In: Alte Uhren, Heft 1, 25–31.
- Wenzel, Johann. 1983b. Das Getriebe im Deckel einer Tischuhr von Jost Bürgi. In: Alte Uhren, Heft 2, 153–162.
- Wenzel, Johann. 1996. Das Astrolab an Uhren. In: Klassik-Uhren, 19, 6, 37–46.
- Wenzel, Johann. 2002. Aequationsuhren. In: Deutsche Gesellschaft für Chronometrie, Jahresschrift 2002. Band 41, 81–96. Stuttgart: Deutsche Gesellschaft für Chronometrie.
- Wenzel, Johann. 2003. Aequationsuhren. In: Deutsche Gesellschaft für Chronometrie, Jahresschrift 2003. Band 42, 27–38. Stuttgart: Deutsche Gesellschaft für Chronometrie.
- Westman, Robert S. 1975. The Melanchthon Circle, Rheticus, and the Wittenberg Interpretation of the Copernican Theory. In: Isis, 66, 2, 165–193. doi 10.1086/351431.
- Wolf, Rudolf. 1858. Biographien zur Kulturgeschichte der Schweiz. Zürich: Orell, Füssli. url: https://archive.org/details/biographienzurk03wolfgoog (10 February 2024).
- Wydra, Stanislav. 1778. Historia matheseos in Bohemia et Moravia cultae. Pragae. (Google Books, 11 November 2022).
- Zinner, Ernst. 1941. Geschichte und Bibliographie der Astronomischen Literatur in Deutschland zur Zeit der Renaissance. Leipzig: Karl W. Hiersemann.
- Zinner, Ernst. 1957. Deutsche und Niederländische astronomische Instrumente des 11. - 18. Jahrhunderts, 2. Auflage. München: Beck.

Manuscripts

- Leovitius, Ciprianus. Tabulae ascensionum obliquarum cum latitudine Meridionali, ad eleuationem poli grad. Biblioteca Apostolica Vaticana.
- Pal. Lat. 1429. url: https://digi.vatlib.it/view/MSS_Pal.lat.1429.
- Pal. Lat. 1431. url: https://digi.vatlib.it/view/MSS_Pal.lat.1431.
- Pal. Lat. 1432, Part A. url: https://digi.vatlib.it/view/MSS_Pal.lat.1432.pt.A

This page is intentionally left blank.

8

Huygens and Flamsteed

Here take notice, that the Sun or the Earth paſſeth the 12. Signes, or makes an entire revolution in the Ecliptick in 365 days, 5 hours, 49 min, or there about, and thoſe days, reckon'd from noon to noon, are of different lengths; as is known to all, that are verſ'd in Astronomy.

Christiaan Huygens, Instructions Concerning the Use of Pendulum-Watches, for Finding the Longitude at Sea (Philosophical Transactions, 1669, 4, p.939)

8.1 Between Kepler and Huygens

8.1.1 Wilhelm Schickard

Wilhelm Schickard (1592–1635), although less well-known, is considered the father of the calculating machine and contributed numerous works in the fields of oriental studies and geography, as well as optics and astronomy.[1] Apparently, with the help of one of the professional authorities in Tübingen, we assume it could have been Besold, Mästlin or Hafenreffer, he came into contact with Kepler, 20 years his senior.[2] Schickard was certainly well acquainted with ancient astronomy, both Ptolemy's *Almagest* and al-Farghānī's *Elements of Astronomy*. He was also well acquainted with the works of Copernicus, Tycho and Kepler. One of his most important contributions was the lunar theory[3], based on the work of Copernicus and Tycho.[4] Schickard apparently used the equation of time in a similar way to Tycho. In a letter from Schickard to Galileo's correspondent Elio Diodati, dated 10 March 1635, Schickard is said to have discovered a new way of using the equation of time, but it is not clear from the available sources what this new way was.[5]

Schickard also used geometry to solve the problem of spherical trigonometry.[6] A manuscript with projection sketches resembling the base of the (Vitruvian) analemma is preserved, but containing projections of the ortive amplitude, right ascension, oblique ascension and ascensional difference.[7]

8.1.2 Horrocks' and Curtz's equation of the centre

Jeremiah Horrocks (1618–1641), the father of English astronomy, was one of the first astronomers to adopt Kepler's laws. Despite his very short life, his contribution to astronomy is very important. He predicted the transit of Venus in 1639 and observed it together with William Crabtree (1610–1644).[8] Among other things, he studied the motion of the Moon and found that it moves in an orbit roughly corresponding to an ellipse.

1 Seck, 1978, Wilhelm Schickard (1592–1635), Astronom, Geograph, Orientalist, Erfinder der Rechenmaschine.

2 Seck, 1978, p.21. Christoph Besold (1577–1638), German jurist and publicist. His work is an important source for describing the causes of the Thirty Years' War. Michael Mästlin (1550–1631), who studied under Philipp Apian, son of Peter Apian, was a proponent of the Copernican model and later Kepler's teacher. Matthias Hafenreffer (1561–1619), German Lutheran theologian.

3 Seck, 1978, pp.199–253. For Kepler's lunar theory, see pp.218–226.

4 Seck, 1978, p.212.

5 Seck, 2002, Wilhelm Schickard Briefwechsel, II, p.323. Seck, 1978, p.241. note 240 also p.286 incl. note 342. For comments on the equation of time in Schickard's work, see Seck, 1978, pp.195, 198, 207, 241 and 286. Élie Diodati (Elia, Elias) (1576–1661), Swiss jurist, correspondent of Galileo.

6 Seck, 1978, pp.153–164.

7 Wiener Handschrift 10565, Seck, 1978, p.154. The geometrical constructions of the right ascension and oblique ascension are incorrect according to the source, see Seck, 1978, pp.157, 160.

8 For more information on the transits of Venus, see Sheehan, Westfall, 2004, The Transits of Venus and Meeus, 1989, Transits.

From HORROCKS' more or less fragmentary legacy, it is possible to reconstruct his equation of the centre:[9]

$$\nu_H - M = 2e \sin M + \frac{5}{4}e^2 \sin 2M + 3e^3 \sin M - \frac{25}{6}e^3 \sin^3 M \ldots \qquad [8.1]$$

Plugging $\sin 3M = 3 \sin M - 4 \sin^3 M$ into KEPLER's equation of the centre (Equation 15.31) yields:

$$\nu - M = 2e \sin M + \frac{5}{4}e^2 \sin 2M + 3e^3 \sin M - \frac{13}{3}e^3 \sin^3 M \ldots$$

and the difference between HORROCKS' and KEPLER's equation of the centre (Equation 15.31) is:

$$\nu_H - \nu = \frac{1}{6}e^3 \sin^3 M \qquad (8.2)$$

ALBERT CURTZ Lat. CURTIUS (1600–1671), a Jesuit astronomer, was a great supporter of the work of TYCHO BRAHE. In 1666, he published *Historia coelestis*, a collection of TYCHO's notes and results of astronomical observations.[10] He also corresponded with KEPLER and apparently accepted KEPLER's first law. He published his ideas on the elliptic orbit, supposedly the first written publication of its kind, in 1626 in *Novum coeli systema*.[11] CURTZ's equation of the centre is:[12]

$$\nu_C - M = 2e \sin M + e^2 \sin 2M + 2e^3 \sin M - \frac{8}{3}e^3 \sin^3 M \qquad [8.3]$$

The difference between CURTZ's and KEPLER's equation of the centre is obtained in the same way as for HORROCKS' equation:

$$\nu_C - \nu = -\frac{1}{4}e^2 \sin 2M - e^3 \sin M + \frac{5}{3}e^3 \sin^3 M \ldots \qquad [8.4]$$

9 Gaythorpe, 1925, On Horrocks' treatment of the evection and the equation of the centre, with a note on the elliptic hypothesis of Albert Curtz and its correction by Boulliau and Newton, p.861.
10 The work was heavily criticised for numerous typographical errors. Jensen, 2007, p.267.
11 It is a very rare work, known from CURTZ's letter to KEPLER of 19 July 1627. See Gaythorpe, 1925, p.861.
12 Gaythorpe, 1925, p.862.

8.1.3 Ismaël Boulliau

Ismaël Boulliau Lat. Ismael Bullialdus (1605–1694), who was in contact with Pascal, Mersenne, Gassendi and other well-known names of the time, and corresponded with Huygens, left a very extensive correspondence. He was well-educated and had a wide range of knowledge, including classics, which was useful for his work as a librarian, which he did, among other things, in Paris. The equation of time was described and tabulated by Boulliau in *Astronomia philolaica*, which is considered one of the most important astronomical works of the period between Kepler and Newton. The table *Tabvla dvplex æqvationvm diervm civilivm* lists both contributions to the equation of time – due to the irregular motion and due to the obliquity of the ecliptic. In both cases, the values are given in minutes and seconds. The tabulated values of the contribution due to the obliquity of the ecliptic are a function of longitude, and the contribution due to irregular motion is a function of the true anomaly. In the *Tabula æquationis dierum ciuilium*, the sum of the two contributions as a function of longitude is given. The contribution from the uneven movement is smaller than it should be, which is why the values of the Boulliau equation of time are smaller than expected.[13] Boulliau had no reputation as an observational astronomer, and the inaccuracies in his equation of time probably account for this. Following Seth Ward's criticisms in 1653 in *In Ismaelis Bullialdi astronomiae philolaicae fundamenta, inquisitio brevis*, Boulliau corrected his calculations of the equation of the centre and published the corrections in 1657 in *Astronomiae philolaïcae fundamenta clarius explicata et asserta*.[14]

Boulliau's corrected equation of the centre, which he derived in his own peculiar but incorrect way, corresponds to the expression:[15]

(8.5)
$$\nu_B - M = 2e \sin M + \frac{5}{4} e^2 \sin 2M + 3e^3 \sin M - \frac{11}{3} e^3 \sin M \ldots$$

e	eccentricity
M	mean anomaly
ν	true anomaly

The difference between Boulliau's and Kepler's equation of the centre is:[16]

(8.6)
$$\nu_B - \nu = \frac{2}{3} e^3 \sin^3 M$$

13 The contribution values correspond reasonably well to the expected values due to the obliquity of the ecliptic.

14 Seth Ward (1617–1689) mathematician and astronomer, founder member of the Royal Society and the first to teach the Copernican system at Oxford. For some explanations about Boulliau's work see Wilson, 1989, Predictive astronomy in the century after Kepler.

15 Gaythorpe, 1925, p.862. Nicholas Greenwood also reached the same result. Greenwood is the author of *Astronomia anglicana* of 1689. See also Wilson, 1989, Predictive astronomy in the century after Kepler, pp.173–174.

16 Wilson, 1989, p.173. Gaythorpe, 1925, p.863.

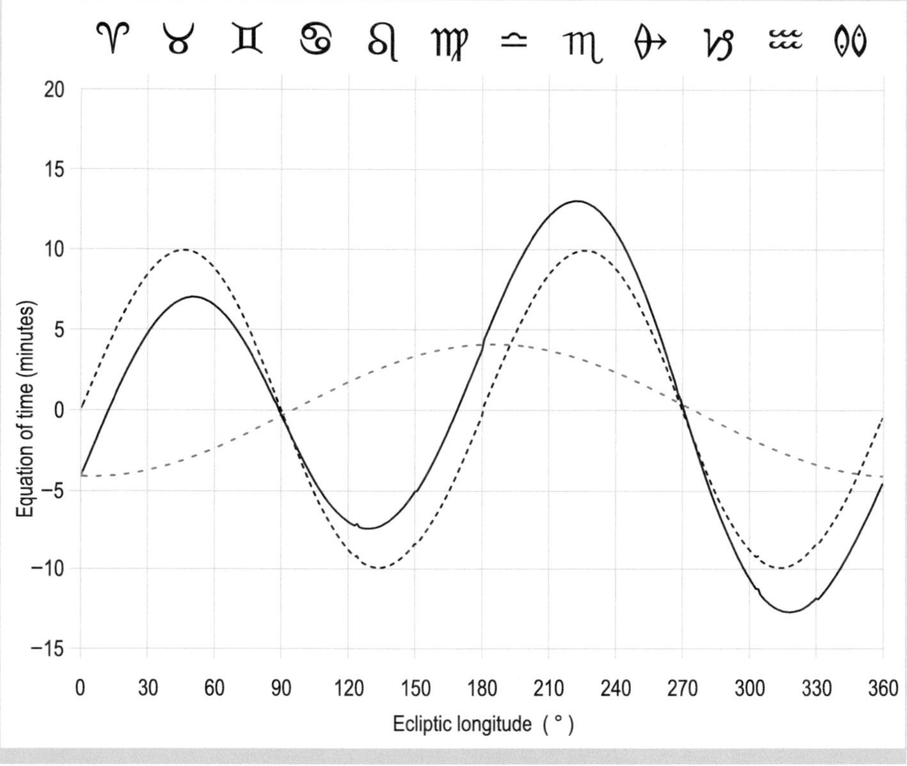

Figure 8.1 BOULLIAU's equation of time. Reference: Boulliau, 1645, Astronomia philolaica, *Tabvla dvplex aeqvationvm diervm civilivm* and *Tabula æquationis dierum ciuilium*, Tables, pp.8–9.

In addition to astronomical works, BOULLIAU's legacy also includes mathematical treatises. He was the first to adapt THEON OF SMYRNA's treatise on arithmetic in 1644, and is also the author of a work on the nature of light, which was clearly influenced by GASSENDI.

The BOULLIAU's values of the equation of time were later also used by SHAKERLEY, see also Chapter 14, Table 14.2/2. VINCENT WING published the explanation of the equation of time in *Astronomia instaurata*.[17] Geometric derivation of the equation of the centre is also given in this treatise. Mathematical procedures of VINCENT WING are also based on BOULLIAU's work.[18] He published a revised model for calculating the true anomaly in *Astronomia Britannica* in 1669.[19]

17 Wing, 1656, pp.48–50. VINCENT WING (1619–1668), self-taught astronomer. WING follows BOULLIAU's approach.
18 Wilson, 1989, p.177.
19 Wilson, 1989, p.177.

8.1.4 Jeremy Shakerley

JEREMY SHAKERLEY (1626–1655?) was an astrologer, self-taught in astronomy and known for some of his astronomical observations. The main source of information about his life and work is his correspondence, specifically the nine letters he wrote to the London astrologer WILLIAM LILLY, HENRY OSBORNE and JOHN MATTESON between 1648 and 1650. He stayed for a time with CHRISTOPHER TOWNELEY, where he apparently also had access to HORROCKS' manuscripts or copies of them. He was the first person known to have recognised the value of HORROCKS' work, and he had also studied, or at least was reasonably familiar with, the work of COPERNICUS, BRAHE, KEPLER and BOULLIAU. He also accepted KEPLER's findings on the elliptical orbits of the planets. He was obviously well acquainted with the ephemerides of the time and their authors, as he mentions the tables of LANSBERGE, LONGOMONTANUS, ARGOLI, KEPLER and ERASMUS REINHOLD's *Prutenic Tables* in his works. He was also a harsh critic, for example criticising the work of WING and LEYBOURN in his *Anatomy of urania practica* in 1649.[20] Among other things, he published the Almanac for 1651 and the *Tabulae Brittanica* of 1653, where he gives a brief description of the equation of time and an example of its use in Chapter 6 on pages 27 and 28. The values of the equation of time in minutes and seconds are also tabulated and given for each degree of longitude in *A compounded table of the æquation of time, which without any ſenſible errour may ſerve for this age*. The values of the time equation are consistent with those of BOULLIAU in the *Tabula æquationis dierum ciuilium...* from the 1645 *Astronomia philolaica*.[21]

SHAKERLEY's work was more or less forgotten after his untimely death, and WALLIS, in a letter to COLLINS dated 13 February 1671, states that he was not familiar with SHAKERLEY's tables. However, SHAKERLEY's work apparently came into the hands of FLAMSTEED, who recognised its value, as is evident from FLAMSTEED's letter to COLLINS on 13 August 1672:[22]

> *The precepts I found translated by the ingenious Mr Shakerley, which I transcribed from him, because I thought them clearer expressed than the English ones in Crabtree's letter, though they are in the substance the very same.*

20 WILLIAM LEYBOURN (1626–1716), mathematician and surveyor.

21 BOULLIAU describes the equation of time on pages 88 to 91 of *Astronomia philolaicae* in Chapter VI, *De vera dierum ciuilium æquatione*.

22 Rigaud and Rigaud (eds.), 1841, Correspondence of Scientific Men of the Seventeenth Century, II, pp.157, 531.

8.1.5 Some other works before Huygens

Philipp Müller (1585–1659), a German physicist, mathematician and physician, described the length of days, albeit without a table of the equation of time, in his 1621 treatise on the unequal length of days, *Ēmerologia seu problema astronomico-geographicum de aequalitate et inaequalitate dierum artificialium & noctium*, which was co-authored by Valerius Schütz.[23] From the treatise of the day as the time interval between sunrise and sunset, or the time interval between two middays, it is clear that the authors were familiar with Geminus' *Phenomena*.

Müller was professor of physics at the University of Leipzig from 1614, and for a long time was also dean and vice-rector. Among his students was Erhard Weigel (1625–1699), who received his doctorate in 1652.[24] Weigel was a professor of mathematics at the University of Jena from 1653, where he mentored several famous mathematicians, including Leibniz, and his influence and intellectual legacy stretched back to Lagrange, Euler, Poisson and others.

In 1628, Matthias Buchold published a treatise on the unequal lengths of days, *Problema astronomicum, de inaequalitate dierum naturalium*.[25] This is a short and general philosophical rather than astronomical treatise. The author clearly had at least a passing familiarity with the work of Copernicus and Kepler, as the text also refers to Copernicus' teachings and Kepler's book *Epitome*. He was also familiar with the Greek term for the length of the day, *νυχθήμερον* (*nychthemeron*).

> *Aſcenſio & deſcenſio ſunt arcus Æquatoris (hoc loco diſcreti tantum,) qui cum dato arcu Eclipticæ, quem ſcilicet Sol motu proprio ſingulis diebus confecit, oriuntur vel occidunt, aut etiam prætvehuntur meridianum.*
>
> *Matthias Buchold, 1628, Problema astronomicum, de inaequalitate dierum naturalium, XX*

Andreas Goldmayer, aka Geelmeyer (1602–1665), German astronomer, mathematician and calendar maker, described the equation of time in 1657 in *Computus directionum astronomicus*, in Chapter 7.[26] Goldmayer's equation of time is rather unusual, somewhat reminiscent of Brahe's equation of time but with different values, and the exact theoretical basis of how the author arrived at it is not clear from the text. Goldmayer, at least according to the sources, must have had a relatively good knowledge of the field of astronomy in which he was involved, but he also seems to have been a rather strange personality. Perhaps due to this reason he can also be attributed with a kind of "equation of time".[27]

23 Valerius Schütz (1601–1632), German teacher, writer and organist.

24 Döring, 1999, Erhard Weigels Zeit an der Universität Leipzig (1647–1653). In: Acta Historica Astronomiae 7, pp.75–84.

25 Hammel, 2015, p.78. Buchold, 1628, Problema astronomicum, de inaequalitate dierum naturalium.

26 Goldmayer, 1657, pp.12–16.

27 For his other works, see Pilz, 1977, pp.276–278. Pilz describes Goldmayer as a fantasist and a dreamer. For more information on his life and work, see Gaab, 2011, pp.419–420 and especially 459–465.

The calendar maker and astronomer EBERHARD WELPER (1590–1664) briefly described the equation of time in Chapter III of his treatise *Brevis instructio de erigenda figvra coeli.*[28]

GILLES LACCARY (1605–1684), a French Jesuit born in the diocese of Castres in southern France, gives an equation of time corresponding to TYCHO's equation of time in his 1657 work *Mathematica tabulae logarithmeticae ad triangulorum rectilineorum & sphaericorum analysin*, Table XII.[29]

ANDREA ARGOLI (1570–1657) composed ephemerides, initially based on the *Prutenic Tables* and later on the model of TYCHO BRAHE. From 1632 he taught mathematics in Padua and wrote other astronomical works. In 1629, in *Astronomicorum libri tres*, he presented his own planetary system, which was geocentric, with Mercury and Venus orbiting the Sun, and the Moon, Sun, Mars and Saturn orbiting Earth, which was similar to that of MARTIANUS CAPELLA, except that ARGOLI also advocated for the rotation of Earth on its own axis. The equation of time is given in the table on page 106 of the 1638 *Ephemerides*, the values are in whole minutes. In the 1648 edition, the table is on page 108.

8.2 Christiaan Huygens

HUYGENS was born into a reputable family in The Hague. At an early age, he showed an interest in and talent for natural sciences and technical fields. His father Constantijn was a diplomat, poet and a supporter of art. RENÉ DESCARTES and MARIN MERSENNE frequented his home. DESCARTES in particular made a great impression on the young HUYGENS since his influence can be observed in HUYGENS' later achievements. With the help of his brother (also, like his father, named) CONSTANTIJN and a special grinding and lens polishing technique, HUYGENS manufactured a telescope with which he resolved the shape of Saturn's rings and its moon Titan in 1655.[30] He was aware of the meaning of accurate time measurement in astronomy, which also influenced the development of the pendulum clock. Although GALILEO GALILEI sketched the design of a pendulum clock in around 1635, he did not put it into practice. GALILEO's son Vincenzio took it up in 1649, but was overtaken by death.[31]

HUYGENS' achievements did not go unnoticed outside his homeland. In 1663 he was nominated a member of the English Royal Society. He spent a lot of time in Paris where in 1676 he met LEIBNIZ, who also made a huge impression upon him, and he was one of the first members of

28 Second edition of 1700, pp.12–13.
29 For other works by LACCARY, see Watt, 1824, II, 580n.
30 For more information see Bedini, 1966, Lens making for scientific instrumentation in the seventeenth century.
31 Repsold, 1908, p.34.

the French Academy of Sciences. For a long time, HUYGENS was in close contact with ISMAEL BOULLIAU, however, they ceased their correspondence after 1662 for unknown reasons; differing opinions about the equation of time might have been a cause.[32] He published his findings on clocks in 1673 in the work *Horologium oscillatorium* and tried to manufacture a marine chronometer; however, the corrosion of the spring and the dependence on temperature caused problems he could not resolve. One of HUYGENS' greatest achievements is his explanation of the wave theory of light.[33]

The tabulated values of the equation of time became an indispensable tool for calibrating the first clocks and instead of the equation of (natural) days term, a new equation of time, Lat. aequatio temporis began to be used. Since some astronomers disagreed with the explanations of the causes for the equation of time from antiquity, HUYGENS was encouraged by ISMAEL BOULLIAU and other astronomers to start studying the phenomenon in more detail.[34] After two years of work, HUYGENS circulated a table with the equation of time in the letters to his correspondents in 1662.[35] In 1669, the values were published in *Instructions Concerning the Use of Pendulum-Watches, for Finding the Longitude at Sea* in their unamended form, with HUYGENS' equation of time being positive. Concurrently, he also confirmed that the equation depends on the eccentricity of Earth's orbit and the obliquity of the ecliptic.[36] The instructions do not contain detailed explanations of the methods used; they merely provide instructions for the correction of apparent solar time.[37] In 1673, the values of the equation were also tabulated in *Horologium oscillatorium*.[38] These values are shifted by 10 days relative to those published in the *Philosophical Transactions*.[39]

These tables made it possible to calibrate mechanical clocks by measuring the distance and orientation of the shadow, which offered a great advantage over the prior method that required the positions of stars. The values stated by HUYGENS (Table 8.1 and Figure 8.2) were sufficient for calibrating clocks, which was also the fundamental purpose of the first tabled values of the equation; however, they were not accurate enough for astronomical observations.

32 HUYGENS also disagreed with BOUILLAU's value of parallax and the equation of the centre. Taton, Wilson (eds.), 2003, Planetary Astronomy from the Renaissance to the Rise of Astrophysics, Part A, p.176.

33 Huygens, 1690, Traité de la lumière.

34 Mahoney, 1980, Christian Huygens: The Measurement of Time and of Longitude at Sea. Bos et al. (eds.), pp.234–270 and 247, note 36. For HUYGENS' early work, see Christiaan Huygens, Oeuvres complètes, HOC, 15, 527ff. For BOULLIAU's explanation of the phenomenon see Boulliau to Huygens, 27 February 1660, HOC, 3, 29, and Boulliau's note, 26 March 1660, HOC, 3, 51–55.

35 Mahoney, 1980, p.247, notes 36 and 40. Huygens to Lodewijk Huygens, 15 February 1662, HOC, 4, 54–57, Huygens to Moray, 17 February 1662, HOC, 4, 60, and Huygens to Petit, 25 April 1662, HOC, 4, 138–142. PIERRE PETIT (1594/98?–1677), French instrument maker, mathematician, and army engineer. He belonged to the circle of scholars that also included DESCARTES, MERSENNE, ÉTIENNE and BLAISE PASCAL.

36 Christiaan Huygens, 1665, Kort Onderwys aengaende het gebruyck der Horologien tot het vinden der Lenghten van Oost en West. A reprint in English was published four years later, in 1669.

37 The tables were, as a rule, produced by clockmakers and were intended to be used for clock calibration.

38 Mahoney, 1980, p.247, note 40. Huygens, 1673, Horologium oscillatorium, pp.151–154. HOC, 18, 112–113. The values in the *Horologium oscillatorium* are equal, but offset, see Figure 8.2.

39 Heckscher, Oettinger (trans., eds.), 1913, Die Pendeluhr, Horologium oscillatorium von Christiaan Huygens, pp.18–19.

	Jan		Feb		Mar		Apr		May		Jun		Jul		Aug		Sep		Oct		Nov		Dec	
Day	min	sec	min	sec	min	sec	min	sec	min	sec	min	sec	min	sec	min	sec	min	sec	min	sec	min	sec	min	sec
1	6	10	0	0	4	46	14	23	19	25	16	24	10	45	11	7	19	41	29	16	31	13	21	14
2	5	47	0	2	5	3	14	39	19	28	16	13	10	38	11	16	20	1	29	30	31	3	20	44
3	5	24	0	4	5	21	14	55	19	29	16	1	10	31	11	25	20	22	29	43	30	53	20	14
4	5	2	0	8	5	39	15	10	19	29	15	49	10	25	11	36	20	43	29	56	30	43	19	44
5	4	41	0	12	5	57	15	25	19	29	15	37	10	19	11	48	21	4	30	9	30	32	19	14
6	*4*	*21*	*0*	*16*	*6*	*15*	*15*	*39*	*19*	*28*	*15*	*24*	*10*	*13*	*12*	*1*	*21*	*25*	*30*	*22*	*30*	*20*	*18*	*44*
7	4	2	0	21	6	33	15	53	19	26	15	11	10	7	12	14	21	47	30	34	30	8	18	14
8	3	44	0	26	6	51	16	7	19	24	14	58	10	2	12	28	22	9	30	45	29	55	17	44
9	3	27	0	32	7	9	16	21	19	21	14	45	9	58	12	42	22	31	30	55	29	40	17	14
10	3	11	0	40	7	27	16	34	19	18	14	32	9	54	12	57	22	52	31	4	29	23	16	44
11	2	55	0	48	7	45	16	47	19	15	14	19	9	51	13	12	23	13	31	12	29	6	16	14
12	2	39	0	57	8	3	16	59	19	11	14	6	9	49	13	27	23	33	31	19	*28*	*48*	*15*	*44*
13	2	23	1	6	8	22	17	11	19	7	13	53	9	47	13	43	23	53	31	26	28	30	15	14
14	2	7	1	16	8	41	17	22	19	2	13	40	9	46	13	59	24	13	31	32	28	8	14	43
15	1	52	1	26	9	1	17	33	18	57	13	27	9	46	14	16	24	33	31	38	27	45	14	12
16	1	38	1	37	9	21	17	43	18	51	13	15	9	46	14	33	24	53	31	43	27	22	13	41
17	1	25	1	49	9	41	17	53	18	45	13	3	9	47	14	50	25	13	31	47	27	8	13	10
18	*1*	*13*	*2*	*2*	*10*	*1*	*18*	*3*	*18*	*39*	*12*	*52*	*9*	*49*	*15*	*8*	*25*	*33*	*31*	*50*	*26*	*45*	*12*	*40*
19	1	2	2	15	10	21	18	13	18	33	12	41	9	52	15	26	25	52	31	53	26	22	12	10
20	0	51	2	28	10	40	18	23	18	26	12	30	9	56	15	45	26	11	31	55	25	58	11	40
21	0	41	2	42	10	59	18	32	18	18	12	19	10	0	16	4	26	30	31	55	25	34	11	10
22	0	32	2	56	11	18	18	39	18	10	12	8	10	4	16	23	26	49	31	55	25	10	10	40
23	0	24	3	11	11	37	18	46	18	3	11	58	10	8	16	42	27	8	31	55	24	45	10	10
24	0	18	3	26	11	56	18	53	17	53	11	48	10	13	17	1	27	26	31	54	*24*	*20*	*9*	*41*
25	0	13	3	41	12	15	18	59	17	41	11	38	10	18	17	21	27	43	31	52	23	55	9	13
26	0	9	3	56	12	34	19	4	17	30	11	28	10	23	17	41	28	0	31	50	23	30	8	45
27	0	6	4	12	12	53	19	9	17	19	11	18	10	28	18	1	28	16	31	47	23	4	8	17
28	0	3	*4*	*29*	13	12	19	14	17	8	11	9	10	34	18	21	28	32	31	43	22	38	7	50
29	0	1			13	31	19	18	16	57	11	0	10	41	18	41	28	47	31	37	22	11	7	23
30	0	0			13	49	*19*	*22*	*16*	*46*	⋆*11*	*52*	10	49	19	1	*29*	*2*	31	30	*21*	*43*	6	58
31	*0*	*0*			14	6			16	35			10	58	19	21			31	22			6	34

Table 8.1 Tabulated values of CHRISTIAAN HUYGENS' equation of time from 1665. Source: Philosophical Transactions, 1669, vol. 4, issue 47, pp.940–941. ⋆ The value for 30 June seems to be wrong. It was supposed to be 10 min 52 s, which is also taken into consideration in Figure 8.2, where the values in italics are represented.

For the starting date where the value of the equation of time was zero, HUYGENS assumed 10 February. The equation of time reaches its local maxima in May, July and November. The highest value – 31 minutes and 55 seconds – occurs between 30 October and 2 November, on the Gregorian calendar.[40]

40 Mahoney, 1980, p.247, note 39, and Huygens, 1669, Instructions Concerning the Use of Pendulum-Watches, for Finding the Longitude at Sea. Philosophical Transactions, 1 January 1669, 4, 45–56, pp.937–976.

Huygens demonstrates by example how to use the equation of time to determine how much a clock is ahead or behind, and hence how to set it. To find out, we need to compare two different days – one when we set a clock using the sundial, and one when we want to find out the exact hour.[41]

Suppose that we want to check a clock on 20 March, given that it matched sundial time on 5 March. The equation of time for 5 March is 3 minutes 11 seconds. The equation of time on 20 March is 7 minutes and 27 seconds or 4 minutes and 16 seconds longer than on 5 March. The sundial must be ahead of the pendulum clock by this same difference if the pendulum clock is accurate. If not, the calculated difference can be used to determine how far ahead or behind the pendulum clock is. Huygens understood the equation of time as *"real time – mean time"*. Today, we think of the equation of time as the difference between *"mean time – real time"*.[42]

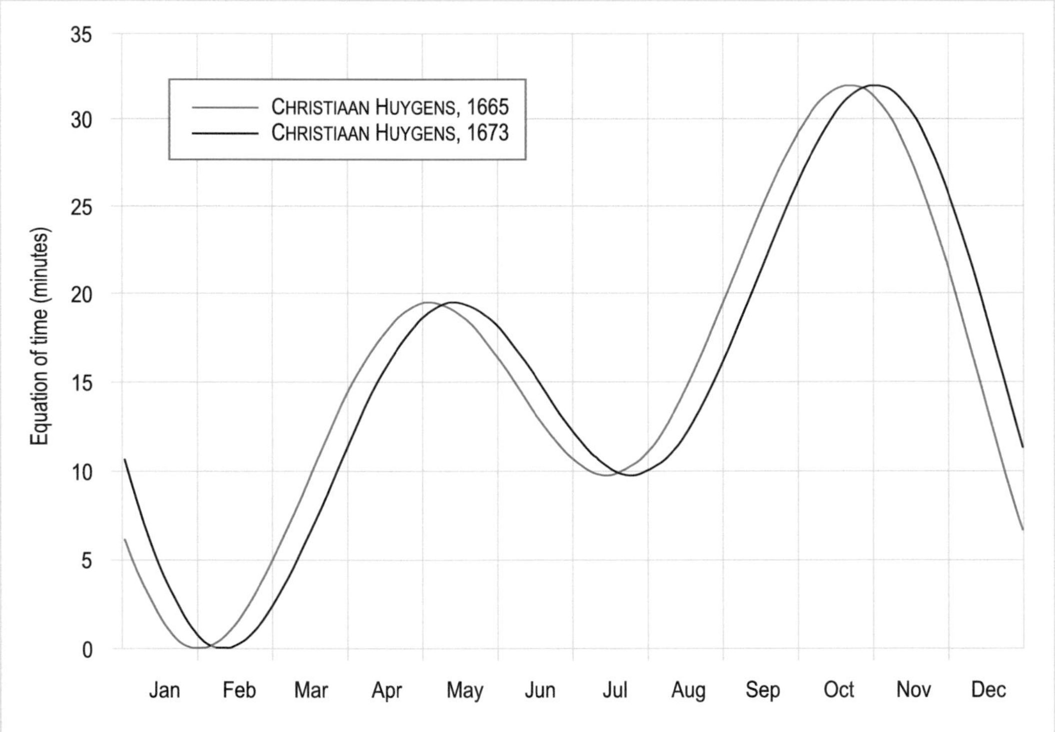

Figure 8.2 Huygens' equation of time from 1665 and 1673. References: Huygens, 1669, Instructions Concerning the Use of Pendulum-Watches, for Finding the Longitude at Sea. Philosophical Transactions, vol. 4, issue 47, pp.940–941, or Table 8.1, and Huygens, 1673, Horologium Oscillatorium, p.15.

41 Heckscher, Oettinger (trans., eds.), 1913, p.17. The values of this equation of time are shifted by 10 days compared to the one originally published in *Philosophical Transactions*.

42 Heckscher, Oettinger (trans., eds.), 1913, p.197, note 17.

The first attempts to measure longitude using a HUYGENS pendulum clock were carried out during a sea voyage of Captain ROBERT HOLMES from Lisbon to Guinea, and the second attempt was the sea voyage of from Guinea via Cape Verde to Barbados.[43] During the voyage to Barbados, where HOLMES wanted to restock on drinking water, the clock proved to be very useful.[44]

HUYGENS also invented and built a clock mechanism for displaying the equation of time. His correspondence shows that he was initially not keen on the idea because of its complexity, preferring instead use of tables.[45] NICOLAUS MERCATOR also modified one of HUYGENS' pendulum clocks to show the equation of time, and presented it to the members of the Royal Society in 1666.[46] The phaenomena determining the equation of time were first substantially detailed in JOHN WALLIS' *Hypothesis about the Flux and Reflux of the Sea*, published in 1666.[47] The treatise describes the influence due to the non-uniform movement of Earth around the Sun, as well as the influence due to the angle of Earth's axis; however, it was not entirely clear back then whether any other phenomena influence the equation. The ideas necessary to calculate the equation of time in this way had been previously published in 1661 in the *Astronomia Carolina* by THOMAS STREETE.[48]

♈	0	7;51
	10	4;40
	20	1;36
♉	30	1; 3
	40	2;58
	50	3;58
♊	60	3;56
	70	2;56
	80	1;10
♋	90	0;59
	100	3;07
	110	4;46
♌	120	5;37
	130	4;12
	140	2; 0
♍	150	0;57
	160	4;23
	170	4;43
♎	180	7;53
	190	11;11
	200	13;52
♏	210	15;39
	220	16;16
	230	15;34
♐	240	13;29
	250	10; 8
	260	5;49
♑	270	0;57
	280	3;56
	290	8;20
♒	300	11;49
	310	14;06
	320	15; 2
♓	330	14;40
	340	13;11
	350	11;49
	360	7;51

Table 8.2 Excerpt from FLAMSTEED's table with the equation of time. Source: Flamsteed, 1678, Horrox, Opera Posthuma, pp.460–461. See also Kitto, 1999, p.181 or Kitto, 2001, p.64.

43 Mahoney, 1980, p.252, note 54. The first voyage took place between 29 April and 4 September 1663, HOC, 4, 446–451. HOLMES' comment on the second voyage, Moray to Huygens 23 January 1665, HOC, 5, 204–206.

44 Mahoney, 1980, pp.252–254, notes 54–56. The initial success may have been followed by disappointments, partly because marine officers did not follow HUYGENS' instructions to operate the clocks. Mahoney, 1980, pp.253–254, note 60.

45 Huygens to Colbert 13 March 1669, HOC, 6, pp.378–379. Huygens to Constantijn Huygens 4 March 1695, HOC, 10, pp.708–710.

46 JOHN EVELYN's diary of 1666 contains an account of *"a visit to the Royal Society in London, where Mercator displayed a curious clock with a time-balance"*. Von Bertele, 1957, Zur Geschichte der Äquationsuhren-Entwicklung, p.84. Moray to Huygens 26 April 1669, HOC, 6, pp.423–425. Also Bertele, 1959, p.46.

47 JOHN WALLIS (1616–1703), English mathematician, Hypothesis about the Flux and Reflux of the Sea, Philosophical Transactions, 6 August 1666, pp.263–281, particularly pp.278–279. WALLIS is also partially responsible for the development of calculus. He was one of the pioneers of analytical geometry and introduced marks for powers and expanded their definition from integers to rational numbers.

48 THOMAS STREETE (1621–1689), English astronomer. His work *Astronomia Carolina*, or a new theory of celestial motions, was used by many, including FLAMSTEED and ISAAC NEWTON. Newton published his tables in the Mathematical Principles of Natural Philosophy.

8.3 John Flamsteed

Poor childhood health frequently kept JOHN FLAMSTEED (1646–1719) from school. He also home-schooled himself for a couple of years before going to Cambridge University. His first known astronomical observation was of the solar eclipse on 12 September 1662, when he was aged 16. Based on daily observations of the passage of the star Sirius, FLAMSTEED induced that Earth evenly rotates around its axis, which helped him draft tables on the values of the equation of time with an accuracy of 12 seconds, a significant improvement on the values used by THOMAS STREETE and others before him. The finding that Earth rotates evenly was published in 1673 in *De inaequalitate dierum solarium dissertatio astronomico*[49] as an addendum to HORROCKS' *Opera posthuma* published by JOHN WALLIS. In 1681, FLAMSTEED's *Doctrine of the Sphere* included an equation of the centre for elliptical motion. The equation was derived from a solution to KEPLER's equation, thus making him the first astronomer to use KEPLER's second law. He also established that KEPLER's third law applies to Jupiter's moons. Based on his observations, he estimated the precession of the equinox to be 1° in 72 years. FLAMSTEED is considered the leading celestial cartographer of his time and he doubled the number of known stars. From 1675 to 1719, he was the first head or royal astronomer at Greenwich Observatory. He estimated the Greenwich latitude as 51°28′34″, which meant an accuracy of 4″.

FLAMSTEED produced the first truly reliable tabulation of the equation of time (Table 8.2).[50] His results, which can also be used for exact astronomical observations, were highly valued among experts. In a letter that FLAMSTEED sent to JOHN COLLINS in 1670, he claims to have written the equation of time some five years before.[51] COLLINS adopted FLAMSTEED's tables as an addendum to JEREMIAH HORROCKS' treatises, which he was working on at that time.[52] It is presumed that FLAMSTEED used the results of HORROCKS' astronomical observations when determining the value of the equation of time since he was impressed by his work. It is hard to imagine the mathematical effort put into determining the value since, for example, FLAMSTEED did not have a reliable pendulum clock as they were still very rare at that time. Therefore, he had no means to precisely verify the results of his work. The lack of temperature-compensation on pendulum length influenced the accuracy of astronomical observations. Phenomena such as aberration and nutation were still unknown, and during their observations astronomers did not record other parameters, such as temperature and air pressure.

49 This treatise (London, 1672) is considered the first work to describe the equation of time in a contemporary way. See An Account of the Revd John Flamsteed, 1835, p.23.

50 His values of the equation of time are published in the *Notes from the Observations with the Mural Arc*, in the catalogue of fixed stars, between 1695 and 1719 with addenda up to 1719, *Zodiacus Britannicus*, as well as in the collection of tables created between 1683 and 1705.

51 Forbes, Murdin, Willmoth (eds.), 1995, The Correspondence of John Flamsteed, the First Astronomer Royal, Volume 1, p.88, Flamsteed to Collins, 3 May 1671, letter 52. In addition to FLAMSTEED, some other astronomers also contributed to the contents. For a summary of the sources of his correspondence, see Sellers, 2012, In Search of William Gascoigne, Chapter 11, pp.103–110, for the contents of the correspondence, see Forbes, Murdin, Willmoth (eds.), 1995–2002, The Correspondence of John Flamsteed, the First Astronomer Royal, volume 1–3. FLAMSTEED's early work on the equation of time is available in the archive of the Cambridge University Library, Royal Greenwich Observatory RGO 1/32A. Flamsteed, Baily, 2011, p.23.

52 Flamsteed, 1678, De Temporio Aequatione Diatriba, pp.460–461.

Nor need we ſcruple about the Æquation of Natural days: I have fully demonſtrated the Æquations, ſo that am perſwaded, no one hereafter will controvert them; and I ſhall ere long, if God will, commit them to your and the publick cenſure.

John Flamsteed, Philosophical Transactions, 17 January 1670, 4, 55, p.1104

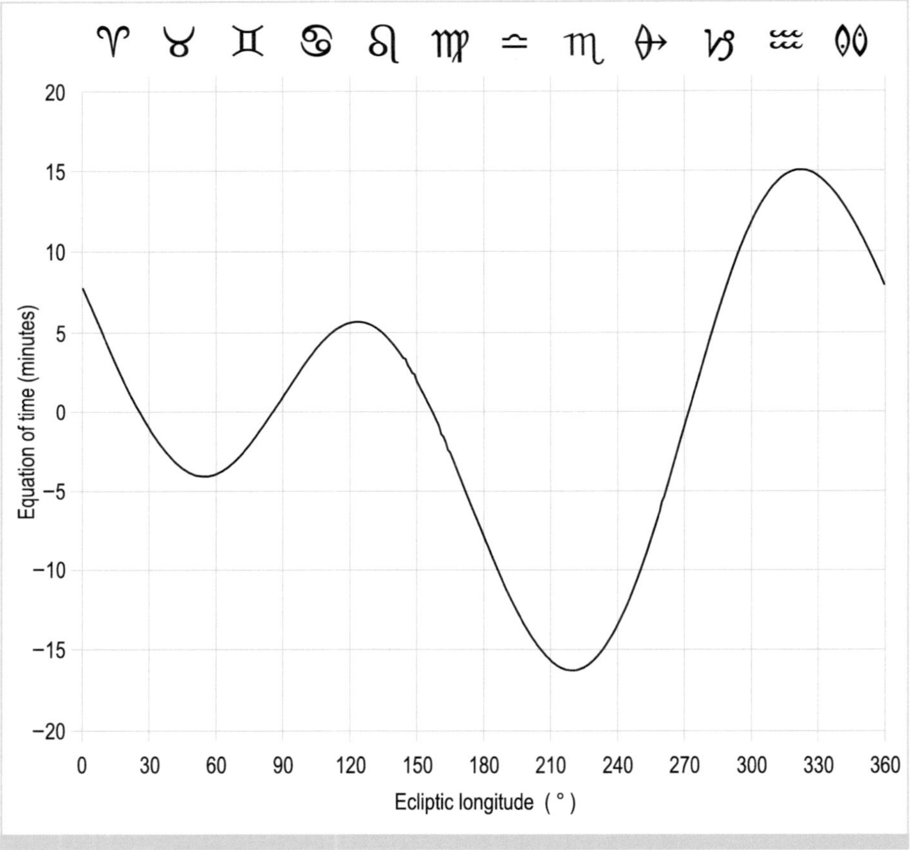

Figure 8.3 Flamsteed's equation of time from 1672. Data source: Flamsteed, 1678, pp.460–461.

Clockmaker John Smith produced several flyers with the equation of time in the period 1678–1694 for equipping his clocks and which we presume were adopted from values published by Huygens. Smith was not alone in preparing and printing tables with the equation of time. Thomas Tompion[53] published equation of time tables for the Julian calendar to be used with the clocks that he produced in 1683. Tompion is considered the father of English clock making. His success was, beside high-

53 Thomas Tompion (1639–1713) trained a range of apprentices who later also became renowned clockmakers. For more information see Symonds, 1969, Thomas Tompion: His Life and Works.

quality products from his workshop, also a result of his knowing ROBERT HOOKE, who supported him. With regards to the available data, we can presume that HOOKE provided the values of the equation of time since he was then the curator of experiments at the Royal Society. It is unknown who actually calculated the values on TOMPION's tables; however, it is presumed that the tables HOOKE provided to TOMPION were drafted by FLAMSTEED.[54] TOMPION otherwise contributed several important improvements for the further development of clocks. As per HOOKE's idea, TOMPION produced two clocks with a long pendulum moving in a small arc for the Royal Observatory, which was established in 1676. The clocks also contained an improved TOWNELEY's deadbeat escapement and were highly accurate, which allowed for even more precise astronomical observations.

TOMPION also made clocks showing the equation of time. The first clock of this type was made in 1696 and had a ring around the dial that bore minute markers and turned automatically.[55]

RICHARD TOWNELEY, a mathematician and astronomer, was the nephew of CHRISTOPHER TOWNELEY who was in contact with the most renowned astronomers of the time – HORROCKS, CRABTREE, GASCOIGNE and JOHN STEPHENSON. He helped FLAMSTEED with his measurements and observations linked to the equation of time, which is also recorded in an extensive collection of preserved correspondence. He was active in meteorology and is known for the first systematic records of precipitation in the British Isles. Additionally, he contributed to BOYLE's establishing the connection between pressure and gas volume, which we know today as BOYLE's law.

FLAMSTEED's values for the equation of time soon came to be well known in other countries. For instance, the *Neu-vermehrte Welperische Gnomonica*, edited by JOHANN GABRIEL DOPPELMAYR, contains tables with FLAMSTEED's values for four consecutive years in a leap year period.[56]

Die Gröſſe von dieſer Variation kan aus denen nechſtfolgenden Tabellen / die ich in dem sehr accuraten und offt bemeldten Aſtronomo Herrn Flamſteed zu dancken habe / auff alle Tag des Jahrs gefunden werden.

Johann Gabriel Doppelmayr, 1708, Neu-vermehrte Welperische Gnomonica, Appendix, p.33.

In 1698, instrument maker THOMAS TUTELL, one of HENRY WYNN's interns, published a table with the equation of time that was most probably the work of FLAMSTEED and calculated for the year 1697. A special aspect is the fact that the column headers contain the terms for acceleration or loss of the pendulum, rather than the customary terms for speeding or lagging of the clock.[57]

54 Davis, 2003, The Equation of Time as shown on Sundials, Bulletin of the British Sundial Society, vol. 16, issue 4, pp.135–144, and Davis, 2005, More on the Equation of Time on Sundials, Bulletin of the British Sundial Society, vol. 17, issue 2, pp.66–74.

55 Von Bertele, 1957, p.88.

56 For additional information about DOPPELMAYR see Pilz, 1977, pp.312–318.

57 Tutell, 1698, The description and uses of a new contriv'd eliptical double dial. See Davis, 2005, More on the Equation of Time on Sundials, Bulletin of the British Sundial Society, vol. 17, issue 2, pp.66–74.

8.4 Gabriel Mouton

At the same time as Huygens and Flamsteed, Gabriel Mouton was also dealing with astronomical issues and thus also time measurement and the equation of time in France.[58] Mouton, who is considered the forefather of the decimal measurement system as we know it today, drafted tables that were comparable to the values of the equation of time as defined by Huygens and Flamsteed, whereas each of the astronomers expressed the values in his own way.[59]

Mouton gives examples of the calculation of the equation of time in *Observationes diametrorum solis et lunae apparentium, meridianarumque aliquot altitudinum solis & paucarum fixarum (Observations of the Sun and Moon's apparent diameter)*.[60] Mouton used tables by Philip van Lansberge.[61] The tables contain the data for the calculation according to the Nabonassar epoch and for the epoch of 0 January of year 1.[62] The time period in van Lansberge's tables is calculated as the sum of the individual sexagesimal contributions, which is the same as in the *Parisian Alfonsine Tables*.

Æquatio temporis est differentia, qua datum temporis æqualis intervallum ab intervallo temporis apparentis superatur: & contrà.

Gabriel Mouton, 1670, Observationes diametrorum solis et lunae, p.398

58 Gabriel Mouton (1618–1694), French theologist and scientist. See Montucla, 1798, Histoire des mathématiques, p.640. Jean-Étienne Montucla (1725–1799), French mathematician, is the author of the first very extensive historical mathematical overview.

59 The need for a universal measurement system was expressed by John Wilkins (1614–1672) as early as 1668 in the treatise An Essay Towards a Real Character and Philosophical Language. Wilkins, otherwise a philosopher and theologist, was also one of the founders of the Royal Society.

60 Mouton, 1670, *Observationes diametrorum*, pp.398–404.

61 Philip van Lansberge (1561–1632), Dutch astronomer. Advocate of the Copernican model.

62 The epoch of day 0 year 1 is also used in later editions of the *Alfonsine Tables*. See Wegener, 1905b, p.9.

Example 8.1 Calculate the mean anomaly κ_M at 5 h 58 min on 29 November 1660 (Mouton, 1670, p.403). The epoch is day 0, the year 1 at noon, and the values are from VAN LANSBERGE's tables of 1662. As in later editions of the *Alfonsine Tables*, VAN LANSBERGE uses 60-degree signs (S = sign) in addition to the zodiac circle signs (each 30°).

Mean anomaly at epoch $\kappa_{M0} = 4^S\ 38°36'34'' = 278°36'34''$ (van Lansberge, 1662, pp.3–4).

The mean anomaly for a given moment is determined according to the number of days since the epoch (van Lansberge, 1662, Tables, p.6):

Number of days from the epoch to 1660

Year	1660	
	1600	$2 \times 60^3 + 42 \times 60^2 + 20 \times 60$
	60	$6 \times 60^2 + 5 \times 60 + 15$
		= 606,315 days

The mean anomaly is determined by the number of days and hours elapsed since the epoch (van Lansberge, 1662, Tables, p.15).

Days	The mean anomaly (partial contributions)
1660 years or 606,315 days	
2×60^3	$16^S\ 39°29'58''32'''$
48×60^2	$18^S\ 39°47'59''24'''$
25×60	$24^S\ 38°28'13''44'''$
15	$0^S\ 14°47'\ 4''56'''$
	$0^S\ 12°33'16''36'''$[63]
29 November or 332 days	
300 = 5 × 60	$4^S\ 55°41'38''44'''$
32	$0^S\ 31°32'26''31'''$
	$5^S\ 27°14'\ 5''15'''$

The mean anomaly for the contribution of hours and minutes is calculated by taking the fraction of hours and minutes in a day (van Lansberge, 1662, Tables, p.7) from the value for the mean anomaly in a day (van Lansberge, 1662, Tables, p.15), which is:

One day mean anomaly	$0^S\ \ 0°\ 59'8''19'''$
Share of hours and minutes	
5 hours = $^{12}/_{60} + {}^{30}/_{3600}$ of a day	$0^S\ \ 0°12'19''14'''$
58 minutes = $^{2}/_{60} + {}^{25}/_{3600}$ days	$0^S\ \ 0°\ \ 2'22''55'''$

The mean anomaly is the sum of the above partial contributions and the mean anomaly at the epoch:

epoch	$4^S\ 38°36'34''$
years	$0^S\ 12°33'16''36'''$
days	$5^S\ 27°14'\ 5''15'''$
hours	$0^S\ \ 0°12'19''14'''$
minutes	$0^S\ \ 0°\ \ 2'22''55'''$
	$4^S\ 18°38'38''\ \ 0'''$

63 This partial result deviates slightly from the value given by MOUTON in the example. Possibly a typo in LANSBERGE's tables or in MOUTON's example.

8.5 William Molyneux

MOLYNEUX is considered one of the major scientific names in Ireland. After graduating Bachelor of Arts in Dublin in 1674, he studied law in London for three years and then returned to Ireland.[64] His notes on the lunar eclipse in 1681 also reached FLAMSTEED, sparking their multiannual correspondence. He founded the Dublin Philosophical Society and published several works in various fields. His most important is *Dioptrica nova*, which deals with optics and lenses, and was published in 1692. The book also contains some of FLAMSTEED's texts, which, although MOLYNEUX stated their source, clearly contributed to ending their friendship. MOLYNEUX invented the sciothericum telescopicum, a precise sundial with two gnomons. An inheritance made him financially independent but nevertheless filled several paying roles. Among other things, he was a construction supervisor between 1684 and 1687, and was also involved in politics.

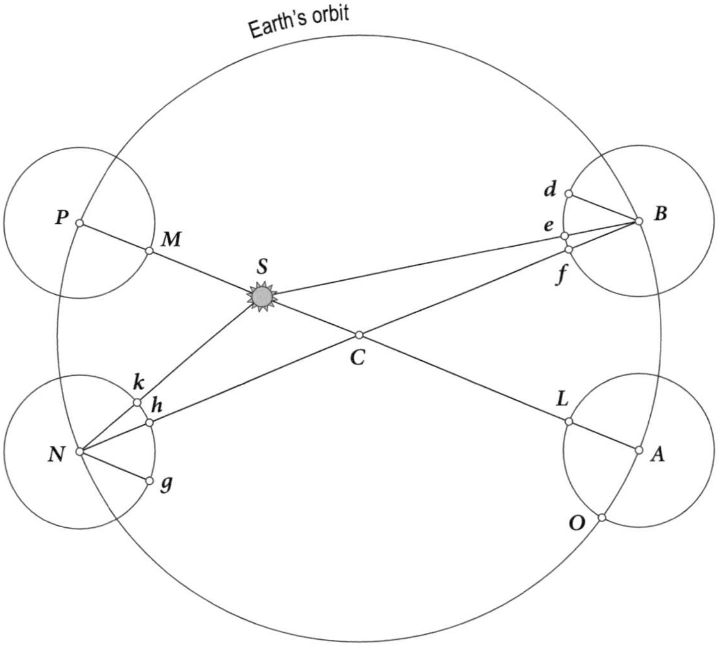

Figure 8.4 The interpretation of FLAMSTEED's equation of time in WILLIAM MOLYNEUX's *Sciothericum telescopicum*, Chapter X, Figure 13, explains the equation of centre. A very similar representation is also found in *Uranoscopia* by CHARLES LEADBETTER, 1735. The source of the original: Niedersächsische Staats- und Universitätsbibliothek Göttingen.

64 For more information on his life and work, see Abrahams, 2007, Acta Historica Astronomiae, 33, 229–246.

MOLYNEUX's explanation of the equation of time, which was based on FLAMSTEED's work, was published in Chapter 10 of *Sciothericum telescopicum*[65] in 1686; this is shown in Figure 8.4. Earth revolves around the Sun in the orbit *ABPN* centred on point *C*. Points *A* and *P* denote Earth's position at aphelion and perihelion. At noon when Earth is at aphelion[66] *AL* is the meridian, and *B* is Earth's position at noon on the following day, exaggerated for clarity. The arc $\widehat{AB}$ or the angle $\widehat{ACB}$ represents the mean movement of Earth from noon at aphelion to noon on the following day.

The Earth rotates, taking point *L* on Earth toward *O* so, the following day, when Earth reaches *B*, *L* passes through point *d* to reach *e* at noon. The arc $\widehat{df}$ or the angle $\widehat{dBf}$ is equal to the angle $\widehat{ACB}$ or the mean daily rotation of Earth. The arc $\widehat{de}$, which Earth's meridian (point *L*) must additionally pass to reach noon, is equal to the angle $\widehat{ASB}$ or the angle corresponding to the path that Earth would travel in one day. The arc $\widehat{ef}$, which is the difference between the mean and the actual day, is equal to the equation of centre[67] or angle $\widehat{SBC}$. Since the movement of Earth around the Sun is uneven, the value of the equation of centre changes over the year. The situation at aphelion applies also to the perihelion. If Earth is at its perihelion at point *P* and reaches point *N* at noon on the following day, the arc $\widehat{PN}$ or the angle $\widehat{PCN}$ is the mean daily motion from noon to noon. The meridian *PM* is the next day at noon when Earth has made one rotation, in position *ng*, which is parallel to the position on the previous day. The true noon occurs when point *g* rotates to point *h*, and the arc $\widehat{gh}$ or the angle $\widehat{gNh}$ represents the mean daily rotation of Earth. The arc $\widehat{hk}$ or the angle $\widehat{CNS}$ is the equation of the centre.

Figure 8.5 shows the quadrant *PCF* with a solstice *C*, *P* is the pole, *AF* is the radius of the equator, and *CA* is the radius of the ecliptic. Point *A* is the equinox, or the position of the Sun at 0° in Aries, on the day of the equinox at noon. The position of the Sun at noon the following day is *S*. The arc $\widehat{PSB}$ is perpendicular to the equator, $\widehat{AS}$ is the daily motion of the Sun, and the equator's arc $\widehat{AB}$ is the right ascension. The arc $\widehat{AS}$ is one of the sides of the spherical right triangle ΔAPB and is not equal to the hypotenuse, that is, the Sun's motion *AS*. The uniform motion along the ecliptic, therefore, implies uneven increments of the arc of the equator $\widehat{AB}$, which implies that solar days are of uneven lengths. To get the mean time, this difference in the second and fourth quadrants should be added to the actual time or deducted in the first and third quadrants.

The first part of MOLYNEUX's interpretation corresponds to the modern contribution of the equation of time due to uneven circulation, where the mean Sun moves along the ecliptic, while the other part where the mean Sun crosses the equator is the contribution due to the obliquity of the ecliptic.

65 Molyneux, 1686, Sciothericum Telescopicum, Chapter X, pp.36–42.
66 MOLYNEUX states the date June 18 as the aphelion.
67 In the *Sciothericum telescopicum*, the equation of the orbit.

His work was continued by his son Samuel Molyneux, who worked with Bradley. The zenith telescope used by Samuel Molyneux was the work of George Graham and was specially made for him in 1725. A smaller version of this instrument had been used by Bradley to detect aberration and nutation.[68]

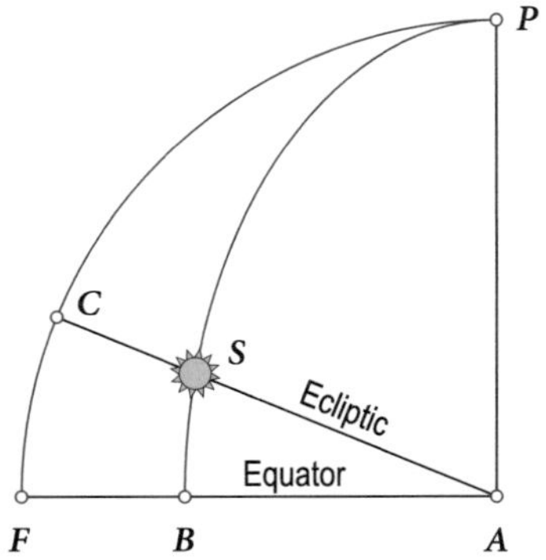

Figure 8.5 The explanation of Flamsteed's equation of time by William Molyneux's, *Sciothericum telescopicum*, Figure 14, is related to the obliquity of the ecliptic. The source of the original: Niedersächsische Staats- und Universitätsbibliothek Göttingen.

8.6 Rømer and Cassini

8.6.1 Speed of light

Shortly discovering Jupiter's four largest moons, Io, Europa, Ganymede and Callisto in 1610, Galileo suggested that the motions of the moons could serve as a natural clock. His proposals were rejected, not least because they were impracticable for use at sea.[69] Galileo published the discovery, along with other astronomical findings, in *Siderius nuncius* in 1610, the first book to discuss astronomical discoveries using a telescope.[70]

The use of the position of Jupiter's moons for timing was made possible by improved ephemerides, which could determine the moons' positions relatively accurately, and by their short orbital periods. For example, Io orbits Jupiter in less than two days.

68 Royal Observatory Greenwich, Telescope: Bradley's 12½-foot Zenith Sector (1727), http://www.royalobservatorygreenwich.org/articles.php?article=1065.

69 Buyse, 2018. Especially when used at sea, this process has some drawbacks, which is why Jupiter's moons have never been widely used for positioning in navigation. The problem is that it is very difficult to make precise observations from a telescope on a restless ship, and Jupiter is invisible for about 10 weeks when it is close to the Sun and it is not possible to use this procedure.

70 For a commentary on Galileo's discovery, see Meeus, 2004, Mathematical astronomy Morsels III, pp.347–361.

The first ephemerides of Jupiter's moons were published in 1668 by CASSINI in *Ephemerides bononienses mediceorum syderum*, who assumed a circular and uniform orbits, though he recognises disturbances in the motion of the second and third moons.[71] In 1692, CASSINI wrote that the moons could be used to determine longitude quickly and reliably for geographical measurements or navigation. CASSINI gave a historical overview of the monitoring of the motion of Jupiter's moons to determine longitude in *Les hypothèses et les tables des satellites de Jupiter, reformées sur de nouvelles observations* in 1693.[72] In addition to GALILEO, PEIRESC and KEPLER, other lesser-known astronomers have also worked on the problem.

NICOLAS-CLAUDE FABRI DE PEIRESC (1580–1637) was a French antiquarian, humanist and influential patron, who inspired HUGO GROTIUS to study law.[73] On a trip to Italy, he met GALILEO, an acquaintance who sparked his interest in numismatics and astronomy. He discovered the Orion Nebula in 1610 and was among the first to study ancient coins in a historical sense. From 1605 he was Senator of Aix in Provence. He corresponded with many famous names of the time. For example, the painter PETER PAUL RUBENS was one of his correspondents. While his written works are not extant, his breadth of expertise is evident in his extant correspondence. We know that NEWTON used his work on optics.

The discovery of Jupiter's moons and their properties, together with more accurate ephemerides, led to another important discovery – they were key to RØMER's estimation of the speed of light, which he succeeded in making in 1673.[74] Observing the beginning and end of eclipses of Jupiter's moons has been crucial for determining the speed of light, where you can observe the beginning of an eclipse before the opposition[75] of Jupiter and the end of an eclipse after the opposition. Among the lesser-known accounts of the measurement of the speed of light is LE VERRIER's 1862 account of LEON FOUCAULT's first precise measurement of the speed of light.[76] LE VERRIER also notes that JEAN PICARD[77] went to Denmark in 1671 to measure the difference in longitude between the location of the TYCHO's old observatory and that of the Observatoire de Paris. OLE RØMER (1644–1710) was assigned as his assistant. Very little information is available about RØMER's youth, but we do know that he completed his education at the Cathedral School

71 Taton, Wilson (eds.), 2003, Planetary Astronomy from the Renaissance to the Rise of Astrophysics, part A. Calculating the ephemerides of Jupiter's moons is not straightforward; for a modern calculation, see Lieske, 1977, also Sampson, 1921. Sampson is also the author of the 1910 work Tables of the Four Great Satellites of Jupiter. Eclipses of Jupiter's moons are part of the tables in the *Nautical Almanac* or *Astronomical Almanac* (section F in modern editions).

72 Bobis, Lequeux, 2008, p.98. Cassini, 1693a, Les hypothèses et les tables des satellites de Jupiter, reformées sur de nouvelles observations. Histoire de l'Académie Royale des Sciences, avec les Mémoires de Mathématiques, t.8; Mémoires (print 1730), pp.317–392.

73 HUIGH DE GROOT (1583–1645), Dutch lawyer. His 1625 work *De jure belli ac pacis* is the foundation of international law in the modern sense. He was also a statesman and diplomat.

74 For a historical overview, see Boyer, 1941, and Römer, Cohen, 1940.

75 In astronomy, opposition means the opposite position of celestial bodies relative to a third celestial body, in this case the Sun.

76 For extant correspondence and written sources on the subject, see Bobis, Lequeux, 2008. URBAIN JEAN JOSEPH LE VERRIER (1811–1877), a French astronomer and mathematician, specialised in celestial mechanics. He is best known for mathematically predicting the position of the planet Neptune. FOUCAULT (1819–1868), a French physicist, is best known for the Foucault Pendulum, which he used to prove the rotation of Earth.

77 JEAN PICARD (1620–1682), a French priest and astronomer, famous for calculating the size of Earth. He had already worked with a HUYGENS pendulum clock.

in Aarhus in 1662 and then studied medicine in Copenhagen with the brothers THOMAS and ERASMUS BARTHOLIN. During his studies, he had the opportunity to study the works of TYCHO BRAHE. After finishing his work in Denmark, PICARD took RØMER with him to Paris, where he later became a member of the Royal Academy. He stayed in Paris for nine years, working in various capacities. In 1679 he visited England and met NEWTON, HALLEY and FLAMSTEED, and two years later, at the request of King CHRISTIAN V, he returned to Denmark, where he became a royal adviser and director of the observatory in Copenhagen.

The best description of RØMER's work is in PEDER HORREBOW's *Basis astronomiae sive astronomiae pars mechanica* of 1735.[78] Letters from GOTTFRIED KIRCH to RØMER are also preserved, giving insights into astronomy at the beginning of the 18th century.[79] From a manuscript where RØMER documented his observations of Io at the beginning and end of the eclipse[80], we can reconstruct his process.[81]

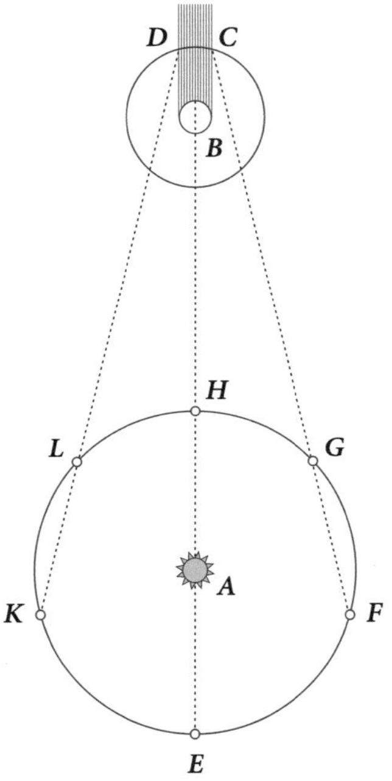

Figure 8.6 A sketch of the observation of eclipses of Jupiter's moon by RØMER. Circle ○EGL is Earth's orbit, C and D are immersion or emersion points.

RØMER observed the emersion on 18 April 1673 at 9 hours 22 minutes 0 seconds and recorded a similar observation on 4 August 1673 at 8 hours 30 minutes 41 seconds. The time is given in local solar time. RØMER is thought to have converted apparent solar time into local clock time using a table with the equation of time, calculated by CASSINI, who later published the values in 1693. The mean solar time on 18 April 1673 corresponds to 9 hours 21 minutes 10 seconds and

78 PEDER NIELSEN HORREBOW (1679–1764), Danish astronomer. From 1703 to 1707, he was RØMER's assistant. He published the work after RØMER's notes and legacy were destroyed by fire in Copenhagen in 1728. He also left several written works of his own on astronomy and navigation.

79 Herbst, 1999.

80 The beginning and end of an eclipse are immersion (eclipse begins, celestial body vanishes) and emersion (celestial body reappears).

81 Backhaus, Quast, 1987. See also Römer, 1978, Der Weg der Physik, Sambursky (ed.), pp.371–373.

on 4 August 1673 to 8 hours 36 minutes 5 seconds. In this time interval, Io made 61 orbits, so one orbit lasted 1 day 18 hours 28 minutes 46 seconds. From the observation of the Io's immersions from 28 November 1672 to 24 March 1673, we calculate an orbital period of 1 day 18 hours 27 minutes 27 seconds, which is 1 minute 19 seconds shorter than calculated from the emersions. As the observations of immersion and emersion are made from different positions on Earth, this difference is the first indication of the finite speed of light. The true orbital period, which is not subject to unwanted influences (the finite speed of light), is obtained if the observation is made over a time interval approximately equal to the synodic period of Jupiter, which is 399 days. At this time, Earth and Jupiter return to the same relative positions as at the start of the observation, so that the unwanted effects of the finite speed of light are compensated for. For example, this was the interval between the observations of 14 March 1672 and 18 April 1673. Io's orbital period is 1 day 18 hours 28 minutes 31 seconds, fifteen seconds shorter than for the period described above. The 61 orbits of Io between 18 April 1672 and 4 August 1672 should therefore have taken 107 days 22 hours 59 minutes 31 seconds if the final speed of light has no effect on the observation. In fact, the time interval was 107 days 23 hours 14 minutes 55 seconds, corresponding to a delay of 15 minutes 24 seconds.

The *Rudolphine Tables* show that the distance between Earth and Jupiter has increased by 1.38 Earth orbital radii during this period. Thus, takes 15 minutes 24 seconds / 1.38 = 11 minutes 9 seconds to travel a distance equal to the radius of Earth's orbit. This result differs significantly from the modern value of 8 minutes 19 seconds. There are probably several reasons for the inaccurate measurement, ranging from disturbances in the motion of Jupiter's moons, for which astronomers at the time did not know the cause, to the shape and position of Jupiter's shadow. The clocks of the time were also not the most accurate, and timekeeping to the second in the second half of the 17th century was still more of a wish than a reality.

Example 8.2 On 17 August 1796, at a solar time of 12 h 30 min 20 s, at a presumed longitude of 88°30′ west of Greenwich, we observe the immersion of Jupiter's first moon, Io (the example is from Hamilton Moore, 1800, p.256). The exact longitude can be calculated using the following simple procedure:

The immersion of the moon **Io** is visible in Greenwich on this day at	18 h 20 min 30 s
Observed immersion	12 h 30 min 20 s
Time difference	+5 h 50 min 10 s
Convert the time difference to degrees (1 h ≡ 15°) and obtain	87°32′½ W

If the time difference is positive, meaning that the Greenwich immersion occurs earlier, then the location is west of Greenwich. The difference between the assumed longitude of 88°30′ and the calculated longitude of 58′ is approximately 58.5 nautical miles (measured at the equator).[82]

To be accurate, the time at the observation site had to be set or synchronised with the solar time before the observation (for example, one day before).

82 The value of 58.5 miles is historic. Considering that a modern nautical mile = 1852 m, the distance is approximately 58 nautical miles. This example clearly shows why accurate time-keeping was so important in maritime navigation.

8.6.2 Eclipses and longitude

Cassini's tables allowed more accurate calculations of eclipses of Jupiter's moons and were an important tool for astronomical observations. The times of the eclipses of Jupiter's moons, or the times of immersion and emersion, could be calculated using the Cassini's tables, which contain, among other things, the following parameters:[83]

- Number of revolutions of the moon since Jupiter was last in perihelium (parameter Num. I)
- Number of revolutions of the moon since Jupiter was last in opposition[84] (parameter Num. II)

The values are summed for each point in time (years, days, etc.), similar to the ancient mean anomaly tables. Cassini tabulates values for the period 1660 to 1720.[85] Example of determining the exact time of immersion or emersion (for practical use, see Example 8.2):

- From the tables, read the values for the year (p.240) and the day or calendar date (pp.241–244). In the case of a leap year, add one day[86] for dates in January or February and calculate the sum.
- From the same tables, we can also get the values of Num. I and Num. II and calculate the sum in the same way as above.
- If Num. I is less than 1224 or greater than 2448, calculate the difference Num. I – 2448. For the absolute value of the result, in the *Tabula primae aequationis conjuctionum primi satellitis cum Jove* (i.e for conjunctions of Io) on p.245, find (interpolate) the corresponding value or the first equation to be added to the sum calculated above.
- If Num. I is between 1224 and 2448, subtract the value from 2448 and find (interpolate) the value of the first equation subtracted from the above sum in the table on page 245.
- Divide the minutes of the first equation by 11, or more precisely by 3⁴⁄₃, then:
  Subtracting the first equation, we add to the sum of Num. II, or
  if the first equation is added, we subtract from the sum of Num. II.
- If the resulting value for Num. II exceeds 225.4, subtract 225.4 from it.
- If the difference is less than 113, find or interpolate the corresponding value (or second equation) in the table on page 246. If the difference is greater than 113, find (interpolate) the corresponding value of the second equation in the same table up to 225.4.
- Add the resulting value to the time or sum calculated above to obtain the time at the mid-point of the eclipse, or halfway between immersion and emission (semimora).

83 Monsieur Cassini his New and Exact Tables for the Eclipses of the First Satellite of Jupiter, Reduced to the Julian Stile, and Meridian of London, Phil. Trans., 1694, vol. 18, pp.237–256.

84 The opposition is the position of the celestial body opposite the position of the Sun, as seen from Earth.

85 Cassini, 1694, p.240.

86 Cassini, 1694, p.250, The Use of the Tables, I.

- In the table on page 247, using the value Num. I, find half the value of the total eclipse, or the semimora,[87] which:
  if it is less than 113 or greater than 225.4 and less than 338, add the values of the emersion,
  if it is greater than 113 or 338, it is subtracted from the immersion value.
- Use the equation of time (pp.248–249) to determine the true solar time of the observation, which, if added to or subtracted from the calculated time, gives the solar time of immersion or emersion.[88]

Example 8.3 FLAMSTEED observed the emersion of Io in Greenwich on 17 September 1677 at 8 h 9 min 40 s. The observation time corresponds to 8 h 9 min 20 s for the meridian through London. The example is from Cassini, 1694, p.251.

The following values can be seen from the table:

		Num. I	Num. II		table, p.
1677	0 d 3 h 14′36″	2028	102.5		p.240
September 17	4 h 4′12″	147	145.4		p.243
Σ	7 h 18′48″	2175	248.0		
Equation 1	26′11″	2448	2.3+	2.3 = 2 6/11	
	6 h 52′37″	273	250.3		
			225.4		
Equation 2	+ 1′39″	interpolated value	24.9		p.246
Semimora	+1 h 7′ 0″	interpolated value for the Num. I = 2175			p.247
Mean time	8 h 1′16″				
Equation of time	9′ 25″	longitude of the Sun ≏ 5°			p.249
Solar time	8 h 10′41″	actual time			
Observation time	8 h 9′20″	for London meridian			
Difference	−1′21″				

For example, the difference could be due to a poorly set clock in the observation. The accuracy of the observation is very important in determining longitude, and, if the observation is not accurate, the error can be significant.

87 See the examples in Cassini, 1694, pp.251–252.
88 For the equation of time, see Cassini, 1694, pp.248–249.

8.7 Astronomers after Flamsteed

8.7.1 Philippe de la Hire

PHILIPPE DE LA HIRE (1650–1718), although he had no formal mathematical-astronomical training, attained an academic level of knowledge. He clearly inherited his father's gift for art and geometry. He left no original achievements in mathematics or astronomy, but his contribution to the experimental sciences and geometry is more than noteworthy. In mathematics, for example, he published several works on conics. in 1687, he published the first part of his *Tabularum astronomicarum*. In 1702, a collection of astronomical tables, *Tabulae astronomicae*, was published, including the equation of time. A second edition of the tables was published after his death in 1727. The second edition also includes a *Tabula aequationis meridiei*, as well as a procedure for calculating the so-called equation of noon, which is the difference between the local solar time and the local clock time at noon.[89] His equation of noon was later improved mathematically by EULER.[90]

A simple procedure for determining the time of noon is given in the 1727 edition (Example 8.4).

Example 8.4 Calculate the time of true noon on 29 October 1726.
On a given day, we measure the following heights of the Sun in the morning and afternoon:

time in the morning	altitude of the Sun	time in the afternoon
9h 33′ 4″	21°0′	2h 7′ 49″
9h 36′ 27″	21°20′	2h 4′ 26″

When calculating the time of true noon, we need to get the same result for both observations. For the observation where we measured the height of the Sun of 21°20′:

time of observation afternoon	2h 4′ 49″
add 12 hours	12h
	14h 7′ 49″
subtract the observation time in the morning	
	9h 33′ 4″
difference	4h 34′ 45″
half of the difference	2h 17′ 22½″
add the observation time in the morning	
	9h 33′ 4″
true noon	11h 50′ 26½″

89 De La Hire, 1727, Tabulæ astronomicæ Ludovici Magni, section with Tables, p.81. Values are given for each degree of declination for the longitude of Paris for 4- to 10-hour time intervals between the observations in seconds and 60th of seconds. For the use and calculation of the true noon by means of spherical trigonometry, see the explanation after the table. The theory (in Latin) is also discussed on pages 75–83 in the section *Usus tabularum*.

90 See Chapter 9.

for the observation where we measured the height of the Sun of 21°0′:

time of observation afternoon	2h 4′ 26″
add 12 hours	12h
	14h 4′ 26″

subtract the observation time in the morning

	9h 36′ 27″
difference	4h 27′ 59″
half of the difference	2h 13′ 59½″

add the observation time in the morning

	9h 36′ 27″
true noon	11h 50′ 26½″

8.7.2 William Whiston

William Whiston (1667–1752) studied mathematics at Cambridge, graduating in 1693. After his studies, he was chaplain at Norwich, and from 1701, Newton's assistant professor of mathematics at Cambridge. After Newton's retirement, he succeeded him as professor of mathematics, but was dismissed in 1710 for religious reasons. Although he made no original mathematical or physical contributions, his contribution to the popularisation of physics is important. He was championed Newton's principles and also distinguished himself by his work and lectures in experimental physics. He also worked with Cotes for a long time. He is also considered one of the fathers of the *Board of Longitude*. In a letter to a newspaper dated 14 July 1713, Whiston and Humphrey Ditton attributed the wreck of Admiral Clowdisley Showel's fleet to an inability to accurately determine its longitude. On the advice of Newton and Halley, a committee called the Board of Longitude was later set up to award a prize for the process of determining longitude to within 30 miles (48 km) (see Chapter 9).

Whiston's scholarly work includes, for example, the Latin edition of Euclid's works published in Cambridge in 1703 and the astronomical treatise *Praelectiones astronomicae* in 1707, which was translated into English a few years later.[91] The reasons for the equation of time, or the explanation for the causes of the inequality of days in the year, as Whiston called the phenomenon, are described in Chapter 10.[92] Less well-known is his work on solar eclipse observations in antiquity.[93]

91 Whiston, 1715, Astronomical Lectures Read in the Publick Schools at Cambridge.
92 Whiston, 1715, pp.115–120.
93 Steele, 2012, Ancient Astronomical Observations and the Study of the Moon's Motion (1691–1757), pp.23–36.

8.7.3 Some other works after Flamsteed

Immanuel Hammer was a grammar school teacher, born in Wittenberg, where he died in 1702.[94] Together with Thomas Helbig, he published in 1667 a treatise on the unequal length of days, *Disputationem mathematicam, dierum aeqvalitatem ac inqvalitatem*.[95] He also contributed to at least two other written works.

Friedrich Krebs (1651–1721) was a student of Weigl, and in 1675 he was called to the rectorship of the University of Jena. In 1672 he published a treatise on the unequal length of days, *Disertatione geographica aestumatas dierum inaequalitates.*

Heinrich Schurzfleisch, also Schurtzfleisch (1664–1722), was a German jurist, historian and librarian. He mentions the equation of time in the *Schurzfleishii annus romanorum iulianus* of 1704.

David Gottfried Häppen, Lat. David Gothofriedus Haeppen defended his dissertation *Disputatio cosmologica de qvantitate dierum* in Riga in 1688.[96]

Johannes Gaupp (1667–1738), a theologian, was also a student of Weigl. He served as parish priest in Lindau on Lake Constance. He is also known as a mathematician and astronomer, and was also a librarian. His legacy is preserved in the Lindau City Archives.[97] Gaupp tabulates Flamsteed's equation of time in the 1708 *Tabulae gnomonicae*, page 76.

Michael Adelbulner (1702–1779), German mathematician, physicist and astronomer, born in Nuremberg, was the son of the famous bookbinder Johann Ernst Adelbulner. After finishing grammar school, he studied bookbinding in Leipzig, in addition to his formal studies. In 1733 he returned to Nuremberg and continued his studies. When Michael Kelsch died in 1742, he succeeded him and became Professor of Mathematics and Physics at the University of Altdorf in 1743 and Professor of Logic from 1766. He was also Rector of the University and Dean of the Faculty of Philosophy. He published an annual calendar. Some of his letters to Christfried Kirch, Christoph Gottlieb von Murr and others have been preserved.[98] In 1733, together with Anders Celsius, who was then in Nuremberg, he started the astronomical journal *Commercium litterarium ad astronomiae incrementum inter huius scientiae amatores*

94 Spitzner, 1830, p.73.

95 Hammel, 2020, p.228.

96 Gottzmann, Hörner, 2007, p.925. A few manuscripts relating to him have been preserved in Latvia. Garber (ed.), 2004, p.1130. Häppen, David Gottfried, 1688, Disputatio cosmologica de qvantitate dierum, qvam in Almo Rigensium Athenæo præside Joh. Paulo Möllero ... publicè ventilandam proponit David Gothofridus Hæppen reg. Pruss. ... Ad diem [XXI] junii anni M DC LXXXVIII.

97 Stadtarchiv Lindau, https://stadtlindau.de/kunst-kultur/stadtarchiv/index.php.

98 Christfried Kkirch (1694–1740), Berlin astronomer and author of astronomical annals. Christoph Gottlieb von Murr (1733–1811), historian, art historian and polyglot. He corresponded with many well-known personalities of the time, edited professional journals and published books in various fields, and left an extensive library. See https://www.e-manuscripta.ch/name/view/2603442?offset=11 also https://www.astronomie-nuernberg.de/index.php?category=rost&page=Adelbulner-Briefwechsel.

communi consilio institutum or *Commerce on the increase in astronomical literature between lovers of that science and academic institutions.*[99] He was admitted as an external member of the Royal Prussian Academy of Sciences, and in 1741 he was also admitted to the Leopoldina Academy of Sciences. He published several treatises on important subjects, including a treatise or dissertation on the unequal length of days, *De inequalitate dierum naturalium*, in 1745. In 1736 he published *De methodo, qua solis observatae eclipses, itemque stellarum par lunam occultationes ad differentias meridianorum investigandas usurpari optime queant*, and in 1761 he wrote *Quo ad observationem rarissimi coeli phenomeni transitus scil. Veneris per Solem die 6. Juni a. c. habendam invitat.*

NIECKE, GOTTLIEB JOHANN (1700–1763), also known as NIKE/NIEKE/NIKIUS, was a theologian. He earned his master's degree and worked as a preacher in Gera from 1736–1739. Under the supervision of FRIEDRICH WEIDLER[100], he wrote a treatise on the equation of time, *De æqvatione temporis observationes selectae,*[101] and also signed the treatise on astrometeorology *Astrologiam meteorologicam systemati mundi physico-geometrico esse contrariam.*[102]

> *Claudius Ptolemaeus magnae compositionis L. III. c. X. primus perspexit inaequalitatis dierum naturalium, et partium eorundem, ueras rationes : ita enim logitur.*
>
> *Weidler, Nikius, 1722, De æqvatione temporis observationes selectae, IV, pp.5–6.*

A 1726 treatise by JOHANN CASPAR FUNCK and GOTTFRIED SCHMIDT, *De tempore horologiorum solarium et automatorum discrepante, ejvsqve aeqvatione*, contains tables with the time at solar noon.[103] FUNCK aka FUNCKE (1680–1729) was a parish priest and a grammar school teacher. He has also authored or co-authored works on history and geography. In 1716, he published the *Liber de coloribus coeli*, eng. *A book about the colours of the sky.*

99 MacTutor, https://mathshistory.st-andrews.ac.uk/Biographies/Celsius/.

100 JOHANN FRIEDRICH WEIDLER (1691–1755), professor of mathematics and chair of mathematics at the University of Wittemberg.

101 Struve, 1845, p.149.

102 Astrometeorology was part of astrology in the Middle Ages, and its purpose was to predict the weather using astrological calculations. Jenks, 1983.

103 Treatise is written in Latin, Ger. title *Warum die Pendul- Schlag- und Sack-Uhren die Zeit nach mit denen Sonnen-Uhren nicht überein kommen, und vvie sie zu vergleichen.* The time difference between sundials and clocks at midday is given in the table *Tabula æquationis pro horologiis.*

Bibliography

- ABRAHAMS, Peter. 2007. When an Eye is armed with a Telescope: The Dioptrics of William and Samuel Molyneux. In: Acta Historica Astronomiae, 33, 229–246. Frankfurt am Main: Harri Deutsch. ISBN 978-3-8171-1804-5.
- ADELBULNER, Michael. 1745. Dissertatio astronomica de inaequalitate dierum naturalium. Altdorf: Georg Meyer. URL: https://www.digitale-sammlungen.de/de/view/bsb10821383 (24 march 2024).
- ADELBULNER, Michael. Correspondence [online source]. URL: https://www.e-manuscripta.ch/name/view/2603442?offset=11 (30 October 2022).
- APPLEBAUM, Wilbur. 1972. Horrocks, Jeremiah. In: Dictionary of Scientific Biography, Gillispie, Charles Coulston (ed.), VI, 514–516. New York: Charles Scribner's Sons.
- APPLEBAUM, Wilbur. Horrocks, Jeremiah. In: Encyclopedia.com, updated May 17 2018 [online source]. URL: https://www.encyclopedia.com/people/science-and-technology/astronomy-biographies/jeremiah-horrocks (16 December 2023).
- ARGOLI, Andreae. 1638. Ephemerides. (Google Books, 18 September 2022).
- ARGOLI, Andrea. 1648. Exactissimae caelestium motuum ephemerides ad longitudinem almae vrbis. Padova: Paolo Frambotti. (Google Books, 18 September 2022).
- Astronomie in Nürnberg, Briefwechsel von Michael Adelbulner [online source]. URL: https://www.astronomie-nuernberg.de/index.php?category=rost&page=Adelbulner-Briefwechsel (30 October 2022).
- BACKHAUS, Udo and Ulrich QUAST. 1987. Die Bestimmung der Lichtgeschwindigkeit nach Rømers Verfahren mit Hilfe eines astronomischen Kalenders. In: Naturwissenschaften im Unterricht Physik/Chemie 35, 27, 35–39.
- BEDINI Silvio A. 1966. Lens making for scientific instrumentation in the seventeenth century. In: Applied Optics, 5, 5, 687–694. DOI 10.1364/AO.5.000687.
- BERNARDI, Gabriela. 2017. Giovanni Domenico Cassini, A Modern Astronomer in the 17th Century. Cham: Springer. DOI 10.1007/978-3-319-63468-5. ISBN 978-3-319-63467-8.
- Biobibliographisches Handbuch der Kalendermacher von 1550 bis 1750. Goldmayer Andreas. URL: https://www.presseforschung.uni-bremen.de/dokuwiki/doku.php?id=goldmayer_andreas (30 October 2022).
- Biobibliographisches Handbuch der Kalendermacher von 1550 bis 1750. Welper Eberhard. URL: https://www.presseforschung.uni-bremen.de/dokuwiki/doku.php?id=welper_eberhard (16 June 2023).
- BOBIS, Laurence and James LEQUEUX. 2008. Cassini, Rømer and the Velocity of Light. In: Journal of Astronomical History and Heritage, 11, 2, 97–105. URL: https://ui.adsabs.harvard.edu/abs/2008JAHH...11...97B/abstract (10 August 2024).
- BOS, H. J. M. 1972. Huygens, Christiaan. In: Dictionary of Scientific Biography, Gillispie, Charles Coulston (ed.), VI, 597–613. New York: Charles Scribner's Sons.
- BOYER, Carl B. 1941. Early Estimates of the Velocity of Light. In: Isis, 33, 1, 24–40. DOI 10.1086/358523. URL: https://www.jstor.org/stable/330649.
- BOULLIAU, Ismael. 1645. Astronomia Philolaica. Paris: Simeonis Piget. (Google Books, 12 March 2023).
- BOULLIAU, Ismael. 1657. Astronomiae Philolaicae fundamenta clarius explicata et asserta adv. Sethi Wardi Oxoniensis Professoris impugnationem. Paris: Gabriel Cramoisy. URL: https://www.digitale-sammlungen.de/en/view/bsb10053335 (24 March 2024).
- BRUHNS, Christian. 1875. Adelbulner, Michael. In: Allgemeine Deutsche Biographie 1, 72 [online version]. URL: https://www.deutsche-biographie.de/pnd116007427.html#adbcontent (30 October 2022).
- BUCHOLD, Matthias. 1628. Problema astronomicum, de inaequalitate dierum naturalium. Leipzig: Gregor Ritzsch.
- BUYSE, Filip A. A. 2018. Galileo Galilei, Holland and the pendulum clock. In: O que nos faz pensar, 26, 41, 9–43.
- CASSINI, Jean-Dominique. 1668. Ephemerides Bononienses Mediceorum syderum. Bologna. (Google Books, 1 May 2023).
- CASSINI, Jean-Dominique. 1692. Nouvelles découvertes de diverses Périodes de mouvement dans la Planete de Jupiter, depuis le mois de Janvier 1691 jusqu'au commencement de l'année 1692. In: Mémoires de Mathématique et de Physique, tirez des Registres de l'Académie Royale des Sciences, Année 1692, t. 10 (printed in 1730), 1–9. URL: https://archive.org/details/mmoiresdemathma00fragoog/mode/2up (10 August 2024).
- CASSINI Jean-Dominique. 1693. Les hypothèses et les tables des satellites de Jupiter, reformées sur de nouvelles observations. In: Histoire de l'Académie Royale des Sciences, avec les Mémoires de Mathématiques, t. 8; Mémoires (printed in 1730), 317–392. URL: https://gallica.bnf.fr/ark:/12148/bpt6k1510925k.image (24 March 2024).
- CASSINI Jean-Dominique. 1694. Monsieur Cassini his New and Exact Tables for the Eclipses of the First Satellite of Jupiter, Reduced to the Julian Stile, and Meridian of London. In: Philosophical Transactions, 31 December 1694, 18, 214, 237–256. DOI 10.1098/rstl.1694.0048.

- Crombie, A. C. 1974. Mersenne, Marin. In: Dictionary of Scientific Biography, Gillispie, Charles Coulston (ed.), IX, 316–322. New York: Charles Scribner's Sons. ISBN 0-684-10120-3.
- Curtz, Albert. 1666. Historia Cælestis. Augsburg. URL: https://archive.org/details/HistoriaCoelestisTichonisBrahe (17 December 2023).
- Datenbank von Astronomen der Frühen Neuzeit 1473 – 1727. Ein frei verfügbares Online-Angebot der Abt. für Geschichte der Naturwissenschaften und Technik an der Universität Stuttgart [online source]. URL: https://www.hi.uni-stuttgart.de/gnt/afn/index.html (30 October 2022).
- Davis, John. 2003. The Equation of Time as shown on Sundials. In: Bulletin of the British Sundial Society, 16, 4, 135–144.
- Davis, John. 2005. More on the Equation of Time on Sundials. In: Bulletin of the British Sundial Society, 17, 2, 66–74.
- Döring, Detlef. 1999. Erhard Weigels Zeit an der Universität Leipzig (1647–1653). In: Acta Historica Astronomiae, 7, 69–90. Frankfurt am Main: Harri Deutsch. ISBN 3-8171-1600-4.
- Dreyer, J. L. E. 1922. Flamsteed's letters to Richard Towneley. In: The Observatory, 45, 280–294.
- Erhard Weigel Gesselschaft [online source]. URL: http://www.erhard-weigel-gesellschaft.de/ (30 October 2022).
- Farell, Maureen. 1976. William Whiston: The longitude man. In: Vistas in Astronomy, 20, 1, 131–134. DOI 10.1016/0083-6656(76)90022-2.
- Farell, Maureen. 1981. William Whiston. New York: Arno Press. ISBN 9780405138546.
- Flamsteed, John. 1669. An accompt of such of the more notable celestial phænomena of the Year 1670, as will be conspicuous in the English horizon. Philosophical Transactions, 4, 55, 1099–1112. DOI 10.1098/rstl.1669.0057.
- Flamsteed, John. 1672. De inaequalitate dierum solarium. London. URL: https://www.e-rara.ch/zut/content/zoom/855272?lang=en (1 May 2023).
- Flamsteed, John. 1678. De Temporis Aequatione Diatriba. In: Jerremiah Horrox, Opera Posthuma, 441–464. London. URL: https://www.e-rara.ch/zut/content/pageview/854825 (24 March 2024).
- Flamsteed, John and Francis Baily. 2011. An Account of the Revd. John Flamsteed, the First Astronomer-Royal To Which Is Added, his British Catalogue of Stars, Corrected and Enlarged. Cambridge: Cambridge University Press. ISBN 9781108038003.
- Force, James E. 2002. William Whiston: Honest Newtonian. Cambridge: Cambridge University Press. ISBN 9780521524889.
- Forbes, Eric Gray, Lesley Murdin and Frances Wilmoth (eds.). 1995. The Correspondence of John Flamsteed, the First Astronomer Royal, Volume 1. Boca Raton, FL: CRC Press. ISBN 978-0750301473.
- Forbes, Eric Gray, Lesley Murdin and Frances Wilmoth (eds.). 1997. The Correspondence of John Flamsteed, the First Astronomer Royal, Volume 2. Boca Raton, FL: CRC Press. ISBN 978-0750303910.
- Forbes, Eric Gray, Lesley Murdin and Frances Wilmoth (eds.). 2002. The Correspondence of John Flamsteed, the First Astronomer Royal, Volume 3. Boca Raton, FL: CRC Press. ISBN 978-0750307635.
- Funck, Johann Caspar and Gottfried Schmid. 1726. De tempore horologiorum solarium et automatorum discrepante, ejvsqve aeqvatione. Ulm: Elias Kühn. (Google Books, 9 July 2023).
- Gaab, Hans. 2011. Der Altdorfer Mathematik- und Physikdozent Abdias Trew (1597–1669). Astronom, Astrologe, Kalendermacher und Theologe. Acta Historica Astronomiae, 42. Frankfurt am Main: Harri Deutsch. ISBN 978-3-944913-33-9.
- Galilei, Galileo. 1610. Sidereus Nuncius. (Google Books, 17 December 2023).
- Galilei, Galileo. 2015. Sidereus Nuncius, Albert van Helden (trans.). Second edition. Chicago: University of Chicago Press. ISBN 978-0-226-32009-0.
- Garber, Klaus (ed.). 2004. Handbuch des personalen Gelegenheitsschrifttums in europäischen Bibliotheken und Archiven, Band 15 Riga, Teil 4. Hildesheim: Georg Olms Verlag. ISBN 978-3-487-11406-4.
- Gaupp, Johann, Indexeintrag. In: Deutsche Biographie [online source]. URL: https://www.deutsche-biographie.de/pnd116468645.html (18 September 2022).
- Gaupp, Johannes. 1708. Tabulae Gnomonicae. Lindau: Johannes Gaupp. (Google Books, 26 October 2022).
- Gaythorpe, S. B. 1925. On Horrocks's treatment of the evection and the equation of the centre, with a note on the elliptic hypothesis of Albert Curtz and its correction by Boulliau and Newton. In: Monthly Notices of the Royal Astronomical Society, 85, 858–865. DOI 10.1093/mnras/85.8.858.
- Gingerich, Owen. 1970. Argoli, Andrea. In: Dictionary of Scientific Biography, Gillispie, Charles Coulston (ed.), I, 244–245. New York: Charles Scribner's Sons. SBN 684-10116-5.
- Goldmayer, Andreas. 1657. Computus Directionum Astronomicus. Strassburg: Johann Heinrich Mittel. (Google Books, 18 September 2022).

- Gottzmann, Carola L. and Petra Hörner. 2007. Lexicon der deutschprachigen Literatur des Baltikums und St. Petersburgs, Band 1. Berlin: De Gruyter. ISBN 978-3-11-019338-1.
- Günther, Siegmund. 1896. Weidler, Johann Friedrich. In: Allgemeine Deutsche Biographie, 41, 453–455 [online version]. URL: https://www.deutsche-biographie.de/pnd100696198.html#adbcontent (30 March 2024).
- Hammel, Jürgen. 2020. Bibliographie der astronomischen Drucke bis 1700 [online version]. URL: https://www.astw.de/kram/download.php?file=Bibliographie_der_astronomischen_Literatur_bis_1700_v1506.pdf (18 September 2022).
- Hammer, Immanuel. In: Deutsche Digitale Bibliothek, Kultur und Wissen online [online source]. URL: https://www.deutsche-digitale-bibliothek.de/person/gnd/128374012 (30 October 2022).
- Helbig, Thomas and Immanuel Hammer. 1667. Disputationem mathematicam, dierum aeqvalitatem ac inqvalitatem. Wittenberg: Michael Wendt. (Google Books, 24 March 2024).
- Hamilton Moore, John. 1800. The New Practical Navigator, 14th Edition. London: J. Crowder. (Google Books, 27 August 2023).
- Heckscher, August and Arthur Joachim von Oettingen (trans., eds.). 1913. Die Pendeluhr, Horologium oscillatorium von Christiaan Huygens. Leipzig: W. Engelmann.
- Herbst, Klaus-Dieter. 1999. Astronomie um 1700. Kommentierte Edition des Briefes von Gottfried Kirch an Olaus Römer vom 25. Oktober 1703. Acta Historica Astronomiae, 4. Frankfurt am Main: Harri Deutsch. ISBN 3-8171-1589-X.
- Horrebow, Peder. 1735. Basis astronomiae sive astronomiae pars mechanica. Copenhagen: Christian Paulli. URL: https://archive.org/details/b30512451 (24 March 2024).
- Houzeau, Jean-Charles. 1889. Bibliographie générale de l'astronomie, 1-2. Bruxelles: Academie Royale de Belgique. (Google Books, 18 September 2022).
- Huygens, Christiaan. 1665. Kort Onderwys aengaende het gebruyck der Horologien tot het vinden der Lenghten van Oost en West. The Hague. (Google Books, 25 February 2020).
- Huygens, Christiaan. 1669. Instructions Concerning the Use of Pendulum-Watches, for Finding the Longitude at Sea. In: Philosophical Transactions, 1 January 1669, 4, 47, 937–976. DOI 10.1098/rstl.1669.0013.
- Huygens, Christiaan. 1673. Horologium oscilatorium. Paris. URL: https://archive.org/details/B-001-004-158/page/n25/mode/2up (24 March 2024).
- Huygens, Christiaan. 1890. Oeuvres complètes de Christiaan Huygens, 3. URL: https://archive.org/details/oeuvrescomplte03huyg (24 March 2024).
- Huygens, Christiaan. 1891. Oeuvres complètes de Christiaan Huygens, 4. URL: https://archive.org/details/oeuvrescomplte04huyg (24 March 2024).
- Huygens, Christiaan. 1893. Oeuvres complètes de Christiaan Huygens, 5. URL: https://archive.org/details/oeuvrescomplte05huyg (24 march 2024).
- Huygens, Christiaan. 1895. Oeuvres complètes de Christiaan Huygens, 6. URL: https://gallica.bnf.fr/ark:/12148/bpt6k77858m.r=completes%20huygens?rk=150215;2 (24 March 2024).
- Huygens, Christiaan. 1905. Oeuvres complètes de Christiaan Huygens, 10. URL: https://gallica.bnf.fr/ark:/12148/bpt6k77858m.r=completes%20huygens?rk=150215;2 (24 March 2024).
- Huygens, Christiaan. 1925. Oeuvres complètes de Christiaan Huygens, 15.
- Huygens, Christiaan. 1934. Oeuvres complètes de Christiaan Huygens, 18.
- Jenks, Stuart. 1983. Astrometeorology in the Middle Ages. In: Isis, 74, 2, 185–210. DOI 10.1086/353243.
- Jensen, Derek. 2007. Curtz, Albert. In: Biographical Encyclopedia of Astronomers, Hockey, Thomas, Virginia Trimble and Thomas R. Williams (eds.), 267. New York: Springer. ISBN 9780387310220.
- Kargon, Robert H. 1974. Molyneux, William. In: Dictionary of Scientific Biography, Gillispie, Charles Coulston (ed.), IX, 464–466. New York: Charles Scribner's Sons. ISBN 0-684-10120-3.
- Kitto, Tony. 2000. John Flamsteed, Richard Towneley and the Equation of Time. In: Antiquarian Horology and Proceedings of the Antiquarian Horological Society, 25, 180–184.
- Kitto, Tony. 2001. Richard Towneley and the Equation of Natural Days. In: Bulletin of the British Sundial Society, 13, 2, 60–65.
- Krebs, Friedrich Johann. 1672. Disertatione Geographica Aestumatas Dierum Inaequalitates. Jena (Google Books, 18 September 2022).
- Laccary, Gilles. 1657. Mathematica Tabulae Logarithmeticae ad Triangulorum Rectilineorum & Spaericorum Analysin. Paris. URL: https://archive.org/details/bub_gb_JD8gYLU--m4C (24 March 2024).
- Leadbetter, Charles. 1735. Uranoscopia or, the contemplation of the heavens. London: J. Wilcox. URL: https://archive.org/details/BUSA297_083/page/n5 (24 March 2024).

- Lieske, Jay H. 1977. Theory of Motion of Jupiter's Galilean Satellites. In: Astronomy and Astrophysics, 56, 333–353. doi 10.1007/978-94-009-9809-4_24.
- Lieske, Jay H. 1981. Catalogue of Eclipses of Jupiter's Galilean Satellites, 1610–2000. In: Astronomy and Astrophysics Supplement Series, 44, 209–216. url: https://www.researchgate.net/publication/23825951_Catalogue_of_eclipses_of_Jupiter's_Galilean_satellites_1610-2000 (10 August 2024).
- Mahoney, Michael S. 1980. Christian Huygens: The Measurement of Time and of Longitude at Sea. In: Studies on Christiaan Huygens, Bos, H.J.M. et al. (eds.), 234–270. London: Routledge. isbn 978-9026503337.
- Meeus, Jean. 1989. Transits. Richmond, VA: Willmann Bell. isbn 978-0943396262.
- Meeus, Jean. 2004. Mathematical astronomy Morsels III. Richmond, VA: Willmann Bell. isbn 0-943396-81-6.
- Molyneux, William. 1686. Sciothericum Telescopicum. Dublin. Andrew Crook and Samuel Helsham. (Google Books, 25 February 2020).
- Montucla, Jean-Étienne. 1798. Histoire des mathématiques. Paris: Henri Agasse. url: https://gallica.bnf.fr/ark:/12148/bpt6k1076512.texteImage (24 March 2024).
- Mouton, Gabriel. 1670. Observationes diametrorum solis et lunae apparentium. Lugduni: Ex Typographiâ Matthæi Liberal. url: https://www.e-rara.ch/zut/collections/content/titleinfo/4006538. doi 10.3931/e-rara-13131.
- Müller, Philipp and Valerius Schütz. 1621. Ēmerologia seu problema astronomico-geographicum de aequalitate et inaequalitate dierum artificialium & noctium. Leipzig: Justus Jansonius Danus. url: https://katalog.ub.uni-leipzig.de/Record/0-1650489951 (30 October 2022).
- Nike, Johann Gottlieb, Indexeintrag. In: Deutsche Biographie [online source]. url: https://www.deutsche-biographie.de/pnd104145951.html (13 November 2022).
- Nikius, Gottlieb Johannes. 1722. De Aeqvatione Temporis Observationes Selectae. Wittemberg: Creusig. url: https://www.digitale-sammlungen.de/en/view/bsb11210650?page=,1 (9 July 2023).
- Nikius, Johann Gottlieb and Georg Philipp Sauber. 1722. Astrologiam meteorologicam systemati mundi physico-geometrico esse contrariam. Wittemberg: Creusig. url: https://www.digitale-sammlungen.de/de/view/bsb10818370?page=1 (30 October 2022).
- Pilz, Kurt. 1977. 600 Jahre Astronomie in Nürnberg. Nürnberg: Hans Carl. isbn 9783418004471.
- Rigaud, Stephen Peter and Stephen Jordan Rigaud (eds.). 1841. Correspondence of Scientific Men of the Seventeenth Century, II. Oxford: Oxford University Press. url: https://archive.org/details/correspondences01rigagoog/mode/2up (24 March 2024).
- Rohlfs, K. 1974. Olaf Römer und die Lichtgeschwindigkeit. In: Sterne und Weltraum, 10, 311–314.
- Römer, Ole. 1978. Eine Demonstration der Bewegung des Lichtes. German translation of original from 1676. In: Sambursky, Shmuel (ed.), Der Weg der Physik, 2500 Jahre des physikalischen Denkens, 371–373. München: Deutscher Taschenbuch Verlag. isbn 3-7608-0384-9.
- Romer, M. and Bernard I. Cohen. 1940. Roemer and the First Determination of the Velocity of Light (1676). In: Isis, 31, 2, 327–379. doi 10.1086/347594. url: https://www.jstor.org/stable/225757.
- Sampson, Ralph Allen. 1921. Theory of the Four Great Satellites of Jupiter. In: Memoirs of the Royal Astronomical Society, 63. url: https://ui.adsabs.harvard.edu/abs/1921MmRAS..63....1S/abstract (24 March 2024).
- Schorn, Ronald A. 2007. Römer, Ole. In: Biographical Encyclopedia of Astronomers, Hockey, Thomas, Virginia Trimble and Thomas R. Williams (eds.), 983. New York: Springer. isbn 9780387310220.
- Schurzfleisch, Heinrich Leonhard, Indexeintrag. In: Deutsche Biographie [online source]. url: https://www.deutsche-biographie.de/pnd117647853.html (18 September 2022).
- Schurzfleisch, Heinrich Leonhard. 1704. Annus Romanorum Iulianus. Wittemberg: Christian Schroedter. (Google Books, 24 March 2024).
- Schütz, Valerius, Indexeintrag. In: Deutsche Biographie [online source]. url: https://www.deutsche-biographie.de/pnd132153998.html (18 September 2022).
- Scriba, Christoph J. 1976. Wallis, John. In: Dictionary of Scientific Biography, Gillispie, Charles Coulston (ed.), XIV, 146–155. New York: Charles Scribner's Sons. isbn 0-684-12926-4.
- Seck, Friedrich. 1978. Wilhelm Schickard (1592–1635), Astronom, Geograph, Orientalist, Erfinder der Rechenmaschine. Tübingen: J. C. B. Mohr. isbn 3-16-939772-9.
- Seck, Friedrich. 2002. Wilhelm Schickard Briefwechsel, II. Stuttgart: Friedrich Frommann Verlag, Günther Holzboog. isbn 3-7728-2162-6.
- Sellers, David. 2012. In Search of William Gascoigne. New York: Springer. isbn 978-1-4614-4096-3.

- Shakerley, Jeremy. 1649. A short Mathematicall Discourse; Laying open the Errors and impertinencies delivered in a Treatise lately published by Mr. Vincent Wing, and Mr. William Leybourne, under the title of Urania practica. By Jeremy Shakerley philomath. London: Thomas Brudenell.
  URL: http://name.umdl.umich.edu/A93028.0001.001 (12 March 2023).
- Shakerley, Jeremy. 1653. Tabulæ Britanicæ: The British Tables. London: R. & W. Leybourn. (Google Books, 13 May 2023).
- Sheehan, William and John Westfall. 2004. The Transits of Venus. Amherst, NY: Prometheus Books.
  ISBN 978-1591021759.
- Simms, J. G. 1982. William Molyneux of Dublin. Dublin: Irish Academic Press. ISBN 978-0716500964.
- Smith, Alan. 2000. A Reconstruction of the Tompion/Towneley/Flamsteed "Great Clocks" at Greenwich. In: Antiquarian Horology and Proceedings of the Antiquarian Horological Society, 25, 185–190.
- Snobelen, Stephen D. 2023. William Whiston, Experimental Lecturing and the Royal Society of London. In: Collected Wisdom of the Early Modern Scholar, Marie Roos, Anna and Gideon Manning (eds.), 349–374. Cham: Springer. DOI 10.1007/978-3-031-09722-5. ISBN 978-3-031-09721-8.
- Speziali, Pierre. 1974. Mouton, Gabriel. In: Dictionary of Scientific Biography, Gillispie, Charles Coulston (ed.), IX, 554–555. New York: Charles Scribner's Sons. ISBN 0-684-10120-3.
- Spitzner, Franz. 1830. Geschichte des Gymnasium uns der Schulanstalten zu Wittenberg. Leipzig: Hartmann. (Google Books, 18 September 2022).
- Stadtarchiv, Lindau. URL: https://stadtlindau.de/kunst-kultur/stadtarchiv/index.php?la=1&object=tx,2412.2208.1&kat=&kuo=2&sub=0 (30 October 2022).
- Steele, John M. 2012. Ancient Astronomical Observations and the Study of the Moon's Motion (1691–1757). New York: Springer. DOI 10.1007/978-1-4614-2149-8. ISBN 978-1-4614-2148-1.
- Streete, Thomas, Johann Gabriel Doppelmayr and Jean-Baptiste Morin. 1705. Astronomia Carolina. Noribergae: Andreæ Ottonis. URL: https://echo.mpiwg-berlin.mpg.de/MPIWG:EEPDNFXR (30 March 2024).
- Struve, Friedrich Georg Willhelm. 1845. Description de l'Observatoire astronomique central de Poulkova. St.Petersburg: Academie Imperiale des Sciences. (Google Books, 18 September 2022).
- Symonds, R. W. 1951. Thomas Tompion: His Life and Works. London: B. T. Badsford.
- Symonds, R. W. 1969. Thomas Tompion: His Life and Works. London: Spring Books. ISBN 978-0600386025.
- Taton, René and Curtis Wilson (eds.). 2003. Planetary Astronomy from the Renaissance to the Rise of Astrophysics, Part A. Cambridge: Cambridge University Press. ISBN 9780521542050.
- Thoren, Victor E. 1972. Flamsteed, John. In: Dictionary of Scientific Biography, Gillispie, Charles Coulston (ed.), V, 22–26. New York: Charles Scribner's Sons. SBN 684-10116-5.
- Tikkanen, Amy. Nicolas-Claude Fabri de Peiresc. In: Encyclopedia Britannica [online version].
  URL: https://www.britannica.com/biography/Nicolas-Claude-Fabri-de-Peiresc (27 October 2022).
- Truffa, Giancarlo. 2007. Andrea Argoli. In: Biographical Encyclopedia of Astronomers, Hockey, Thomas, Virginia Trimble and Thomas R. Williams (eds.), 59. New York: Springer. ISBN 9780387310220.
- Van Lansberge, Johan Philip. 1662. Tabula motuum caelestium perpetuae. Middelburg: Zachariam Roman. URL: https://archive.org/details/bub_gb_Y1L75RRCdkIC/mode/2up (24 March 2024).
- Von Bertele, Hans. 1957. Zur Geschichte der Äquationsuhren-Entwicklung. In: Nagler, Josef (ed.), Blätter für Technikgeschichte, 19, 78–121. Technisches Museum für Industrie und Gewerbe in Wien, Forschungsinstitut für Technikgeschichte. Wien: Springer. DOI 10.1007/978-3-7091-5534-9_5. ISBN 978-3-7091-5534-9.
- Von Bertele, Hans. 1959. The Development of Equation Clocks, Part 1. In: La Suisse Horlogerie, International Edition, 3, 39–46.
- Wallis, John. 1666. Hypothesis about the Flux and Reflux of the Sea. Philosophical Transactions, 1665-1666, 6 August 1666, 1, 16, 263–281. DOI 10.1098/rstl.1665.0108.
- Ward, Seth. 1653. In Ismaelis Bullialdi astronomiae philolaicae fundamenta, inquisitio brevis. Oxford. DOI 10.3931/e-rara-8728.
- Watt, Robert. 1824. Bibliotheca Britannica; Or, A General Index to British and Foreign Literature. Part II Authors. Edinburgh: Archibald Constable and Co. London: Longman, Hurst, Rees, Orme, Brown & Green. (Google Books, 18 September 2022).
- Webster, Charles. 1976. Towneley, Richard. In: Dictionary of Scientific Biography, Gillispie, Charles Coulston (ed.), XIII, 444–445. New York: Charles Scribner's Sons. ISBN 0-684-12925-6.
- Wegener, Alfred. 1905. Die Alfonsinischen Tafeln für den Gebrauch eines modernen Rechners. Dissertation. Berlin: Friedrich-Wilhelms-Universität zu Berlin. URL: https://www.digi-hub.de/viewer/!fulltext/BV041981724/45/ (24 March 2024).

- Welper, Eberhard. 1700. Brevis Instructio de Erigenda Figvra Coeli, second edition. Strasbourg: Johann Albert Dolhopff. (Google Books, 16 June 2023).
- Welper, Eberhard G. 1708. Neu-vermehrte Welperische Gnomonica, Doppelmayr, Johann Gabriel (ed.). Nürnberg: Johann Christoph Weigel. URL: https://digital.ub.uni-duesseldorf.de/urn/urn:nbn:de:hbz:061:1-174115 (24 March 2024).
- Westermayer, Georg. 1876. Curtz, Albert. In: Allgemeine Deutsche Biographie, 4, 654–655 [online version]. URL: https://www.deutsche-biographie.de/pnd124643221.html#adbcontent (24 March 2024).
- Wilkins, John. 1668. An Essay Towards a Real Character and Philosophical Language. London. (Google Books, 25 February 2020).
- Wilson, Curtis. 1989. Predictive astronomy in the century after Kepler. In: Planetary astronomy from the Renaissance to the rise of astrophysics, Volume 2, Part A: Tycho Brahe to Newton, Taton, René and Curtis Wilson (eds.). Cambridge: Cambridge University Press. ISBN 0 521 24254 1.
- Wing, Vincent and William Leybourn. 1649. Ens fictum Shakerlæi or the annihilation of Mr. Jeremie Shakerley, his in-artificiall anatomy of Urania practica. Wherein his falacies or ignorance, are demonstratively detected his malice in its groundlesse colours display'd, and the authors of the said Urania practica justly vindicated from his unjust aspersions. London: Robert Leybourn. URL: http://name.umdl.umich.edu/A96673.0001.001 (12 March 2023).
- Wing, Vincent. 1656. Astronomia Instaurata. London: R. and W. Leybourn. (Google Books, 13 May 2023).
- Whiston, William. 1707. Prælectiones Astronomicæ Publicis Habitæ. London: Benjamin Tooke. (Google Books, 9 July 2023).
- Whiston, William. 1715. Astronomical Lectures Read in the Publick Schools at Cambridge. London: R. Senex, W. Taylor. (Google Books, 11 September 2021).

This page is intentionally left blank.

9

Euler and Tobias Mayer

Ad verum meridiei tempus inueniendum inter ali Aſtronomi vti ſolent binis obſeruationibus aequalium ſolis altitudinum, quarum altera ante altera poſt meridiem ſunt factae.

Leonhard Euler, 1735, Methodus Compvtandi Aeqvationem Meridiei, §.1

9.1 Leonhard Euler

9.1.1 Euler's solution to Kepler's equation

LEOHARD EULER (1707–1783) is one of the greatest names in mathematics. Despite his gilding nearly every branch of mathematics that he touched, contemporary sources describe him as a very modest personality. He is known for his iterative solution to KEPLER's equation, which was already known before him, and may have originated with KEPLER himself. EULER's work also extends to spherical trigonometry, he published a treatise on the equation of noon, as he called it, and his now-forgotten derivation of the equation of the centre is also interesting. He also laid the modern foundations of complex analysis, an especially important contribution being EULER's formulae from which we can derive many further expressions, several of which are also used in the mathematical derivations in this book. His publications also deal with perturbation theory, and his contribution to the motion of the Moon is important.

In 1740, EULER published his *Orbitae solaris determinatio* (*Determination of orbits around the Sun*), in which he worked out the orbits of the planets from FLAMSTEED's data.[1] In his *De motu planetarum et orbitarum determinatione* (*On the motion of planets and orbits*), published in the same year, EULER also proposed a new solution to the KEPLER equation. He had written this treatise in 1735. EULER's solution to KEPLER's equation has the form:[2]

[9.1] $$E = M + e(\sin M + e\sin(M + e\sin M + \cdots$$

e	eccentricity
M	mean anomaly

which corresponds to the sum of successive approximations:

$$E_n = M + e\sin E_{n-1} \qquad n = 1, 2, 3, \ldots \qquad E_0 = M$$

This iterative procedure, often attributed to EULER, was used by PEDER HORREBOW to solve KEPLER's equation in his 1717 work, and may have roots in KEPLER's own work.[3]

1 Euler, 1740b, Orbitae solaris determinatio, p.86. He used data from *Historia coelestis Britannica*, 1725, a collection of the observations of JOHN FLAMSTEED.

2 Euler, 1740a, De motu planetarum et orbitarum determinatione, pp.70–71, §.9.

3 Collwell, 1993, Solving Kepler's Equation over three Centuries, p.19. See also Jørgensen, 1974, Peter Horrebow's solution of the Kepler Problem. HORREBOW (1679–1764) was Professor of Astronomy at the University of Copenhagen from 1714, and is credited with documenting the work of CHRISTIAN RØMER.

9.1.2 The equation of noon

The EULER equation of noon E_{en}, which is the difference between the true solar noon and the noon of the mean Sun or clock, is given by Expression 9.2 (values are in degrees) for positive values of declination or north declination (see also Figure 9.1):[4]

$$E_{en} = \frac{\widehat{SPT}}{720°} \mathrm{d}t \left(\frac{\tan \widehat{ZE}}{\sin \frac{\widehat{SPT}}{2}} - \frac{\tan \widehat{OE}}{\tan \frac{\widehat{SPT}}{2}} \right) \quad [9.2]$$

For negative values of declination (south declination) it applies:

$$E_{en} = \frac{\widehat{SPT}}{720°} \mathrm{d}t \left(\frac{\tan \widehat{ZE}}{\sin \frac{\widehat{SPT}}{2}} + \frac{\tan \widehat{OE}}{\tan \frac{\widehat{SPT}}{2}} \right) \quad [9.3]$$

Multiplying the above two expressions by 4 gives the result in seconds:

$$E_{en} = \frac{\widehat{SPT}}{180°} \mathrm{d}t \left(\frac{\tan \widehat{ZE}}{\sin \frac{\widehat{SPT}}{2}} \mp \frac{\tan \widehat{OE}}{\tan \frac{\widehat{SPT}}{2}} \right) \quad [9.4]$$

P	celestial pole
$\widehat{PZE}$	meridian at desired location
$\widehat{PZ}$	latitude complement
$\widehat{ZE}$	latitude
$\widehat{ZS}$	complement of the altitude of the Sun
$\widehat{PS}$	complement of the declination
CD, cd	parallels
$\widehat{OE}$	declination
$\mathrm{d}t$	daily increment of declination
$\widehat{SPT}$	angle between two moments of observation, morning and afternoon, when the Sun is at the same altitude. Determined by observation time (clock time) and converted to degrees

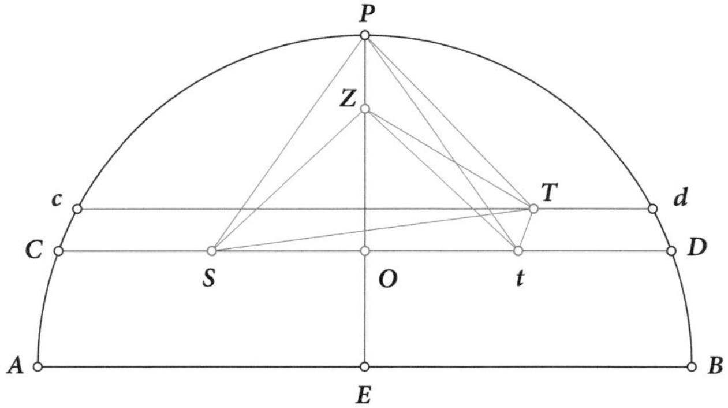

Figure 9.1 EULER'S sketch for the equation of noon. Source: Euler, 1741, Methodus computandi aequationem meridiei, pp.53–54, Figure 2.

4 Euler, 1741, Methodus computandi aequationem meridiei, §.16, pp.59–60 also Headley, 2022, A Method for Calculating the Equation of Noon, p.73.

The whole derivation process, which is not easy to understand, is described in EULER's 1741 treatise *Methodus computandi aequationem meridiei*. The procedure contains elements of spherical trigonometry, calculus and logarithms.[5] EULER was clearly motivated by a desire for simpler calculations than those offered in other works, for example by DE LA HIRE, whose tables EULER also used.[6] The basic geometrical construction on which EULER built the further derivation is shown in Figure 9.2.

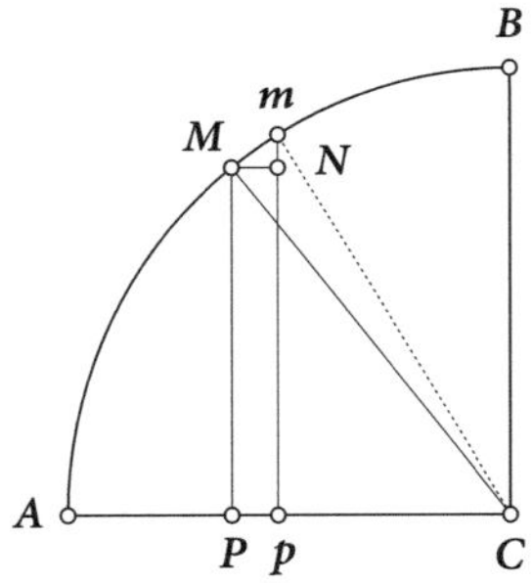

Figure 9.2 EULER'S sketch to understand the basic concepts, the figure is a projection of Figure 9.3 on the equatorial plane. Source: Euler, 1741, Methodus computandi aequationem meridiei, Figure 1.

The following shows EULER's procedure for deriving the equation of noon using the original (Eulerian) notation. If we consider that the radius of the circle $R = 1$ and $PM = A$, for any arc $\widehat{AM}$ (see Figures 9.2 and 9.3) applies:[7]

$$\sin \widehat{PM} = \sin \widehat{ACE} = A$$

or:

$$\cos \widehat{CP} = \cos \widehat{ACE} = a = \sqrt{1 - A^2}$$

We assume that the arc $\widehat{Mm}$ is a small increment of the arc $\widehat{AM}$. Since the increment is small, the arc extension or increment can be approximated as a straight line. For arc $\widehat{Am}$:

$$\sin \widehat{Am} = pm = PM + mN = A + mN$$

5 EULER used various tables of trigonometric function values in the procedure, for example for radii 10^7 and 10^{10}.

6 EULER seems to have relied heavily on DE LA HIRE's data, or astronomical tables, according to the references in his works. For DE LA HIRE's example of the calculation of the true noon, which is methodologically similar to EULER's, see De la Hire, 1727, explanation of the use of table *Tabula æquationis meridiei*.

7 Euler, 1741, §.8, p.53. EULER is using A in two diferent ways, both as a point and as the length of the line segment PM in Figure 1. Headley, 2022, p.70, note 8. In the tables used by EULER, the radius was not always equal to 1.

The triangles $\triangle CPM$ and $\triangle mNM$ are similar triangles for which:[8]

$$CM : CP = Mm : mN \qquad \text{or} \qquad 1 : a = Mm : mN$$

therefore:

$$mN = a \cdot Mm \quad \text{and} \quad pm = A + a \cdot Mm$$

the following also applies:

$$Pp = MN = A \cdot mM$$

The cosine of the extended arc $\widehat{Am}$ is:

$$\cos \widehat{Am} = Cp = CP - Pp = a - Pp$$

Using the above equations, the expressions for the sine and cosine of the extended arcs are:[9]

$$\sin \widehat{Am} = \sin \widehat{ACm} = A + a \cdot \underbrace{\widehat{Mm}}_{=\widehat{MCm}} \tag{9.5}$$

$$\cos \widehat{Am} = \cos \widehat{ACm} = a - A \cdot \underbrace{\widehat{Mm}}_{=\widehat{MCm}} \tag{9.6}$$

The daily increment of declination was determined by EULER using the daily motion of the Sun along the ecliptic (Figure 9.3), both parameters in arc seconds.

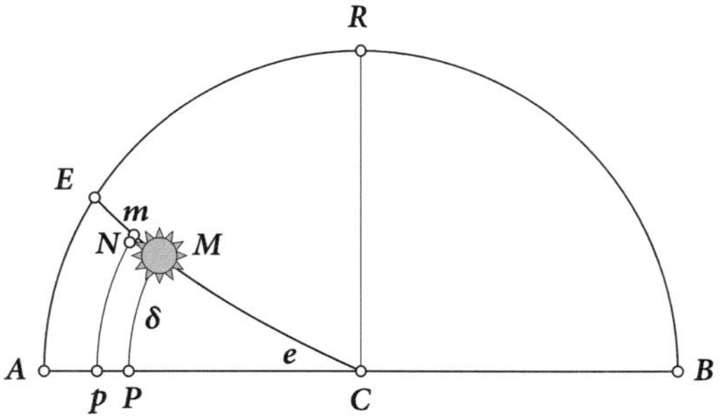

Figure 9.3 EULER'S sketch for determining the daily increment of declination. Source: Euler, 1741, Methodus computandi aequationem meridiei, p.62 or Figure 4.

R	North pole
AB	Celestial equator
$ARBA$	northern hemisphere
EC	ecliptic
e	obliquity of the ecliptic or $\widehat{ACE} = 23°29'$.

8 Euler, 1741, §.8, p.53.
9 Euler, 1741, §.8, p.53.

In Figure 9.3 the Sun is at point M, the arc $\widehat{PM}$ is the declination. In one day, the Sun travels a distance $\widehat{Mm}$ on an arc of the ecliptic equal to the increment of longitude, denoted by dk. The declination increment dt is:10

$$[9.7] \qquad \widehat{mp} - \widehat{MP} = \widehat{mN} = \mathrm{d}t$$

If we assume that the following applies $\quad \sin\widehat{MP} = b \quad$ and $\quad \cos\widehat{MP} = B$

and considering the arc $\widehat{mN} \ll 1$, using the expression for the sine of the sum of the angles in Equation 9.7, it follows that:

$$[9.8] \qquad \sin\widehat{mp} = b + B\mathrm{d}t$$

The sine of the angle $\widehat{ACE}$ (Figure 9.3), which is the obliquity of the ecliptic, is denoted by e or $\sin\widehat{ACE} = e$ by EULER, where:

$$e : 1 = \sin\widehat{PM} : \sin\widehat{CM} = b : \sin\widehat{CM}$$

therefore: $\sin\widehat{CM} = \dfrac{b}{e}$

We then take into account

$$\cos\widehat{CM} = f \qquad \text{and} \qquad \sin\widehat{Cm} = \frac{b}{e} + f\mathrm{d}k$$

Considering that the sine of 90° or the sinus totus = R:[11]

$$\sin\widehat{ACE} : \text{sinus totus} = e : 1 = \sin\widehat{pm} : \sin\widehat{Cm} = (b + B\mathrm{d}t) : \left(\frac{b}{e} + f\mathrm{d}k\right)$$

or: $\qquad ef\mathrm{d}k = B\mathrm{d}t$

This gives the expression for the (daily) increment of declination dt:[12]

$$[9.9] \qquad \mathrm{d}t = \frac{ef\mathrm{d}k}{B}$$

For the obliquity of the ecliptic, we assume that $e = \sin 23°29'$, f is equal to the cosine of the Sun's distance from the equinox, B is equal to the cosine of the declination, and dk is known or inferred from the ephemeris. The expression dk can be converted to seconds if necessary, and the declination increment dt can be expressed in seconds.

10 Euler, 1741, §.19, p.62.
11 For sinus totus see Chapter 2, note 12.
12 Euler, 1741, §.19, p.62.

Example 9.1 The following is an example of the calculation of the equation of noon as given by EULER in his work (Euler, 1741, §.20–22, pp.62–65).[13] Calculation results are in sixtieths of a second. EULER's mathematical notation is not always the strictest, for example he used the letter l for the decimal logarithm, t or tan for the tangent trigonometric function, etc.

The example below, based on EULER's original notation, therefore requires a little more attention from the reader. The specific features or differences with modern mathematical notation are explained in the text.

At latitude 52°27′, the Sun's altitude is measured at 8:21 am and the Sun reaches the same altitude again in the afternoon at 3:49 pm. The longitude of the Sun is ♉16°35′6″ and the declination is 16°49′. We want to determine the (clock) time of day when the Sun reaches noon or solar noon occurs. The daily motion of the Sun, as seen from the ephemeris, is 57′4″ or 3424″.

Longitude increment $\mathrm{d}k = 3424''$,
the distance of the Sun from the equinox 46°35′ or $f = \cos 46°35'$,
obliquity of the ecliptic $e = 23°29'$,
declination or $B = \cos 16°49'$.

For the logarithm of the obliquity of the ecliptic e, EULER's working shows:

$$\log e = \log(\sin 23°29') = 9.6004090$$

The result is obtained by considering the expression

$$\log e = \log(10^{10} \cdot \sin 23°29') = \log 10^{10} + \log(\sin 23°29') = 9.6004090$$

This expression is obviously a simplification following from the radius or maximum value of the sine function, which, in the tables used by EULER, is not 1 but 10^{10}. The expressions in the following example from EULER are:

$$\log f = \log(10^{10} \cdot \cos 46°35') = \log 10^{10} + \log(\cos 46°35') = 9.8371456$$

$$\log \mathrm{d}k = \log 3424 = 3.5345338$$

$$\log(ef\mathrm{d}k) = 22.9720884$$

This expression is, by the definition of logarithms, the sum of

$$\log(ef\mathrm{d}k) = \log e + \log f + \log \mathrm{d}k$$

It is similar also for declination:

$$\log B = \log(10^{10} \cdot \cos 16°49') = \log 10^{10} + \log(\cos 16°49') = 9.9810187$$

The result for the declination increment (Equation 9.9) again follows from the definition of logarithms:

$$\log \mathrm{d}t = \log e + \log f + \log \mathrm{d}k - \log B = 12.9910697$$

For the sake of consistency of the calculation, we must also take into account the radius or $\log 10^{10} = 10$, so that for the logarithm of the increment of the declination $\mathrm{d}t$ we get:

$$\log \frac{\mathrm{d}t}{10^{10}} = 12.9910697 - 10 = 2.9910697$$

13 Also Headley, 2022, pp.75–77.

For the daily increment of the declination dt, the value 979″ follows from the above expression ($10^{2.9910697} \approx 979$), which corresponds to 16′19″. Multiplying 979″ by 4 gives the result in seconds (Equation 9.4).

Converting the time difference between the two observations, which is 7 hours and 28 minutes, into degrees (1 hour ≡ 15°) gives an angle of $\widehat{SPT}$, which is 112° and a half angle $\frac{\widehat{SPT}}{2}$ of 56°.

We then use these values to continue the calculation:

$$\log \widehat{SPT} = \log 112 = 2.0492180$$

$$\log \mathrm{d}t = 2.9910697 \qquad \text{(corrected value taking into account radius} = 10^{10}\text{)}$$

The sum of the above logarithms, corresponding to the expression $\log \widehat{SPT}\,\mathrm{d}t = 5.0402877$

$$\log 180 = 2.2552725$$

For the expression $\log \frac{\widehat{SPT}}{180}\mathrm{d}t$ or the expression before the bracket in Equation 9.4, we get

$$\log \frac{\widehat{SPT}}{180}\mathrm{d}t = \log \widehat{SPT} + \log \mathrm{d}t - \log 180 = 2.7850152$$

The second part of Equation 9.4 (the bracketed terms) is calculated as follows:

$$\log (\tan \widehat{ZE}) = \log (10^{10} \cdot \tan 52°27') = \log 10^{10} + \log (\tan 52°27') = 10.1142350$$

$$\log \left(\sin \frac{\widehat{SPT}}{2} \right) = \log (10^{10} \cdot \sin 56°) = \log 10^{10} + \log (\sin 56°) = 9.9185742$$

Subtracting the second equation from the first (using the rules of logarithm arithmetic) gives:

$$\log \left(\frac{\tan \widehat{ZE}}{\sin \frac{\widehat{SPT}}{2}} \right) = 10.1142350 - 9.9185742 = 0.1956608$$

The ratio $\frac{\tan \widehat{ZE}}{\sin \frac{\widehat{SPT}}{2}}$ is therefore $= 10^{0.1956608} = 1.569$.

Similarly, calculate the second fraction in brackets (Equation 9.4):

$$\log (\tan \widehat{OE}) = \log (10^{10} \cdot \tan 16°49') = \log 10^{10} + \log (\tan 16°49') = 9.4803451$$

$$\log (\tan \frac{\widehat{SPT}}{2}) = \log (10^{10} \cdot \tan 56°) = \log 10^{10} + \log (\tan 56°) = 10.1710126$$

Using the logarithm rules, we again get:

$$\log \left(\frac{\tan \widehat{OE}}{\tan \frac{\widehat{SPT}}{2}} \right) = 9.4803451 - 10.1710126 = -0.6406675$$

This result corresponds to $-1+0.3593325$, or -1.3593325 in EULER's original notation.[14]

The ratio $\dfrac{\tan\widehat{OE}}{\sin\dfrac{\widehat{SPT}}{2}}$ is therefore $= 10^{-0.6406675} = 0.229$.

The expression in brackets (Equation 9.4) is therefore equal to $1.569 - 0.229 = 1.340$ or $\log 1.340 = 0.1271048$.

Add to this the result obtained above for $\log \dfrac{\widehat{SPT}}{180}\mathrm{d}t$

and for the logarithm of the equation of noon we get $\log E_{en} = 2.9121200$. The equation of noon is, then,

$E_{en} = 10^{2.9121200} \approx 817'''$ or $13''37'''$ in time units.

Since the Sun is in ♉ in this case, the declination is increasing. To determine the time of solar noon, subtract the resulting equation for noon from half the time difference between the two observations.

The median time interval between observations we calculate as follows:

8 hours 21 minutes
15 hours 49 minutes[15]

24 hours 10 minutes
half = 12 hours 5 minutes

Subtract the equation of noon to get the clock time of solar noon as calculated by EULER:
12 hours 5 minutes – 13 seconds $37'''$ = 12 hours 4 minutes 46 seconds $23'''$.[16]

14 This is another peculiarity of EULER's mathematical notation, if it is not taken into account we get the wrong result.
15 EULER, in his account, instead of 15 hours 49 minutes, counts from 3 hours 49 minutes and adds 12 hours.
16 Euler, 1741, §.22, p.65.

9.1.3 Equation of the centre

EULER derived his, now forgotten, equation of the centre in a 1778 treatise, *Nova methodus motum planetarum determinandi* (*New method to determine the motion of planets*).[17]

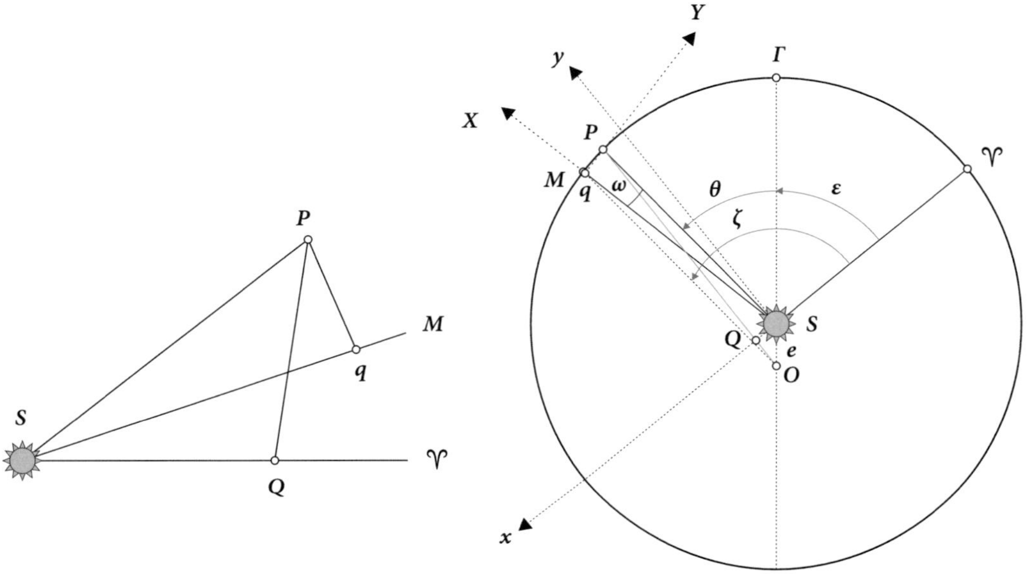

Figure 9.4 EULER'S sketch for the derivation of the equation of the centre (left) and the coordinate system he used for the derivation (right). Source: Euler, 1781, Nova methodus motum planetarum determinandi.

S	Sun
P	planet (Earth)
S♈	axis pointing in the direction of the 0° of ♈
M	mean planet (opposite of the mean Sun)
$\widehat{♈SM}$	longitude of the mean planet, ζ
$\widehat{MSP} = \omega$	equation of the centre

The planet passes through ♈ 0° (Figure 9.4) and then after a time τ reaches a point P, from where we draw a perpendicular line on the S♈ axis, thus defining two coordinate axes or coordinates $x = SQ$ and $y = QP$, where the distance between the planet P and the Sun is S:

$$v = \sqrt{x^2 + y^2} \quad [9.10]$$

Using the laws of motion, EULER wrote the following two equations:[18]

$$\frac{1}{\Delta}\frac{\mathrm{dd}x}{\mathrm{d}\tau^2} = -\frac{x}{v^3} \quad \text{and} \quad \frac{1}{\Delta}\frac{\mathrm{dd}y}{\mathrm{d}\tau^2} = -\frac{y}{v^3} \quad [9.11]$$

17 Very good summary with some additional comments is Bistafa, 2021, Euler's Forgotten Equation of the Center.
18 Euler, 1781, Nova methodus motum planetarum determinandi p.278.

The position of a planet P can also be determined using the following two coordinates (see Figure 9.4 for a better understanding):[19]

$$X = Sq \quad \text{and} \quad Y = qP, \quad \text{where} \quad \upsilon^2 = X^2 + Y^2$$

The coordinate transformation gives the relationship between the two coordinate systems:[20]

[9.12] $$x = X\cos\zeta - Y\sin\zeta \quad \text{and} \quad y = X\sin\zeta + Y\cos\zeta$$

After differentiating Equations 9.12 twice, EULER obtained:[21]

$$\mathrm{dd}x = \mathrm{dd}X\cos\zeta - \mathrm{dd}Y\sin\zeta - 2\mathrm{d}\zeta(\mathrm{d}X\sin\zeta + \mathrm{d}Y\cos\zeta) - \mathrm{d}\zeta^2(X\cos\zeta - Y\sin\zeta)$$

$$\mathrm{dd}y = \mathrm{dd}X\sin\zeta + \mathrm{dd}Y\cos\zeta + 2\mathrm{d}\zeta(\mathrm{d}X\cos\zeta - \mathrm{d}Y\sin\zeta) - \mathrm{d}\zeta^2(X\sin\zeta + Y\cos\zeta)$$

Multiplying the left-hand sides of the equations by $\cos\zeta$ and $\sin\zeta$ respectively and adding and subtracting the equations gives:[22]

[9.13] $$\mathrm{dd}x\cos\zeta + \mathrm{dd}y\sin\zeta = \mathrm{dd}X - 2\mathrm{d}\zeta\mathrm{d}Y - X\mathrm{d}\zeta^2$$

[9.14] $$\mathrm{dd}y\cos\zeta - \mathrm{dd}x\sin\zeta = \mathrm{dd}Y + 2\mathrm{d}\zeta\mathrm{d}X - Y\mathrm{d}\zeta^2$$

Euler then inserted these two expressions into Equation 9.11 and obtained:

$$\frac{\mathrm{dd}x\cos\zeta + \mathrm{dd}y\sin\zeta}{\Delta\mathrm{d}\tau^2} = \frac{-x\cos\zeta - y\sin\zeta}{\upsilon^3} = -\frac{X}{\upsilon^3}$$

$$\frac{\mathrm{dd}y\cos\zeta - \mathrm{dd}x\sin\zeta}{\Delta\mathrm{d}\tau^2} = \frac{-y\cos\zeta + x\sin\zeta}{\upsilon^3} = -\frac{Y}{\upsilon^3}$$

or:

[9.15] $$\frac{\mathrm{dd}X\cos\zeta - 2\mathrm{d}\zeta\mathrm{d}Y - X\mathrm{d}\zeta^2}{\Delta\mathrm{d}\tau^2} = -\frac{X}{\upsilon^3}$$

[9.16] $$\frac{\mathrm{dd}Y\cos\zeta + 2\mathrm{d}\zeta\mathrm{d}X - Y\mathrm{d}\zeta^2}{\Delta\mathrm{d}\tau^2} = -\frac{Y}{\upsilon^3}$$

19 Euler, 1781, p.279.
20 See Chapter 20 for an explanation of coordinate transformation.
21 EULER's notation $\mathrm{dd}x = \mathrm{d}^2x$ in modern notation. The rules for differentiating in modern notation are:
$\mathrm{d}y = y'(x)\mathrm{d}x \qquad \mathrm{d}^2y = y''(x)\mathrm{d}x^2 \qquad \mathrm{d}^3y = y'''(x)\mathrm{d}x^3 \ldots$
22 Euler, 1781, p.280. In EULER's treatise, only Equations 9.13 and 9.14 appear, not the derivation.

Using the equations above, we can determine the X and Y coordinates from the longitude of the planet (Earth). If the Earth P orbits the Sun at a constant speed (mean Earth motion) then $X = 1$ and $Y = 0$, which implies $v = 1$ (see equations above). In this case, the first equation implies:[23]

$$-\frac{d\zeta^2}{\Delta d\tau^2} = -1 \qquad \text{or} \qquad d\zeta^2 = \Delta d\tau^2$$

Δ is thus: $\Delta = \dfrac{d\zeta^2}{d\tau^2}$

Where ζ is the mean longitude of the Earth or planet, which can be expressed in days (or degrees) at time τ and denoted by t,

where $\Delta = \dfrac{dt^2}{d\tau^2}$ and Δdt^2 can be written instead of $\Delta d\tau^2$. Since the time τ corresponds to

the mean motion of Earth, for one day or $59'8''19'''$ and considering this (for any planet) we can note (Equations 9.15 and 9.16):

$$\frac{ddX \cos\zeta - 2d\zeta dY - Xd\zeta^2}{dt^2} = -\frac{X}{v^3} \tag{9.17}$$

$$\frac{ddY \cos\zeta - 2d\zeta dX - Yd\zeta^2}{dt^2} = -\frac{Y}{v^3} \tag{9.18}$$

Since the mean longitude ζ is a function of the mean daily path of the planet or t, where $\zeta = nt$ and $d\zeta = ndt$, respectively, and the above two expressions (Equations 9.17 and 9.18) become:

$$\frac{ddX \cos\zeta}{dt^2} - \frac{2ndY}{dt} - n^2X = -\frac{X}{v^3} \tag{9.19}$$

$$\frac{ddY \cos\zeta}{dt^2} - \frac{2ndX}{dt} - n^2Y = -\frac{Y}{v^3} \tag{9.20}$$

Given that the mean distance from the Sun to the planet is a, and assuming a circular orbit or an eccentricity $e = 0$, then $Y = 0$ and $X = v = a$ and from the first of the above expressions it follows

that the $n^2 = \dfrac{1}{a^3}$ second expression is 0.

If the planet's orbit is eccentric, then $Sq = X$, where $X = a(1 + x)$ and $qP = Y = ay$ and the following applies:

$$v = a\sqrt{(1 + x)^2 + y^2}$$

23 Euler, 1781, p.280.

Where x and y are new coordinates, not the ones from the start of the derivation. Inserting these expressions into Equations 9.19 and 9.20 gives:[24]

$$\frac{\mathrm{d}\mathrm{d}x}{\mathrm{d}t^2} - \frac{2n\mathrm{d}y}{\mathrm{d}t} - n^2(1+x) = -\frac{-(1+x)}{a^3((1+x)^2+y^2)^{\frac{3}{2}}} \tag{9.21}$$

$$\frac{\mathrm{d}\mathrm{d}y}{\mathrm{d}t^2} - \frac{2n\mathrm{d}x}{\mathrm{d}t} - n^2 y = \frac{-y}{a^3((1+x)^2+y^2)^{\frac{3}{2}}} \tag{9.22}$$

Because $n^2 = \frac{1}{a^3}$,

$$\frac{\mathrm{d}\mathrm{d}x}{\mathrm{d}t^2} - \frac{2n\mathrm{d}y}{\mathrm{d}t} - n^2(1+x) = -\frac{-n^2(1+x)}{((1+x)^2+y^2)^{\frac{3}{2}}}$$

$$\frac{\mathrm{d}\mathrm{d}y}{\mathrm{d}t^2} - \frac{2n\mathrm{d}x}{\mathrm{d}t} - n^2 y = \frac{-n^2 y}{((1+x)^2+y^2)^{\frac{3}{2}}}$$

Given that $y^2 \ll (1+x)^2$ Euler expanded the right-hand sides into series and got:[25]

$$\frac{\mathrm{d}\mathrm{d}x}{\mathrm{d}t^2} - \frac{2n\mathrm{d}y}{\mathrm{d}t} - n^2(1+x) = \frac{-n^2}{(1+x)^2} + \frac{3n^2y^2}{2(1+x)^4} - \frac{3\cdot 5n^2y^4}{2\cdot 4(1+x)^6} + \frac{3\cdot 5n^2y^6}{2\cdot 4\cdot 6(1+x)^8}\cdots$$

$$\frac{\mathrm{d}\mathrm{d}y}{\mathrm{d}t^2} - \frac{2n\mathrm{d}x}{\mathrm{d}t} - n^2 y = \frac{-n^2 y}{(1+x)^3} + \frac{3n^2y^3}{2(1+x)^5} - \frac{3\cdot 5n^2y^5}{2\cdot 4(1+x)^7} + \frac{3\cdot 5\cdot 7n^2y^7}{2\cdot 4\cdot 6(1+x)^9}\cdots$$

The individual terms on the right-hand side of the two resulting equations were again expanded into series and, taking into account the longitude of the planet or the expressions $nt = \zeta$, $n\mathrm{d}t = \mathrm{d}\zeta$ and $n^2\mathrm{d}t^2 = \mathrm{d}\zeta^2$, he obtained:[26]

$$\begin{aligned} \frac{\mathrm{d}\mathrm{d}x}{\mathrm{d}\zeta^2} - \frac{2\mathrm{d}y}{\mathrm{d}\zeta} &= 3x - 3x^2 + 4x^3 - 5x^4 + 6x^5 - 7x^6 \ldots \\ &+ \frac{3}{2}y^2 - 6xy^2 + 15x^2y^2 - 30x^3y^2 + \frac{105}{2}x^4y^2 \ldots \\ &- \frac{15}{8}y^4 + \frac{45}{4}xy^4 - \frac{315}{8}x^2y^4 \ldots \\ &+ \frac{35}{16}y^6 \ldots \end{aligned} \tag{9.23}$$

24 Euler, 1781, §.10, p.281, see also Bistafa, 2021.

25 For expressions for the series expansion, see for example Bronshtein, Semendyayev, 1985, Handbook of Mathematics, p.23.

26 Euler, 1781, §.13, p.282.

[9.24] $$\frac{\mathrm{dd}y}{\mathrm{d}\zeta^2} - \frac{2\mathrm{d}x}{\mathrm{d}\zeta} = 3xy - 6x^2 + 10x^3y - 15x^4y + 21x^5y \ldots$$

$$+\frac{3}{2}y^3 - \frac{15}{2}xy^3 + \frac{45}{2}x^2y^3 - \frac{105}{2}x^3y^3 \ldots$$

$$-\frac{15}{8}y^5 + \frac{105}{8}xy^5 \ldots$$

For small eccentricity e, the following convergent series can be used to describe the x and y coordinates:[27]

[9.25] $$x = eP + e^2Q + e^3R + e^4S + e^5T \ldots$$

[9.26] $$y = ep + e^2q + e^3r + e^4s + e^5t + e^6u \ldots$$

Because the terms P, p, Q, q, etc., are not functions of eccentricity, the following pairs of differential equations follow from Expressions 9.23 and 9.24 for the individual contributions to Equations 9.25 and 9.26:[28]

for terms with e:

[9.27] $$\frac{\mathrm{dd}P}{\mathrm{d}\zeta^2} - \frac{2\mathrm{d}p}{\mathrm{d}\zeta} = 3P$$

[9.28] $$\frac{\mathrm{dd}p}{\mathrm{d}\zeta^2} + \frac{2\mathrm{d}P}{\mathrm{d}\zeta} = 0$$

for terms with e^2:

$$\frac{\mathrm{dd}Q}{\mathrm{d}\zeta^2} - \frac{2\mathrm{d}q}{\mathrm{d}\zeta} = 3Q - 3P^2 + \frac{3}{2}p^2 \quad \text{and} \quad \frac{\mathrm{dd}q}{\mathrm{d}\zeta^2} + \frac{2\mathrm{d}Q}{\mathrm{d}\zeta} = 3Pp$$

for terms with e^3:

$$\frac{\mathrm{dd}R}{\mathrm{d}\zeta^2} - \frac{2\mathrm{d}r}{\mathrm{d}\zeta} = 3R - 6PQ + 4P^3 + 3pq - 6p^2P$$

$$\frac{\mathrm{dd}r}{\mathrm{d}\zeta^2} + \frac{2\mathrm{d}R}{\mathrm{d}\zeta} = 3pQ + 3qP - 6P^2 + \frac{3}{2}p^3$$

27 Euler, 1781, p.283.
28 Euler, 1781, pp.283–287.

for terms with e^4:

$$\frac{ddS}{d\zeta^2} - \frac{2ds}{d\zeta} = 3S - 6PR - 3Q^2 + 3pr + \frac{3}{2}q^2 - 6p^2Q + 12P^2 - 12pqP - 5P^4 + 15p^2P^2 - \frac{15}{8}p^4$$

$$\frac{dds}{d\zeta^2} + \frac{2dS}{d\zeta} = 3pR + 3qQ + 3pP - 12pPQ - 6qP^2 + \frac{9}{2}p^2q + 10pP^3 - \frac{15}{2}p^3P$$

EULER gave the solution to these differential equations in terms of a new variable θ, or mean anomaly:[29]

$$P = \cos\theta \quad \text{and} \quad p = -2\sin\theta$$

$$Q = -\frac{1}{2} + \frac{1}{2}\cos 2\theta \quad \text{and} \quad q = \frac{1}{4}\sin 2\theta$$

$$R = -\frac{3}{8}\cos 3\theta \quad \text{and} \quad r = \frac{9}{8}\sin\theta - \frac{7}{24}\sin 3\theta$$

$$S = \frac{407}{8} - \frac{209}{24}\cos 2\theta + \frac{219}{965}\cos 4\theta \quad \text{and} \quad s = \frac{355}{48}\sin 2\theta + \frac{49}{480}\sin 4\theta$$

Plugging these solutions into the expressions for the x and y coordinates gives the following equations, which are functions of the eccentricity e and the mean anomaly θ:[30]

[9.29] $$x = e\cos\theta + e^2\left(\frac{1}{2} - \frac{1}{2}\cos 2\theta\right) - \frac{3}{8}e^3\cos 3\theta + e^4\left(\frac{407}{3} - \frac{209}{24}\cos 2\theta + \frac{719}{960}\cos 4\theta\right)$$

[9.30] $$y = -2e\sin\theta + \frac{1}{4}e^2\sin 2\theta + e^3\left(\frac{9}{8}\sin\theta - \frac{7}{24}\sin 3\theta\right) + e^4\left(\frac{355}{48}\sin 2\theta + \frac{49}{480}\sin 4\theta\right)$$

The calculated coordinates are used to determine the equation of the centre, or $\widehat{MSP}$, and the planet's position, or true longitude is then calculated as the difference between the mean longitude and the equation of the centre. True longitude is (see Figure 9.4):[31]

[9.31] $$\widehat{\Upsilon SP} = \zeta + \omega$$

The equation of the centre ω or angle $\widehat{MSP}$ is:[32]

[9.32] $$\omega = \operatorname{atan}\frac{y}{1+x}$$

29 Euler, 1781, p.286.
30 Euler, 1781, p.301.
31 Euler, 1781, p.301. For the sign of the equation of the centre see also Chapter 1.
32 Euler, 1781, p.302.

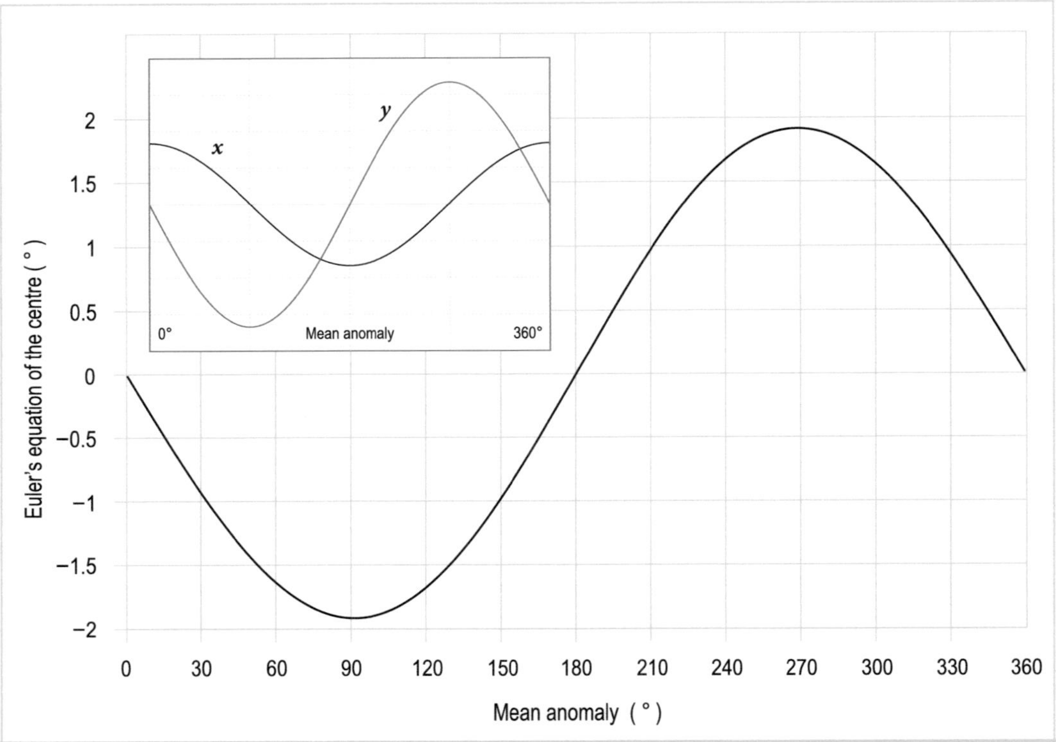

Figure 9.5 EULER'S equation of the centre and coordinates x and y, Equations 9.29 and 9.30, as function of the mean anomaly, calculated with the modern value of eccentricity, $e = 0.01671022$.

Example 9.2 Differential equations of this type (e.g., Equations 9.27 and 9.28) are today solved by transforming them into linear homogeneous differential equations of higher order with constant coefficients. The first two differential equations (Equations 9.27 and 9.28) written in modern form are:

$$y'' - 2x' = 3y \qquad\qquad x'' + 2y' = 0$$

Differentiating the first equation gives this system:

$$y''' - 2x'' = 3y'$$

$$x'' + 2y' = 0$$

We multiply the second equation by 2, rearrange the two equations, add them together and get the following differential equation:

$$y''' + y' = 0$$

which can be solved using the characteristic equation, which is:[33]

$$r^3 + r = 0 \qquad \text{or} \qquad r(r^2 + 1) = 0$$

The characteristic equation has one real and two complex conjugate roots:

$$r_1 = 0 \qquad r_{2,3} = \pm i$$

In this case, the general solution of the equation takes the form:

$$y = C_1 e^{ax} + C_2 e^{ax} \cos bx + C_3 e^{ax} \sin bx$$

Since we are dealing with periodic motion, the equation must have a final solution such that $a = 0$ and $C_1 = 0$. The general solution for the given example is therefore:

$$y = C_2 \cos bx + C_3 \sin bx$$

The solution is further simplified by introducing new variables C and D:

$$C_2 = C \cos D \qquad C_3 = C \sin D$$
$$y = C \cos D \cos bx + C \sin D \sin bx$$

it follows that:[34]

$$y = C \cos(bx - D)$$

This is also the form corresponding to EULER's solution of the first equation, where, assuming $C = 1$ and introducing a new variable θ, EULER obtained:

$$P = \cos\theta \qquad \text{or} \qquad p = -2 \sin\theta$$

Where:

$\theta = \zeta - \varepsilon$

ζ	mean longitude (accompanying part, orbital motion)
ε	longitude of the apogee (constant)

We then proceed to solve the other differential equations in a similar way.

33 Tenenbaum, Pollard, 1963, pp.217–220.

34 If $C_2 = C \sin D$ and $C_3 = C \cos D$ we get solution $y = C \sin(bx + D)$. Both solutions are equivalent, see Tenenbaum, Pollard, 1963, p.218.

9.1.4 Euler's contribution to spherical trigonometry

When it came to the historical literature on spherical trigonometry, EULER was familiar with THEODOSIUS' *Sphaerics*, but he was not familiar with MENELAUS' *Sphaerics*, nor with its Arabic adaptations. In his 1755 *Principes de la trigonométrie sphérique tirés de la méthode des plus grands et des plus petits*, where EULER derived the expressions of spherical geometry by means of the calculus of variations, he derived a complete set of expressions for spherical triangles for the triangle $\triangle APM$ (Figure 9.6).

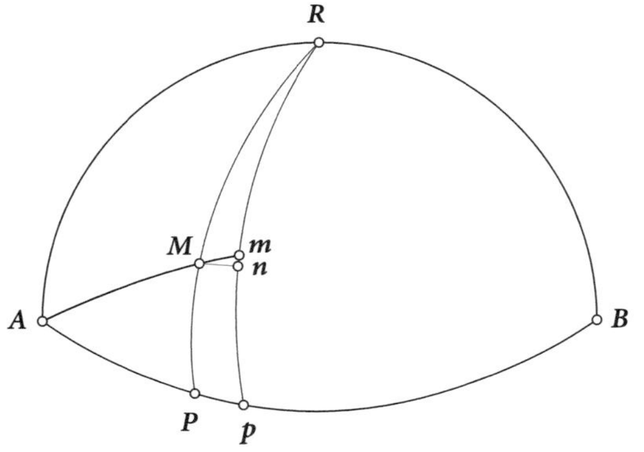

$\widehat{AB}$	equator
$\widehat{AM}$	horizon
$\widehat{PM}$	meridian

Figure 9.6 EULER'S sketch to determine the spherical trigonometry expressions for the triangle $\triangle APM$. Source: Euler, 1755, Principes de la trigonometrie spherique tires de la methode des plus grands et plus petits, Fig. 2.

Denoting the arc angles as $\widehat{AP} = x$, $\widehat{PM} = y$ and $\widehat{AM} = s$, EULER derived the expression known as the Pythagorean theorem for a spherical triangle:

[9.33] $$\cos s = \cos x \cos y$$

The spherical form of the Pythagorean theorem can also be derived using NAPIER's rules, see Apendix D, Section D.5. The above expression thus corresponds to the expressions D.35 for the spherical triangle in Figure D.6.

He also derived and published cosine theorems in the treatise. The 1755 spherical trigonometry treatise was followed in 1782 by a second *Trigonometria sphaerica universa, ex primis principiis breviter et dilucide derivata*, where, unlike the earlier publication, EULER derived the expressions using geometry.

In his 1755 treatise, EULER systematically derived the following expressions, among others, for spherical trigonometry:

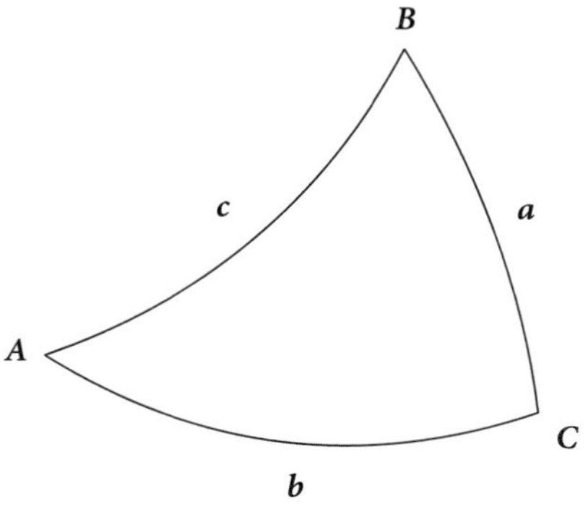

Figure 9.7 Spherical triangle from EULER'S treatise. Source: Euler, 1755, Tab. V, Figure 2.

$\cos c = \cos a \cos b$	$\tan A = \dfrac{\tan a}{\sin b}$	$\tan B = \dfrac{\tan b}{\sin a}$
$\cos b = \dfrac{\cos c}{\cos a}$	$\sin A = \dfrac{\sin a}{\sin c}$	$\cos B = \dfrac{\tan a}{\tan c}$
$\cos a = \dfrac{\cos c}{\cos b}$	$\cos A = \dfrac{\tan a}{\tan c}$	$\sin B = \dfrac{\sin a}{\sin c}$
$\cos a = \dfrac{\cos c}{\cos b}$	$\cos A = \dfrac{\tan b}{\tan c}$	$\sin B = \dfrac{\sin b}{\sin c}$
$\sin b = \dfrac{\tan a}{\tan A}$	$\sin c = \dfrac{\sin a}{\sin A}$	$\sin B = \dfrac{\cos A}{\cos a}$
$\tan b = \sin a \tan B$	$\tan c = \dfrac{\tan a}{\cos B}$	$\cos A = \cos a \sin B$
$\tan a = \sin b \tan A$	$\tan c = \dfrac{\tan b}{\cos A}$	$\cos B = \cos b \sin A$
$\sin a = \dfrac{\tan b}{\tan B}$	$\sin c = \dfrac{\sin b}{\sin B}$	$\sin A = \dfrac{\cos B}{\cos b}$
$\sin a = \sin c \sin A$	$\tan b = \tan c \cos A$	$\tan B = \dfrac{1}{\cos c \tan A}$
$\sin b = \sin c \sin B$	$\tan a = \tan c \cos B$	$\tan A = \dfrac{1}{\cos c \tan B}$
$\cos a = \dfrac{\cos A}{\sin B}$	$\cos b = \dfrac{\cos B}{\sin A}$	$\cos c = \dfrac{1}{\tan A \tan B}$

Examples of expressions derived by EULER using the calculus of variations and reported in his 1755 publication. Source: Euler, 1755, Principes de la trigonometrie spherique tires de la methode des plus grands et plus petits, p.232.

Euler also derived or gave expressions for:

- cosine law for sides
- cosine law for angles
- relationships between two sides and two angles
- sines of half angles
- cosines of half angles
- tangent of half angles
- the product, sum and difference of the two tangents of the half angles
- sines of the half sides
- cosines of half sides
- tangent of half sides
- the sum and difference of the two tangents of the half sides
- the tangent of half of the sum and difference of two angles
- the tangent of half of the sum and difference of two sides and others.

9.1.5 Euler's formula

Euler's formula, or equation as we call it today

$$e^{i\alpha} = \cos\alpha + i\sin\alpha \quad [9.34]$$

was published by Euler in his 1748 treatise, *Introductio in analysin infinitorum*, wherein he did not yet use the symbol i for the imaginary unit, but $\sqrt{-1}$.[35] A somewhat similar expression, but only in the form of text, had already been published by Cotes in 1714 (see Figure 9.8):[36]

$$OE\sqrt{-1} \propto \frac{CX + EX\sqrt{-1}}{R} \qquad \text{or} \qquad \alpha\sqrt{-1} \propto \cos\alpha + \sin\alpha\sqrt{-1} \quad [9.35]$$

In the same publication, Cotes also gave a fairly good approximation of e.[37] Powers of e, including $e^{i\alpha}$, of course obey the ususal rules for exponents, for example:

$$e^{-i\alpha} = e^{i(-\alpha)} \qquad \text{and} \qquad e^{-i\alpha}e^{i\alpha} = 1$$

35 Euler, 1748, p.104.

36 Cotes, Halley, 1714, Logometria auctore Rogero Cotes, p.32.

37 The constant e = 2.7182818… or Euler's number is base of the natural logarithm. Sometimes it is also called Napier's constant. The value of e was published to 23 decimal places in 1748 Euler's treatise *Introductio in Analysin infinitorum*, volume 2, p.288, but without information about the calculation details. See also volume 1, p.318, Exemplum III. Number **e** can be defined as infinite series or integral:

$$e = \sum_{n=0}^{\infty}\frac{1}{n!} = 1 + \frac{1}{1} + \frac{1}{1\cdot 2} + \frac{1}{1\cdot 2\cdot 3} + \cdots \qquad \text{or} \qquad \int_1^e \frac{1}{x}dx = 1$$

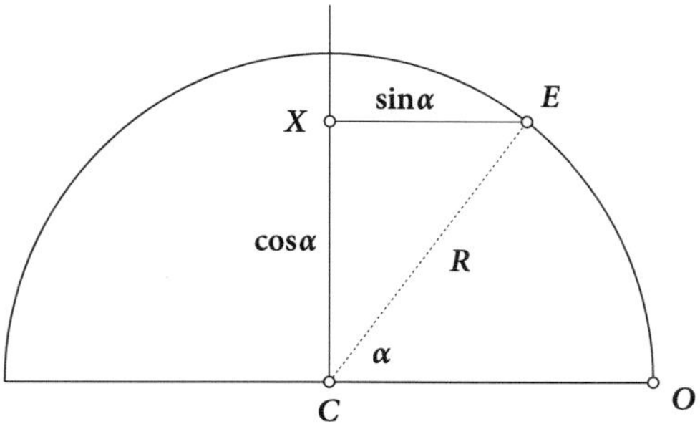

Figure 9.8 Reconstruction of COTES' geometric construction on the basis of which he wrote Equation 9.35 in a textual form. Based on Cotes, 1714.

From Equation 9.34:

$$e^{-i\alpha} = \cos(-\alpha) + i\sin(-\alpha)$$

Because $\sin(-\alpha) = -\sin\alpha$ and $\cos(-\alpha) = \cos\alpha$

we find that: $e^{-i\alpha} = \cos\alpha - i\sin\alpha$

Example 9.3 Several important trigonometric expressions can be derived using EULER's formula:

- Sum of squares

The expression $e^{i\alpha}e^{-i\alpha} = (\cos\alpha + i\sin\alpha)(\cos\alpha - i\sin\alpha)$

is rearranged: $e^{i\alpha}e^{-i\alpha} = \cos^2\alpha - i\cos\alpha\sin\alpha + i\cos\alpha\sin\alpha - i^2\sin^2\alpha$

$$e^{i\alpha}e^{-i\alpha} = \cos^2\alpha + \sin^2\alpha$$

and we get the known formula:[38] $\sin^2\alpha + \cos^2\alpha = 1$

- Sum of angles[39] $e^{i(\alpha+\beta)} = e^{i\alpha}e^{i\beta}$

$$\cos(\alpha+\beta) + i\sin(\alpha+\beta) = (\cos\alpha + i\sin\alpha)(\cos\beta + i\sin\beta)$$

$$\cos(\alpha+\beta) + i\sin(\alpha+\beta) = \cos\alpha\cos\beta + i\cos\alpha\sin\beta + i\sin\alpha\cos\beta + i^2\sin\alpha\cos\beta$$

Combine elements with real and imaginary parts

$$\cos(\alpha+\beta) = \cos\alpha\cos\beta - \sin\alpha\sin\beta$$

$$\sin(\alpha+\beta) = \cos\alpha\sin\beta + \sin\alpha\cos\beta$$

38 Euler, 1748a, p.97.
39 Euler, 1748a, p.98.

Equating the real and imaginary parts on the left and right gives

$$\cos(\alpha + \beta) = \cos\alpha \cos\beta - \sin\alpha \sin\beta$$
$$\sin(\alpha + \beta) = \cos\alpha \sin\beta + \sin\alpha \cos\beta$$

In the same way, expressions for difference of angles can be derived.

- Double angle functions $e^{i2\alpha} = (e^{i\alpha})^2$

By considering EULER's formulae on both sides

$$\cos 2\alpha + i\sin 2\alpha = (\cos\alpha + i\sin\alpha)^2$$
$$\cos 2\alpha + i\sin 2\alpha = \cos^2\alpha + 2i\cos\alpha\sin\alpha + i^2\sin^2\alpha$$

Combining real and imaginary parts

$$\cos 2\alpha + i\sin 2\alpha = (\cos^2\alpha - \sin^2\alpha) + 2i\cos\alpha\sin\alpha$$

Equate the real and imaginary parts

$$\cos 2\alpha = \cos^2\alpha - \sin^2\alpha$$
$$\sin 2\alpha = 2\cos\alpha\sin\alpha$$

- In the same way, we can derive expressions for the formulae of multiple angles

$\sin n\alpha$ or $\cos n\alpha$

In this case, we can use the binomial theorem:[40]

$$(a+b)^n = a^n + {}^nC_1 a^{n-1}b + {}^nC_2 a^{n-2}b^2 + \cdots + {}^nC_k a^{n-k}b^k + \cdots + {}^nC_{n-1} ab^{n-1} + b^n$$

where

$${}^nC_k = \frac{n!}{k!\,(n-k)!} \quad \text{or} \quad \binom{n}{k}$$

Example expressions for the sine and cosine of sextuple angles, $\sin 6\alpha$ and $\cos 6\alpha$: $e^{i6\alpha} = (e^{i\alpha})^6$

Using EULER's formula, we can write: $\cos 6\alpha + i\sin 6\alpha = (\cos\alpha + i\sin\alpha)^6$

Using the binomial theorem:

$$\cos 6\alpha + i\sin 6\alpha = \cos^6\alpha + 6i\cos^5\alpha\sin\alpha + 15i^2\cos^4\alpha\sin^2\alpha$$
$$+20i^3\cos^3\alpha\sin^3\alpha + 15i^4\cos^2\alpha\sin^4\alpha$$
$$+6i^5\cos\alpha\sin^5\alpha + i^6\sin^6\alpha$$

40 Euler, 1748a, p.98.

or:

$$\cos 6\alpha + i\sin 6\alpha = \cos^6\alpha + 6i\cos^5\alpha\sin\alpha - 15\cos^4\alpha\sin^2\alpha$$

$$-20i\cos^3\alpha\sin^3\alpha + 15\cos^2\alpha\sin^4\alpha$$

$$+6i\cos\alpha\sin^5\alpha - \sin^6\alpha$$

Equate the real and imaginary parts:

$$\cos 6\alpha = \cos^6\alpha - 15\cos^4\alpha\sin^2\alpha + 15\cos^2\alpha\sin^4\alpha - \sin^6\alpha$$

$$\sin 6\alpha = 6\cos^5\alpha\sin\alpha - 20\cos^3\alpha\sin^3\alpha + 6\cos\alpha\sin^5\alpha$$

From the expressions for sine and cosine, we can also calculate the expressions for the tangent of multiple angles.

- EULER's formulae can also be used to derive polar coordinates and the derivatives and integrals of trigonometric functions, and series expansion is also useful.
- Expressions for trigonometric functions are also very useful, and can be derived using EULER's formula:

$$\sin\alpha = \frac{e^{i\alpha} - e^{-i\alpha}}{2i} \qquad \cos\alpha = \frac{e^{i\alpha} + e^{-i\alpha}}{2} \qquad \tan\alpha = \frac{e^{i\alpha} - e^{-i\alpha}}{i(e^{i\alpha} + e^{-i\alpha})}$$

9.1.6 Other works by Euler

In 1745, EULER compiled the *Tabulae astronomicae solis et lunae*, where, among other things, the values of the mean longitude, the mean solar anomaly, the longitude of the apogee, etc., are given for each year. His tables are also designed that the final result of the individual parameters is calculated as the sum of the individual contributions. A special feature is the period values, which are not only in Julian calendar years, but also for years in a Julian period.[41]

Among the more important works, at least as far as the subject matter covered in this book is concerned, are:

- *Theoria motuum planetarum et cometarum*, 1744, (*Theory of the planets and comets*).
- *Theoria motus lunae exhibens omnes eius inaequalitates*, 1753, (*Theory of the motion of the Moon which exhibits all its irregularities*).
- *Theoria motum Lunae*, 1772, (*Theory of lunar motion*).

41 Euler, 1746, p.141. See Appendix A for an explanation of the **Julian period** or the **Julian day number**.

9.2 Tobias Mayer

9.2.1 Mayer's life and work

TOBIAS MAYER (1723–1762), born in Marbach near Stuttgart, grew up in modest circumstances. He received a classical education where he distinguished himself as an outstanding student, while in mathematics and astronomy he was mostly self-taught. Among other things, he studied with shoemaker GOTTLIEB DAVID KANDLER, whose world view clearly exceeded the local environment since he was interested in mathematics and bought mathematical books from which MAYER also learned. Even as a child, MAYER was a talented painter. Before his premature death, he worked as a professor in Göttingen and his tables on the movement of the Moon and the Sun, which surpassed the precision of his predecessors, were published in 1753.[42] Despite being largely self-taught in the field of science, he was indisputably capable. He corresponded with EULER, who appreciated his work, and his tables were later published in the *Nautical Almanac* as their value was recognised by MASKELYNE.[43] With his tables, he competed for a cash prize from the English Board of Longitude for a solution that would allow the easy determination of longitude at sea. As acknowledgment, his widow received a partial prize of £3,000 after his death.[44]

9.2.2 Tables on the movement of the Sun and Moon

It was not easy to prove that the movement of celestial bodies changes over time because the results of previous astronomical observations, sometimes several centuries old, are needed as evidence. In book III of the *Almagest*, PTOLEMY cites HIPPARCHUS' assumption that the length of the year may not be constant; TYCHO and LONGOMONTANUS also later considered that the movement of celestial bodies changes over time.

KEPLER, despite all his great achievements, failed to determine a model for the movement of the Moon, which was one of the more important issues in astronomy in the 17th and 18th centuries. In this regard, it is worth mentioning HORROCKS' contribution, but his work remained almost completely unknown for several decades after his death. Towards the end of the 17th century, EDMOND HALLEY discovered the Moon's secular acceleration. On the basis of NEWTON's work and his own observations, TOBIAS MAYER compiled some useful tables on the movement of the

42 For some comments see also Pilz, 1977, p.321.

43 For the correspondence with EULER, see Forbes, 1971, The Euler-Mayer Correspondence; for EULER's recognition of MAYER's work, see, for example, p.79, Euler to Mayer, 26 February 1754. 12 letters from MAYER's correspondence with LACAILLE are also, at least partially, preserved. Forbes, Gapaillard, 1996.

44 Mayer, 1770, Tabulæ Motuum Solis et Lunæ Novæ et Correctæ, p.92.

Moon and Sun.[45] Entitled *Novae tabulae motuum Solis et Lunae*, the tables were published in 1753 in the second part of the *Commentarii societatis regiae scientiarum gottingensis*. Unfortunately, a lack of evidence prevent us from knowing exactly how MAYER calculated the values.[46] His tables allow, various parameters of the motion of the Sun to be determined, such as its true longitude, right ascension and apogee, and the tables contain corrections for the effects of the Moon ☽, Venus ♀, Jupiter ♃ and nutation ☊.[47]

9.2.3 Example of using Mayer's tables

The following example presents the process of determining the orbital parameters of the Sun and the equation of time for any day of the year.[48] MAYER's tables accommodate the period from 1600 to 1807. We can calculate the equation of time using the following procedure:

- First, we determine the mean longitude for the desired day and time using the equation of time for the same day, but another year – for example, using the values published in the *Nautical Almanac* (Mayer, 1770, Tables XII, XIV, XIX).
- The true longitude is determined from the mean longitude for the beginning of the selected year, adding the values from the beginning of the year to the particular day of the year, and incorporating corrections due to the influences of the Moon, Jupiter, Venus and nutation (Mayer, 1770, Table XXII).
- The apogee is also determined from the tables; the initial values are added to the correction.
- We determine the correction to the equation of the centre; the argument of the equation of the centre in MAYER's tables is the mean anomaly, which is calculated as the difference between the mean longitude and the apogee (Mayer, 1770, Tables XX, XXI).
- To calculate the right ascension, we require the obliquity of the ecliptic (Mayer, 1770, problem VI, Table XXV).
- Then we calculate the right ascension (Mayer, 1770, problem VII, Tables XXVIII, XIX).
- To the mean longitude, we add the correction to the equation of equinoxes.
- The equation of time is equal to the difference of the mean longitude, taking into account the correction of the equation of equinoxes and right ascension.

45 For more information see Steele, 2012, Ancient Astronomical Observations and the Study of the Moon's Motion (1691–1757).

46 Due to a lack of theoretical foundations, MAYER was criticised by D'ALEMBERT. MAYER may have determined certain parameters on the basis of observational results. More data including mathematical procedures are available for calculating the tables on the movement of the Moon. See Steele, 2012, pp.95–122 and particularly Wepster, 2010, Chapter 5.4, pp.66–87.

47 Mayer, 1770, p.101.

48 Vince, 1797, A Complete System of Astronomy, Chapter IV, pp.45–46.

Example 9.4 Using MAYER's tables, determine the value of the equation of time on 1 July 1792 for the meridian through Greenwich.

We use MAYER's tables, *Tabulæ motuum Solis et Lunæ novæ et correctæ* (1770). For the approximate value of the equation of time for July 1, 1788, we find the value of 3′28″ in the *Nautical Almanac*.

First, define the true longitude – see Mayer, 1770, Example III, pp.101–105 and Table XII. In Table XII, we find the initial value of the mean longitude for 1792, which is $9^S\ 10°50'0.''7$ = ♑ $10°50'0.''7$ or $280°50'0.''7$, the count of signs begins with 0, the ninth sign is the Sea goat (Capricorn) ♑. From the same table comes the value of the apogee longitude $3^S\ 9°23'46''$ or $99°23'46''$, and the number of corrections № 1 – № 4 due to the influence of the Moon, Venus, Jupiter and nutation are 241, 227, 123 and 478, respectively.

Corrections for July 1 are given in Table XVI:

mean longitude	$5^S\ 29°23'16.''2 = 129°23'16.''2$
apogee	$33''$

Arguments for corrections № 1 – № 4 due to the effects of:

Moon	☽	163
Venus	♀	456
Jupiter	♃	312
nutation	☊	27

For the value of the equation of time, we also determine the correction of the mean longitude $7.''4$ and $1.''1$, which follows from Table XIX.

The corrected mean longitude is $3^S\ 10°13'25.''4$.

For the correction of the equation of the centre, $-1'37.''1$, the value is interpolated, the argument is the mean anomaly, which is the difference between the mean longitude and the longitude of the apogee, from Tables XX and XXI.

Corrections due to the Moon, Venus, Jupiter and nutation are given in Table XXII - intermediate values are also interpolated:

Moon	☽	$+4.''5$
Venus	♀	$-4.''7$
Jupiter	♃	$+3.''65$
nutation	☊	$-0.''6$

We obtain the true longitude from the mean longitude, taking into account the above corrections, amounting to $3^S\ 10°11'51.''15$.

The obliquity of the ecliptic for 1792 is determined as follows. We use the value for 1756, which is $23°28'16''$ (Mayer, 1770, Problem IV, page 106). Subtract the value $16.''6$, i.e. the change in the obliquity of the ecliptic in 35 years, obtained from Table V. The value for 35 years must be interpolated.

We must also consider the change in nutation. The value of argument №4 for changes due to nutation is:

for year 1792	487	(Table VII)
for September	36	(Table XVII)

The above sum of argument № 4 = 523 follows the value of the correction of the obliquity of the ecliptic due to the nutation $-9.''5$, obtained from Table XXV.

The obliquity of the ecliptic is calculated by taking into account the corrections:

for year 1756	$23°28'16''$
change in 35 years	$-16.''6$
correction due to nutation	$-9.''5$

The obliquity of the ecliptic is therefore $23°27'49.''9$.

Right ascension is defined as follows (Mayer, 1770, Problem VII, pp.109–110):

From Tables XXVIII and XIX, for the true longitude value 3S $10°11'51.''15$, we can obtain or interpolate the values of the reduction $+2'30''$ and the change of $4.''9$. The obliquity of the ecliptic is compared with the value $23°28'15''$. The difference is $25.''1$.

Ratio $60'' : 4.''9$ (change) = 12.2.
The same ratio applies to the difference: $25.1 : 12.2 \approx 2$.

We add the result to the value of the ecliptic if the latter is greater than $23°28'15''$ or subtract if the value of the ecliptic is lower.

Thus the reduction $+0°53'52'' - 2'' = +0°53'50''$.

We add the resulting true longitude and get the true right ascension: $3^S\ 10°11'51.''15 + 0°53'50'' = 3^S\ 11°5'41.''15$.

We must also take into account the correction due to the equation of the equinoxes:

correction due to nutation $^{11}/_{12} \times (-0.''6) = -0.''55$

For the explanation of the value $^{11}/_{12}$ see Chapter 13.1.

The equation of time is the difference between the mean longitude, taking into account the correction for the equation of the equinoxes, and the right ascension:

$$3^S\ 10°13'24.''85 - 3^S\ 11°5'41.''15 = -52'16.''3$$

The result is then converted to time units ($1° \equiv 4$ min = 3 min 29 s).

	Mean longitude	Longitude of the apogee	№ 1	№ 2	№ 3	№ 4
Initial value for year 1792	9ˢ 10°50′0.″7 [♑ 10°50′0.″7 280°50′0.″7]	3ˢ 9°23′46″ [♋ 9°23′46″ 99°23′46″]	241	227	123	478
Mean motion July 1	5ˢ 29°23′16.″2 [♍ 29°23′16.″2 179°23′16.″2]	33″	163	456	312	27
Time (equation of time for the year 1788) 3′28″	7.″4 1.″1					
Mean longitude with corrections	3ˢ 10°13′25.″4 [♋ 10°13′25.″4 100°13′25.″4]	Longitude of the apogee with corrections 3ˢ 9°24′19″ [♋ 9°24′19″ 99°24′19″]	404	683	435	505
Equation of the centre	−1′37.″1 ⋆	Mean longitude − apogee = mean anomaly 3ˢ 10°13′25.″4 − 3ˢ 9°24′19″ = 49′6.″4 In MAYER's tables, the mean anomaly is the argument of the equation of the centre.				
☽	+4.″5 ⋆					
♃	−4.″7 ⋆					
♀	+3.″65⋆					
☊	−0.″6 ⋆					
True longitude	3ˢ 10°11′51.″15 [♋ 10°11′51.″15 100°11′51.″15]					

Table 6.1 The process of calculating true longitude for a given day using MAYER'S tables as in Example 9.4. Summarised from: Vince, 1797, *A Complete System of Astronomy*, vol. II, p.46. ⋆ Values are interpolated.

Tobie Mayer est universellement considéré comme l'un des plus grands astronomes, non-seulement du 18e siècle, mais de tous les temps et de tous les pays.

Jean-Baptiste Delambre, 1827,
Histoire de l'Astronomie aux Dix-Huitième Siècle, p.429

9.3 Roger Boscovich

ROGER BOSCOVICH or RUĐER BOŠKOVIĆ (1711–1787), one of the most versatile scholars of his time, published his treatise on trigonometry, *Trigonometria sphæricæ constructio*, where he also used an analemma, in 1735.[49] He published his next trigonometric treatise, *Trigonometria sphærica*, in 1745. In this paper, he also referred to terms descriptively, as was the custom in professional literature at the time. In 1754, in his first book *Elementorum universæ matheseos*, he treated trigonometry in detail. He did not use the formulae until his treatise on spherical trigonometry in 1785. The treatise, which includes differential formulae of spherical trigonometry, may have been written under the influence of EULER's treatises, and was published in the 4th volume of his *Opera pertinentia*.[50]

In the expressions below, x, y and z are the sides and p, q and r are the corresponding angles. Differential terms, as derived by BOSCOVICH, are:

$$\begin{aligned} & dx - dy\cos r - dz\cos q - dp\sin z\sin q = 0 \\ & dx\sin q - dy\cos z\sin p - dp\sin z - dr\sin x\cos q = 0 \\ & dx\cot x - dy\cot y - dp\cot p + dq\cot q = 0 \\ & dx\sin r\sin y - dp - dq\cos z - dr\cos y = 0 \end{aligned} \qquad [9.36]$$

BOSCOVICH's discussions of trigonometry were based primarily on the need to solve practical problems. BOSCOVICH's good knowledge of trigonometry also helped him in his pursuit of surveying. For example, in 1750, together with the English Jesuit CHRISTOPHER MAIRE, he measured or determined the meridian through Rome.[51]

49 For an overview of his works, see Proverbio, 2007, Catalogo delle opere a stampa di Ruggiero Giuseppe Boscovich, 1711–1787, Accademia nazionale delle scienze detta dei XL.
50 *Opera pertinentia*, 1785, book 4, p.322.
51 Triplat et al., 2011, Application of Bošković's Geometric Adjustment Method on Five Meridian Degrees.

Bibliography

- ANTHES, Erhard and Armin HÜTTERMANN (eds.). 2013. Tobias Mayers Beiträge zur Wissenschaft des 18. Jahrhunderts im Lichte neuerer Untersuchungen. Acta Historica Astronomiae, 48. Leipzig: Akademische Verlagsanstallt. ISBN 978-3-944913-39-1.
- BISTAFA, Sylvio R. 2021. Euler's Forgotten Equation of the Center. In: Advances in Historical Studies, 10, 44–52. DOI 10.4236/ahs.2021.101005.
- BOSCOVICH, Roger. 1745. Trigonometria Sphærica. Venice: Monaldini. URL: https://www.digitale-sammlungen.de/en/view/bsb10081489?page=,1 (11 February 2024).
- BOSCOVICH, Roger. 1754. Elementorum universæ matheseos, Book 1. Rome: Generoso Salomoni. (Google Books, 29 July 2023).
- BOSCOVICH, Roger. 1785. Opera Pertinentia, Book 4. Venice: Remondini. (Google Books, 29 July 2023).
- BRONSHTEIN, I. N. and K. A. SEMENDYAYEV. 1985. Handbook of Mathematics. Thun and Frankfurt am Main: Harri Deutsch; New York: Van Nostrand Reinhold Company. ISBN 0-442-21171-6.
- COLWELL, Peter. 1993. Solving Kepler's Equation over three Centuries. Richmond, VA: Willmann Bell. ISBN 0-943396-40-9.
- COTES, Roger and Edmond HALLEY. 1714. Logometria auctore Rogero Cotes, Trin. Coll. Cantab. Soc., Astr. & Ph. Exp. Professore Plumiano, & R. S. S. In: Philosophical Transactions, 31 March 1714, 5–45. DOI 10.1098/rstl.1714.0002.
- DE LA HIRE, Philipp. 1727. Tabulae Astronomicae Ludovici Magni. Paris: Montalant. URL: https://archive.org/details/tabulaeastronom00suntgoog/mode/2up (18 February 2023).
- DELAMBRE, Jean-Baptiste. 1827. Historie de l'Astronomie au Dix-Huitième Siècle, reprint 2004. Paris: Éditions Jacques Gabay. ISBN 2-87647-253-8.
- DEL CENTINA, Andrea and Alessandra FIOCCA. 2018. "A masterly though neglected work", Boscovich's treatise on conic sections. In: Archive for History of Exact Sciences, 72, 4, 453–495. DOI 10.1007/s00407-018-0213-3.
- EULER, Leonhard. 1740a. De motu planetarum et orbitarum determinatione. In: Commentarii academiae scientiarum Petropolitanae, Volume 7, 67–85. Opera Omnia, Series 2, 28, 1–16. URL: https://scholarlycommons.pacific.edu/euler-works/37/ (11 February 2024).
- EULER, Leonhard. 1740b. Orbitae solaris determinatio (Determination of orbits around the sun). In: Commentarii academiae scientia-rum Petropolitanae 7, 86–96. Opera Omnia: Series 2, 28, 17–25. Reprinted in Commentarii academiae scientiarum Petropolitanae 7, ed. nova, Bononiae 1748, 78–87. URL: http://eulerarchive.maa.org/docs/originals/E038.pdf (11 February 2024).
- EULER, Leonhard. 1741. Methodus computandi aequationem meridiei. In: Commentarii academiae scientiarum Petropolitanae, 8, 48–65. Opera Omnia Series 2, 30, 13–25. URL: https://scholarlycommons.pacific.edu/euler-works/50/ (11 February 2024).
- EULER, Leonhard. 1743. De summis serierum reciprocarum ex potestatibus numerorum naturalium ortarum dissertatio altera. In: Miscellanea Berolinensia 7, 172–192. URL: https://scholarlycommons.pacific.edu/euler-works/61/ (11 February 2024).
- EULER, Leonhard. 1746. Tabulae astronomicae solis et lunae. In: Opuscula varii argumenti, Volume 1, 137–168. Opera Omnia Series 2, Volume 23, 1–10. URL: https://scholarlycommons.pacific.edu/euler-works/87/ (11 February 2024).
- EULER, Leonhard. 1748. Introductio in Analysin Infinitorum, Volume 1. Lucerne: Bosquet. URL: https://scholarlycommons.pacific.edu/euler-works/101/ (11 February 2024).
- EULER, Leonhard. 1748. Introductio in Analysin Infinitorum, Volume 2. Lausanne: Bosquet. URL: https://scholarlycommons.pacific.edu/euler-works/102/ (11 February 2024).
- EULER, Leonhard. 1755. Principes de la trigonometrie spherique tires de la methode des plus grands et plus petits. In: Memoires de l'academie des sciences de Berlin 9, 223–257. URL: http://eulerarchive.maa.org/docs/originals/E214.pdf (11 February 2024).
- EULER, Leonhard. 1781. Nova methodus motum planetarum determinandi. In: Acta Academiae Scientiarum Imperialis Petropolitanae, II, 277–302. Opera Omnia Series 2, 29, 336–359. URL: https://scholarlycommons.pacific.edu/euler-works/519/ (11 February 2024).
- EULER, Leonhard. 1782. Trigonometria sphaerica universa, ex primis princpiis breviter et dilucide. In: Acta academiae scientarum imperialis Petropolitinae 3, 72–86; Opera Omnia Series 1, 26, 224–236. URL: http://eulerarchive.maa.org//docs/originals/E524.pdf (11 February 2024).
- FLECKENSTEIN, J. O. 1963. Boscovich als Mitbegründer der Sphärischen Trigonometrie. Contributi dell' Osservatorio astronomico di Milano-Merate, Nouva Serie N. 207. Milano: Art Grafiche F. Milli.
- FORBES, Eric Gray. 1967. The Life and Work of Tobias Mayer (1723-62). In: Quarterly Journal of the Royal Astronomical Society, 8, 227–251.
- FORBES, Eric Gray. 1970a. Tobias Mayer (1723-62): A Case of Forgotten Genius. In: The British Journal for the History of Science, 5, 1, 1–20. URL: https://www.jstor.org/stable/4025350.

- Forbes, Eric Gray. 1970b. Tobias Mayer's contributions to the development of lunar theory. In: Journal for the History of Astronomy, 1, 2, 144–154. doi 10.1177/002182867000100205.
- Forbes, Eric Gray. 1971. The Euler-Mayer correspondence (1751-1755); a new perspective on 18th century advances in the lunar theory. New York: American Elsevier. doi 10.1007/978-1-349-01250-3. isbn 0-444-19580-7.
- Forbes, Eric Gray (ed.). 1972. The Unpublished Writings of Tobias Mayer, Volume I, Astronomy and Geography. Arbeiten aus der Niedersächsischen Staats- und Universitätsbibliothek Göttingen, Bd. 9. Göttingen: Vandenhoeck and Ruprecht. isbn 3-525-85257-6.
- Forbes, Eric Gray. 1980a. Tobias Mayer (1723-62): Pioneer of enlightened science in Germany. Göttingen: Vandenhoeck & Ruprecht. isbn 978-3525852682.
- Forbes, Eric Gray. 1980b. Tobias Mayer's Contributions to Observational Astronomy. In: Journal for the History of Astronomy, 11, 1, 28–49. doi 10.1177/002182868001100104.
- Forbes, Eric Gray and Jacques Gapaillard. 1996. La correspondance astronomique entre l'abbé Nicolas-Louis de Lacaille et Tobias Mayer. In: Revue d'histoire des sciences, 49, 4, 483–541. url: https://www.jstor.org/stable/23633769, https://www.persee.fr/doc/rhs_0151-4105_1996_num_49_4_1267 (19 February 2023).
- Headley, Patrick T. (trans.). 2022. A Method for Calculating the Equation of Noon (an English translation of Methodus Computandi Aequationem Meridiei). In: Euleriana 2, 2, 67–78. doi 10.56031/2693-9908.1034.
- Jørgensen, Niels Terkel. 1974. Peter Horebow's solution of the Kepler's Problem. In: Centaurus, 18, 3, 173–183. doi 10.1111/j.1600-0498.1974.tb00296.x.
- Lapaine, Miljenko (ed.). 2016. Ruđer Bošković i geoznanosti / Ruđer Bošković and the Geosciences. Zagreb: Geodetski fakultet Sveučilišta u Zagrebu, Državna geodetska uprava, Hrvatska komora ovlaštenih inženjera geodezije. isbn 978-953-6082-14-8.
- Mayer, Tobias. 1753. Novae Tabulae Motuum Solis et Lunae. In: Commentarii Societatis Regiae Scientiarum Gottingensis, 383–430. (GoogleBooks, 7 April 2020).
- Mayer, Tobias. 1770. Tabulæ Motuum Solis et Lunæ Novæ et Correctæ. Explanation and use of the Solar and Lunar tables. Londini: Gulielmi et Johannis Richardson. doi 10.3931/e-rara-467.
- Mayer, Tobias. 1770. Tabulæ Motuum Solis et Lunæ Novæ et Correctæ. Londini: Gulielmi et Johannis Richardson. url: https://archive.org/details/bim_eighteenth-century_tabulae-motuum-solis-et-_mayer-tobias_1770 (11 February 2024).
- Papadopoulos, Athanase. 2014. On the works of Euler and his followers on spherical geometry. HAL Id: hal-01064269. url: https://hal.archives-ouvertes.fr/hal-01064269 (version 16 September 2014).
- Pilz, Kurt. 1977. 600 Jahre Astronomie in Nürnberg. Nürnberg: Hans Carl. isbn 9783418004471.
- Proverbio, Edoardo. 1988. R. G. Boscovich's determination of instrumental errors in observation. In: Archive for History of Exact Sciences, 38, 2, 135–152. doi 10.1007/BF00348455.
- Proverbio, Edoardo. 2007. Catalogo delle opere a stampa di Ruggiero Giuseppe Boscovich, 1711–1787. Rome: Accademia nazionale delle scienze detta dei XL. url: http://www.brera.inaf.it/boscovich/progetto-sito/opere_a_stampa.pdf (11 February 2024).
- Proverbio, Edoardo. 2013. Boscovich: scientist and man of letters. In: Memoire Societa Astronomica Italiana, Supplement, 22, 34–54.
- Rickey, V. Frederick. 2017. Did Cotes Anticipate Euler? [online note], May 11, 2017. url: https://faculty.washington.edu/etou/eulersoc/documents/EulerCotes_Rickey.pdf (11 February 2024).
- Sheynin, Oscar. B. 1984. On the history of the statistical method in astronomy. In: Archive for History of Exact Sciences, 29, 2, 151–199. doi 10.1007/BF00348245.
- Steele, John M. 2012. Ancient Astronomical Observations and the Study of the Moon's Motion (1691–1757). New York: Springer. isbn 978-1-4614-2148-1.
- Triplat Horvat Martina, Miljenko Lapaine and Dražen Tutić. 2011. Application of Bošković's Geometric Adjustment Method on Five Meridian Degrees. In: 15th Scientific-Professional Colloquium on Geometry and Graphics, Došlić, Tomislav and Ema Jurkin (eds.), 67–74. Zagreb: Hrvatsko društvo za geometriju i grafiku.
- Vince, Samuel. 1797. A Complete System of Astronomy. Volume I. Cambridge: J. Burges. (GoogleBooks, 30 May 2020).
- Vince, Samuel. 1823. A Complete System of Astronomy. Volume III, 2nd Edition. London: J. Mawman; Baldwin, Cradock & Joy; W. Baynes and Son; T. Hamilton; Simpkin and Marshall. (GoogleBooks, 30 May 2020).
- Wepster, Steven. 2010. Between Theory and Observations; Tobias Mayer's Explorations of Lunar Motion, 1751–1755. New York: Springer. isbn 978-1-4419-1313-5. doi 10.1007/978-1-4419-1314-2.

Internet links

- Publications by Leonhard Euler available on archive.org. url: https://lambentresearch.com/eulerindex/
- Euler Archive. url: https://scholarlycommons.pacific.edu/euler-monographs/

This page is intentionally left blank.

10 Equation of time in the 18th and 19th centuries

The Aſtronomers are the bold Men who tell us so; for they, by their nice Search into Things, have found, that the Sun's apparent Motion is no ways equal; they obſerve, that he now and then ſlackens his Pace, and afterwards quickens it again: And therefore Equal Time, which goes on always at the ſame Rate, cannot truly be measured by the Sun's Motion.

John Keill, 1739, An introduction to the true astronomy: or, Astronomical Lectures, chapter XXV, p.313

10.1 Uppsala University in the 17th and 18th centuries

10.1.1 Samuel Krook

Uppsala University, the oldest of the Nordic universities, was founded in 1477; the first professors were graduates from Germany. Astronomy lectures were held at the Faculty of Philosophy at least as early as 1480. Records of the professors survive since 1593, when the university was revived after the Reformation.[1] The main source for astronomical lectures was SACROBOSCO's *De sphaera*. Around 1650, the university had an observatory on the roof of the professor's house.[2] Prof. ANDERS SPOLE also built his own observatory in 1680, but it was destroyed in the fire that struck Uppsala in 1702. The first truly modern observatory was built in Uppsala by ANDERS CELSIUS in 1741.[3]

SAMUEL KROOK (1661–1716), who later became a professor of mathematics and lectured at Pärnu University in Estonia, defended his doctoral thesis on unequal day length at Uppsala University in 1688.[4] KROOK's thesis cites several reasons for the variable length of the so-called unequal days. In addition to irregular motion due to eccentricity and the contribution of the obliquity of the ecliptic, it also cites the changing longitude of the apogee or perigee over time as a reason.[5] It also mentions parallax and the obliquity of the horizon, or oblique ascension, and vapours in the atmosphere.[6]

KROOK seems to have drawn a lot of help from RICCIOLI's *Almagestum novum*, from which he quotes more data.[7] The contribution from the Sun's irregular motion as described by KROOK is listed below (Figure 10.1).[8]

The circle ○*BCDE*, which is the ecliptic with the Earth at the centre *A*, has diameter *BD* and *CE* drawn on it. The eccentric point *K* represents the centre of the eccentric circle of ○*GHI*, or the orbit that represents the Sun's annual path. Draw a line *ML* through the centres of the two circles, points *M* and *N* being the perigee and apogee respectively. If we observe the Sun from point *A* as it travels along the eccentric circle ○*FGH*, it appears to be travelling along the

1 University of Uppsala, Professors of astronomy, https://www.astro.uu.se/history/proflist_eng.html (24 November 2022).

2 From 1648 to1657 the Professor of Astronomy was MARTINUS OLAI NYCOPENSIS (1596–1657).

3 University of Uppsala, A brief history of astronomy at Uppsala University, https://www.astro.uu.se/history/ (24 November 2022).

4 Piirimäe, 1999, p.210.

5 Krook, 1688, for the longitude of the apogee see §VIII.

6 Krook, 1688, §X, p.19.

7 KROOK, for example in §IV, p.7, gives values for the Sun's path lengths in the individual quadrants derived from RICCIOLI's work: between the winter solstice and the spring equinox 92 days 23 hrs 52 min 45 sec, between the spring equinox and the summer solstice 93 days 14 hrs 56 min 0 sec, between the summer solstice and the autumn equinox 89 days 12 hrs 30 min and between the autumn equinox and the winter solstice 89 days 2 hrs 30 min (see also Figure 0.2).

8 Krook, 1688, §VI.

southern semicircle of the ecliptic *BCD*, and later the semicircle *DEB*. Because the arc $\widehat{FGH}$ is longer than arc $\widehat{HIF}$, the Sun takes longer to travel along the longer arc than the shorter arc, and dwells longer around the centre of the northern part of the ecliptic *BCD* than around the southern part of *DEB*. In terms of observed angles, the Sun appears from Earth to be travelling irregularly. From any point other than the centre, a uniform circular motion appears irregular.

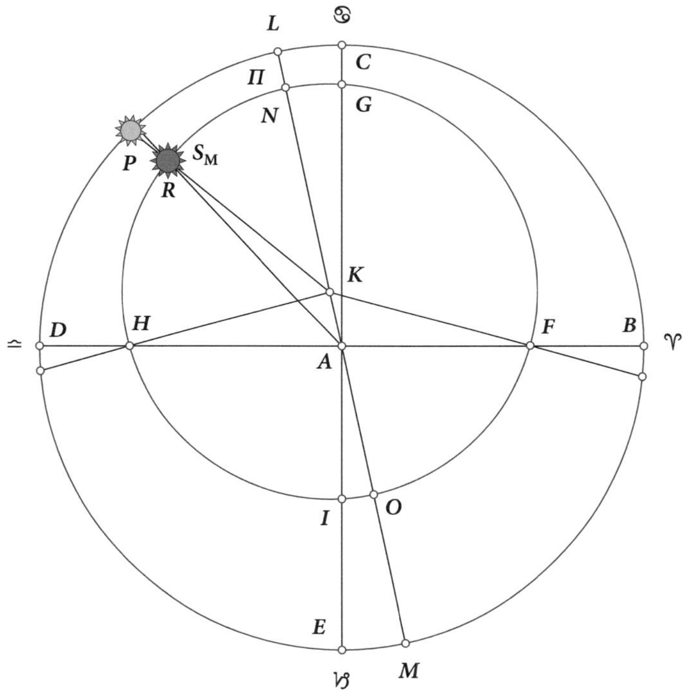

Figure 10.1 Samuel Krook's explanation of non equal day lengths. Source: Krook, 1688, Disputatio astronomica de inæqualitate dierum civilium, p.10. An explanation of the figure is given in § VI, pages 9 to 11. It is summarised by the original; the source of the original: DiVA portal.

In order to explain the eccentricity *KA* in the sketch, points *KH* and *KF* must be connected by straight lines, with the mean Sun at point *R*. Point *P* on the left is the apparent Sun, connected to the centre by the line *AP*, and *KR* is the line between the centre and the mean Sun. Krook also lists the contribution to the equation of time due to the obliquity of the ecliptic.[9] Krook's treatise explains the arc of the ecliptic - for the meridian to be in the same place every day, or for the days to have the same duration, the Sun would have to travel the same arc length of the ecliptic every day.[10] The obliquity of the ecliptic is given as 23°30′, which is derived from the observations of Tycho Brahe on 11 June 1593, when Tycho measured the altitude of the Sun at the equinox of 57°35′, the complement of Uraniborg's latitude being 34°5′30″.

9 Krook, 1688, §IX.

10 The explanation of this phenomenon is the same as for Geminus, see Chapter 1.

10.1.2 Johan Thun

JOHAN THUN's doctoral *Dissertatio de inæqualitate dierum qvam* from 1712 lists PER ELVIUS as the author, though he was apparently the mentor.[11] THUN also drew on RICCIOLI's work. On pages 15 and 16, of his thesis, THUN summarises the reasons for the uneven length of days: parallax and refraction, the obliquity of the ecliptic, and the uneven motion or uneven obliquity[12] of the horizon, as THUN calls it.

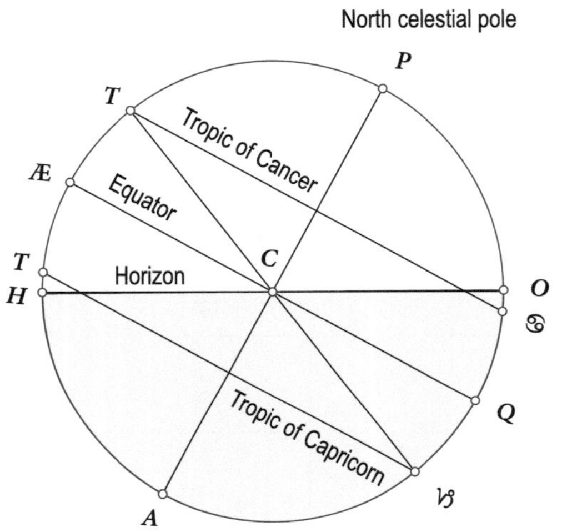

Figure 10.2 JOHAN THUN'S presentation of the uneven obliquity of the horizon as he calls it. Source: Thun, 1712, *Dissertatio de inæqualitate dierum qvam*, p.20. Source of the original: DiVA portal.

> *Diversa horizontis obliquitas, ob quam etiamsi Sol in eodem constitutus sit parallelo, tamen dies unius Horizontis non est aequalis ulli diei in alus horizontibus.*
>
> *Johan Thun, 1712, Dissertatio de inæqualitate dierum qvam, pp.15–16*

11 PER (PETRUS) ELVIUS (1660–1718), a Swedish mathematician, mentored many PhD students in astronomy and produced almanacs and calendars. His textbooks were still in use after his death. Source: https://www.astro.uu.se/history/p_elvius.html (24 November 2022). Struve, 1845, p.149, and Houzeau, 1889, p.1163, cite Elvius, 1712, De inaequalitate dierum, Strengnesiae. Strengnesiae = Strengnäs am Mälar See.

12 Oblique ascension.

10.1.3 Andreas Boström

ANDREAS BOSTRÖM's 1749 doctoral thesis *Specimen academicum de æquatione temporis, quod præside …* at Uppsala University also contains a table on pages 21 to 23 with the equation of time for 1750 based on FLAMSTEED's values.[13] The values are adjusted to Uppsala's meridian and are given for each day of the month in minutes, seconds and sixtieth of seconds.[14] On pages 25 to 27, the values for 1802 are given in the same way.

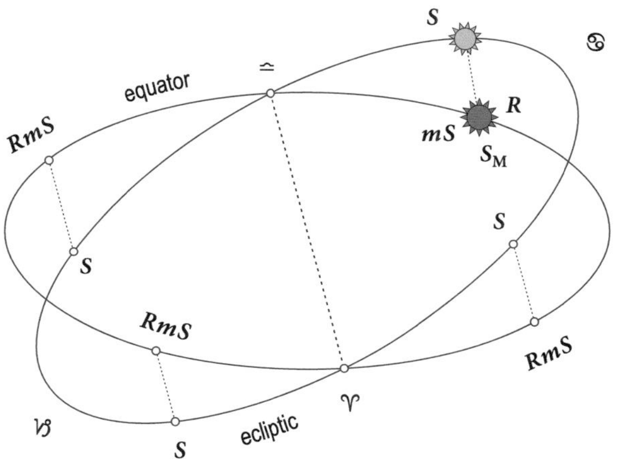

Figure 10.3 shows that the position of the Sun and the position of the mean Sun coincide four times a year, i.e. at times when the value of the equation of time is zero.

mS	mean Sun
S	Sun
R	right ascension or projection of the Sun on the equator

Figure 10.3 Presentation of the equation of time from BOSTRÖM'S 1749 dissertation, *Specimen academicum de æquatione temporis…*, p.14. Source of the original: DiVA portal.

Sit porro m locus solis ſicti, loco solis veri S respondens & manifestum eſt arcum mR in tempus converſum elle differentiam temporis inter tranſitum utriusque ſideris per meridianum, h. e. inter tempus verum & medium.[15]

Boström, 1749, Specimen academicum de æquatione temporis…, p.18

13 STRUVE mentions two parts: Boström, Andreas, 1749–1752, Specimen academicum de aequatione temporis. Part I, Holmise. Part II, Upsalla. See Struve, 1845, p.149. BOSTRÖM's mentor was Prof. MÅRTEN STRÖMER (1707–1770), one of Sweden's most important mathematicians. STRÖMER is best known for his 1744 edition of EUCLID's *Elements* and for his works *Inledning till trigonometria plana* or *Introduction to Planimetry* (1749) and *Läran om klotet och sphæeriska trigonometrien* or *A Doctrine on Spherical Trigonometry* (1759). He made astronomical and meteorological observations and was also a cartographer.

14 See also Chapter 3, note 57.

15 Approximate translation of the original: *"Let it be further that the* **m** *place of the mean Sun corresponds to the place of the true Sun S and it is clear that the arc mR is reversed in time, or the difference in time between the transit of both stars through the meridian, henceforth the true & mean time".*

10.1.4 Frederick Gabriel Craelius

In 1792, CRAELIUS defended his thesis *Examen controversi cujusdam experimenti physici,...* under the tutelage of ZACHARIAS NORDMARK.[16] He is also the author of *Disputatio de incrementis et decrementis dierum*, a 1794 treatise on the inequality of the length of days, written under the tutelage of NICOLAI LANDERBECK.[17] This treatise is interesting because CRAELIUS uses trigonometry to deal with the inequality of the length of days, in a similar way to SAMUEL VINCE, but in a much more mathematically sophisticated way. In this work he also derived differential expressions for declination, true anomaly, ortive amplitude and ascensional difference, using plane and spherical trigonometry, but without using MENELAUS' theorem.

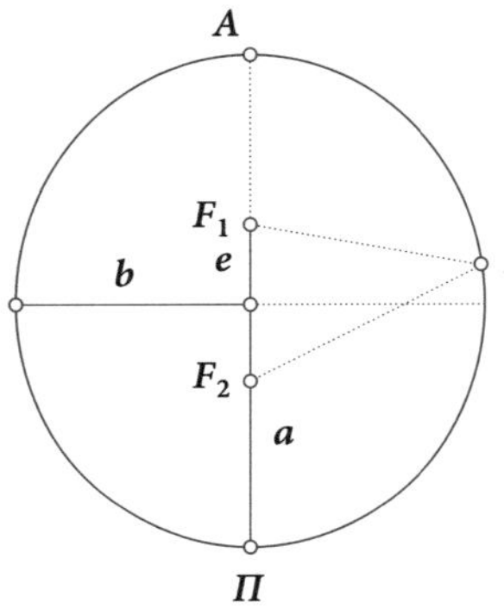

CRAELIUS derived his expression for an elliptic orbit using the following procedure:

Figure 10.4 CRAELIUS' elliptical orbit. Based on Craelius, 1794, Fig. 1.

Elliptic orbits exhibit the usual properties of an ellipse (Figure 10.4):

[10.1] $$A\Pi = 2a \qquad a^2 = b^2 + e^2 \qquad F_1P + F_2P = 2a$$

a, b	axes of the ellipse
e	linear eccentricity $= \sqrt{a^2 - b^2}$
$F_{1,2}$	foci of the ellipse

The following definitions follow from Figure 10.5:

$AVPTH$	ecliptic
C	centre
S	Sun (focal point)
$CP = a$	see Figure 10.4
$SC = e$	linear eccentricity
$ST = R$	in modern terms radii vectors R, V
$SH = V$	
$\widehat{VSH}$	M, true anomaly when the Sun is at H
$\widehat{VST}$	z, the longitude measured from the solstice to the moment when the Sun is at T
$\widehat{VSP}$	v, longitude of the perihelion or $\widehat{PST} = z - v$,

16 ZACHARIAS NORDMARK, aka NORDMARCK (1751–1828), Swedish physicist and astronomer.
17 NILS LANDERBECK (1735–1810), Swedish mathematician, also rector of Uppsala University.

According to CRAELIUS' interpretation and definitions, its parameters z, υ and M can be compared with the modern parameters' longitude, longitude of the perihelion and true anomaly. The only difference is in the starting point or origin of these parameters.[18]

Point V is the winter solstice; in CRAELIUS' time, the winter solstice and perihelion were closer than they are today. In the 13th century, they coincided, but today the difference is about 14 days.

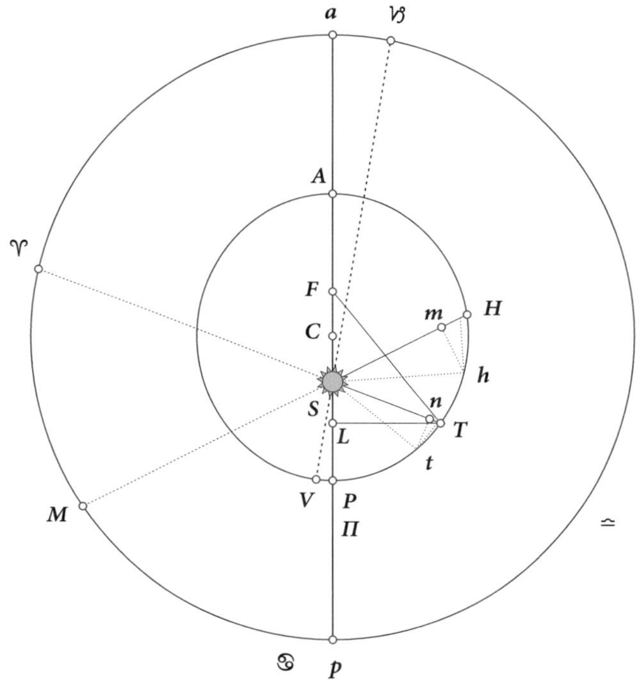

Figure 10.5 Approximate reconstruction of CRAELIUS' diagram used to derive the expressions below. Craelius, 1794, Fig. 1. Source of the original: DiVA portal.

CRAELIUS derived the expression for the radius vector R or V as follows:

From the triangle $\triangle FLT$ (Figure 10.5), we get the following two expressions:

$$FT = 2a - R$$

[10.2] $$FT^2 = FL^2 + LT^2$$

From the triangle $\triangle SLT$:

$$ST^2 = SL^2 + LT^2$$

$$LT^2 = ST^2 - SL^2$$

18 For a modern definition, see Chapter 15.

The last equation is substituted into Equation 10.2 to give:

$$FT^2 = FL^2 + ST^2 - SL^2$$

The equation is rearranged:

$$(FT^2 - ST^2) = FL^2 - SL^2$$

$$(FT + ST)(FT - ST) = (FS + SL)^2 - SL^2$$

$$(FT + ST)(FT - ST) = FS^2 + 2FSSL + SL^2 - SL^2$$

$$(FT + ST)(FT - ST) = FS^2 + 2FSSL$$

Then insert the equations below into the equation:[19]

$$SL = R\cos(z - v) \qquad FS = 2e \qquad FT = 2a - R \qquad ST = R$$

We find that:

$$(2a - R + R)(2a - 2R) = 4e^2 + 4eR\cos(z - v)$$

$$4a^2 - 4e^2 = 4aR + 4eR\cos(z - v)$$

express the radius vector R and get:

[10.3] $$R = \frac{a^2 - e^2}{a + e\cos(z - v)}$$

If we replace the distance $SC = e$, as denoted by CRAELIUS, where e is the linear eccentricity, by $a{\cdot}e$, where e is the eccentricity of the ellipse:

$$e = \sqrt{1 - \left(\frac{b}{a}\right)^2}$$

we obtain the polar equation of the ellipse relative to the focal point (see Chapter 15, section 15.2.2). The same applies to the SH or the radius to point V:

[10.4] $$V = \frac{a^2 - e^2}{a + eR\cos M}$$

The differential expression for the true anomaly follows from KEPLER's second law - the radius from Sunt to planet sweeps over the same area at equal intervals. For the sectors ◔tST and ◔hSH described by the radii ST and SH:

[10.5] $$\frac{hm}{tn} = \frac{ST}{SH}$$

Taking infinitesimally small values into account, CRAELIUS apparently obtained equations from the sine law for the areas and triangles $\triangle tST$ and $\triangle hSH$, respectively (see also Appendix D):[20]

19 The first is derived using trigonometry and the others follow from the properties of the ellipse.
20 The notation radius = 1 is very common in older literature and replaces the expression sin 90° = 1. See also Section 10.6.

$$\frac{\overbrace{\text{radius} = 1}^{\sin \widehat{StT}=1}}{dz} = \frac{ST}{tn} \quad \text{and} \quad \frac{dM}{\underbrace{\text{radius} = 1}_{\sin \widehat{ShH}=1}} = \frac{hm}{SH} \tag{10.6}$$

From Equation 10.5:

$$hm = \frac{tn \cdot ST}{SH} \quad \text{and} \quad tn = \frac{hm \cdot SH}{ST} \tag{10.7}$$

From Equation 10.6:

$$dz = \frac{tn}{ST} \quad \text{or} \quad tn = ST dz$$

$$dM = \frac{hm}{SH} \quad \text{or} \quad hm = SH dM$$

Insert into Equation 10.5 and we get:

$$SH dM = \frac{tn \cdot ST}{SH} \quad \text{or} \quad ST dz = \frac{hm \cdot SH}{ST}$$

$$tn = \frac{SH^2 dM}{ST} \quad \text{or} \quad hm = \frac{ST^2 dz}{SH}$$

Rearranging

$$\frac{hm}{tn} = \frac{\frac{ST^2 dz}{SH}}{\frac{SH^2 dM}{ST}} = \frac{ST^2 ST dz}{SH^2 SH dM} \quad \text{or} \quad \frac{hm dM}{tn dz} = \frac{ST^2}{SH^2}\frac{ST}{SH}$$

Taking Equation 10.5 into account, we get:

$$\frac{hm dM}{tn dz} = \frac{ST^2}{SH^2}\frac{hm}{tn} \quad \text{or} \quad \frac{dM}{dz} = \frac{ST^2}{SH^2}$$

Using the expressions for the radius vector and Equations 10.3 and 10.4, we get:

$$\frac{dM}{dz} = \frac{\left(\frac{a^2 - e^2}{a + e\cos(z - v)}\right)^2}{\left(\frac{a^2 - e^2}{a + e\cos M}\right)^2}$$

Rearrange the above expression:[21]

$$dz = \frac{(a + e\cos M)^2 dM}{(a + e\cos(z - v))^2} \tag{10.8}$$

21 Craelius, 1794, p.4.

Other expressions were also derived by CRAELIUS using the rules of spherical trigonometry. Figure 10.6 shows:

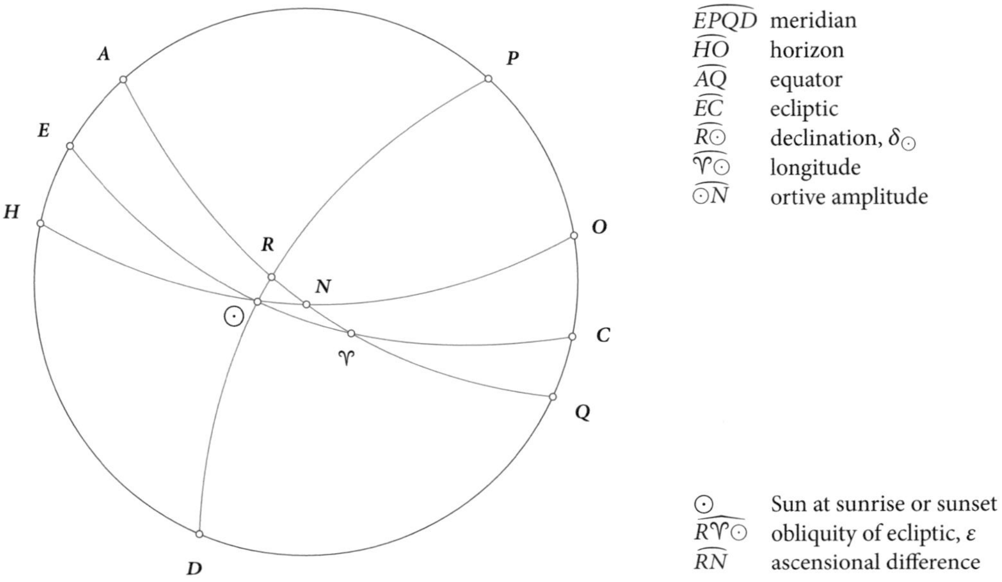

Figure 10.6 CRAELIUS' sketch from which he derived several expressions using spherical trigonometry. Summarised from: Craelius, 1794, Fig. 2. Source of the original: DiVA portal.

The expressions for the complement of latitude, the ortive amplitude and the ascensional difference were derived by CRAELIUS, using the sine law, from the spherical triangle $\triangle R\odot\Upsilon$:

[10.9] $$\frac{\sin \widehat{\odot R}}{\sin \widehat{R\Upsilon\odot}} = \frac{\sin \widehat{\Upsilon\odot}}{\underbrace{\sin \widehat{SR\odot}}_{\sin 90°=1}}$$

Putting this into a modern form, we get the familiar equation for declination (Equation D.37):

[10.10] $$\sin \delta_{\odot} = \sin \varepsilon \sin \lambda_{\odot}$$

or

[10.11] $$\cos \delta_{\odot} = \sqrt{1 - \sin^2 \varepsilon \sin^2 \lambda_{\odot}}$$

From the triangle $\triangle R\odot N$, CRAELIUS derived an expression for the angle $\widehat{R\odot N}$ or F:[22]

[10.12] $$\frac{\sin \widehat{R\odot}}{\sin \widehat{RN\odot}} = \frac{\sin \widehat{RN}}{\sin \widehat{R\odot N}} \quad \text{or} \quad \frac{\sin \delta_{\odot}}{\sin(90° - \varphi)} = \frac{\sin n}{\sin F}$$

22 Craelius, 1794, p.5. See also Figure 0.13.

by applying Equation 0.15 we get:

$$\sin F = \frac{\sin \varphi}{\cos \delta_{\odot}} = \frac{\sin \varphi}{\sqrt{1 - \sin^2 \varepsilon \sin^2 \lambda_{\odot}}} \tag{10.13}$$

$$\cos F = \frac{\sqrt{\cos^2 \varphi - \sin^2 \varepsilon \sin^2 \lambda_{\odot}}}{\sqrt{1 - \sin^2 \varepsilon \sin^2 \lambda_{\odot}}} \tag{10.14}$$

In a similar way, he derived the expression for the ortive amplitude, or arc $\widehat{\odot N}$ or H:[23], from the same triangle:

$$\frac{\sin \widehat{\odot N}}{\underbrace{\sin \widehat{NR\odot}}_{\sin 90^\circ = 1}} = \frac{\sin \widehat{\odot R}}{\sin \widehat{RN\odot}} \quad \text{or} \quad \frac{\sin \delta_{\odot}}{\cos \varphi} = \sin H \tag{10.15}$$

therefore:

$$\sin H = \frac{\sin \varepsilon \sin \lambda_{\odot}}{\cos \varphi} \tag{10.16}$$

$$\cos H = \frac{\sqrt{\cos^2 \varphi - \sin^2 \varepsilon \sin^2 \lambda_{\odot}}}{\cos \varphi} \tag{10.17}$$

And the ascensional difference, or arc or K:[24]

$$\frac{\sin \widehat{RN}}{\sin \widehat{R\odot N}} = \frac{\sin \widehat{\odot R}}{\sin \widehat{RN\odot}} \quad \text{or} \quad \sin K = \frac{\sin \delta_{\odot}}{\cos \varphi} \sin F \tag{10.18}$$

therefore:

$$\sin K = \frac{\sin \varepsilon \sin \varphi \sin \lambda_{\odot}}{\cos \varphi \sqrt{1 - \sin^2 \varepsilon \sin^2 \lambda_{\odot}}} \tag{10.19}$$

$$\cos K = \frac{\sqrt{\cos^2 \varphi - \sin^2 \varepsilon \sin^2 \lambda_{\odot}}}{\cos \varphi \sqrt{1 - \sin^2 \varepsilon \sin^2 \lambda_{\odot}}} \tag{10.20}$$

CRAELIUS also derived differential expressions for the above expressions using the laws of elliptic orbit and spherical trigonometry, respectively, which can be found in the literature.[25]

23 Craelius, 1794, p.6. See also Figure 0.13 and Equation 0.18.
24 Craelius, 1794, p.6.
25 Craelius, 1794, see p.3 for commentary.

10.2 Connoissance des temps

The German astronomer JOHANNES HECKER (1625–1675), a relative of HEVELIUS, published several editions of the ephemeris, apparently highly regarded in astronomical circles, based on KEPLER's *Rudolphine Tables*. In the 1662 edition of HECKER's *Ephemerides* there is a table with TYCHO's equation of time, *Tabvla æqvationis diei*, in the introductory part. After HECKER's death, there was a void in the field of literature, but with the support of King LOUIS XIV and the newly founded Académie des Sciences in Paris, a new publication of annual ephemerides was launched shortly afterwards.[26] The King granted the licence to publish to JOACHIM DALENCÉ, a consultant and librarian. DALENCÉ published the first edition of the *Connoissance des temps* in March 1679 with the help of astronomer JEAN PICARD. The calculations for the first edition were contributed by JEAN PICARD, assisted by JEAN LE FÈVRE and PHILIPPE DE LA HIRE.[27] The tables with the equation of time have been part of the *Connoissance des temps* since the first edition, except that modern editions give the times of the Sun's meridian transit.[28] *Connoissance des temps* is also the oldest publication of its kind still in print.[29]

Connoissance des temps was edited in the first decades of its publication by astronomers JEAN PICARD, JEAN LE FÈVRE, JACQUES LIETAUD, LOUIS GODIN and GIOVANNI DOMENICO MARALDI.[30]

26 For a historical overview, see Boistel, 2022.

27 Boistel, 2022, p.30. JEAN PICARD (1620–1682) collaborated with GASSENDI for a time and his work was also appreciated by HUYGENS. Among other things, he made measurements of the Earth using the SNELL triangulation method. JEAN LE FÈVRE (1652–1706), maker of ephemerides, later became involved in a bitter dispute with DE LA HIRE, which may have led to his expulsion from the Academy. He apparently later published *Ephemerides* under the pseudonym J. DE BEAULIEU.

28 Bureau de longitude, 2022, pp.26–29, temps de passage.

29 Modern title is *Connaissance des temps*.

30 JACQUES LIEUTAUD (1660–1730) was DE LA HIRE's student and later assistant. LOUIS GODIN (1704–1760) was one of the members of an expedition sent to Peru in 1735 to determine the shape of the Earth through measurements. For historical data on the expedition, see Ferreiro, 2011, Measure of the Earth. GIOVANNI DOMENICO MARALDI (1709–1788) is best known for his work on fixed stars.

10.3 John Keill

In 17th century English writers used the term *equation of natural days*,[31] eventually adopting *equation of time* in the early 18th century. In 18th century, we can already find a formal mathematical definition of the equation of time solvable with trigonometry.

JOHN KEILL (1671–1721) studied at Edinburgh, where he completed his studies in mathematics and physics in 1692. He continued his education at Oxford, graduating after two years. For some time, he worked at Oxford University's mathematical institute and was then employed as treasurer for a fund for immigrants fleeing the attacks of LOUIS XIV. He was an advocate of NEWTON's principles and later an ardent supporter of NEWTON in his dispute with LEIBNIZ over the discovery of calculus. In his *An Introduction to the True Astronomy*, KEILL describes both contributions to the equation of time, as he calls it (Figure 10.7):[32] The mean Sun m is uniformly circling along the equator Æ, and the true Sun S along the ecliptic E. The arc of the meridian $\widehat{SA}$, which connects the position of the Sun to the equator, is a declination.

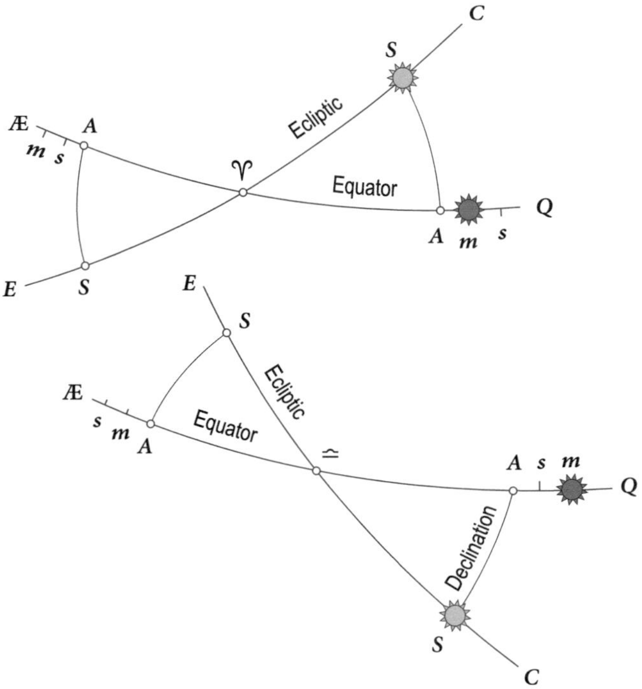

Figure 10.7 Explanation of the equation of time from JOHN KEILL'S *An Introduction to the True Astronomy: or Astronomical Lectures of 1739*. Summarised from Figures 3 and 4, plate XXIV, and the description on page 318. Source of the original: Staats- und Stadtbibliothek Augsburg.

31 The term *æquation of natural days* was used by JOHN FLAMSTEED.

32 Keill, 1739, An Introduction to the True Astronomy or Astronomical Lectures Read in the Astronomical School of the University of Oxford, XXV, pp.312–326. The explanation below and the figures are summarised according to the data on p.318 and Figures 3 and 4, Plate XXIV.

The meridian AS passes through the true Sun, where the mean Sun or m can be left or right from A, but according to the ecliptic longitude, where $\widehat{Am}$, converted into time, is the equation of time. Let arc $\widehat{♈s}$ or $\widehat{♎s}$ be equal to $\widehat{♈S}$ or $\widehat{♎S}$ on the ecliptic. The arc $\widehat{sm}$ thus corresponds to the difference between the mean and the true Sun, and depends on the anomaly of the Sun. The arc $\widehat{As}$ is the difference between the hypotenuse $\widehat{♈S}$ of the right triangle $\Delta ♈SA$ and the side $\widehat{♈A}$, which can be calculated. The arc $\widehat{Am}$ or the equation of time is equal to the sum or difference of the arcs $\widehat{As}$ and $\widehat{sm}$, and can be calculated.

Pierre le Monnier (1715–1799), known for his observations, brought English astronomical ideas and instrumentation to France. His book *Institutions astronomiques*, published in 1746, is an improved translation of Keill's *An Introduction to the true Astronomy.*

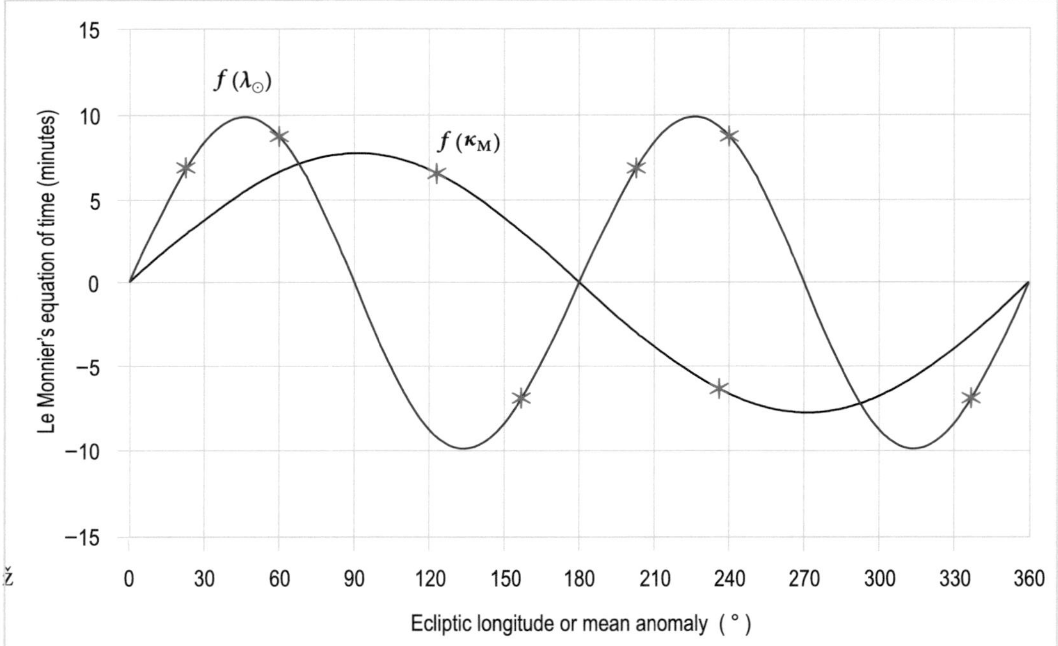

Figure 10.8 Le Monnier's equation of time, contribution of unequal motion or equation of the centre as a function of the mean anomaly and contribution due to the obliquity of the ecliptic as a function of the true ecliptic longitude. The places marked with the symbol (✶) contain obvious errors in the original and the diagram has corrected values. Data for plotting the diagram: Le Monnier, 1746, Institutions Astronomiques, p.154, Tables I, II.

10.4 Nevil Maskelyne

NEVIL MASKELYNE (1732–1811), the fifth Astronomer Royal, headed the observatory at Greenwich between 1765 and 1811, where he succeeded NATHANIEL BLISS. Following an order from the *Royal Society*, which was adopted in 1758, he travelled to St. Helena in 1761 to observe the transit of Venus. The expedition was not the most successful since the crucial moments were compromised by cloud. However, while navigating, he tried to determine the longitude from the Moon's position. This method had been argued for by ISAAC NEWTON and EDMOND HALLEY. A journey to Barbados in 1763 was also aimed at studying navigation at sea, and MASKELYNE tested H4, the fourth chronometer of JOHN HARRISON, while on the trip. He published the results of his experiments in 1763 in *The British Mariner's Guide*. This book also constituted one of the bases for the launch of the *Nautical Almanac*, whose first edition, for 1767, was published in 1766. He then supervised the almanac for a further 50 years until his death. The *Nautical Almanac*, where values of the equation of time were also published, later became a fundamental authority whose values of the equation were applied by others.

In a treatise published in 1764, MASKELYNE described the calculation of the equation of time in the same form as we know it today, at least in principle (Figure 10.9). He defined the equation as proportional to the differences in the right ascension of the true Sun travelling along the ecliptic $\overparen{AQS}$ and the equatorial mean Sun $\overparen{APU}$, whereby the connections follow from Figure 10.9:[33]

[10.21] $$E = \alpha_{\odot} - \alpha_{\mathrm{M}} \equiv \overparen{AQS} - \overparen{APU} = \overparen{BQS} - \overparen{BPU}$$

In response to the false assumptions by DE LALANDE[34], MASKELYNE evaluated the very small contribution of nutation to the equation of time:[35]

[10.22] $$\frac{\sin \overparen{PTQ}}{\sin \overparen{TPQ}} = \frac{\overparen{PQ}}{TQ}$$

33 Maskelyne, 1764, On the Equation of Time and the True Manner of Computing it, Philosophical Transactions, No. 54, p.342.

34 For LALANDE's work see Section 10.5.

35 Maskelyne, 1764, p.340.

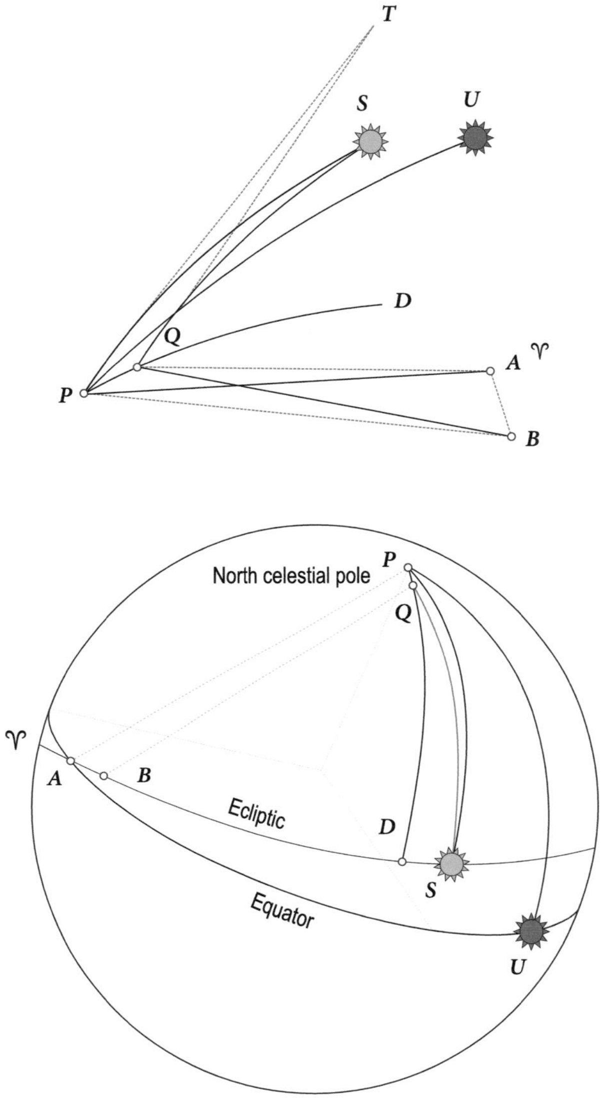

T	the point at which the tangents to the meridians coincide
S	true Sun
U	equatorial mean Sun
P	north celestial pole
A	vernal equinox ♈
BA	correction due to precession
$\widehat{PD}$, $\widehat{PS}$, $\widehat{PU}$	celestial meridians

Figure 10.9 The definition of the equation of time of NEVILLE MASKELYNE from 1764. Summarised by *On the Equation of Time and the True Manner of Computing it.* Source: Philosophical Transactions, 1764, vol. 54, p.337.

10 Jan	8 min 5 s
20 Jan	11 min 34 s
30 Jan	13 min 49 s
10 Feb	14 min 42 s
20 Feb	14 min 9 s
28 Feb	12 min 55 s
10 Mar	10 min 35 s
20 Mar	7 min 42 s
30 Mar	4 min 36 s
10 Apr	1 min 19 s
20 Apr	1 min 12 s
30 Apr	0 min 3 s
10 May	3 min 56 s
20 May	3 min 55 s
30 May	2 min 59 s
10 Jun	1 min 9 s
20 Jun	0 min 56 s
30 Jun	3 min 3 s
10 Jul	4 min 47 s
20 Jul	5 min 50 s
30 Jul	5 min 59 s
10 Aug	4 min 58 s
20 Aug	3 min 5 s
30 Aug	0 min 25 s
10 Sep	3 min 10 s
20 Sep	6 min 39 s
30 Sep	10 min 0 s
10 Oct	12 min 56 s
20 Oct	15 min 5 s
30 Oct	16 min 10 s
10 Nov	15 min 51 s
20 Nov	14 min 4 s
30 Nov	10 min 58 s
10 Dec	6 min 47 s
20 Dec	1 min 57 s
30 Dec	3 min 2 s

Table 10.1 Excerpt from the table with the equation of time of MASKELYNE from the year 1767. Source: *The Nautical Almanac and Astronomical Ephemeris for the Year 1767.*

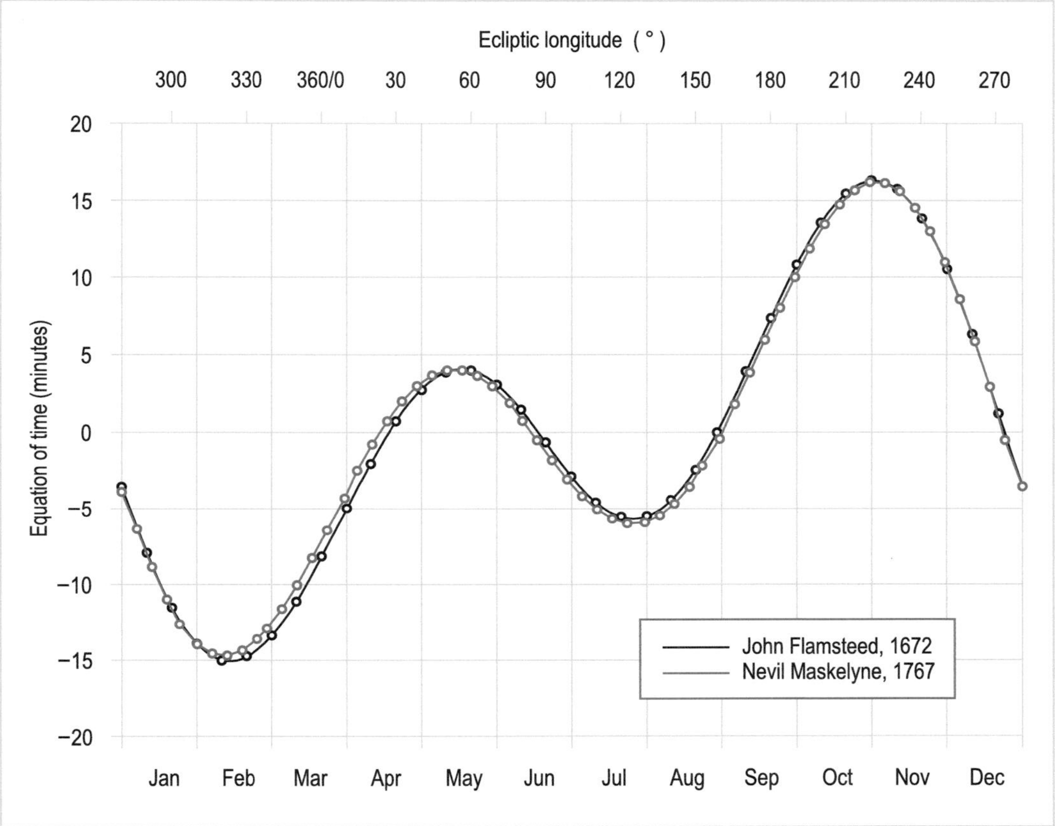

Figure 10.10 The values of FLAMSTEED'S and MASKELYNE'S equation of time, shown in the way the equation of time is presented today. Although the original values do not yet have a sign, the values of both authors are fairly well-matched, with FLAMSTEED'S equation of time showing some deviations for individual values. Sources: Flamsteed, 1678, pp.460–461 and *The Nautical Almanac and Astronomical Ephemeris for the Year 1767.*

Under the assumption $\widehat{PQ} \approx \sin \widehat{PQ}$ and $\widehat{PTQ} \approx \sin \widehat{PTQ}$, the effect of nutation is:[36]

[10.23] $$\widehat{PTQ} = \frac{\widehat{PQ} \cdot \sin \widehat{TPQ}}{TQ} = \widehat{PQ} \sin \widehat{TPQ} \cdot \tan \delta$$

$\sin \widehat{TPQ}$ sine of the right ascension of the true Sun measured from meridian $\widehat{PQD}$ in which the pole moves

whereby the displacement due to the nutation $\widehat{PQ} \leq 9''½$ so value of the angle $\widehat{PTQ}$ cannot be greater than $9''½ \cdot \tan 23°30' = 4.''1$.

36 Maskelyne, 1764, p.340.

10.5 Joseph-Jérôme Lefrançais de Lalande

De Lalande (1732–1807) was a student of Joseph-Nicolas Delisle, and later also took up his teaching position and taught many astronomers. Delisle's achievements have also been recognised abroad. Peter the Great thus proposed to him the establishment of an observatory in Russia. The St Petersburg Observatory, where Delisle worked from 1725 to 1747, trained several astronomers, and published observations of eclipses of Jupiter's satellites in the first six volumes of the *Commentarii academiae imperialis scientiarum petropolitanae.* He left a large correspondence, which is only partly edited or printed.[37] An example of the use of the equation of time can be found, for example, in a letter from Navarre to Delisle dated 3 May 1721.[38]

De Lalande in 1760 succeeded Giovanni Domenico Maraldi as the editor of the astronomical almanac *Connoissance des temps* and Delisle as the Royal Professor of Astronomy, and in 1768 he became the director of the Paris Observatory. Together with Alexis-Claude Clairaut, he used the solution to the problem of three bodies to precisely determine when Halley's Comet would return. In 1753, according to Delisle's orders, he observed along with other astronomers the passage of Mercury, and he made his name by, among other things, documenting the transit of Venus, a very rare astronomical phenomenon that he observed from the Luxembourg palace in 1769. He needed the parallax for observing this phenomenon, and estimated it at $8.''5$ to $8.''75$. Twenty years later, he watched the Sun at the solstice and, with the help of 40-year-old observation notes from the same site, realised that the obliquity of the ecliptic shrank by $38''$ per century. He also compiled a comprehensive catalogue of 50,000 stars, which was published as *Histoire celeste* in 1801. The work is one of the most important achievements of astronomy in the 18th century. In Lalande's notes, American astronomer Samuel C. Walker found a description of a star unknown until then, which later turned out to be the planet Neptune, which eventually came to be discovered on the basis of astronomical calculations. Lalande's works in the field of astronomy history are still an important source of data today.

Lalande describes the equation of time as the difference between the longitude of the equatorial mean Sun and the right ascension of the true Sun:[39]

$$E = \lambda_{M} - \alpha_{\odot} \qquad [10.24]$$

37 Doublet (ed.), 1910, Correspondance échangée de 1720 a 1739 entre L'Astronome J.-N. Delisle et M. de Navarre and Omont, H. (ed.), 1919, Lettres de J.-N. Delisle au comte de Maurepas et à I'abbé Bignon sur ses travaux géographiques en Russie (1726–1730).

38 Doublet (ed.), 1910, Navarre à Delisle, le 3 may 1721, p.62.

39 Lalande, 1774, Abrégé d'Astronomie, p.142. Longitude of the equatorial mean Sun is the same as its right ascension. With the exception of the sign, Equation 10.24 is identical to Maskelyne's (Equation 10.21).

In Part 1 of *Astronomie*, LALANDE lists the tables with the equation of time separately for each of the contributions.[40] The contribution due to the obliquity of the ecliptic is tabulated as a function of the mean anomaly, and the contribution due to uneven circular motion or eccentric orbit as a function of true longitude. In Figure 10.11, summarising LALANDE's work, $\widehat{SA}$ is the meridian, and $\widehat{EQA}$ the equator. The highest value of the contribution due to the obliquity of the ecliptic is 7 minutes and 42 seconds. The highest value of the contribution due to the uneven circular motion is 9 minutes and 53 seconds, occurring at the true ecliptic longitude of 46¼° measured from the equinox. The highest value of the LALANDE equation of the centre is 1°55′31.″6, which is about 7 minutes 42 seconds. In LALANDE's interpretation of the equation of time, the Sun is at point O, where the path to S requires 4 minutes for each degree of the arc $\widehat{QA}$ on the equator. The mean Sun moves along the equator at the even speed of 59′8″ per day starting at the equinox. If we take the value of 23°28′20″ as the obliquity of the ecliptic, which is LALANDE's value, then $\widehat{EA}$ = 2°28′24.″8, or 9 min 53.7 sec in time units, applies for the ecliptic longitude $\lambda_\odot = \widehat{ES}$ = 46¼°. The combination of both contributions gives an equation of time with a maximum value of approximately 16 min 12 sec.[41]

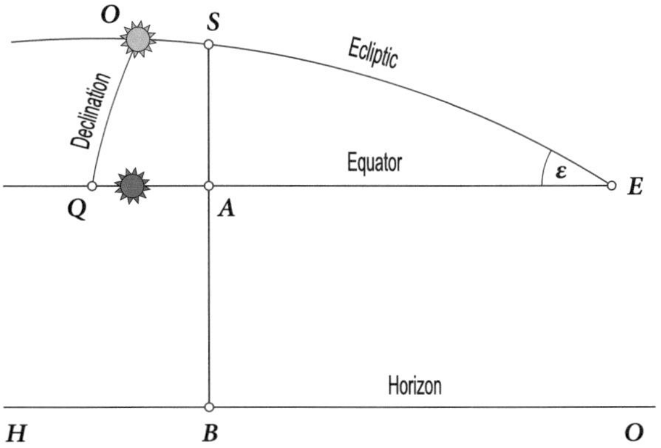

Figure 10.11 DE LALANDE'S geometric model for the description of the equation of time. Summarised from: Lalande, 1774, Abrégée d'Astronomie, plate IV, Figure 21.

0	0 min 0 s
10	3 min 15 s
20	6 min 9.1 s
30	8 min 22.8 s
40	9 min 39.7 s
50	9 min 48.4 s
60	8 min 45.0 s
70	6 min 34.3 s
80	3 min 31.5 s
90	0 min 0 s
100	3 min 31.5 s
110	6 min 34.3 s
120	8 min 45.0 s
130	9 min 48.4 s
140	9 min 39.7 s
150	8 min 22.8 s
160	6 min 9.1 s
170	3 min 15 s
180	0 min 0 s
190	3 min 15 s
200	6 min 9.1 s
210	8 min 22.8 s
220	9 min 39.7 s
230	9 min 48.4 s
240	8 min 45.0 s
250	6 min 34.3 s
260	3 min 31.5 s
270	0 min 0 s
280	3 min 31.5 s
290	6 min 34.3
300	8 min 45.0 s
310	9 min 48.4 s
320	9 min 39.7 s
330	8 min 22.8 s
340	6 min 9.1 s
350	3 min 15 s
360	0 min 0 s

Table 10.2 Contribution to the equation of time due to the obliquity of the ecliptic as a function of the mean anomaly. Source: Lalande, 1764, Astronomie, Table XIV.

40 Lalande, 1774, pp.142–143, for the tables see Lalande, 1764, Astronomie, Book 1, Tables XIII and XIV. The tables are also in subsequent editions, for example, Lalande, 1771, Second Edition, Tables XV and XVI or Lalande, 1792, Third Edition, Tables XIX and XX for 1800. Here, too, both contributions to the equation of time are given separately. The arguments, in this case, are also the mean longitude or mean anomaly.

41 Lalande, 1774, p.143.

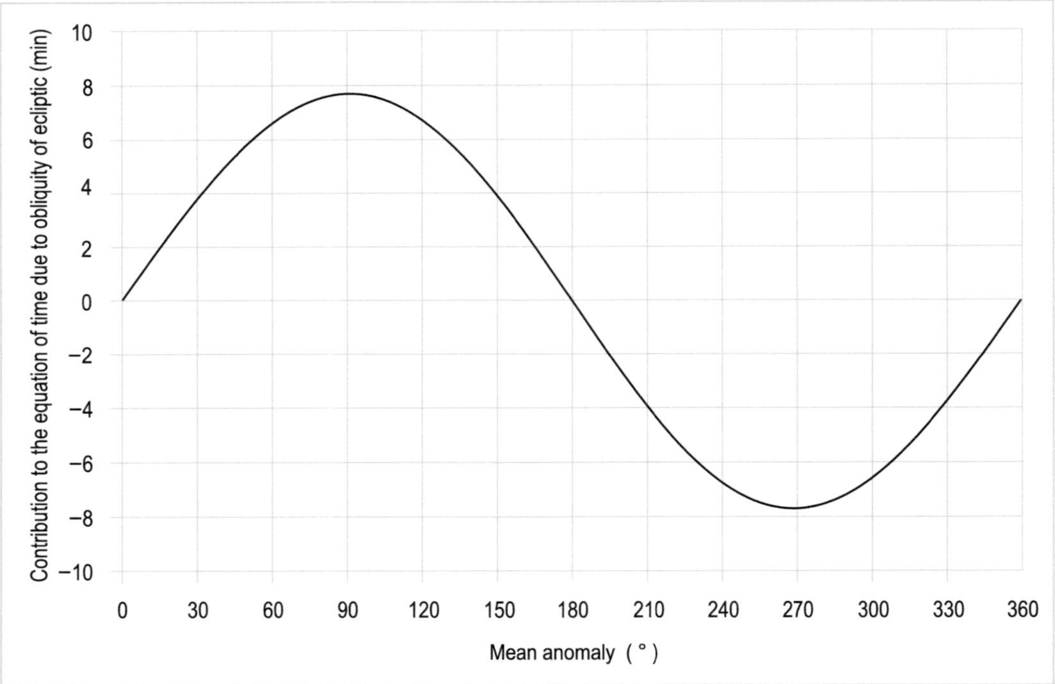

Figure 10.12 Contribution due to the obliquity of the ecliptic as a function of the mean anomaly. Data for plotting the diagram: Lalande, 1764, Astronomie, Book 1, Table XIV.

... l'équation du tems est la différence entre la longitude moyenne et l'ascension droite vraie du soleil convertie en tems.

Jérôme de Lalande, 1774, Abrégé d'Astronomie, Liv. II, p.142

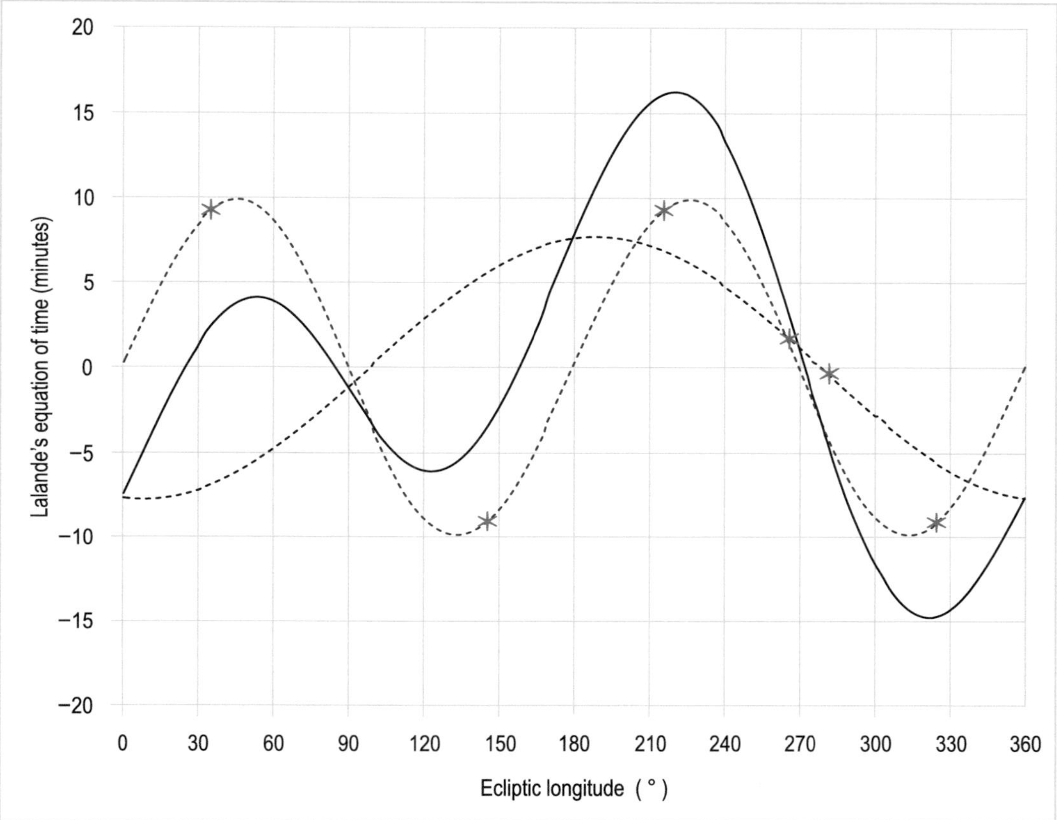

Figure 10.13 DE LALANDE'S equation of time as a function of true ecliptic longitude. The values of both contributions are interpolated linearly. The places marked with the symbol (∗) contain obvious errors in the original and the diagram has corrected values. Data for plotting the diagram: Lalande, 1764, Astronomie, Book 1, Tables XIII, XIV.

10.6 Samuel Vince

In his book *A Plan of a Course of Lectures on the Principles of Natural Philosophy* from 1793, Samuel Vince (1749–1821), an English cleric, mathematician and astronomer, described an equation of time that also includes an equation of equinoxes.[42] In Figure 10.14, P is the Earth's pole, $vwyz$ is the equator, where Earth rotates around its own axis in the direction $vwyz$. ΥDLE is the celestial equator and ΥCL is the ecliptic. At noon, the Sun is located for two consecutive days at points a and m. $Pvae$ and $Prmh$ are perpendicular to the equator and connectors between the Earth and the Sun. At noon, the observer is at point s on the meridian Pv, and the Sun is also at the meridian. At one turn of Earth around its axis, Psv is in the same position, while the Sun has moved to point m. In order to have the Sun again on the observer's meridian, Earth must move additionally by the angle $\widehat{vPr}$. Angle $\widehat{vPr}$ is equal to the arc $\widehat{eh}$, which is the increment to the right ascension of the Sun in one day. The solar day is therefore equal to the amount of time Earth needs for a rotation around its axis and an increment to the right ascension. Since the Sun progresses unevenly along the ecliptic, the daily increment $\widehat{eh}$ is not uniform but varies and is dependent on the obliquity of the ecliptic, as well as on unequal circular motion. The difference in the transition of the Sun through the local meridian compared to the local clock time is the equation of time. The clock time coincides with the apparent mean Sun, which circulates along the equator with a right ascension increment and an angular velocity of $59'8.''2$ per day.

> *... hence, the length of a true solar day is equal to the time of the earth's rotation about its axis + the time of its describing an angle equal to the increase of the sun's right ascension in a true solar day.*
>
> *Samuel Vince, 1797, A Complete System of Astronomy, Chapter IV, p.43*

A theoretical explanation of the spherical trigonometry required to understand the derivation of the equation of time was given by Vince's *A Treatise of Plane and Spherical Trigonometry*, published in 1800.

42 See pp.130–132. A more extensive description of the equation of time is also found in other works of Vince – see for example Vince, 1810, The Elements of Astronomy, Third Edition, Chapter 4. Vince, 1797, A Complete System of Astronomy, Part 1, Chapter 4, pp.42–49; a more precise calculation of the equation of time based on Laplace's work, presented with tables in Part 3, see for example the 1823 edition, explanation pp.17–18, Table XXIX, XXX, pp.90–94.

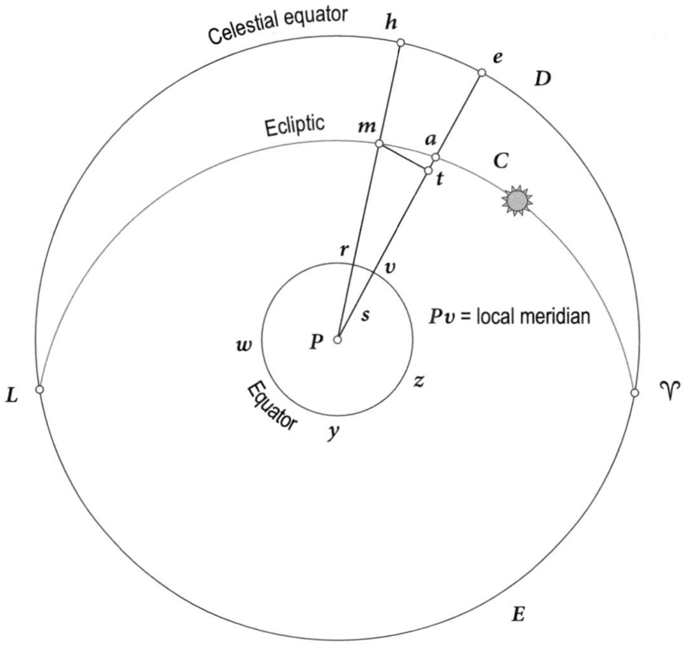

Figure 10.14 Interpretation of uneven lengths of days during the year. Summarised from: Vince, 1797, A Complete System of Astronomy, plate II, Figure XXIII. Summarised from the original; source of the original: Bayerische Staatsbibliothek München.

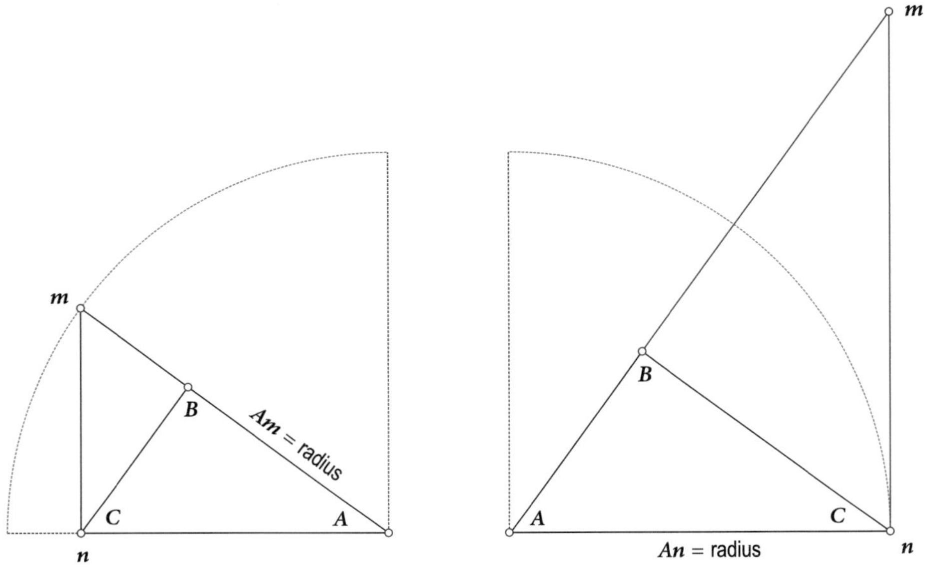

Figure 10.15 Geometric constructions summarised from the text in Vince, 1800, A Treatise on Plane and Spherical Trigonometry, propositions XII and XIII (§123 and 125), pp.63–64.

The terms below are derived from the geometric constructions in Figure 10.15 using the rules that apply to similar triangles. For the sine and cosine functions, the following apply (Figure 10.15, left):

[10.25] $$\frac{AC}{CB} = \frac{Am}{nm} = \frac{\text{radius}}{\sin \widehat{BAC}}$$

$$\frac{AC}{AB} = \frac{Am}{An} = \frac{\text{radius}}{\cos \widehat{BAC}}$$

For the tangent function (Figure 10.15, right):

[10.26] $$\frac{AB}{BC} = \frac{An}{nm} = \frac{\text{radius}}{\tan \widehat{mAn}}$$

We can also use the geometric construction in Figure 10.16.

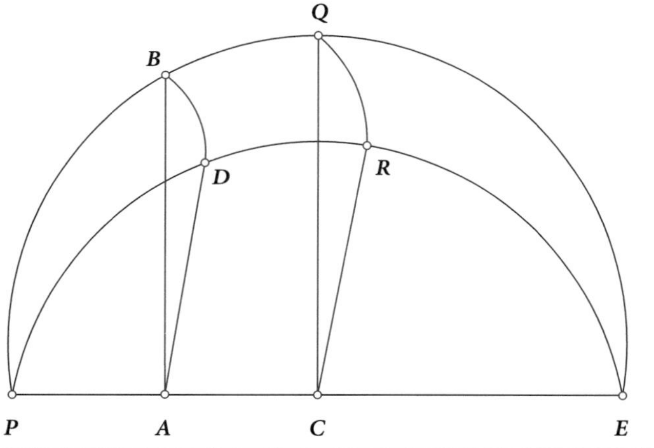

Figure 10.16 This geometric construction helps to derive an equation of time, as VINCE did. Image source: Vince, 1797, A Complete System of Astronomy, plate I, Figure I. Summarised from the original; source of the original: Bayerische Staatsbibliothek München.

Therefore, from Figure 10.14, taking into account *ma* infinitely small, the following equalities based on the rules of plane trigonometry follow:[43]

[10.27] $$\frac{\widehat{ma}}{mt} = \frac{\text{radius}}{\sin \widehat{mat}} = \frac{\text{radius}}{\sin \widehat{\Upsilon ae}}$$

43 Vince, 1800, A Treatise on Plane and Spherical Astronomy, definition 65 and proposition XIII (§125), pp.44, 64. See also Figure 10.15.

From Figure 10.16, VINCE wrote the following relation:[44]

$$[10.28] \quad \frac{\overset{\frown}{QR}}{\overset{\frown}{BD}} = \frac{QC}{BA} = \frac{\text{radius}}{\cos \overset{\frown}{BQ}}$$

Or, from Figure 10.14:

$$[10.29] \quad \frac{mt}{\overset{\frown}{eh}} = \frac{\cos \overset{\frown}{ae}}{\text{radius}}$$

VINCE applied NAPIER's rules to the spherical triangle $\triangle ea♈$ (Figure 10.14) formed by the vernal equinox, the Sun, and the projection of the Sun onto the celestial equator. From the rules of spherical trigonometry (Appendix D) it follows that:[45]

$$\sin \overset{\frown}{♈ae} = \frac{\cos \overset{\frown}{a♈e}}{\cos \overset{\frown}{ae}} \quad \text{or} \quad \sin \overset{\frown}{♈ae} = \frac{\cos \varepsilon}{\cos \delta_{\odot}}$$

where ε is the obliquity of the ecliptic and $\delta_{\odot}$ is declination of the Sun. From Equations 10.27 and 10.29, he obtained:[46]

$$[10.30] \quad \frac{\overset{\frown}{ma}}{\overset{\frown}{eh}} = \frac{\cos \overset{\frown}{ae}}{\sin \overset{\frown}{♈ae}} = \frac{\cos^2 \overset{\frown}{ae}}{\cos \overset{\frown}{a♈e}}$$

From Equation 10.30, it follows that the ratio changes unevenly, which means that the path of the Sun is uneven.

In Figure 10.17, *APLS* is the ecliptic, *ALv* the equator, *A* is 0° ♈, or the equinox, *P* is the apogee of the Sun and S is the Sun. *Sv* is the perpendicular to the equator, whereby $An = AP$. If the Sun is located in *P*, the mean Sun circling the equator is at point *n*. When the mean Sun crosses the meridian, the mean solar time (clock) is noon. If $nm = Ps$ and the mean Sun enters position *m*, the position of the true Sun if the motion was uniform would be at point *s*, but because of the uneven circular motion, the actual position of the Sun is *S*. *S* and *v* are on the meridian, *m* is the position of the mean Sun, where *mv* is the equation of time. $An = AP$, $nm = Ps$ and $Am = APs$ all apply, from which follow the equivalence $mv = Av - Am = Av - APs$. From the mean equinox at point *a*, we draw the perpendicular *az* to *AL*, where $Am = Az + zm = Aa \times \cos a\ Az + zm = {}^{11}/_{12} \times Aa + zm$ applies, therefore, $mv = Av - zm - {}^{11}/_{12} \times Aa$.[47] The right ascension of the Sun is *Av*, *zm* is the mean right ascension or mean longitude, and the expression ${}^{11}/_{12} \times Aa$ (*Az*) is the equation of equinoxes. The equation of time is equal to the difference between the right ascension of the Sun and the mean longitude corrected by the equation of equinoxes.

44 Vince, 1810, §13, p.4. VINCE's proof is based on EUCLID's *Elements*, Book 10, Proposition 10.
45 See also Appendix D, Section D.1 and D.6.1. Vince, 1823, p.44. For NAPIER's rules see Vince, 1800, § 212, pp.121–123.
46 Vince, 1823, p.44.
47 For the explanation of the value ${}^{11}/_{12}$ see Chapter 13, Delambre.

When the distance *Am* is smaller than *Av*, the local clock time overtakes the true time, and when it is greater, the local clock time falls behind.

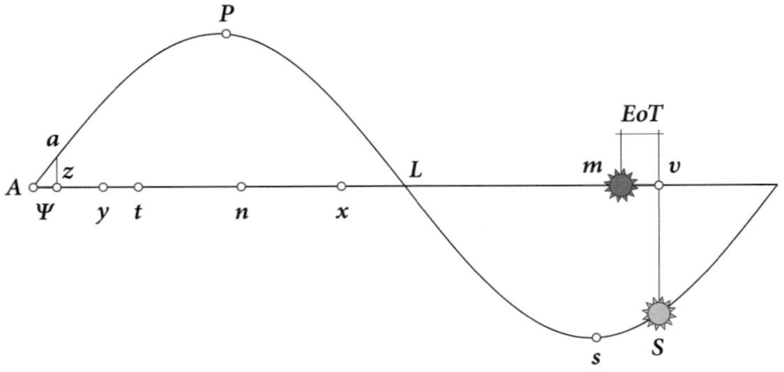

Figure 10.17 Explanation of the equation of time and the equation of equinoxes. Summarised from: Vince, 1797, A Complete System of Astronomy, plate II, Figure XXIV. Summarised from the original; source of the original: Bayerische Staatsbibliothek München.

VINCE also dealt with logarithms and complex numbers in mathematics, and his work on the resistance or movement of bodies through fluids was also important for the subsequent development of aviation.

... hence the equation of time is equal to the difference of the sun's true right ascension, and its mean longitude corrected by the equation of the equinoxes in right ascension.

Samuel Vince, 1797, A Complete System of Astronomy, Chapter IV, pp.44–45

Bibliography

- Anthes, Erhard and Armin Hüttermann (eds.). 2013. Tobias Mayers Beiträge zur Wissenschaft des 18. Jahrhunderts im Lichte neuerer Untersuchungen. In: Acta Historica Astronomiae, 48. Leipzig: Akademische Verlagsanstallt. ISBN 978-3-944913-39-1.
- Boistel, Guy. 2022. "Pour la Gloire de M. La Lande" Une histoire matérielle, scientifique, institutionnelle et humaine de la Connaissance des temps, de 1679 à 1920. Paris: IMCCE. ISBN 978-2-910015-87-9. (Google Books, 15 February 2024).
- Boström, Andreas. 1749. Specimen academicum de æquatione temporis, quod ... præside ... Mag. Martino Strömer, ... publice ventilandum sistit Andreas Boström Vestergothus in Auditorio Carol. Majori die XX Dec. MDCCXLIX. H. A. M. S. Holmiae (Stockholm): Laurenth Salvii. URL: https://www.diva-portal.org/smash/get/diva2:1452934/FULLTEXT01.pdf (15 February 2024).
- Bureau des longitudes. 2022. Éphémérides Astronomiques 2023, Connaissanse des Temps. Paris: IMMCE. ISSN 2259-4191. URL: https://www.imcce.fr/publications/publications-institutionnelles/ (15 February 2024).
- Craelius, Fredrick Gabriel. 1792. Examen controversi cujusdam experimenti physici, quod, ... præside Mag. Zach. Nordmark, ... publicæ censuræ modeste defert Fredericus Gabr. Crælius, smolandus. In Audit. Gust. Maj. d. IV April. MDCCXCII. URN: urn:nbn:se:uu:diva-399564, URL: https://www.diva-portal.org/smash/get/diva2:1378473/FULLTEXT01.pdf (15 February 2024).
- Craelius, Fredrick Gabriel. 1794. Disputatio de incrementis et decrementis dierum, quam ... sub præsidio mag. Nicolai Landerbeck, ... pro gradu publice proponit Fredericus Gabr. Crælius, Smolandus. In Audit. Gust. Maj. d. XXX Aprilis MDCCXCIV. H. A. M. C. Uppsala: Johannes Edmann. URL: https://www.diva-portal.org/smash/get/diva2:1344349/FULLTEXT01.pdf (15 March 2024).
- De Lalande, Joseph Jérôme Le Français. 1762. Mémoire sur les Équations Séculaires, Et sur les moyens mouvemens du Soleil, de la Lune, de Saturne, de Jupiter, & de Mars, Avec les observations de Tycho-brahé, faites sur Mars en 1593, tirées des manuscripts de cet Auteur. In: Histoire de L'Académie Royale des Sciences, Anne 1757, 411–482. Paris: Imprimerie Royale. (Google Books, 9 April 2020).
- De Lalande, Joseph Jérôme Le Français. 1764. Astronomie, Book 1. Paris: Desaint & Saillant, Libraires, rue S. Jean-de-Beauvais. (Google Books, 15 March 2020).
- De Lalande, Joseph Jérôme Le Français. 1771. Astronomie, Book 1. Paris: Desaint & Saillant, Libraires, rue S. Jean-de-Beauvais. URL: https://archive.org/details/gri_astronomiexx01lala/mode/2up (15 February 2024).
- De Lalande, Joseph Jérôme Le Français. 1774. Abrégé d'Astronomie. Amsterdam: Barthelemi. URL: https://gallica.bnf.fr/ark:/12148/bpt6k9611142j/f1.item.texteImage (15 February 2024).
- De Lalande, Joseph Jérôme Le Français. 1792. Astronomie, Book 1, 3rd Edition. Paris: P. Didot. (Google Books, 26 June 2021).
- Delambre, Jean-Baptiste. 1827. Historie de l'Astronomie au Dix-Huitième Siècle, reprint 2004. Paris: Éditions Jacques Gabay. ISBN 2-87647-253-8.
- Flamsteed, John. 1678. De Temporis Aequatione Diatriba. In: Jerremiah Horrox, Opera Posthuma, 441–464. London. URL: https://www.e-rara.ch/zut/content/pageview/854825 (15 February 2024).
- Ferreiro, Larrie D. 2011. Measure of the Earth. New York: Basic Books. ISBN 978-0-465-01723-2.
- Gingerich, Owen. 1973. Lacaille. In: Dictionary of Scientific Biography, Gillispie, Charles Coulston (ed.), VII, 542–545. New York: Charles Scribner's Sons. SBN 684-10118-1.
- Hecker, Johannes. 1662. Johannis Heckeri Motuum caelestium ephemerides ab anno ae. v. 1666 ad 1680. Gdansk: Simon Reininger. (Google Books, 5 August 2023).
- Keill, John. 1718. Introductio ad Veram Astronomiam, seu Lectiones Astronomicae Habitæ in Schola Astronomica Acadamiæ Oxoniensis. London: Hen. Clements. URL: https://archive.org/details/b30537241/page/n4/mode/2up (15 February 2024).
- Keill, John. 1730. An Introduction to the True Astronomy or Astronomical Lectures Read in the Astronomical School of the University of Oxford. 2nd Edition. London: Bernard Lintot. (Google Books, 7 April 2020).

- Krook, Samuel. 1688. Disputatio astronomica de inæqualitate dierum civilium, ... in regia acad. Upsal. præside ... Joanne Bilberg ... publico bonorum examini subjicitur a Samuele Krook Smoland. In audit Gustav. majore horis ante merid. consuetis die 5. Decemb. ann. 1688. Upsala: Henrici Keysers. URL: https://www.diva-portal.org/smash/get/diva2:1540374/FULLTEXT01.pdf (15 March 2024).
- La Connoissance des Temps. Ou Calendrier et Ephemerides du lever & coucher du Soleil, de la Lune, & des autres Planetes. 1679. Paris: Jean Baptiste Coignard. URL: https://gallica.bnf.fr/ark:/12148/bpt6k6506709d/f1.item (15 February 2024).
- Maskelyne, Nevil. 1764. On the Equation of Time and the True Manner of Computing it. In: Philosophical Transactions, 54, 336–347. DOI 10.1098/rstl.1764.0054.
- Maskelyne, Nevil. 1767. The Nautical Almanac and Astronomical Ephemeris for the Year 1767. London: Printed by W. Richardson and S. Clark. (Google Books, 7 April 2020).
- McKeon, Robert. 1973. Le Fèvre. In: Dictionary of Scientific Biography, Gillispie, Charles Coulston (ed.), VIII, 131–132. New York: Charles Scribner's Sons. SBN 684-10119-X.
- Nitschelm, Christian. 2007. Monnier, Pierre-Charles le. In: Biographical Encyclopedia of Astronomers. Hockey, Thomas, Virginia Trimble and Thomas R. Williams (eds.), 798–799. New York: Springer. ISBN 9780387310220. DOI 10.1007/978-0-387-30400-7_972.
- Piirimäe, Helmut. 1999. Pärnu - the university town. Pärnu: Tartu Ülikooli Kirjastus. ISBN 9789985400852.
- Przypkowski, Tadeusz. 1969. Hecker, Johannes. In: Neue Deutsche Biographie 8, 182 [online version]. URL: https://www.deutsche-biographie.de/pnd128706112.html#ndbcontent.
- Quill, H. 1963. John Harrison, Copley Medallist, and the £20,000 Longitude Prize. In: Notes and Records of the Royal Society, 18, 2, 146–160. DOI 10.1098/rsnr.1963.0018.
- Quill, H. 1976. John Harrison Copley Medallist and the £20,000 Longitude Prize. Reprint, first published in Notes and Records of the Royal Society, 18, 2, December 1963. Wadhurst: The Antiquarian Horological Society.
- Steele, John M. 2012. Ancient Astronomical Observations and the Study of the Moon's Motion (1691–1757). New York: Springer. ISBN 978-1-4614-2148-1.
- Thun, Johan. 1712. Dissertatio de inæqualitate dierum qvam ... præside viro amplissimo mag. Petro Elvio ... Publico examini sistit Johannes Thun O. Gothus. In aud. Gust. maj. ad d. 24 Maji anno 1712. horis ante meridiem solitis. Strengnesiae: Widmanniana. URL: https://www.diva-portal.org/smash/get/diva2:1544853/FULLTEXT01.pdf (15 February 2024).
- Vince, Samuel. 1793. A Plan of a Course of Lectures on the Principles of Natural Philosophy. Cambridge: J. Archdeacon. (Google Books, 30 May 2020).
- Vince, Samuel. 1797. A Complete System of Astronomy. Volume I. Cambridge: J. Burges. (Google Books, 30 May 2020).
- Vince, Samuel. 1800. A Treatise on Plane and Spherical Trigonometry; with an Introduction, Explaining the Nature and use of Logarithms. Cambridge: J. Burgess. (Google Books, 30 May 2020).
- Vince, Samuel. 1810. The Elements of Astronomy: Designed for the use of Students in the University, 3rd edition. Cambridge: J. Smith. URL: https://archive.org/details/elementsofastron00vincrich (15 February 2024).
- Vince, Samuel. 1823. A Complete System of Astronomy. Volume III, 2nd Edition. London: J. Mawman; Baldwin, Cradock & Joy; W. Baynes and Son; T. Hamilton; Simpkin and Marshall. (Google Books, 30 May 2020).
- Widmalm, Sven. Zacharias Nordmark. In: Svenskt biografiskt lexikon, [online source]. URL: https://sok.riksarkivet.se/sbl/Presentation.aspx?id=8314 (13 November 2022).

11

Tables of solar altitude

Wenn man demnach die Tageslänge weiß, so läßt sich die Höhe der Sonne für jeden Augenblick durch die bloße Addition dreyer Logarithmen finden.

Johann Heinrich Lambert, 1765,
A Beyträge zum Gebrauche der Mathematik und deren Anwendung, §. 80., p.418

11.1 Johann Heinrich Lambert

Johann Heinrich Lambert (1728–1777), born in Alsace, was a self-taught mathematician, yet his achievements place him alongside the greatest mathematicians of his time. He was active in various cities in Europe, partly because of the situation at the time. His contributions to cartography and probability theory sre very important. He was also among the first to systematically study hyperbolic functions.[1]

Of interest are his continued fractions, for example:[2]

$$\tan x = \cfrac{1}{\cfrac{1}{x} - \cfrac{1}{\cfrac{3}{x} - \cfrac{1}{\cfrac{5}{x} - \cfrac{1}{\cfrac{7}{x} - \cfrac{1}{\cfrac{9}{x} + \cdots}}}}}$$

In 1771 he published the following series, valid for |x|<1:[3]

$$\frac{x}{1-x} + \frac{x^2}{1-x^2} + \frac{x^3}{1-x^3} + \cdots = x + 2x^2 + 2x^3 + 3x^4 + 2x^5 + 4x^6 + 2x^7 + 4x^8 + 3x^9 + \cdots$$

or:

$$\frac{x}{1-x} + \frac{2x^2}{1-x^2} + \frac{3x^3}{1-x^3} + \frac{4x^3}{1-x^4} + \cdots$$

$$= x + 3x^2 + 4x^3 + 7x^4 + 6x^5 + 12x^6 + 8x^7 + 15x^8 + 13x^9 + \cdots$$

He is the subject of various anecdotes. During an opera, for instance, he was reportedly so lost in his thoughts that when the lights were switched off at the end of the performance, he had to find his way out in the dark. Later, he said that during the performance, he had spent the entire time thinking about how to calculate the refraction of the light beams.[4] In the *Beyträge zum Gebrauche der Mathematik und deren Anwendung* of 1765, in the third part of Anmerkungen und Zusätze zur Trigonometrie, he published a method for calculating the altitude of the Sun and the hour angle, for the purpose of setting the exact clock time, in a similar way to that published by Bode a few years later.[5]

1 He published two important treatises on hyperbolic functions. First in 1761, Abhandlung über einige bemerkenswerte Eigenschaften transzedenter kreisförmiger und logaritmischer Grössen (Mémoire sur quelques propriétés des quantités transcedantes circulaires et logarithmiques). The second treatise he published in 1770. Hyperbolic functions were discovered by Vincentio Riccati before Lambert, but Lambert is credited for their systematic treatment.

2 Löwenhaupt, 1943, pp.30–31.

3 Lambert, 1771, Anlage zur Architectonic, Band 2, p.507. Löwenhaupt, 1943, p.32.

4 Löwenhaupt, 1943, p.37.

5 Lambert, 1765, in particular p.418 §.80 to p.423 §.85. Bode, 1776a and 1776b.

LAMBERT's derivation is based on the spherical triangle $\triangle PZS$ between the north celestial pole, the zenith, and the Sun (see Appendix D, Figure D.9). The angles in this spherical triangle run clockwise, the reverse of most of the examples in this book, but this does not affect the result. In addition, LAMBERT uses capital letters for sides or arcs and lower case for angles (Figure 11.1, right image). In the derivations below, the notation is consistent with other parts of this book, with sides in lower case and angles in upper case, where:

$H_\odot, H_{\odot\uparrow\downarrow}$	hour angle, hour angle of sunrise/sunset
b	distance between the pole and the Sun, $\widehat{PS} = 90° - \delta_\odot$
$a_\odot$	altitude of the Sun (°) = $90° - z$
z	zenith angle or complement of the altitude of the Sun
ϕ	distance between the pole and the zenith, $\widehat{PZ} = 90° - \varphi$
q	parallactic angle
β	complement of azimuth, $180° - A_\odot$ if azimuth is measured from South and $360° - A_\odot$ if azimuth is measured from North

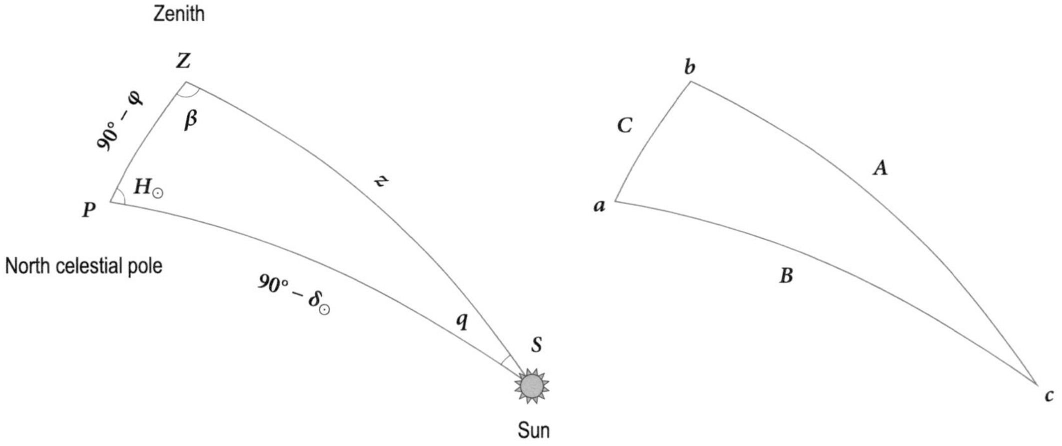

Figure 11.1 Spherical triangle with vertices at the North celestial pole, the zenith and the Sun (left) and same triangle with LAMBERT'S notation (right). After Lambert, 1765, Trigonometria, Pract. Mathem. Tab IV. Fig. X. See also explanations in the text and Appendix D.

For a spherical triangle $\triangle PZS$, the cosine law for sides applies:[6]

[11.1] $$\cos z = \cos b \cos \phi + \sin b \sin \phi \cos H_\odot$$

Taking into account that $\cos H_\odot = 1 - 2\sin^2 \frac{H_\odot}{2}$

we find that: $$\cos z = \cos b \cos \phi + \sin b \sin \phi \left(1 - 2\sin^2 \frac{H_\odot}{2}\right)$$

6 For derivation see Lambert, 1765, pp.416–417. For cosine law see also Appendix D, Section D.2.2. or D.2.3.

Therefore:

$$\frac{\cos z}{2\sin b\sin\phi} = \frac{\cos b\cos\phi}{2\sin b\sin\phi} + \frac{\sin b\sin\phi}{2\sin b\sin\phi} - \sin^2\frac{H_\odot}{2}$$

Given the addition theorem $\cos(b-\phi) = \cos b\cos\phi + \sin b\sin\phi$

we can rewrite as: $\dfrac{\cos z}{2\sin b\sin\phi} = \dfrac{\cos(b-\phi)}{2\sin b\sin\phi} - \sin^2\dfrac{H_\odot}{2}$

In the above equation, substitute $\dfrac{\cos(b-\phi)}{2\sin b\sin\phi} = \sin^2\dfrac{H_{\odot\uparrow\downarrow}}{2}$

so that: $\dfrac{\cos z}{2\sin b\sin\phi} = \sin^2\dfrac{H_{\odot\uparrow\downarrow}}{2} - \sin^2\dfrac{H_\odot}{2} = \left(\sin\dfrac{H_{\odot\uparrow\downarrow}}{2} + \sin\dfrac{H_\odot}{2}\right)\left(\sin\dfrac{H_{\odot\uparrow\downarrow}}{2} - \sin\dfrac{H_\odot}{2}\right)$

By rearranging the above equation LAMBERT got the expression for the zenith angle, from which he derived the altitude:[7]

[11.2] $$\cos z = 2\sin b\sin\phi\sin\frac{(H_{\odot\uparrow\downarrow}+H_\odot)}{2}\sin\frac{(H_{\odot\uparrow\downarrow}-H_\odot)}{2}$$

Meaning of the angle $H_{\odot\uparrow\downarrow}$:

In the equation $\dfrac{\cos z}{2\sin b\sin\phi} = \sin^2\dfrac{H_{\odot\uparrow\downarrow}}{2} - \sin^2\dfrac{H_\odot}{2}$

insert $z = 90°$ or $\cos z = 0°$ to get: $\sin^2\dfrac{H_{\odot\uparrow\downarrow}}{2} = \sin^2\dfrac{H_\odot}{2}$

For $z = 90°$, or when the Sun is on the horizon, then $H_{\odot\uparrow\downarrow} = H_\odot$. The hour angles of sunrise and sunset have the same absolute value but the opposite sign.

The terms $\dfrac{(H_{\odot\uparrow\downarrow}+H_\odot)}{2}$ and $\dfrac{(H_{\odot\uparrow\downarrow}-H_\odot)}{2}$

correspond to the arcs that define the position of the Sun from sunrise and sunset, respectively, and together are the angle between sunrise and sunset, or the length of the day in degrees.

The method is not applicable beyond the polar circle when the Sun does not set. In that case:

[11.3] $$\frac{\cos(b-\phi)}{2\sin b\sin\phi} > 1$$

7 Lambert, 1765, §.76, p.417.

LAMBERT gives the following procedure for such a case:[8]

$$\frac{\cos(b-\phi)}{2\sin b\sin\phi} = \tan^2\psi \qquad \text{and} \qquad \sin^2\frac{H_\odot}{2} = \tan^2\xi$$

Similar to the previous example, we get:

$$\frac{\cos z}{2\sin b\sin\phi} = \tan^2\psi - \tan^2\xi = (\tan\psi + \tan\xi)(\tan\psi - \tan\xi)$$
$$= \frac{\sin(\psi+\xi)\sin(\psi-\xi)}{\cos^2\psi\cos^2\xi}$$

Therefore:

[11.4] $$\cos z = \frac{2\sin b\sin\phi}{\cos^2\psi\cos^2\xi}\sin(\psi+\xi)\sin(\psi-\xi)$$

He derived the expression for the hour angle similarly. For a given triangle $\triangle PZS$, the cosine law for angles applies:[9]

[11.5] $$\cos H_\odot = \sin\beta\sin q\cos z - \cos\beta\cos q$$

Taking into account $\cos z = -1 + 2\cos^2\frac{z}{2}$

we find that: $\cos H_\odot = \sin\beta\sin q\left(-1 + 2\cos^2\frac{z}{2}\right) - \cos\beta\cos q$

Therefore: $\cos H_\odot = 2\sin\beta\sin q\cos^2\frac{z}{2} - \sin\beta\sin q - \cos\beta\cos q$

By applying the addition theorem $\cos(\beta - q) = \cos\beta\cos q + \sin\beta\sin q$

follows: $\cos H_\odot = 2\sin\beta\sin q\cos^2\frac{z}{2} - \cos(\beta - q)$

we rearrange the equation

$$\frac{\cos H_\odot}{2\sin\beta\sin q} = \cos^2\frac{z}{2} - \frac{\cos(\beta-q)}{2\sin\beta\sin q}$$

In the above equation, substitute

$$\frac{\cos(\beta-q)}{2\sin\beta\sin q} = \cos^2\frac{H_{\odot\uparrow\downarrow}}{2}$$

8 Lambert, 1765, §.81, p.419.
9 Lambert, 1765, pp.420–421.

We find that:[10]

$$\frac{\cos H_\odot}{2\sin\beta\sin q} = \cos^2\frac{z}{2} - \cos^2\frac{H_{\odot\uparrow\downarrow}}{2} = \left(\cos\frac{z}{2} + \cos\frac{H_{\odot\uparrow\downarrow}}{2}\right)\left(\cos\frac{z}{2} - \cos\frac{H_{\odot\uparrow\downarrow}}{2}\right)$$

By rearranging the above equation the expression for the hour angle follows:

[11.6] $$\cos H_\odot = 2\sin\beta\sin q\sin\frac{(z + H_{\odot\uparrow\downarrow})}{2}\sin\frac{(z - H_{\odot\uparrow\downarrow})}{2}$$

Here again LAMBERT gives a special procedure for latitudes above the polar circle:[11]

$$\frac{\cos(\beta - q)}{2\sin\beta\sin q} = \tan^2\psi \qquad \text{and} \qquad \cos^2\frac{z}{2} = \tan^2\xi$$

Similar to the previous example, we get:

$$\frac{\cos H_\odot}{2\sin\beta\sin q} = \tan^2\xi - \tan^2\psi = (\tan\xi + \tan\psi)(\tan\xi - \tan\psi)$$
$$= \frac{\sin(\psi + \xi)\sin(\psi - \xi)}{\cos^2\psi\cos^2\xi}$$

Therefore:

[11.7] $$\cos H_\odot = \frac{2\sin\beta\sin q}{\cos^2\psi\cos^2\xi}\sin(\xi + \psi)\sin(\xi - \psi)$$

10 There are some errors in this expression in LAMBERT's article, probably errors in the typesetting, which are corrected here.

11 Lambert, 1765, §.84., p.422.

11.2 Johann Elert Bode

The German astronomer JOHANN ELERT BODE (1747–1826), together with LAMBERT, conceived the *Berliner Astronomisches Jahrbuch*, an annual astronomical journal, in 1774, which he edited for 51 years, so he is credited with the publication of this publication in a similar way to MASKELYNE with the *Nautical Almanac*. The journal is one of the oldest astronomical publications.[12] In 1768 he became head of the Berlin Observatory, where he remained for almost 40 years. He is famous for the Titius-Bode law and for determining the orbit of the planet Uranus. In 1801 he published *Uranographia*, in which he published 20 star charts and catalogued 17,240 stars and nebulae. He supported the theories of LAMBERT, KANT and HERSCHEL, and his correspondence with GAUSS and BESSEL survives.[13] BODE is credited with the publication of BESSEL's *Fundamenta astronomiae*, and he was also instrumental in BESSEL's preparation of star charts. In the *Astronomisches Jahrbuch* for 1778, he published a procedure for determining the altitude of the Sun and the parallactic angle for use with tables of the Sun's altitude, which were useful for setting the exact time of clocks.

Examples of calculations based on data from the *Astronomisches Jahrbuch* for 1776 are given by FRANZ SCHAFGOTSCH in his 1788 treatise *Abhandlung über die Berechnung der Ephemeriden*.[14]

BODE derived the expressions for the Sun's altitude and parallactic angle from the two right-angled spherical triangles $\triangle eox$ and $\triangle pSx$.[15] For the triangle $\triangle eox$ (Figure 11.2):

$$\widehat{eo} = H_{\odot} = \overbrace{6 \text{ hours}}^{=90^\circ} - \widehat{oE}$$

$H_{\odot}$ hour angle

The angle ϕ is the complement of latitude, or in older literature, the altitude of the equator.[16] At the latitude of Berlin:

$$\phi = 90^\circ - \varphi = 37^\circ 27' 30''$$

12 The contents of the *Berliner Astronomisches Jahrbuch*, although no longer under its original name, are still published as two separate publications.

13 Schwemin, 2014 and 2017. BODE's correspondence is mainly concerned with descriptions of novelties, ephemerides and similar topics, and only a small part is mathematical. Complex astronomical calculations, or so the sources say, were not BODE's strong point.

14 Von Schaffgotsch, 1788, p.1. FRANZ VON SCHAFFGOTSCH (1743–1809) was a natural scientist, mathematician and astronomer. He also published astronomical papers in the *Berliner Astronomisches Jahrbuch*.

15 Bode, 1776b, pp.183–185.

16 Äquatorhöhe in old German literature. For a better understanding see also Figure 0.13.

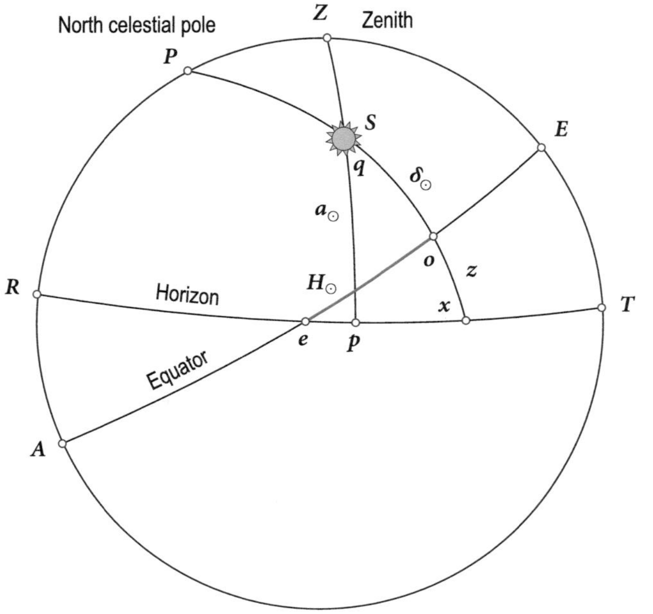

P	north celestial pole
Z	zenith
$\widehat{AE}$	Equator
$\widehat{RT}$	horizon
$\widehat{oex}$	latitude complement, $\phi = 90° - \varphi$
S	Sun
$a_{\odot}$	altitude of the Sun
$\delta_{\odot}$	declination

Figure 11.2 Spherical triangle as used by BODE. After Bode, 1776, Astronomisches Jahrbuch oder Ephemeriden für das Jahr 1778, Tab III, Fig. VIII.

From the laws of spherical trigonometry or NAPIER's rules for the triangle $\triangle eox$ (see Appendix D):

$$\tan z = \sin H_{\odot} \tan \phi$$
$$\cos x = \cos H_{\odot} \sin \phi$$

The angle x is the same for both triangles. For the second triangle $\triangle pSx$, also based on NAPIER's rules:[17]

[11.8] $$\sin x \sin(z + \delta_{\odot}) = \sin \widehat{pS} = \sin a_{\odot}$$ altitude of the Sun

[11.9] $$\tan x \cos(z + \delta_{\odot}) = \cot q$$ parallactic angle

The parallactic angle q is like $\widehat{PSZ}$ or $\widehat{pSx}$ (Figure 11.2) and has nothing to do with the parallax.[18]

17 See Appendix D, Section D.1.
18 See also Appendix D, Section D.6.3. For an explanation, see Smart, 1990, p.49.

For different declinations and time of the day, the Sun's altitude and parallactic angle are given by the expressions below:

- Less than 6 hours:

Altitude of the Sun

[11.10] $$\sin a_{\odot} = \sin x \sin \begin{pmatrix} z + \delta_{\odot} \\ z - \delta_{\odot} \end{pmatrix} \quad \begin{matrix} \text{north declination} \\ \text{south declination} \end{matrix}$$

Parallactic angle:

[11.11] $$\cot q = \tan x \cos \begin{pmatrix} z + \delta_{\odot} \\ z - \delta_{\odot} \end{pmatrix} \quad \begin{matrix} \text{north declination} \\ \text{south declination} \end{matrix}$$

- 6 hours:

Altitude of the Sun:

[11.12] $$\sin a_{\odot} = \sin x \sin \delta_{\odot}$$

Parallactic angle:

[11.13] $$\cot q = \tan x \cos \delta_{\odot}$$

- More than 6 hours, solution is a complement to 12 hours:

Altitude of the Sun:

[11.14] $$\sin a_{\odot} = \sin x \sin(\delta_{\odot} - z)$$

Parallactic angle:

[11.15] $$\cot q = \tan x \cos(\delta_{\odot} - z)$$

The parameters x and z are tabulated for the latitude of Berlin in the literature.[19]

19 See Bode, 1776a, pp.177–182.

Example 11.1 Determine the altitude of the Sun in Berlin on 16 April at 3 h 13 min and 46 seconds in the afternoon.

The latitude and declination on 16 April for Berlin are:[20]

$$\varphi = 52°32'30''$$
$$\delta_\odot = 10°11'12''$$

	z	x
3 hours 13 minutes	27° 1′47″	63° 0′58″
3 hours 14 minutes	26°54′58″	62°54′10″

The values are from BODE's tables, and the intermediate value for 46 seconds has to be interpolated.

For z and x, we interpolate the values from the resulting data:

$$z = 26°56'35''$$
$$x = 62°55'44''$$

$$z + \delta_\odot = 37°\ 7'47'' \qquad \text{(north declination)}$$

$$\sin a_\odot = \sin x \sin(z + \delta_\odot)$$
$$a_\odot = \arcsin\left(\sin x \sin(z + \delta_\odot)\right)$$

The altitude of the Sun is:

$$a_\odot = 32°30'47''$$

20 For parameters $\boldsymbol{x}$ and $\boldsymbol{z}$ see Bode, 1776a, table on p.180.

11.3 Friedrich Christoph Müller

Friedrich Christoph Müller (1751–1808) studied theology, mathematics and astronomy. He was a teacher at the Prussian Military Cadet-Corps or Cadet School (Corps des Cadets), later worked as a priest in various parishes and was also active as a cartographer. The Berlin Academy awarded him the title of Corresponding Member. His astronomical tables of 1791 and 1792 are well known and were used to set the exact time of mechanical clocks or watches. The 1791 tables contain the time for individual altitudes of the Sun for places between 47° and 54° in 1° increments. We know that he worked with Mayer's tables.[21] The 1792 tables give values for 52° N or Wittemberg. These tables include, for consecutive days of the year:

- time of day for the Sun's azimuth in 5° increments in the morning and after-noon from 5° to 125°,
- the time for each Sun altitude in 1° increments in the morning and afternoon,
- time of the rising and setting of the stars,
- the contribution of the equation of time at different times of the day for each day in seconds,
- a table showing the solar time at midnight (0:00h).

The work also mentions Lambert's[22] and Bode's[23] methods for determining the position and altitude of the Sun in the sky, respectively. Similar tables were also published later by other authors.[24]

21 Müller, 1791, p.18.
22 Müller, 1791, p.10.
23 Müller, 1792, pp.32, 34.
24 See for instance Göring, 1862, p.VII, values of the equation of time is summarised by *Berliner astronomisches Jahrbuch*. Also Hammer, 1893.

11.4 Johann Samuel Halle

Johann Samuel Halle (1727–1810) was a Prussian historian and toxicologist. He is the author of several books on history of natural sciences, was active as a translator, and was also a copper cutter. He studied theology, philosophy and mathematics in Königsberg. He came to Berlin in 1753 as an educator, and in 1760 took up a post as professor of history at the Prussian Military Cadet-Corps or Cadet School.

In *Werkstäte der heutigen Künste oder die neue Kunsthistorie*, he gave a comprehensive and detailed account of the handicraft art of his time in six books.

In *Magie oder die Zauberkräfte der Natur, so auf den Nutzen und die Belustigung angewandt worden* he described various experiments and topics in the field of natural science, in the 1796 edition (8th book). In the chapter on table and wall clocks, on page 127, we also find the equation of time or Aequationstabelle, which gives the values for the local clock time at solar noon for the six days of each month, and later in the book, Halle also describes Müller's tables of the altitude of the Sun.

Bibliography

- Bode, Johann Elert. 1776a. Herrn Bode Tafel, aus der gegebenen Zeit und Abweichung , die Höhe und den parallactifchen Winkel der Sonne, des Mondes und der Sterne zu finden; für die Polhöhe von 52°32′30″. In: Astronomisches Jahrbuch oder Ephemeriden für das Jahr 1778, 177–182. Berlin: Haude und Spenersche Buchhandlung. (Google Books, 25 September 2022).
- Bode, Johann Elert. 1776b. Erklärung und Gebrauch vorstehender Tafel. In: Astronomisches Jahrbuch oder Ephemeriden für das Jahr 1778, 183–185. Berlin: Haude und Spenersche Buchhandlung. (Google Books, 25 September 2022).
- Bullynck, Maarten. Johann Heinrich Lambert (1728–1777), Collected Works – Sämtliche Werke Online. url: http://www.kuttaka.org/~JHL/WerkeThem.html (30 October 2022).
- Clemens, Hugo. 1902: Die älteren Ephemeridenausgaben der Berliner Akademie und die Begründung des Astronomischen Jahrbuches. In: Veröffentlichungen des Königlichen Astronomischen Rechen-Instituts zu Berlin, 20, 171–196. Berlin.
- Frommert, Hartmuth. 2007. Lambert, Johann Heinrich. In: Biographical Encyclopedia of Astronomers. Hockey, Thomas, Virginia Trimble and Thomas R. Williams (eds.), 671–672. New York: Springer. isbn 9780387310220.
- Göring, Heinrich. 1862. Der Zeitmesser: Tafeln der Mittagsverbesserung für die Breitengrade 30 bis 60 und der Zeitgleichung für alle Tage; nebst einer populären Anleitung, Sonnenhöhen zu bestimmen, die Zeit zu berechnen, eine Mittagslinie zu ziehen (Paderborn: Schöningh). url: https://sammlungen.ulb.uni-muenster.de/hd/content/titleinfo/2407888?lang=en (17 February 2024).
- Günther, Siegmund. 1885. Müller, Friedrich Christoph. In: Allgemeine Deutsche Biographie 22, 530 [online version]. url: https://www.deutsche-biographie.de/pnd121786951.html#adbcontent (30 October 2022).
- Halle, Johann Samuel. 1796. Fortgesetzte Magie, oder, Die Zauberkräfte der Natur: so auf den Nutzen und die Belustigung angewandt worden. (Google Books, 19 September 2022).
- Hammer, Ernst H. 1893. Zeitbestimmung (Uhr-Kontrole) ohne Instrumente durch Benützung der Ergebnisse einer Landvermessung allgemein verständlich dargestellt. Mit Schautafeln der Sonnen-Deklination und der Zeit-Gleichung für 1893 bis 1896. Stuttgart: Metzler.
- Lambert, Johann Heinrich. 1765. Beyträge zum Gebrauche der Mathematik und deren Anwendung. Berlin: Buchladen der Realschule. url: https://www.deutsche-digitale-bibliothek.de/item/PUK353PD4ZZBSIBNKBGKDD5KGOBGOQVK (30 October 2022).
- Lambert, Johann Heinrich. 1771. Anlage zur Architectonic, oder Theorie des Ersten und des Einfachen in der philosophischen und mathematischen Erkenntniß, Band 2. Riga: Johann Friedrich Hartknoch. url: https://www.deutschestextarchiv.de/book/view/lambert_architectonic02_1771?p=1 (30 September 2023).
- Löwenhaupt, Friedrich. 1943. Johann Heinrich Lambert. Leistung und Leben. Mülhausen/Elsass: Braun.
- Kokott, Wolfgang. 2007. Bode, Johann Elert. In: Biographical Encyclopedia of Astronomers. Hockey, Thomas, Virginia Trimble and Thomas R. Williams (eds.), 140–142. New York: Springer. isbn 9780387310220.
- Kraus, Andreas. 1982. Lambert, Johann Heinrich. In: Neue Deutsche Biographie, 13, 437–439 [online version]. url: https://www.deutsche-biographie.de/pnd118568876.html#ndbcontent (30 October 2022).
- Müller, Friedrich Christoph. 1791. Friedrich Christoph Müllers Tafeln der Sonnenhöhen für ganz Deutschland und dessen westlich und östlich benachbarte Länder. Leipzig: Siegfried Lebrecht Crusius. doi 10.3931/e-rara-1194.
- Müller, Friedrich Christoph. 1792. Friedrich Christoph Müllers Gemeinnützige Astronomische Tafeln (hauptsächlich zur richtigen Stellung der Uhren) für alle Oerter Deutschlandes und der [...] : nebst einem in Kupfer gestochenen Sextanten und einer Anweisung ein solches Werkzeug zu verfertigen und damit zu beobachten. Leipzig: Siegfried Lebrecht Crusius. doi 10.3931/e-rara-1190.
- Schwemin, Friedhelm. 2006. Der Berliner Astronom; Leben und Werk von Johann Elert Bode (1747–1826). Acta Historica Astronomiae, 30. Frankfurt am Main: Harri Deutsch. isbn 978-3-944913-21-6.

- SCHWEMIN, Friedhelm. 2014. Der Briefwechsel zwischen Carl Friedrich Gauß und Johann Elert Bode. Acta Historica Astronomiae, 53. Leipzig: Akademische Verlagsanstalt. ISBN 978-3-944913-43-8.
- SCHWEMIN, Friedhelm. 2017. Der Briefwechsel zwischen Friedrich Wilhelm Bessel und Johann Elert Bode. Acta Historica Astronomiae, 61. Leipzig: Akademische Verlagsanstalt. ISBN 978-3-944913-52-0.
- SCRIBA, Christoph J. 1973. Lambert, Johann Heinrich. In: Dictionary of Scientific Biography, Gillispie, Charles Coulston (ed.),V II, 595–600. New York: Charles Scribner's Sons. SBN 684-10118-1.
- SMART, William Marshall. 1990. Textbook on Spherical Astronomy. Reprint. Cambridge: Cambridge University Press. ISBN 0 521 29180 1.
- STICKER, Bernhard. 1970. Bode, Johann Elert. In: Dictionary of Scientific Biography, Gillispie, Charles Coulston (ed.), II, 220–221. New York: Charles Scribner's Sons.
- VON BRAUNMÜHL, Anton. 1891. Christoph Schreiner als Mathematiker, Physiker und Astronom. Bamberg: Buchnersche Verlagsbuchhandlung.
  URL: https://archive.org/details/christophschein00braugoog/mode/2up (17 February 2024).
- VON SCHAFFGOTSCH, Franz. 1788. Abhandlung über die Berechnung der Ephemeriden. Dresden: Walterische Hofbuchhandlung. DOI 10.25673/49815.
  URL: https://opendata.uni-halle.de//handle/1981185920/51766 (17 February 2024).

12 Lagrange, Schulze and power series

Die mittlere Länge der Sonne = ⊙
Die mittlere Anomalie derſelben = a
Die mittlere ſtündliche Bewegung der Sonne = h
Die Zeitgleichung ſo zur Zeit T ſtimmet = x

Dieſes voraus geſetzt, ſo kann die mittlere Zeit, welche zur wahren Zeit T ſtimmet, durch T + x ausgedruckt werden, und zu dieſer mittlern Zeit wird die mittlere Länge der Sonne = ⊙ + hx ſeyn.

Johann Schulze, 1778, Ueber ein Mittel die Zeitgleichung unmittelbar zu beſtimmen, Aſtronomiſches Jahrbuch oder Ephemeriden für das Jahr 1781, Theil 2, p.139

12.1 Power series expansion

12.1.1 The Kerala - Mādhava School

The expansion of trigonometric functions into power series was already known to Indian mathematician MĀDHAVA OF SAṄGAMAGRAMMA (1340/50?–1425?), a representative of the Kerala School of Mathematics in the 14th century.[1] We know almost nothing about MĀDHAVA's life. He was born in Cochin on the coast of Kerala in south-eastern India. All his mathematical works have been lost but some of his astronomical records are preserved. His works are known from the works of other mathematicians of the Kerala School. He developed a series for the arc tangent[2] function, wrote the number π to 11 decimals, and compiled sine function tables.

MĀDHAVA knew about expression as a series as early as 1400, more than 200 years before a similar discovery in Europe. It is possible that his work on series was known to the Jesuits who worked in India before NEWTON and LEIBNIZ developed the foundation of calculus. MĀDHAVA used the series to calculate the length of the arc of a circle; his construction for calculating the length of the arc $s = R$, or the angle $\measuredangle EP$ can be seen in Figure 12.1. If we divide $EB = t$ into n equal parts and construct triangles as shown in Figure 12.1, the values of k or the hypotenuse of the triangles are given by the expression:

$$k_i^2 = R^2 + \left(\frac{it}{n}\right)^2$$

The expression for the approximation of the length of the arc s is:[3]

$$s = \measuredangle EP \approx \left(\frac{t}{n}\right)\left[\left(\frac{R^2}{k_0^2}\right) + \left(\frac{R^2}{k_1^2}\right) + \left(\frac{R^2}{k_2^2}\right) + \cdots + \left(\frac{R^2}{k_n^2}\right)\right] \tag{12.1}$$

1 For the achievements of the Kerala School mathematicians see Sarma, 1972, A History of the Kerala School of Hindu Astronomy. Also Gheverghese, 2005, Kerala Mathematics, History and its Possible Transmission to Europe. Van Brummelen, 2009, The Mathematics of the Heavens and the Earth, pp.113–121. Sarma et al., 2008, Gaṇita-Yukti-Bhāṣā, vol. I, chapters 7.5 and 7.6, pp.221–234. For the work of LEIBNIZ, GREGORY and NĪLAKAṆTHĀ, see Roy, 1990, The Discovery of the Series Formula for π by Leibniz, Gregory and Nilakantha.

2 Inverse tangent function, atan or arctan.

3 For the process leading to the Equation 12.1, see Sarma et al., 2008, volume 1, chapter 6 - in particular section 6.3.1.

After rearrangement we get a power series:[4]

[12.2] $$s = \frac{R \cdot \mathrm{Sin}\,\alpha}{\mathrm{Cos}\,\alpha} - \frac{R \cdot \mathrm{Sin}^3\,\alpha}{3 \cdot \mathrm{Cos}^3\,\alpha} + \frac{R \cdot \mathrm{Sin}^5\,\alpha}{5 \cdot \mathrm{Cos}^5\,\alpha} - \frac{R \cdot \mathrm{Sin}^7\,\alpha}{7 \cdot \mathrm{Cos}^7\,\alpha} + \cdots$$

α	angle, see Figure 12.1
R	radius
Sin	Hindu sine
Cos	Hindu cosine
$s = R\alpha$	arc length in radians

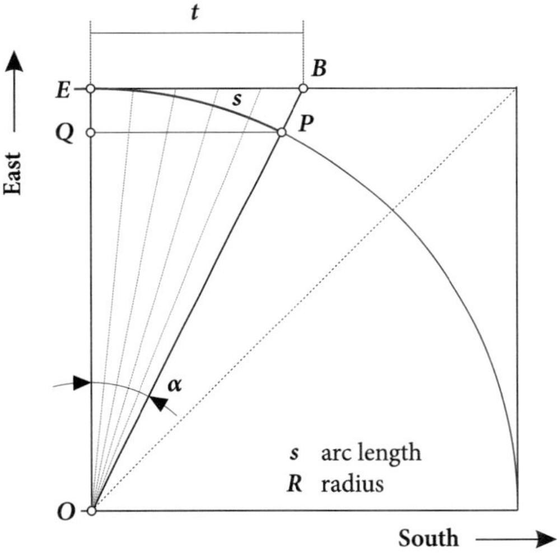

Figure 12.1 MĀDHAVA'S geometric construction for calculating the length of the arc or the angle. Summarised from: Sarma et al., 2008, volume I, p.199, Figure 6.6.

If the Hindu sine, which is the product of the trigonometric function and the radius (see chapter 2), is replaced by the familiar sine and cosine, we get:

$$R\alpha = \frac{R \cdot \sin\alpha}{\cos\alpha} - \frac{R \cdot \sin^3\alpha}{3 \cdot \cos^3\alpha} + \frac{R \cdot \sin^5\alpha}{5 \cdot \cos^5\alpha} - \frac{R \cdot \sin^7\alpha}{7 \cdot \cos^7\alpha} + \cdots$$

If $\frac{\sin\alpha}{\cos\alpha}$ is replaced by $\tan\alpha$ and R is cancelled out, we get:

[12.3] $$\alpha = \tan\alpha - \frac{\tan^3\alpha}{3} + \frac{\tan^5\alpha}{5} - \frac{\tan^7\alpha}{7} + \cdots$$

or: $$\arctan\alpha = \alpha - \frac{\alpha^3}{3} + \frac{\alpha^5}{5} - \frac{\alpha^7}{7} + \cdots$$

4 Sarma et al., 2008, volume 1, chapter 6.6, translation p.68, commentary pp.198–200, equation 6.75, p.198. See also Van Brummelen, 2009, pp.113–121.

MĀDHAVA's legacy also includes precise tables of the sine function for arcs, multiples of 225′, up to an angle of 90°.[5] Basically, this is the half value of the chord function or the modern sine, multiplied by the radius. It is very likely that MĀDHAVA arrived at the exact values by expanding the sine and cosine functions into a power series. In modern notation, the series expansions of the sin, cos, and arctan functions are written as:

[12.4] $$\sin x = x - \frac{x^3}{3!} + \frac{x^5}{5!} - \frac{x^7}{7!} + \cdots = \sum_{n=0}^{\infty} \frac{(-1)^n x^{2n+1}}{(2n+1)!}$$

$$\cos x = 1 - \frac{x^2}{2!} + \frac{x^4}{4!} - \frac{x^6}{6!} + \cdots = \sum_{n=0}^{\infty} \frac{(-1)^n x^{2n}}{(2n)!}$$

$$\arctan x = x - \frac{x^3}{3} + \frac{x^5}{5} - \frac{x^7}{7} + \cdots = \sum_{n=0}^{\infty} \frac{(-1)^n x^{2n+1}}{2n+1}$$

12.1.2 Gregory and Taylor

In Europe, a series of discoveries paved the way to power series thanks to the Scottish mathematician JAMES GREGORY (1638–1675).[6] GREGORY studied theology, but initially worked on optics and his book *Optica promota* also contains a description of the telescope design, now known as Gregorian. He studied power series in Padua, Italy, in 1664, where he collaborated with ANGELI. In the summer of 1668 he was probably already familiar with series for functions sin, cos and arctan. Before leaving Padua, he also wrote a book on the development of calculus, *Geometriae pars universalis*, which was the first systematic book on the topic. He worked on calculus at the same time as NEWTON, although independently. His work on differential geometry, drew him into a dispute with HUYGENS. Later he worked in St. Andrews, which he apparently left due to unfavourable circumstances, and went to Edinburgh where he died at a young age.

In the 1667 treatise *Vera circuli et hyperbolae quadratura*, GREGORY mentioned the concept of a convergent series, although his notion differs from the modern definition. He used the concept of a convergent series to describe a sequence of two values that correspond to the following conditions:[7]

$$a_n = f(a_{n-1}, b_{n-1}) \qquad \text{and} \qquad b_n = g(a_{n-1}, b_{n-1}), \qquad n > 1$$

$$|a_n - b_n| < |a_{n-1} - b_{n-1}|$$

5 Ramasubramanian, Sriram, 2011, Tantrasaṅgraha of Nīlakaṇṭha Somayājī, p.64.

6 Malet, 1993, James Gregorie on tangents and the "Taylor" rule for series expansions, Archive for History of Exact Sciences, 46, 2, pp.97–137.

7 Ferraro, 2008, The Rise and Development of the Theory of Series up to the Early 1820s, p.21.

He used the definition in the study of the quadrature of a circle by means of right n-polygons inscribed and surrounding a conic section. In a letter to COLLINS on November 23, 1670, GREGORY also quoted the interpolation formula, which is today named after him and NEWTON.[8] The expansion into a series was finally completed by BROOK TAYLOR.[9] TAYLOR was born into a wealthy family who could afford a private teacher, and afterwards he left to study at Cambridge. He corresponded with MACHIN[10] and KEILL, and his first important mathematical publication was in 1714. TAYLOR was the first to publish the series in 1715, in *Methodus incrementorum directa & inversa*.[11] He had already arrived at the theorem some years before then, as is evident from his letter to JOHN MACHIN on 6 July 1712.[12]

TAYLOR's father instilled in him a love of music and painting, which were later also fields of his mathematical involvement; he studied both vibrating strings and perspective from a mathematical point of view. The importance of his work on series for the development of calculus was only recognised by LAGRANGE in 1772, forty years after TAYLOR's death. A similar approach as used by TAYLOR was surely also known by NEWTON, BERNOULLI, LEIBNIZ and DE MOIVRE, and possibly by GREGORY.[13] The name TAYLOR's series appears to have been first used by SIMON L'HUILIER in a publication from 1786.[14]

The TAYLOR series of the function $f(x)$ around $x = a$ is:

[12.5] $$f(x) = f(a) + (x-a)f'(a) + \frac{(x-a)^2}{2!}f''(a) + \frac{(x-a)^3}{3!}f'''(a) + \cdots + \frac{(x-a)^n}{n!}f^n(a) + \cdots$$

where f^n is the n-th derivative of f with respect to x. In essence, MĀDHAVA's notation is the same as the MACLAURIN[15] series, which is a special case of the TAYLOR series with $a = 0$ and is written in contemporary notation as:

[12.6] $$f(x) = f(0) + xf'(0) + \frac{x^2}{2!}f''(0) + \frac{x^3}{3!}f'''(0) + \cdots + \frac{x^n}{n!}f^n(0) + \cdots$$

MACLAURIN published an explanation of the series by means of geometry in 1742 in a treatise on fluxions.

8 Turnbull, 1939, James Gregory Tercentary Memorial Volume. For the formula before Gregory, see also Goldstine, 1977, A History of Numerical Analysis from the 16th Century through to the 19th Century.

9 BROOK TAYLOR (1685–1731), English mathematician, known for his work in the field of power series and calculus.

10 JOHN MACHIN (1680–1751), English mathematician and astronomer, best known for the formula for calculating the number π.

11 Feigenbaum, 1985, Brook Taylor and the method of increments, Archive for History of the Exact Sciences 34, pp.1–140.

12 Bateman, 1905, The correspondence of Brook Taylor, Bibliotheca Mathematica, 3, 7, pp.367–372. He wrote a treatise some years earlier as a result of a disagreement with JOHANN BERNOULLI.

13 NEWTON wrote his form of TAYLOR series in *De quadratura curvarum* from 1692. Newton, Isaac. The Mathematical Papers of Isaac Newton, D. T. Whiteside (ed.), vol. 7, pp.96–98, also Feigenbaum, 1985. Also Enerström, 1911–1912, Zur Vorgeschichte der Entdeckung des Taylorschen Lehrsatzes, Bibliotheca Mathematica, 3, 12, pp.333–336.

14 SIMON ANTOINE JEAN L'HUILIER (1750–1840), Swiss mathematician. L'Huilier, 1786, Exposition élémentaire des principes des calculs supérieurs: qui a remporté le prix proposé par l'Académie royale des sciences et belles-lettres pour l'année 1786.

15 COLIN MACLAURIN (1698–1746), Scottish mathematician. He is known for contributions to the theory of series, elliptic integrals, and the gravity of ellipsoids. Maclaurin, 1742, A Treatise of Fluxions in Two Books. For an explanation of the series, see Proposition 20. See also Cantor, 1901, Vorlesungen über Geschichte der Mathematik, 3. Band, pp.678–686.

12.1.3 Lagrange's contribution

JOSEPH-LOUIS LAGRANGE's works show EULER's influence. As an 18-year-old, LAGRANGE impressed EULER with a letter he sent to him in 1754, in which he described his work before he published his first treatise. The letter referred to EULER's 1744 treatise, *Methodus inveniendi lineas curvas maximi minimive proprietate gaudentes*, which contained procedures for solving problems of calculus of variations.[16] In Berlin, where LAGRANGE arrived in 1766, he was quickly taken under the wings of LAMBERT and JOHANN CASTILLON.[17] LAGRANGE succeeded EULER as the head of the Berlin Academy in the same year.

In the field of equation-solving, LAGRANGE is credited with work on numerical and algebraic solutions for higher degree polynomials, where he continued the work, or rather studied the contributions of TSCHIRNHAUS, EULER and BÉZOUT, which were particularly important for his work, and he was also well-acquainted with the works of VIÈTE, DESCARTES and FERRARI.[18]

An important milestone that enabled a simple formal mathematical notation of trigonometric series expansion is represented by the two works published by LAGRANGE in 1770 and 1771. On 18 January and 5 April 1770, LAGRANGE gave the lecture *Nouvelle méthode pour résoudre les équations littérales par le moyen des séries* to the Berlin Academy; the procedure described in the lectures was probably based on an interview with LAMBERT.[19] On 1 November 1770, LAGRANGE sent a letter to the Academy describing the series as a solution to KEPLER's problem.[20] He had already mentioned the procedure in a letter to D'ALEMBERT on 26 August 1770. The first lecture or publication provides procedures for solving polynomial equations, while the other relates to KEPLER's work and provides the procedure for calculating the mean anomaly when the true anomaly is known, where KEPLER's equation must be solved.[21]

LAGRANGE's work provides a relatively simple solution to KEPLER's equation; it follows from his work that the solution of the equation which has the following general form where z is independent of x:[22]

16 Fraser, 1985, J. L. Lagrange's Calculus of Variations, p.155. For biographical information on Lagrange, see Itard, 1973, Lagrange, Dictionary of Scientific Biography, pp.559–573.

17 JOHANN CASTILLON (1704–1791), an Italian mathematician, is famous for his work on cardioids.

18 TSCHIRNHAUS published his work in this field in 1683, Hamburg, 1976, p.20. Lagrange, Œuvres III, 222-237. EHRENFRIED WALTER VON TSCHIRNHAUS (1651–1708) German mathematician, also known as the father of European porcelain, also worked on experiments with concentrated solar radiation. ÉTIENNE BÉZOUT (1730–1783), a French mathematician known for his work on polynomial equations. LODOVICO FERRARI (1522–1565), a mathematics lecturer in Milan, famous for his algebraic solution of the biquadratic equation. A biquadratic equation is a fourth-order equation involving only even terms, for example $y = a + bx^2 + cx^4$.

19 Itard, 1973, p.566.

20 Itard, 1973, p.567. Lagrange, Œuvres complètes, III, 113–138.

21 Lagrange, 1770, Nouvelle méthode pour résoudre les équations littérales pour le moyen des séries. Mémoires de l'Académie Royale des Sciences (Berlin), 24 also Œuvres, III, 5–73. Lagrange, 1771, Sur le Proléme de Kepler. Mémoires de l'Académie Royale des Sciences (Berlin), 25, pp.204–233 also Oeouvres Completes, III, 113–138.

22 KEPLER's equation takes the form $E = M + e\sin E$. See also Lagrange, 1869, Œuvres, 3, Mémoires de Berlin (VII à XXII), vol. 1–2. Reprint, 2004, Paris, Jacques Gabay, pp.25–26. Bottazzini, 1989, Lagrange et le problème de Kepler. Revue d'histoire des sciences, 42, 1/2, pp.27–42. For an example of calculating the true anomaly from an eccentric anomaly, see Klügel, 1789, Formel aus der anomalia excentri die mittlere anomalie zu finden, Astronomisches Jahrbuch für das Jahr 1789, pp.191–193.

[12.7] $y = z + x\phi(y)$

can be written as series, which converges for sufficiently small values of x.[23]

Equation 12.7 is written as a MACLAURIN series (Equation 12.6):

[12.8] $$f(y) = f(0) + x\left(\frac{\partial f(z)}{\partial z}\right)_0 + \frac{x^2}{2!}\left(\frac{\partial^2 f(z)}{\partial z^2}\right)_0 + \frac{x^3}{3!}\left(\frac{\partial^3 f(z)}{\partial z^3}\right)_0 + \cdots$$

For function (y):

[12.9] $$f(y) = f(0) + x\left(\frac{\partial f(z)}{\partial z}\right)_0 + \frac{x^2}{2!}\left(\frac{\partial^2 f(z)}{\partial z^2}\right)_0 + \frac{x^3}{3!}\left(\frac{\partial^3 f(z)}{\partial z^3}\right)_0 + \cdots$$

Differentiating Equation 12.7 gives the following expressions:

[12.10] $$\frac{\partial y}{\partial z} = 1 + x\frac{\partial\phi(y)}{\partial y}\frac{\partial y}{\partial z} \qquad \frac{\partial y}{\partial x} = \phi(y) + x\frac{\partial\phi(y)}{\partial y}\frac{\partial y}{\partial x}$$

From Equations 12.10:

$$\frac{\partial y}{\partial x}\left(1 - x\frac{\partial\phi(y)}{\partial y}\right) = \phi(y) \qquad \frac{\partial y}{\partial z}\left(1 - x\frac{\partial\phi(y)}{\partial y}\right) = 1$$

or

[12.11] $$\frac{\partial y}{\partial x} = \phi(y)\frac{\partial y}{\partial z}$$

Substituting $\frac{\partial y}{\partial x}$ for Equation 12.11 in Equations 12.9, we get:

$$\frac{\partial f(y)}{\partial x} = \frac{\partial f(y)}{\partial y}\phi(y)\frac{\partial y}{\partial z}$$

or

[12.12] $$\frac{\partial f(y)}{\partial x} = \phi(y)\frac{\partial f(y)}{\partial z}$$

For $x = 0$, $y = z$ or $(y) = (z)$, the zero of the first derivative is:

$$\left(\frac{\partial f(y)}{\partial x}\right)_0 = \phi(z)\frac{\partial f(z)}{\partial z}$$

23 The following procedure is summarised from Ottley, 1838, A Treatise on Differential Calculus, pp.108–110. See also Williamson, 1899, An Elementary Treatise on Differential Calculus, Ball, 1908, A Treatise on Spherical Astronomy, pp.155–156, Smart, 1960, Celestial Mechanics, pp.27–28, or Schaub, 1950, Vorlesungen über Sphärische Astronomie, pp.35–36.

If we introduce a new function $F_1(y)$, Equation 12.12 can be written as:

$$\frac{\partial f(y)}{\partial x} = \phi(y)\frac{\partial f(y)}{\partial z} = \frac{\partial F_1(y)}{\partial z} \tag{12.13}$$

From the above equation, we calculate the second derivative and, taking Equation 12.12 into account, we get:

$$\frac{\partial^2 f(y)}{\partial x^2} = \frac{\partial^2 F_1(y)}{\partial x \partial z} = \frac{\partial}{\partial z}\frac{\partial F_1(y)}{\partial x} = \frac{\partial}{\partial z}\phi(y)\frac{\partial F_1(y)}{\partial z}$$

Or by taking into account Equation 12.13:

$$\frac{\partial^2 f(y)}{\partial x^2} = \frac{\partial}{\partial z}\left((\phi(y))^2\frac{\partial f(y)}{\partial z}\right) \tag{12.14}$$

Equation 12.14 implies the zero of the second derivative:

$$\left(\frac{\partial^2 f(y)}{\partial x^2}\right)_0 = \frac{\partial}{\partial z}\left((\phi(z))^2\frac{\partial f(z)}{\partial z}\right)$$

When calculating the third derivative, we introduce the function $F_2(y)$:

$$(\phi(y))^2\frac{\partial f(y)}{\partial z} = \frac{\partial F_2(y)}{\partial z}$$

Or from Equation 12.14:

$$\frac{\partial^2 f(y)}{\partial x^2} = \frac{\partial^2 F_2(y)}{\partial z^2}$$

Then calculate the third derivative:

$$\frac{\partial^3 f(y)}{\partial x^3} = \frac{\partial^3 F_2(y)}{\partial x \partial z^2} = \frac{\partial^2}{\partial z^2}\frac{\partial F_2(y)}{\partial x} = \frac{\partial^2}{\partial z^2}\left(\phi(y)\frac{\partial F_2(y)}{\partial z}\right) = \frac{\partial^2}{\partial z^2}\left((\phi(y))^3\frac{\partial f(y)}{\partial z}\right) \tag{12.15}$$

For the zero of the third derivative, we get:

$$\left(\frac{\partial^3 f(y)}{\partial x^3}\right)_0 = \frac{\partial^2}{\partial z^2}\left((\phi(z))^3\frac{\partial f(z)}{\partial z}\right)$$

Similarly, we also obtain expressions for higher-order derivatives, for which the general expression applies:

$$\frac{\partial^{n-1} f(y)}{\partial x^{n-1}} = \frac{\partial^{n-2}}{\partial z^{n-2}}\left((\phi(y))^{n-1}\frac{\partial f(y)}{\partial z}\right)$$

When calculating the next element, we again introduce a new function $F_{n-1}(y)$:

$$\frac{\partial^{n-1} f(y)}{\partial x^{n-1}} = \frac{\partial^{n-1} F_{n-1}(y)}{\partial z^{n-1}}$$

For the n-th derivative, we get:

[12.16] $$\frac{\partial^n f(y)}{\partial x^n} = \frac{\partial^{n-1}}{\partial z^{n-1}} \frac{\partial F_{n-1}(y)}{\partial x} = \frac{\partial^{n-1}}{\partial z^{n-1}} \left(\phi(y) \frac{\partial F_{n-1}(y)}{\partial z} \right) = \frac{\partial^{n-1}}{\partial z^{n-1}} \left((\phi(y))^n \frac{\partial f(y)}{\partial z} \right)$$

The zero of the n-th derivative is:

$$\left(\frac{\partial^n f(y)}{\partial x^n} \right)_0 = \frac{\partial^{n-1}}{\partial z^{n-1}} \left((\phi(z))^n \frac{\partial f(z)}{\partial z} \right)$$

Plugging the expressions for the zeros of the individual derivatives into Equation 12.8 gives:

[12.17] $$f(y) = f(z) + \frac{x}{1!} \phi(z) \frac{\partial f(z)}{\partial z} + \frac{x^2}{2!} \frac{\partial}{\partial z} \left((\phi(z))^2 \frac{\partial f(z)}{\partial z} \right) + \frac{x^3}{3!} \frac{\partial^2}{\partial z^2} \left((\phi(z))^3 \frac{\partial f(z)}{\partial z} \right) + \cdots + \frac{x^n}{n!} \frac{\partial^{n-1}}{\partial z^{n-1}} \left((\phi(z))^n \frac{\partial f(z)}{\partial z} \right) + \cdots$$

or $$f(y) = f(z) + \sum_{n=1}^{\infty} \frac{x^n}{n!} \frac{\partial^{n-1}}{\partial z^{n-1}} \left((\phi(z))^n \frac{\partial f(z)}{\partial z} \right)$$

Equation 12.17 is also known as the LAGRANGE *reversion theorem* or the *expansion theorem of* LAGRANGE.[24]

With the help of a series, LAGRANGE also expressed a formula that links the ecliptic longitude and the right ascension of the Sun, and that also contributes to the equation of time due to the obliquity of the ecliptic:[25]

[12.18] $$\lambda_\odot - \alpha_\odot = \tan^2 \frac{\varepsilon}{2} \sin 2\lambda_\odot - \frac{1}{2} \tan^4 \frac{\varepsilon}{2} \sin 4\lambda_\odot + \frac{1}{3} \tan^6 \frac{\varepsilon}{2} \sin 6\lambda_\odot - \frac{1}{4} \tan^8 \frac{\varepsilon}{2} \sin 8\lambda_\odot + \frac{1}{5} \tan^{10} \frac{\varepsilon}{2} \sin 10\lambda_\odot - \cdots$$

For this expression in the historical literature, the term reduction of the ecliptic to the equator is often used. An example of the derivation for a circular eccentric orbit is given in Chapter 14 and for a KEPLER orbit in Chapter 15.

Mr Lagrange is here, and I am happy to have seen him and spoken with him. This is a quite singular young man of great talent and modesty. His character is sweet and melancholy. He appears to know no other pleasure than just being and that he is motivated more by indolence than ambition.

Clairaut to Daniel Bernoulli, 27 December 1763
(Crépel, Jouve, 2014, Lettera Mathematica, 2, 1–2, p.18)

24 For more information see Weiss, 1885, Entwickelungen zum Lagrange'schen Reversionstheorem. For the solution of KEPLER's equation by applying the LAGRANGE reversion theorem see Chapter 15.

25 Lagrange, 1776, Positionswinkel und Abweichung für alle Grade der Breite, in Formeln ausgedrückt, von Hrn. de la Grange. Sammlung Astronomischer Tafeln, pp.226–227. For power series expansion see Chapter 15, derivation of the Equation 15.7.

12.2 Equation of time in power series

12.2.1 Johann Karl Schulze

It was not long before the equation of time was addressed with the help of trigonometric series. JOHANN KARL SCHULZE (1749–1790) published a solution for the equation in 1778 using the power series. From 1793 SCHULZE assisted JOHANN BERNOULLI III (1744–1807) at the astronomical observatory in Berlin.[26] He also contributed a great deal of content to the astronomical almanac *Berliner Astronomisches Jahrbuch* between 1777 and 1781. We have little information about him, but the main source of information about his life and work is BODE's notes, with whom, he was not on the best of terms.[27] SCHULZE was a student of LAMBERT and a member of the Berlin Academy of Sciences when the Academy's mathematical section was directed by LAGRANGE, who succeeded EULER.

SCHULZE had already prepared material for tables containing seven-digit values of trigonometric functions and logarithms for sexagesimal seconds, but did not publish the material because of LAGRANGE's proposal. LAGRANGE proposed a decimal division of the quadrant instead of the sexagesimal division, which was subsequently published in tables by HOBERT and IDELER in 1799.[28]

For the mean time, or increment $t + x$, where t is the actual time and x is the equation of time, the mean longitude at the increment of longitude h is, as expressed by SCHULZE is:[29]

$$\overline{\lambda}_{\Delta} = \overline{\lambda} + hx \qquad [12.19]$$

$\overline{\lambda}$ longitude of the mean Sun
$\overline{\lambda}_{\Delta}$ longitude of the mean Sun after time increase, $t + x$
h the path or longitude that the mean Sun travels in one hour, SCHULZE's value = 147.″8[30]
x equation of time

In a similar way, SCHULZE also wrote the increment of the mean anomaly, $M + hx$. By considering the equation of the centre expanded into a series and the relation between the true and mean longitudes, he obtained the following expression:[31]

26 Wielen and Wielen, 2010, pp.43, 66.
27 Wielen and Wielen, 2011, p.265 also Wielen and Wielen, 2010.
28 Hobert, Ideler, 1799, Neue trigonometrische tafeln für die decimaleintheilung des quadranten. French mathematician and naval officer JEAN-CHARLES DE BORDA (1733–1799) also developed decimal trigonometric tables, which were published posthumously in 1801. See Danjon, 1980, p.20.
29 Schulze, 1778b, Ueber ein Mittel die Zeitgleichung unmittelbar zu bestimmen, Astronomisches Jahrbuch or Ephemeriden für das Jahr 1781, part II, p.139.
30 The value 147.″8 is obtained by dividing the number of arc seconds in a year by the number of hours in a year $= (360 \times 60 \times 60)/(365.25 \times 24)$.
31 For the equation of time expanded into a series, see Schulze, 1778b, p.147.

[12.20] $$\lambda_{\odot} = \bar{\lambda} + hx - 6931'' \sin(M + hx) + 73'' \sin 2(M + hx) - 1'' \sin 3(M + hx)$$

$\bar{\lambda}$ longitude of the mean Sun
M mean anomaly
h the longitude that the mean Sun travels in one hour = 147.″8
x equation of time

Since the product $h \cdot x$ is small, SCHULZE took into account the simplifications or approximations $\sin(hx) \approx hx$ and $\cos(hx) \approx 1$. With these simplifications:[32]

$$\sin(M + hx) = \sin M \cos hx + \cos M \sin hx \approx \sin M + hx \cos M$$

$$\sin 2(M + hx) = \sin 2M \cos 2hx + \cos 2M \sin 2hx \approx \sin 2M + 2hx \cos 2M$$

$$\sin 3(M + hx) = \sin 3M \cos 3hx + \cos 3M \sin 3hx \approx \sin 3M + 3hx \cos 3M$$

or

[12.21] $$\lambda_{\odot} = \bar{\lambda} + hx - 6931'' (\sin M + hx \cos M) + 73'' (\sin 2M + 2hx \cos 2M) - 1'' (\sin 3M + 3hx \cos 3M)$$

Taking into account the numerical values of the longitude that the mean Sun travels in one hour ($h = 147.''8$) and converted into radians, he obtained:

[12.22] $$\lambda_{\odot} = \bar{\lambda} + 0.00071656 \cdot x - 0.03360244 \sin M - 0.00002408 \cdot x \cos M + 0.00035391 \sin 2M + 0.00000051 \cdot x \cos 2M - 0.00000485 \sin 3M$$

or

[12.23] $$x = \lambda_{\odot} - \bar{\lambda} = 0.00071656 \cdot x - 0.03360244 \sin M - 0.00002408 \cdot x \cos M + 0.00035391 \sin 2M + 0.00000051 \cdot x \cos 2M - 0.00000485 \sin 3M$$

SCHULZE obtained the following expression from Equation 12.23:[33]

[12.24] $$(1.00000000 + 0.00002410 \cos M - 0.00000051 \cos 2M)x = -0.03362653 \sin M + 0.00035416 \sin 2M - 0.00000485 \sin 3M$$

SCHULZE derived the second contribution to the equation of time using the LAGRANGE expression (Equation 12.18), inserting the expression for the Sun's true longitude (Equation 12.21) into the LAGRANGE expression.[34]

32 Schulze, 1778b, p.139.
33 The values differ slightly from the values given in Equation 12.23. SCHULZE does not provide any additional explanation why.
34 Contribution due to obliquity of ecliptic. Schulze, 1778b, p.141. Lagrange, 1776, pp.226–227.

Based on Equation 12.22, he derived the expression for the sine of multiple angles of the ecliptic longitude. For multiples of longitude:[35]

$$\sin n\lambda_{\odot} = \sin n(\bar{\lambda} + x) = \sin n\bar{\lambda}\cos nx + \cos n\bar{\lambda}\sin nx$$

Taking into account the above expression and the equalities $\sin x = -\sin(-x)$ and $\cos x = \cos(-x)$, the following expression obtains:[36]

$$\begin{aligned}\sin n\lambda_{\odot} &= \sin n\bar{\lambda}\cos\left\{\begin{matrix}0.03360244n\sin M + 0.00002408nx\cos M\\ -0.00071656nx + 0.00000485n\sin 3M\\ -0.00035391n\sin 2M - 0.00000051nx\cos 2M\end{matrix}\right\}\\ &- \cos n\bar{\lambda}\sin\left\{\begin{matrix}0.03360244n\sin M + 0.00002408nx\cos M\\ -0.00071656nx + 0.00000485n\sin 3M\\ -0.00035391n\sin 2M - 0.00000051nx\cos 2M\end{matrix}\right\}\end{aligned}$$

Schulze used multiples of longitude up to and including 8 times $\lambda_{\odot}$. For example, for a factor of 8 times $\lambda_{\odot}$, the above expression gives:

$$\begin{aligned}\sin 8\lambda_{\odot} &= \sin 8\bar{\lambda}\cos\left\{\begin{matrix}0.2688195\sin M - 0.0057325x\\ -0.0028313\sin 2M + 0.0001926x\cos M\\ +0.0000388\sin 3M - 0.0000040x\cos 2M\end{matrix}\right\}\\ &- \cos 8\bar{\lambda}\sin\left\{\begin{matrix}0.2688195\sin M - 0.0057325x\\ -0.0028313\sin 2M + 0.0001926x\cos M\\ +0.0000388\sin 3M - 0.0000040x\cos 2M\end{matrix}\right\}\end{aligned}$$

Similarly for other multiples of $\lambda_{\odot}$.

He replaced the terms in brackets by ϕ and expressed $\sin\phi$ and $\cos\phi$ as trigonometric series. For the sine and cosine of multiple angles the following formulas apply:[37]

$$\cos n\theta = \begin{cases}\cos\theta\left(1 - \dfrac{(n^2-1^2)x^2}{2!} + \dfrac{(n^2-1^2)(n^2-3^2)x^4}{4!} - \cdots\right) & n = \text{odd}\\ 1 - \dfrac{n^2x^2}{2!} + \dfrac{n^2(n^2-2^2)x^4}{4!} - \cdots & n = \text{even}\end{cases}$$

35 For an explanation, see Todhunter, 1860, Plane Trigonometry, pp.213–217.

36 There are several typographical errors in section XIV of Schulze's paper, which have been corrected in this text. See Schulze, 1778b, p.141.

37 Bromwich, 1908, An Introduction to the Theory of Infinite Series, pp.177–181. See also application of Euler formulas, Chapter 9, Section 9.1.5.

$$\sin n\theta = \begin{cases} nx - \dfrac{n(n^2-1^2)x^3}{3!} + \dfrac{n(n^2-1^2)(n^2-3^2)x^5}{5!} - \cdots & n = \text{odd} \\ n\cos\theta\left(x - \dfrac{(n^2-2^2)x^3}{3!} + \dfrac{(n^2-2^2)(n^2-4^2)x^5}{5!} - \cdots\right) & n = \text{even} \end{cases}$$

where $x = \sin\theta$.

Taking into account the following general expressions for the powers of trigonometric functions:[38]

$$\sin^2 x = \frac{1}{2}(1 - \cos 2x) \qquad \cos^2 x = \frac{1}{2}(1 + \cos 2x)$$

$$\sin^3 x = \frac{1}{4}(3\sin x - \sin 3x) \qquad \cos^3 x = \frac{1}{4}(3\cos x + \cos 3x)$$

$$\sin^4 x = \frac{1}{8}(\cos 4x - 4\cos 2x + 3) \qquad \cos^4 x = \frac{1}{8}(\cos 4x + 4\cos 2x + 3)$$

$$\vdots \qquad\qquad \vdots$$

Schulze obtained the following expressions for $n = 8$:

$$\begin{aligned}\cos\phi = {} & 0.9820135 + 0.0003783\cos M + 0.0015267x\sin M \\ & - 0.0000162x^2 + 0.0179523\cos 2M - 0.0000418x\sin 2M \\ & - 0.0003771\cos 3M + 0.0000056x\sin 3M \\ & + 0.0000342\cos 4M - 0.0000001x\sin 4M \\ & - 0.0000011\cos 5M + 0.0000011x^2\cos 4M \\ & - 0.0000002x^2\cos 2M\end{aligned}$$

or

$$\begin{aligned}\sin\phi = {} & -0.0056295x + 0.2663984\sin M + 0.0001887x\cos M \\ & - 0.0027803\sin 2M - 0.0001070\cos 2M \\ & + 0.0008437\sin 3M + 0.0000038x\cos 3M \\ & - 0.0000254\sin 4M - 0.0000001x\cos 4M \\ & + 0.0000012\sin 5M + 0.0000043x^2\sin M\end{aligned}$$

38 Expressions for higher order powers are obtained using de Moivre's formula.

Taking into account that[39]

$$n\lambda_{\odot} = n\bar{\lambda} - \frac{n}{8}\phi$$

it follows:

$$\sin n\lambda_{\odot} = \sin\left(n\bar{\lambda} - \frac{n}{8}\phi\right) \quad \text{or} \quad \cos n\lambda_{\odot} = \cos\left(n\bar{\lambda} - \frac{n}{8}\phi\right)$$

by applying addition formulas a general expression that applies to arbitrary multiples of longitude follows:[40]

[12.25]
$$\sin n\lambda_{\odot} = \sin n\bar{\lambda}\left(1 - \frac{n^2}{64}\cdot\frac{\phi^2}{2!} + \frac{n^4}{4096}\cdot\frac{\phi^4}{4!} - \frac{n^6}{262144}\cdot\frac{\phi^6}{6!} + \cdots\right)$$
$$- \cos n\bar{\lambda}\left(\frac{n}{8}\cdot\phi - \frac{n^3}{512}\cdot\frac{\phi^3}{3!} + \frac{n^5}{32768}\cdot\frac{\phi^5}{5!} - \cdots\right)$$

He then inserted this into the LAGRANGE expression, or Equation 12.18. The result is a trigonometric series of multiples of the mean longitude. For $n = 2$, he obtained the following result:[41]

$$\lambda_{\odot} - \alpha_{\odot} = \tan^2\frac{\varepsilon}{2}\sin 2\bar{\lambda}\left\{\begin{array}{c} 0.9988711 + 0.0000962x\sin M - 0.0000010x^2 \\ +0.0000238\cos M - 0.0000026x\sin 2M + 0.0000001x^2\cos M \\ +0.0011284\cos 2M - 0.0000238\cos 3M + 0.0000005\cos 4M \end{array}\right\}$$
$$- \tan^2\frac{\varepsilon}{2}\cos 2\bar{\lambda}\left\{\begin{array}{c} 0.0671670\sin M - 0.0014315x - 0.0000001x\sin M \\ -0.0007078\sin 2M + 0.0000480x\cos M + 0.0000223\sin 3M \\ -0.0000026x\cos 2M - 0.0000004\sin 4M + 0.0000001x\cos 3M \end{array}\right\}$$
$$-\frac{1}{2}\tan^4\frac{\varepsilon}{2}\sin 4\bar{\lambda}\ldots$$

He calculated similar expressions for other multiples of the mean longitude.

In a similar way, SCHULZE also calculated the correction for the obliquity of the ecliptic. For the obliquity of the ecliptic, he took the value 23°28′, which was already used by CASSINI. He also took into account the correction of the ecliptic with $\psi < 2'$.

$$\varepsilon = 23°28' + \psi$$

39 Schulze, 1778b, p.142.
40 Bromwich, 1908, pp.177–182.
41 Schulze, 1778b, p.143.

For example, for twice the mean longitude of the Sun, he derived the following expression:

$$\sin 2\bar{\lambda} \tan^2 \frac{\psi}{2} \left\{ \begin{array}{c} 0.4328236 + 0.0000417x \sin M - 0.0000004x^2 \\ +0.0000103 \cos M - 0.0000011x \sin 2M + 0.0004890 \cos 2M \\ -0.0000103 \cos 3M + 0.0000002 \cos 4M \end{array} \right\}$$

$$- \cos 2\bar{\lambda} \tan^2 \frac{\psi}{2} \left\{ \begin{array}{c} 0.0291043 \sin M - 0.0006202x - 0.0003067 \sin 2M \\ +0.0000208x \cos M + 0.0000097 \sin 3M - 0.0000011x \cos 2M \\ -0.0000002 \sin 4M \end{array} \right\}$$

By taking into account $\tan\frac{\psi}{2} = \tan\frac{1}{2}2'\mu$,

where μ is the ratio of a given number of seconds, but less than 120 seconds and $2'$, Schulze simplified the expressions further. Since $\mu < 1$, he considered an approximation for small values of the tangent function, $\tan x \approx x$. Therefore:

$$\tan\frac{\psi}{2} \approx 1'\mu$$

For the above expression, Schulze took the value 0.0002909, which corresponds to $1'$ converted to radians.[42] We write C for equation of the centre:

[12.26] $$15^\circ\, x = C - (\lambda_\odot - \alpha_\odot)$$

or in radians $$0.2617994\, x = C - (\lambda_\odot - \alpha_\odot)$$

and considering the equation of the centre (see Equation 12.23):[43]

[12.27] $$C = -0.0336024 \sin M - 0.0000241x \cos M + 0.0003539 \sin 2M + 0.0000005x \cos 2M - 0.0000048 \sin 3M$$

42 Schulze, 1778b, p.146. This part of Schulze's explanation is unclear. No expression for the numerical value of the variable μ is given in the text, only a statement about the ratio.

43 See also Schulze, 1778b, p.147.

By writing E instead of x in Equation 12.26, Schulze obtained the following result, in seconds, taking into account both contributions to the equation of time and rearranging the result:[44]

$$\begin{aligned}E = &-592.52 \sin 2\bar{\lambda} - 462.07 \sin M + 0.13 \sin 2\left(M - \bar{\lambda}\right) \\ &+ 12.74 \sin 4\bar{\lambda} + 4.87 \sin 2M - 0.54 \sin 2\left(M + \bar{\lambda}\right) \\ &- 0.36 \sin 6\bar{\lambda} - 0.07 \sin 3M + 0.01 \sin 8\bar{\lambda} \\ &+ 19.93 \sin\left(M + 2\bar{\lambda}\right) + 19.93 \sin\left(M - 2\bar{\lambda}\right) - 0.86 \sin\left(M + 4\bar{\lambda}\right) \\ &+ 0.04 \sin 2\left(M + 2\bar{\lambda}\right) - 0.02 \sin 2\left(M - 2\bar{\lambda}\right) - 0.86 \sin\left(M - 4\bar{\lambda}\right) \\ &+ 0.03 \sin\left(M + 6\bar{\lambda}\right) + 0.04 \sin\left(M - 6\bar{\lambda}\right) + 0.01 \sin\left(3M - 2\bar{\lambda}\right) \\ &- 1.73\mu \sin 2\bar{\lambda} + 0.06\mu \sin\left(M + 2\bar{\lambda}\right) + 0.07\mu \sin 4\bar{\lambda} \\ &+ 0.06\mu \sin\left(M - 2\bar{\lambda}\right)\end{aligned} \tag{12.28}$$

Schulze also gave an expression for calculating the Sun's true longitude in radians:[45]

$$\begin{aligned}\lambda_{\odot} = &\bar{\lambda} - 0.0336944 \sin M - 0.0001179 \sin 2\bar{\lambda} + 0.0003564 \sin 2M \\ &+ 0.0000026 \sin 4\bar{\lambda} - 0.0000051 \sin 3\bar{\lambda} + 0.0000060 \sin\left(M + 2\bar{\lambda}\right) \\ &- 0.0000004 \sin 2\left(M + \bar{\lambda}\right) + 0.0000019 \sin\left(M - 2\bar{\lambda}\right) + 0.0000004 \sin 2\left(M - \bar{\lambda}\right) \\ &- 0.0000001 \sin\left(M + 4\bar{\lambda}\right) - 0.0000001 \sin\left(M - 4\bar{\lambda}\right)\end{aligned} \tag{12.29}$$

$\bar{\lambda}$ longitude of the mean Sun
M mean anomaly
μ = 0.0002909 rad

Or if we ignore the smallest terms and convert from radians to seconds:[46]

$$\begin{aligned}\lambda_{\odot} = &\bar{\lambda} - 6950''.0 \sin M - 24''.3 \sin 2\bar{\lambda} + 1''.2 \sin\left(M + 2\bar{\lambda}\right) \\ &+ 73''.5 \sin 2M + 0''.5 \sin 4\bar{\lambda} - 1''.1 \sin 3M + 0''.4 \sin\left(M - 2\bar{\lambda}\right) \\ &- 0''.1 \sin 2\left(M + \bar{\lambda}\right) + 0''.1 \sin 2\left(M - \bar{\lambda}\right)\end{aligned} \tag{12.30}$$

44 For Schulze's procedure see Schulze, 1778b, pp.147–151, for the equation see p.151. From Schulze's description it is not clear how exactly he treated the equation of the centre or Equation 12.27 in the procedure of calculating the final result of the equation of time. Schulze gives only a brief description and the final result. Schulze's two papers about the equation of time also include many typo errors so a detailed understanding of his procedures would require precise study of his work.

45 Schulze, 1778b, p.153.

46 Schulze, 1778b, p.153.

One year later, in the *Astronomisches Jahrbuch oder Ephemeriden*, he published tables with the values of the equation of time for the mean anomaly and mean longitude arguments.[47] In this paper, he separated the equation of time and the correction due to the obliquity of the ecliptic:

The equation of time as a function of mean longitude in seconds is:[48]

[12.31]
$$\begin{aligned} E = &-592.51 \sin 2\bar{\lambda} - 462.07 \sin M + 0.12 \sin 2(M - \bar{\lambda}) \\ &+ 12.81 \sin 4\bar{\lambda} + 5.04 \sin 2M - 0.56 \sin 2(M + \bar{\lambda}) \\ &- 0.37 \sin 6\bar{\lambda} - 0.07 \sin 3M + 0.01 \sin 8\bar{\lambda} \\ &+ 20.19 \sin(M + 2\bar{\lambda}) + 19.76 \sin(M - 2\bar{\lambda}) - 0.87 \sin(M + 4\bar{\lambda}) \\ &+ 0.04 \sin 2(M + 2\bar{\lambda}) - 0.02 \sin 2(M - 2\bar{\lambda}) - 0.86 \sin(M - 4\bar{\lambda}) \\ &+ 0.03 \sin(M + 6\bar{\lambda}) + 0.04 \sin(M - 6\bar{\lambda}) + 0.01 \sin(3M - 2\bar{\lambda}) \end{aligned}$$

For the equation of time as a function of mean anomaly and longitude of the apogee, the expression in seconds is:[49]

[12.32]
$$\begin{aligned} E = &-592.51 \sin 2\bar{\lambda} - 462.07 \sin M + 20.19 \sin(M + 2\bar{\lambda}) + 12.81 \sin 4\bar{\lambda} \\ &+ 5.04 \sin 2M + 0.04 \sin 2(M + 2\bar{\lambda}) - 0.37 \sin 6\bar{\lambda} - 0.07 \sin 3M \\ &+ 19.76 \sin(M - 2\bar{\lambda}) + 0.01 \sin 8\bar{\lambda} - 0.02 \sin 2(M - 2\bar{\lambda}) \\ &- 0.12 \sin 2\varpi - 0.04 \sin(5\bar{\lambda} + \varpi) - 0.56 \sin(4\bar{\lambda} - 2\varpi) + 0.03 \sin(7\bar{\lambda} - \varpi) \\ &- 0.87 \sin(5\bar{\lambda} - \varpi) + 0.01 \sin(\bar{\lambda} - 3\varpi) + 0.86 \sin(3\bar{\lambda} + \varpi) \end{aligned}$$

Also new is the expression for the correction in the obliquity of the ecliptic for each minute ($1'$):[50]

[12.33]
$$-0.86 \sin 2\bar{\lambda} + 0.03 \sin(M + 2\bar{\lambda}) + 0.04 \sin 4\bar{\lambda} + 0.03 \sin(M - 2\bar{\lambda})$$

Schulze is also the author of a comprehensive two-part handbook on geometry and trigonometry.[51] He also compiled extensive tables of trigonometric functions and logarithms, which were later supplemented or extended by Georg Vega.[52]

47 Schulze, 1779, pp.159–170.

48 Schulze, 1779, p.159.

49 Schulze, 1779, p.160. In the third term the sign seems to be wrong in Schulze's paper.

50 Schulze, 1779, p.160.

51 Schulze, 1782/1783, Taschenbuch für diejenigen, so gründliche Anwendungen der Meßkunst zu machen sich vorsetzen – insbesondere aber zum Gebrauch meiner Vorlesungen abgefasset.

52 Schulze, 1778a, Neue und erweiterte Sammlung logarithmischer, trigonometrischer und anderer zum Gebrauch der Mathematik unentbehrlicher Tafeln.

12.2.2 Georg Vega

GEORG VEGA is best known for his tables of logarithms which were used by mathematicians for almost 200 years, from the end of the 18th century until the advent of electronic calculators. He built on the work of his predecessors ADRIAAN VLACQ and KARL SCHULZE and was also familiar with the work of BARTHOLOMEO PITISCUS. That VEGA was clearly well acquainted with the work of his predecessors is shown by the first edition of his tables in 1783, in which he gives the corrections to the values in SCHULZE's tables, published in 1778. The first edition of VEGA's work also contains instructions for solving plane and spherical triangles[53] and many trigonometric expressions and examples of the expansion of trigonometric functions or expressions into series. The second edition of 1797 also contains practical examples of problems in spherical astronomy, including an example of calculating the difference between mean (clock) time and local solar time from the altitude of the Sun, in VEGA's own way, as well as a table of values of the equation of time.

Example 12.1 From the altitude of the Sun at a given time, calculate the difference between solar time and local time. The example is taken from Vega, 1797, p.XXXVII, Problem II. The data for the calculation are:

Local time	8 h 58 min 32 s in the morning
Altitude of the Sun	$40°28'20''$
Declination, $\delta_\odot$	$15°\ 2'50''$
Latitude of Vienna, φ	$48°12'30''$

Vega calculated the hour angle $\widehat{ZPS} = H_\odot$ using the rule for the sine of half angles of spherical triangles.[54] For the triangle defined by the north celestial pole, zenith and Sun, it follows that (see Appendix D, Figure D.9):

$$\widehat{ZS} = a \qquad \widehat{ZP} = b \qquad \widehat{PS} = c$$

$$\frac{1}{2}(a + b + c) = p$$ (see Appendix D, Section D.3.1)

Sun's altitude complement (zenith angle, z)	$a = 49°31'40''$
Latitude complement $= 90° - \varphi$	$b = 41°47'30''$
Complement of declination $= 90° - \delta_\odot$	$c = 74°57'10''$
Perimeter	$2p = 166°16'20''$
Semiperimeter	$p = 83°8'10''$
	$p - b = 41°20'40''$
	$p - c = 8°11'0''$

53 Vega, 1783, Logarithmische, trigonometrische, und andere zum Gebrauche der Mathematik eingerichtete Tafeln und Formeln, pp.LIV–LXV.

54 Kells et al., 1940, Plane and Spherical Trigonometry, pp.295–298, also Bronshtein, Semendyayev, 1985, Handbook of Mathematics, p.188.

The expression for the sine or cosine of a half angle $\widehat{ZPS}$ is (see Appendix D, Equations D.9 and D.10):

$$\sin\frac{\widehat{ZPS}}{2} = \sqrt{\frac{\sin(p-b)\sin(p-c)}{\sin b \sin c}} \qquad \text{or} \qquad \cos\frac{\widehat{ZPS}}{2} = \sqrt{\frac{\sin p \sin(p-a)}{\sin b \sin c}}$$

$$\tfrac{1}{2}\widehat{ZPS} = 22°28'20''$$

$$\widehat{ZPS} = 44°56'40''$$

The hour angle converted to time (15° ≡ 1 hour) is:

	44° ≡ 2 h 56 min 0 s
	56′ ≡ 3 min 44 s
	40″ ≡ 2 s
Hour angle converted to time	2 h 59 min 46 s
Local solar time	9 h 0 min 14 s
Local (clock) time	8 h 58 min 32 s
Difference or equation of time	– 1 min 42 s

The hour angle can also be converted to time using VEGA's tables, 1797 edition, book 2, p.177.

12.3 Equations of condition

When addressing astronomical problems, variables are common as functions of other quantities, variables or constants and times whose value at a given moment is known and is the result of observations.[55] The variables x, y, z... are functions of time t and the parameters a, b, c... and we write them as:

$$x = f_1(a, b, c, \dots, t)$$
$$y = f_2(a, b, c, \dots, t) \dots$$

From the known or presumed parameter a, b, c... we can determine the variable quantities x, y, z... at a given moment in time.

From the results of the observations of quantities x, y, z... at different moments in time, we can determine the values of parameters a, b, ...

If, for a case of uniform rotation around the centre by the angle of $\widehat{\Upsilon OW}$ for the initial moment, or the origin from which we measure the time (Figure 12.2), we take the value of b, then c is the arc that the celestial body travels in a time unit, so the angle $\widehat{\Upsilon OS}$ is at any time determined by the equation:

$$\widehat{\Upsilon OS} = b + ct \qquad [12.34]$$

The rectangular coordinates of point S (for example the Sun) are:

$$x = a\cos(b + ct)$$
$$y = a\sin(b + ct)$$

- If the values a, b and c are known, we can calculate the values of x and y for various initial moments.
- If the values of coordinates x and y are measured or obtained by observations at different moments in time t_1, t_2 ..., we can calculate the values of x and y at the initial moment if we know the values of a, b, c... If the calculated values do not match the expected result, the cause is an error that could be due to inaccurate observations or false values of parameters a, b, c... Based on the calculation, we can also determine the size of the error if, for example, we have observation values available.

55 For more information and solutions to the equations of condition, see Newcomb, 1960, pp.63–84.

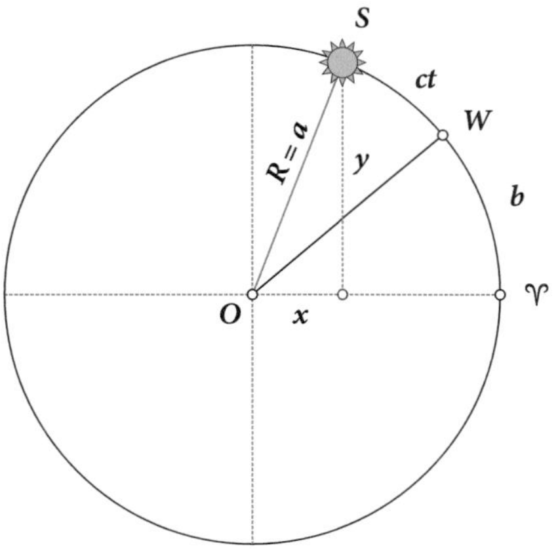

Figure 12.2 Graphical representation of an object moving on a circular path. Adopted from: Newcomb, 1906, p.34, Figure 3.

- In an extreme case, the values a, b and c are unknown. In this case, the values can be determined from the known observation results or the values of x and y at moments t_1 and t_2. For parameters a, b and c, the system of equations applies.

$$\begin{aligned} x_1 &= a\cos(b+ct_1) \\ y_1 &= a\cos(b+ct_2) \\ z_1 &= a\cos(b+ct_3) \end{aligned} \tag{12.35}$$

Such equations are called the equations of condition because the parameters a, b and c must fulfil certain conditions so that the equation system is solvable. Using values from three observations at different points on the circle, sufficiently distant from one another, we can determine adequate values of a, b and c to be used for subsequent calculations.[56]

The equations of condition were used by MAYER, LAPLACE and DELAMBRE.[57] LAPLACE was probably encouraged by MAYER's work *Abhandlung über die Umwälzung des Monds um seine Axe, und die scheinbare Bewegung der Mondsflecken: Worinnen der Grund einer verbesserten Mondbeschreibung aus neuen Beobachtungen gelegt wird*,[58] in which MAYER successfully applied equations of condition to the study of the moon's motion. EULER's work *Recherches sur la question des inégalités du mouvement de Saturne et de Jupiter* (1749) was probably available to MAYER, although we do not know for certain. LAPLACE, however, used LAGRANGE's treatise *Théorie de la libration de la lune et des autres phénomènes qui dépendent de la figure non-sphérique de cette planète*,[59] in which LAGRANGE refers to MAYER's treatise on the moon's motion. LAPLACE's study of the effects of Jupiter and Saturn, *Théorie de Jupiter et de Saturne*, subsequently assisted DELAMBRE in compiling tables that were later classified by DE LALANDE in the third edition of his book *Astronomie*.[60]

56 For the development of statistical methods in astronomy, see Sheynin, 1973, Mathematical Treatment of Astronomical Observations (A Historical Essay), Archive for History of Exact Sciences, vol. 11, no. 2/3, pp.97–126.

57 Wilson, 1980, II, pp.257–268.

58 Published in Kosmographische Nachrichten aus Nuerenberg, 1750.

59 Nouveaux Memoires de l'Academie Royale des Sciences et Belles_Lettres, Berlin, 1780, pp.203–309.

60 Wilson, 1980, II, p.279.

Bibliography

- Amma, Saraswathi T. A. 1963. The development of mathematical series in India after Bhaskara II. In: Bulletin of the National Institute of Sciences in India, 21, 320–343.
- Ball, Robert. 1908. A Treatise on Spherical Astronomy. Cambridge: University Press. URL: https://archive.org/details/atreatiseonsphe00ballgoog (17 February 2024).
- Bateman, H. 1905. The correspondence of Brook Taylor. In: Bibliotheca Mathematica, Zeitschrift für Geschichte der mathematischen Wissenschaften, 3, 7, 367–371. Leipzig: Teubner.
- Bottazzini, Umberto. 1989. Lagrange et le problème de Kepler. In: Revue d'histoire des sciences, 42, 1/2, 27–42. URL: https://www.persee.fr/doc/rhs_0151-4105_1989_num_42_1_4133 (17 February 2024).
- Bromwich, Thomas John. 1908. An introduction to the theory of infinite series. London: Macmillan. URL: https://archive.org/details/introductiontoth00bromuoft/mode/2up (17 February 2024).
- Calinger, Roland. S. 1971. Castillon, Johann. In: Dictionary of Scientific Biography, Gillispie, Charles Coulston (ed.), III, 119–120. New York: Charles Scribner's Sons.
- Cantor, Moritz. 1901. Vorlesungen über Geschichte der Mathematik, 3. Band. Reprint 1965. Stuttgart: Teubner.
- Crépel, Pierre and Gouillaume Jouve. 2014. 1765: A year in the correspondence between d'Alembert and Lagrange. In: Lettera Mathematica, 2, 1–2, 17–24. DOI 10.1007/s40329-014-0039-z.
- Danjon, André. 1980. Astronomie Générale. Reprint 1994. Paris: Albert Blanchard. ISBN 2-85367-068-6.
- De Lalande, Joseph Jérôme Le Français. 1792. Astronomie, Book 1, 3rd Edition. Paris: P. Didot. URL: https://archive.org/details/astronomielalande01lala (17 February 2024).
- Enerström, Gustav. 1911–1912. Zur Vorgeschichte der Entdeckung des Taylorschen Lehrsatzes. In: Bibliotheca Mathematica, Zeitschrift für Geschichte der mathematischen Wissenschaften, 3, 12, 333–336. Leipzig: Teubner.
- Erman, Jean Pierre. 1799. Éloge de Monsieur Schultz. In: Mémoires de l'Académie Royale des Sciences et Belles-Lettres, MDCCXCIV et MDCCXCV (1794 et 1795), 55–70. Berlin: George Decker.
- Feigenbaum, L. 1985. Brook Taylor and the method of increments. In: Archive for History of Exact Sciences, 34, 1–2, 1–140. DOI 10.1007/BF00329903.
- Ferraro, Giovanni. 2008. The Rise and Development of the Theory of Series up to the Early 1820s. New York: Springer. ISBN 978-0-387-73467-5. DOI 10.1007/978-0-387-73468-2.
- Fraser, Craig. 1985. J. L. Lagrange's changing approach to the foundations of the calculus of variations. In: Archive for History of Exact Sciences, 32, 2, 151–191. DOI 10.1007/BF00329871.
- Goldstine, Herman H. 1977. A History of Numerical Analysis from the 16th Century Through the 19th Century. New York: Springer. DOI 10.1007/978-1-4684-9472-3. ISBN 978-0-387-90277-7.
- Gregory, James. 1667. Vera circuli et hyperbola quadratura. Padova: Cadorini. DOI 10.3931/e-rara-4024.
- Gupta, Radha Charan. 1973. The Madhava-Gregory series. In: The Mathematics Education, 7, 3, B67–B70.
- Gupta, Radha Charan. 1991. The Madhava-Gregory Series for $\tan^{-1} x$. In: Indian Journal of Mathematics Education, 11, 3, 107–110.
- Hamburg, Robin Rider. 1976. The Theory of Equations in the 18th Century: The Work of Joseph Lagrange. In: Archive for History of Exact Sciences, 16, 1, 17–36. DOI 10.1007/BF00348304.
- Itard, Jean. 1973. Lagrange. In: Dictionary of Scientific Biography, Gillispie, Charles Coulston (ed.), VII, 582–583. New York: Charles Scribner's Sons. SBN 684-10118-1.
- Itard, Jean. Lagrange. In: Encyclopedia.com [online version]. Update 27 June 2018. URL: https://www.encyclopedia.com/people/history/historians-miscellaneous-biographies/joseph-louis-louis-lagrange (23 August 2023).
- Joseph, George Gheverghese. 2009. Kerala Mathematics, History and its Possible Transmission to Europe. Delhi: BR Publishing Corp. ISBN 978-81-7646-662-2.
- Kells, Lyman M., Willis F. Kern and James R. Bland. 1940. Plane and Spherical Trigonometry. New York, London: McGraw-Hill.
- Klügel, Georg Simon. 1789. Formel aus der anomalia excentri die mittlere anomalie zu finden, Bode, J. E. (ed.). Astronomisches Jahrbuch für das Jahr 1789, 191–193. Berlin: George Jacob Decker.

- Lagrange, Joseph-Louis. 1770. Nouvelle méthode pour résoudre les équations littérales par le moyen des séries. In: Mémoires de l'Académie Royale des Sciences (Berlin), 24, 5–73. URL: https://gallica.bnf.fr/ark:/12148/bpt6k229222d/f6 (17 February 2024).
- Lagrange, Joseph-Louis. 1771. Sur le Problème de Kepler. In: Mémoires de l'Académie Royale des Sciences (Berlin), 25, 204–233. URL: https://gallica.bnf.fr/ark:/12148/bpt6k229222d/f114 (17 February 2024).
- Lagrange, Joseph Louis. 1771. Sur le Problème de Képler. In: Œuvres complètes, III, 113–138.
- Mémoires de l'Académie royale des sciences et belles-lettres de Berlin, t. XXV. URL: https://gallica.bnf.fr/ark:/12148/bpt6k229222d/f114 (17 February 2024).
- Lagrange, Joseph-Louis. 1776. Positionswinkel und Abweichung für alle Grade der Breite, in Formeln ausgedrückt, von Hrn. de la Grange In: Sammlung Astronomischer Tafeln, 226–227. Berlin: George Jacob Decker. (Google Books, 31 May 2020).
- Lagrange, Joseph-Louis. 1869. Œuvres, t. 3 - Mémoires de Berlin (VII à XXII), vol. 1–2, Joseph-Alfred Serret (ed.). Paris: Gauthier-Villars. Reprint 2004. Paris: Jacques Gabay. ISBN 978-2-87647-237-2.
- Laplace, Pierre Simon. 1802. Traité de Mécanique Céleste, Part 3. Paris: J. B. M. Duprat. (Google Books, 31 May 2020).
- L'Huilier, Simon. 1786. Exposition élémentaire des principes des calculs supérieurs: qui a remporté le prix proposé par l'Académie royale des sciences et belles-lettres pour l'année 1786. Berlin: George Jacob Decker. URL: https://catalogue.bnf.fr/ark:/12148/cb30815722d (17 February 2024).
- Maclaurin, Colin. 1742. A Treatise of Fluxions in Two Books. Edinburgh: Ruddimans. URL: https://archive.org/details/atreatisefluxio00conggoog/mode/2up (17 February 2024).
- Malet, Antoni. 1993. James Gregorie on tangents and the 'Taylor' rule for series expansions. In: Archive for History of the Exact Sciences, 46, 2, 97–137. DOI 10.1007/BF00375656.
- Newcomb, Simon. 1960. A compendium of spherical astronomy. Reprint of 1906 edition. New York: Dover.
- Newton, Isaac. 1967–1981. The Mathematical Papers of Isaac Newton, vol. 7, Whiteside, D. T. (ed.). Cambridge: Cambridge University Press. ISBN 9780521045896.
- Ottley, W. C. 1838. Treatise on the Differential Calculus with a Collection of Examples. London: W. P. Grant and Whittaker and Co. URL: https://archive.org/details/atreatiseondiff00ottlgoog (17 February 2024).
- Ramasubramanian, K. and M. S. Sriram. 2011. Tantrasaṅgraha of Nīlakaṇṭha Somayājī. New Delhi: Hindustan Book Agency. ISBN 978-93-80250-09-0.
- Roy, Ranjan. 1990. The Discovery of the Series Formula for π by Leibniz, Gregory and Nilakhanta. In: Mathematics Magazine, 63, 5, 291–306. DOI 10.2307/2690896.
- Sarma, K. V. 1972. A History of the Kerala School of Hindu Astronomy. Hoshiarpur: Vishveshvaranand Institute.
- Sarma, K. V. (transl.). 2008. Gaṇita-Yukti-Bhāṣā of Jyeṣṭhadeva, Volume I, II. With Explanatory Notes by K. Ramasubramanian, M. D. Srinivas and M. S. Sriram. New Delhi: Hindustan Book Agency/Springer. ISBN 978-1-84882-072-2.
- Schaub, Werner. 1950. Sphärische Astronomie. Leipzig: Geest & Portig.
- Schulze, Johann Karl. 1778a. Neue und erweiterte Sammlung logarithmischer, trigonometrischer und anderer zum Gebrauch der Mathematik unentbehrlicher Tafeln. Berlin: A. Mylius. URL: https://archive.org/details/bub_gb_fO42AAAAMAAJ/mode/2up (27 June 2021).
- Schulze, Johann Karl. 1778b. Ueber ein Mittel die Zeitgleichung unmittelbar zu bestimmen. In: Astronomisches Jahrbuch oder Ephemeriden fur das Jahr 1781, Part 2, 135–160. Berlin: George Jacob Decker. URL: https://mdz-nbn-resolving.de/urn:nbn:de:bvb:12-bsb10538408-1 (17 February 2024).
- Schulze, Johann Karl. 1779. Nachtrag zu der im vorigen Jahr dieser Ephemeriden gegebenen Abhandlung uber ein Mittel die Zeitgleichung unmittelbar zu bestimmen. In: Astronomisches Jahrbuch oder Ephemeriden fur das Jahr 1782, Part 2, 159–170. Berlin: George Jacob Decker. URL: https://www.digitale-sammlungen.de/en/view/bsb10538409?page=,1 (17 February 2024).
- Schulze, Johann Karl. 1782. Taschenbuch für diejenigen, so gründliche Anwendungen der Meßkunst zu machen sich vorsetzen – insbesondere aber zum Gebrauch meiner Vorlesungen abgefasset. Heft 1. Berlin: Julius Mylius. (Google Books, 30 September 2023).

- Schulze, Johann Karl. 1783. Taschenbuch für diejenigen, so gründliche Anwendungen der Meßkunst zu machen sich vorsetzen – insbesondere aber zum Gebrauch meiner Vorlesungen abgefasset. Heft 2. Berlin: Julius Mylius. (Google Books, 30 September 2023).
- Sheynin, O. B. 1973. Mathematical Treatment of Astronomical Observations (A Historical Essay). In: Archive for History of Exact Sciences, 11, 2/3, 97–126. doi 10.1007/BF00343531.
- Stigler, Stephen M. 1978. Laplace's Early Work: Chronology and Citations. In: Isis, 69, 2, 234–254. doi 10.1086/352006.
- Taton, René. 1974. Inventaire chronologique de l'œuvre de Lagrange. In: Revue d'histoire des sciences, 27, 1, 3–36. url: https://www.persee.fr/doc/rhs_0151-4105_1974_num_27_1_1044 (17 February 2024).
- Taylor, Brook. 1717. Methodus incrementorum directa & inversa. Londini: impensis Gulielmi Innys. doi 10.3931/e-rara-4090.
- Turnbull, Herbert Westren. 1939. James Gregory Tercentary Memorial Volume. London: G. Bell.
- Todhunter, Isaac. 1860. Plane Trigonometry for the Use of Colleges and Schools with Numerous Examples. 2nd edition. Cambridge: Macmillan. (Google Books. 31 May 2020).
- Van Brummelen, Glen. 2009. The Mathematics of the Heavens and the Earth. Princeton: Princeton University Press. isbn 978-0-691-12973-0.
- Vega, Georg. 1783. Logarithmische, trigonometrische, und andere zum Gebrauche der Mathematik eingerichtete Tafeln und Formeln. Wien: Johann Thomas von Trattner. url: https://www.dlib.si/details/URN:NBN:SI:DOC-IL8YP8WP?&language=eng (26 June 2021).
- Vega, Georg. 1797. Logarithmisch – trigonometrische nebst andern zum Gebrauch der Mathematik eingerichtete Tafeln und Formeln, Band II. 2nd Edition. Leipzig: Weidmanische Buchhandlung. (Google Books, 26 June 2020).
- Weiss, E. 1885. Entwickelungen zum Lagrange'schen Reversionstheorem, und Anwendung derselben auf die Lösung der Keppler'schen Gleichung. Denkschriften der Mathematisch-Naturwissenschaftlichen Classe der Kaiserlichen Akademie der Wissenschaften, XLIX. Sonderabdruck. Wien: Kaiserlich-Königliche Hof- und Staatsdruckerei, Karl Gerold's Sohn.
- Whitaker, Edmund. 1949. Laplace. In: The Mathematical Gazette, 33, 303, 1–12. doi 10.2307/3608408.
- Wielen, Roland and Ute Wielen. 2010. Johann Elert Bodes Geschichte der Berliner Sternwarte bis zum Jahr 1811. Edition der Handschrift. Heidelberg: Astronomisches Rechen-Institut Zentrum für Astronomie Universität Heidelberg. doi 10.11588/heidok.00011523.
- Wielen, Roland and Ute Wielen. 2011. Die Reglements und Statuten des Astronomischen Rechen-Instituts und zugehörige Schriftstücke im Archiv des Instituts. Edition der Dokumente. Heidelberg: Astronomisches Rechen-Institut Zentrum für Astronomie Universität Heidelberg. doi 10.11588/heidok.00013066.
- Williamson, Benjamin. 1899. An Elementary Treatise on the Differential Calculus. New York, Bombay: Longmans, Green and Co. url: https://archive.org/details/rishabh-lib/mode/2up (17 February 2024).
- Wilson, Curtis A. 1980. Perturbations and Solar Tables from Lacaille to Delambre: The Rapprochement of Observation and Theory, Part I. In: Archive for History of Exact Sciences, 22, 1/2, 53–188. doi 10.1007/BF00327870.
- Wilson, Curtis A. 1980. Perturbations and Solar Tables from Lacaille to Delambre: The Rapprochement of Observation and Theory, Part II. In: Archive for History of Exact Sciences, 22, 3, 189–304. doi 10.1007/BF00348584.
- Wollenschläger, K. 1933. Der Mathematische Briefwechsel zwischen Johann I Bernoulli und Abraham de Moivre. In: Verhandlungen der Naturforschenden Gesellschaft in Basel, 43, 151–317.
- Woodhouse, Robert. 1821. A Treatise on Astronomy, Theoretical and Practical, Part 1, Volume 1. Cambridge: J. Deighton & Sons and G. & W. B. Whittaker. url: https://archive.org/details/treatiseonastron11woodrich (17 February 2024).
- Woodhouse, Robert. 1823. A Treatise on Astronomy, Theoretical and Practical, Part 2, Volume 1. Cambridge: J. Deighton & Sons and G. & W. B. Whittaker. url: https://archive.org/details/p2treatiseonastr01wooduoft (17 February 2024).

13 Contemporary notation for the equation of time

As the Equation of time, or difference between the time ſhewn by a well regulated Clock and a true Sun-dial, depends upon two cauſes, namely, the obliquity of the Ecliptic, and the unequal motion of the Earth in its annual Orbit; it will be proper, firſt, to explain the effects of these cauſes ſeparately considered; and then the united effects reſulting from their combination.

James Ferguson, 1756, Astronomy Explained upon Sir Isaac Newton's Principles and made Easy to those who have not studied Mathematics, p.100.

13.1 Jean-Baptiste Joseph Delambre

Delambre, Lalande's student, was a successful astronomer and is also considered one of the fathers of the metre as a unit of measurement. His contribution to geodesy is also important. In 1790, he won the Academy of Sciences and Arts award for calculating the orbit of the newly-discovered planet Uranus. In the same year, he was elected as a foreign member to the Swedish Royal Academy of Sciences, and in 1792 became a member of the French Academy of Sciences. During an expedition to determine the length of one metre on the route between Dunkerque and Barcelona, Cassini was assigned control over measurements of the northern part of the route between Dunkerque and Rodez. He did not accept the post for political reasons, howewer, Delambre was appointed instead. The measurements were completed in 1798 and the results presented at an international conference a year later. Delambre is also the author of many of the more frequently cited works in the field of astronomy history.[1] In 1814, his extensive work Astronomie theorique et pratique was published, which replaced Lalande's teaching materials that had been in use beforehand.

Delambre's first publication, in 1785, gave a mathematical treatment of the Sun's position used Lacaille's mathematical procedures and Mayer's tables.[2] A year later, Delambre published another treatise on the same topic in which the influence of Laplace's work is already recognisable.[3] The equation of time was addressed by Delambre in *Astronomie theoretique et pratique* and *Tables astronomiques*, citing the equation of time as the difference between the right ascensions of the true and the mean Sun:[4]

[13.1] $$E = \alpha_{\odot} - \overline{\alpha}$$

$\alpha_{\odot}$ right ascension
$\overline{\alpha}$ right ascension of the mean Sun

With the derivation, Delambre wrote the equation of time as a function of the longitude of the mean Sun which, in addition to longitude, also contained corrections for the gravitational perturbations due to other celestial bodies.[5]

1 Histoire de l'astronomie ancienne, 2 volumes, 1817, Histoire de l'astronomie du moyen age, 1819, Histoire de l'astronomie moderne, 2 volumes, 1821, Histoire de l'astronomie au dix-huitième siècle, 1827.

2 Delambre, 1787, Mémoire sur les élémens de l'orbite solaire, dans lequel on détermine, par de nouvelles observations, l'apogée, la longitude moyenne & la plus grande équation du soleil. Nouveaux Mémoires de l'Académie Royale des Sciences et Belles-Lettres de Berlin, Année 1785, vol. 16, pp.191–218. See also Wilson, 1980, Perturbations and Solar Tables from Lacaille to Delambre: The Rapprochement of Observation and Theory, Part II, pp.271–272.

3 Delambre, 1788, Second mémoire sur les élémens de l'orbite solaire. Nouveaux Mémoires de l'Académie Royale des Sciences et Belles-Lettres de Berlin, Année 1786, vol. 17, pp.234–248.

4 Delambre, 1814, Astronomie Theoretique et Pratique, II, Chapter 23, pp.205, §30. Also Delambre, Bürg, 1806, Tables astronomiques.

5 Delambre, 1814, p.197, §8.

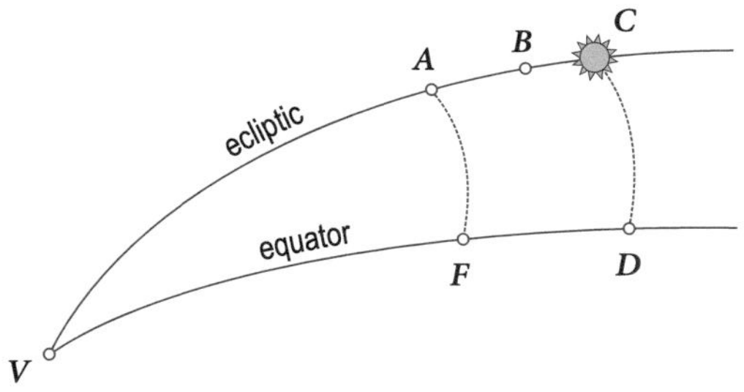

Figure 13.1 DELAMBRE's graphical interpretation of the equation of time. Summarised from: Delambre, 1814, Astronomie Theoretique et Pratique II, planche II, Figure 29.

DELAMBRE's definition of the equation of time can be seen from *Astronomie theoretique et pratique*, where the figure shows the following expressions:

The mean longitude is: $\bar{\lambda} = \overset{\frown}{VA}$

The true longitude, taking into account the equation of the centre, is:[6]

$$\overset{\frown}{VB} = \overset{\frown}{VA} + \overset{\frown}{AB} \qquad \text{or} \qquad \lambda_{\odot} = \bar{\lambda} + C$$

The true longitude, taking into account the influence of the Moon and the planets and the perturbations of P, is:[7]

$$\overset{\frown}{VC} = \overset{\frown}{VA} + \overset{\frown}{AB} + \overset{\frown}{BC} \qquad \text{or} \qquad \lambda_{\odot} = \bar{\lambda} + C + P \tag{13.2}$$

The perturbations are clearly based on the terms LAPLACE published in *Traité de mécanique céleste*.[8] We do not know exactly how DELAMBRE arrived at the results since his working material is not preserved. He also used the results of BRADLEY's observations, published by HORNSBY in 1798, and the results of MASKELYNE's observations.[9]

Right ascension can be determined from the mean longitude directly by spherical trigonometry or from the true longitude by reduction of the ecliptic to the equator, a mathematical process that is more complicated but yields the same result.[10] The values of the reduction of the ecliptic were tabulated in astronomical manuals, making this procedure relatively simple in practice. The reduction of the ecliptic to the equator as given by DELAMBRE is (Figure 13.1 or Equation 12.18):

6 For the explanation or sign of the equation of the centre see also Equation 1.6.

7 Delambre, 1814, pp.196–197, §.7.

8 Wilson, 1980, Part II, p.287. Laplace, 1802, Traité de Mécanique Céleste, volume III, book VI, Chapter X, pp.105–114.

9 Forbes, 1965, Bradley's Astronomical Observations. Bradley and Bliss, 1798, Astronomical Observations, made at the Royal Observatory in Greenwich, from the year 1750 to the year 1764.

10 See Appendix D, Section D.6.1 or Chapter 15.

$$[13.3] \quad R = \widehat{VC} - \widehat{VD} = \lambda_\odot - \alpha_\odot$$
$$= \tan^2\frac{\varepsilon}{2}\sin 2\widehat{VC} - \frac{1}{2}\tan^4\frac{\varepsilon}{2}\sin 4\widehat{VC} + \frac{1}{3}\tan^6\frac{\varepsilon}{2}\sin 6\widehat{VC} - \cdots$$

And the right ascension is:

$$[13.4] \quad \alpha_\odot = \bar{\lambda} + C + P - R$$

In calculating the true longitude, DELAMBRE also took into account, in addition to perturbations, the nutation in the longitude and in the right ascension:

$$[13.5] \quad \lambda_\odot = \bar{\lambda} + C + P + a\sin N$$

For the right ascension corrected for nutation, DELAMBRE gives the following expression:[11]

$$[13.6] \quad \alpha_\odot = \lambda_\odot + R = \bar{\lambda} + C + P + R + a\sin N$$

$\bar{\lambda}$	longitude of the mean Sun
P	perturbations due to the planets
C	equation of the centre
R	reduction of the ecliptic to the equator
$a\sin N$	nutation = $0.''09925 \cdot \sin N$[12]
	$N = 360° -$ ascending node of the Moon's orbit[13]

Nutation, or the equation of equinoxes, has two contributions:[14]

- Nutation in longitude (the equation of the equinoxes in longitude): $n_\lambda = 18'' \sin N$
- Nutation in right ascension:[15] $n_\alpha = 18'' \sin N \cos\varepsilon$

For the nutation, or the equation of the equinoxes, the following relations also apply:

$$n_\lambda = n_\alpha \frac{1}{\cos\varepsilon} = n_\alpha \sec\varepsilon \qquad n_\alpha = n_\lambda \cos\varepsilon$$

ε obliquity of the ecliptic

11 Delambre, Bürg, 1806, p.30 (page sequence, not a given page number). The sign of R is opposite to that in the book *Astronomie Théoretique et Pratique*.
12 Value $0.''09925$ is in seconds of time. For the values of nutation, see Delambre, Bürg, 1806, Table XIII.
13 Delambre, 1814, p.197, §9.
14 For an explanation of nutation, see Woodhouse, 1823, pp.353–383.
15 DELAMBRE uses the value 18″, see Delambre, 1814, p.197. Woodhouse, 1823, for example, states the value $18.''034$ on p.532.

For the difference DELAMBRE gives the following expression:

$$n_\lambda - n_\alpha = 18'' \sin N\,(1 - \cos\varepsilon) = 36'' \sin^2\frac{\varepsilon}{2}\sin N = 1.''4887 \sin N$$

or 0.''09925 in seconds of time.

For an obliquity of the ecliptic of 23°28″, the value of $\sin\varepsilon$ is approximately $^{9173}/_{10000}$ or $^{11}/_{12}$, from which, taking into account the values for the obliquity of the ecliptic, follows the value of the nutation in right ascension $\approx 16.''5 \sin N$.

The equation of time is the difference between the true and the mean right ascension, so that for the value of the equation of time, already converted into time units, from the above expressions and Equation 13.1, it follows that in seconds of time:[16]

[13.7]
$$E = \frac{1}{15}(\overline{\lambda} + C + P + R + a\sin N - \overline{\lambda} - a\cos\varepsilon\sin N)$$
$$= \frac{1}{15}(C + P + R + 2a\sin^2\frac{\varepsilon}{2}\sin N) = \frac{1}{15}(C + P + R) + 0.''09925 \sin N$$

$\overline{\lambda}$	longitude of the mean Sun
P	perturbations due to the planets
C	equation of the centre
R	reduction of the ecliptic to the equator, part of the equation of time due to the obliquity of the ecliptic, $R = \lambda_\odot - \alpha_\odot$
$a \sin N$	contribution due to nutation in seconds of time = $0.''09925 \cdot \sin N$[17]

For the equation of the centre, we can find the following record in DELAMBRE's work:[18]

[13.8]
$$C = a\sin(\overline{\lambda} - \varpi) + b\sin 2(\overline{\lambda} - \varpi) + c\sin 3(\overline{\lambda} - \varpi) + \cdots$$
$$= a\cos\varpi\sin\overline{\lambda} - a\sin\varpi\cos\overline{\lambda} + b\cos 2\varpi\sin 2\overline{\lambda} - b\sin 2\varpi\cos 2\overline{\lambda}$$
$$+ c\cos 3\varpi\sin 3\overline{\lambda} - c\sin 3\varpi\cos 3\overline{\lambda} + \cdots$$

$\overline{\lambda}$	longitude of the mean Sun
ϖ	longitude of perihelion

The coefficients a, b, c are:[19]

$$a = \frac{1}{4}\left(8e - e^3\right) \qquad b = \frac{5}{4}e^2 \qquad c = \frac{13}{12}e^3$$

e eccentricity

16 Delambre, Bürg, 1806, (p.30). For obliquity of the ecliptic DELAMBRE uses the ω sign.

17 For the values of nutation, see Delambre, Bürg, 1806, Table XIII.

18 Delambre, 1814, p.198. Also Delambre, Bürg, 1806. The equation of the centre for the year 1810 is tabulated in Table XII, pp.195–207, and the table with the equation of time in Table VIII, pp.187–190.

19 For DELAMBRE's values see Delambre, 1814, Chapter XX, p.7. See Equation 15.31 also Schaub, 1950, Vorlesungen über Sphärische Astronomie, p.38 or Meeus, 2000, Mathematical Astronomy Morsels, p.341.

The reduction of the ecliptic to the equator as given by DELAMBRE is:[20]

[13.9]
$$\frac{1}{15}R = -\left(\frac{\tan^2\frac{\varepsilon}{2}}{\sin 15''}\right)\sin(2\bar{\lambda}+2C) + \left(\frac{\tan^4\frac{\varepsilon}{2}}{\sin 30''}\right)\cos(4\bar{\lambda}+4C)$$
$$-\left(\frac{\tan^6\frac{\varepsilon}{2}}{\sin 45''}\right)\sin(6\bar{\lambda}+6C) + \cdots$$
$$= -\left(\frac{\tan^2\frac{\varepsilon}{2}}{\sin 15''}\right)\cos 2C \sin 2\bar{\lambda} - \left(\frac{\tan^2\frac{\varepsilon}{2}}{\sin 15''}\right)\sin 2C \cos 2\bar{\lambda}$$
$$+\left(\frac{\tan^4\frac{\varepsilon}{2}}{\sin 30''}\right)\cos 4C \sin 4\bar{\lambda} + \left(\frac{\tan^4\frac{\varepsilon}{2}}{\sin 30''}\right)\sin 4C \cos 4\bar{\lambda}$$
$$-\left(\frac{\tan^6\frac{\varepsilon}{2}}{\sin 45''}\right)\cos 6C \sin 6\bar{\lambda} - \left(\frac{\tan^6\frac{\varepsilon}{2}}{\sin 45''}\right)\sin 6C \cos 6\bar{\lambda}$$

$\bar{\lambda}$ longitude of the mean Sun
ε obliquity of the ecliptic
C equation of the centre

20 Delambre, Bürg, 1806 or Delambre, 1814, p.198. Equation follows from LAGRANGE's equation or Equation 12.18, where the true longitude $\lambda_\odot$ is substituted by $\bar{\lambda} + C$.

From Equation 13.7, we get DELAMBRE's equation of time, including allowance for perturbations:[21]

$$
\begin{aligned}
E = &-\left(\frac{t\sin^2 a}{\sin 30''}\right)\sin\varpi + \left(\frac{bt}{15}\right)\sin 2\varpi + \left(\frac{a}{15}\right)(1+t)\cos\varpi\sin\bar{\lambda} \\
&-\left(\frac{a}{15}\right)(1-t)\sin\varpi\cos\bar{\lambda} - \left(\frac{t\cos^2 a}{\sin 15''}\right)\sin 2\bar{\lambda} + \left(\frac{t\sin^2 a}{\sin 60''}\right)\cos\varpi\sin\bar{\lambda} \\
&+\left(\frac{b}{15}\right)(1-t)\cos 2\varpi\sin 2\bar{\lambda} - \left(\frac{b}{15}\right)(1+t)\sin 2\varpi\cos 2\bar{\lambda} + \left(\frac{t^2\sin^2 a}{\sin 60''}\right)\sin 2\varpi\cos 2\bar{\lambda} \\
&-\left(\frac{at}{15}\right)(1+t)\cos\varpi\sin 3\bar{\lambda} + \left(\frac{at}{15}\right)(1-t)\sin\varpi\cos 3\bar{\lambda} + \left(\frac{t^2\cos^2 2a}{\sin 30''}\right)\sin 4\bar{\lambda} \\
&-\left(\frac{t^2\sin^2 a}{\sin 30''}\right)\cos 2\varpi\sin 4\bar{\lambda} + \left(\frac{t^2\sin^2 2a}{\sin 30''}\right)\sin 2\varpi\cos 4\bar{\lambda} + \left(\frac{bt}{15}\right)\sin 2\varpi\cos 4\bar{\lambda} \\
&-\left(\frac{bt}{15}\right)\cos 2\varpi\sin 4\bar{\lambda} + \left(\frac{at^2}{15}\right)\cos\varpi\sin 5\bar{\lambda} - \left(\frac{at^2}{15}\right)\sin\varpi\cos 5\bar{\lambda} \\
&-\left(\frac{t^3\cos^2 3a}{\sin 45''}\right)\sin 6\bar{\lambda} + \left(\frac{bt^2}{15}\right)\cos 2\varpi\sin 6\bar{\lambda} - \left(\frac{bt^2}{15}\right)\sin 2\varpi\cos 6\bar{\lambda} \\
&+\frac{1}{15}P + 0''\!.09925\sin N + 0''\!.117\sin(2\bar{\lambda}+N+180°) + 0''\!.013\sin(2\bar{\lambda}-N)
\end{aligned}
$$

where: $t = \tan^2\frac{\varepsilon}{2}$

For the year 1810, the values are: apogee at 9°39′ in the ninth sign of the zodiac or $\varpi = 279°39'$, $\varepsilon = 23°27'50''$, $a = 1°55'27''$ and $b = 72''\!.7$, so DELAMBRE's equation of time for the year 1810 is:[22]

[13.10]
$$
\begin{aligned}
E = &\ 0''\!.047 + 79''\!.378\sin\bar{\lambda} + 435''\!.840\cos\bar{\lambda} - 596''\!.878\sin 2\bar{\lambda} + 1''\!.628\cos 2\bar{\lambda} \\
&- 3''\!.424\sin 3\bar{\lambda} - 18''\!.801\cos 3\bar{\lambda} + 12''\!.949\sin 4\bar{\lambda} - 0''\!.073\cos 4\bar{\lambda} \\
&+ 0''\!.142\sin 5\bar{\lambda} + 0''\!.848\cos 5\bar{\lambda} - 0''\!.373\sin 6\bar{\lambda} + 0''\!.003\cos 6\bar{\lambda} \\
&+ \frac{1}{15}P + 0''\!.09925\sin N + 0''\!.117\sin(2\bar{\lambda}+N+500) - 0''\!.013\sin(2\bar{\lambda}-N)
\end{aligned}
$$

For the contribution of nutation, DELAMBRE lists the following simplifications:[23]

$$0''\!.09925\sin N \approx 0''\!.1\sin N \quad \text{and} \quad 0''\!.117\sin(2\bar{\lambda}+N+500) \approx 0''\!.1\sin(2\bar{\lambda}+N+500)$$

N is the longitude of the ascending node of the Moon. The last part is neglected, and for $2\bar{\lambda}+N+500$, he takes into account the value $-3''\!.1$.

21 Delambre, Bürg, 1806, in this edition it seems to be an error in the first term of the equation. Delambre, 1814, p.199. Also the calculation procedure that results in this equation is highly complex. In the source DELAMBRE gives only the final result.
22 Delambre, Bürg, 1806, p.33.
23 Delambre, Bürg, 1806, p.33.

13.2 Franz von Zach

FRANZ XAVER VON ZACH (1754–1832), Austrian-German astronomer, surveyor, mathematician and officer, was born in Pest and studied in Vienna. He worked in several European cities from Ukraine to England, and was in contact with the greatest names in astronomy of the time. He was a great organiser and was most active in Gotha, Saxony, where he arranged the first astronomical congress.[24] The congress was prompted by a visit from LALANDE, and was an opportunity for meeting and sharing knowledge. ZACH also began the first periodical scientific publication in the field of astronomy, *Monatliche Correspondenz zur Beförderung der Erd- und Himmelskunde*, which also covered the field of geography. He also founded the first formal astronomical association and improved the ephemeris.[25] One of his students was GAUSS.

ZACH also relied on the work of his predecessors, in particular CLAIRAUT, MAYER and DELAMBRE, and was probably provided with THOMAS HORNSBY's observation data for the calculation of the tables published in 1792.[26] ZACH states the equation of the centre for every 10′ of the mean anomaly.[27]

Example 13.1 Using ZACH's tables, determine the true longitude on 11 June 1787 for the meridian through Greenwich.

We use ZACH's tables from 1792, *Tabulæ motuum solis et lunæ novæ et correctæ*; for the initial value of the mean longitude for the year 1787, we find in Table III the value of $9^S\,10°1'28.''42$. The correction of the mean longitude for 11 June is $5^S\,9°40'29.''49$ – see Table IV. An additional correction can be applied for a particular time of day - see Table V.

For the longitude of the apogee, the same tables give us an initial value for 1787 of $3^S\,9°14'54''$ and the correction $+27''$ for 11 June. The mean longitude and longitude of the apogee with corrections are given by:

mean longitude, λ_M :	$2^S\,19°41'57.''91$
apogee, λ_Γ :	$3^S\,9°15'21''$

24 For information on his life and work see: Dick, Schwarz, 2016, Franz Xaver von Zach und die Astronomie seiner Zeit; Brosche, 2014, Zach-Spätlese; Balasz, Brosche, Duerbeck, Zsoldos (eds.), 2004, The European Scientist; Brosche, 2000, Der Astronom der Herzogin, Leben und Werk von Franz Xaver von Zach.

25 Delambre, 1806, Tables du Soleil, and improved in Connaisance des Tems, 1820. Zach, 1806–1807, Tabulae Specialis Aberrationis et Nutationis I, II. Zach, 1807, Tabulae Specialis Aberrationis et Nutationis, Monatliche Correspondenz zur Beförderung der Erd- und Himmelskunde, 16. Band, April, pp.365–388. Wurm, 1817, Ueber die Zeit in welcher Beobachtungen angegeben werden. Zeitschrift für Astronomie und verwandte Wissenschaften, May–June, p.305.

26 Von Zach, 1812, Versuch die Berechnung der Zeit-Gleichung zu erleichtern. Monatliche Correspondenz zur Beförderung der Erd- und Himmelskunde, XXV, April, pp.297–321. Monatliche Correspondenz zur Beförderung der Erd- und Himmelskunde, February 1811, Versuch einer Geschichtlichen Darstellung der Fortschritte der Sternkunde im verfloßenen Deccenio, pp.101–150, in particular p.132. Königliche Gesellschaft der Wissenschaften, 1807, Göttingische Gelehrte Anzeigen, 2. Band, pp.937–947, equation of time, pp.941–942. Delambre, Bürg, 1806, Tables astronomiques, publiées par le Bureau des Longitudes de France, Table VIII.

27 Von Zach, 1792, Tabulae motuum solis novae et correctae ex theoria gravitatis et observationibus recentissimis erutae, equation of the centre, Table XXX.

Arguments for corrections № 2 – № 6 (Table III) and corrections (Table IV) are:

Moon	☽	372	486
Jupiter	♃	646	406
Venus	♀	994	277
Mars	♂	44	207
nutation	☊	209	24

We calculate the sum of the above arguments № 2 – № 6 and ignore the thousands:

Moon	☽	858
Jupiter	♃	52
Venus	♀	271
Mars	♂	251
nutation	☊	233

Then we calculate the mean anomaly:

$$\kappa_M = \lambda_M - \lambda_\Gamma = 2^S 19°41'57.''91 - 3^S 9°15'21'' \approx 11^S 10°26'36''$$

From Table VI on page XXXVIII, we interpolate the value of the equation of the centre for the calculated mean anomaly:

$$c = 37'55.''2$$

From the values of the arguments for the Moon, Jupiter, Venus, Mars and for nutation, we found the following interpolated values in Tables VI, VII, IX, XI, XII and XIII:

Moon	☽	−5.''99
Jupiter	♃	+4.''21
Venus	♀	+7.''96
Mars	♂	−0.''36
nutation	☊	+17.''72

Sum of perturbations and nutation = +23.''54

True longitude is the sum of the mean longitude, the equation of the centre and the perturbations:

$$\lambda_\odot = 2^S\ 20°20'16.''65 \approx 2S\ 20°20'16''$$

	Mean longitude	Longitude of the apogee	№ II ☽	№ III ♃	№ IV ♀	№ V ♂	№ VI ☊	
Initial value for the year 1787	$9^S\,10°1'28''\!.42$ [♑ $10°1'28''\!.42$ / $280°1'28''\!.42$]	$3^S\,9°14'54''$ [♋ $9°14'54''$ / $99°14'54''$]	372	646	994	44	209	Table III
Mean longitude 11 June	$5^S\,9°40'29''\!.49$ [♍ $9°40'29''\!.49$ / $159°40'29''\!.49$]	$+27''$	486	406	277	207	24	Table IV
Mean longitude	$2^S\,19°41'57''\!.91$ [♊ $19°41'57''\!.91$ / $79°41'57''\!.91$]	Longitude of the apogee with corrections $3^S\,9°15'21''$ [♋ $9°15'21''$ / $99°15'21''$]	858	52	271	251	233	
Equation of the centre ☽ ♃ ♀ ♂ ☊	$+37'55''\!.2$ ∗ $-5''\!.99$∗ $+4''\!.21$∗ $+7''\!.96$∗ $-0''\!.36$∗ $+17''\!.72$∗	Mean anomaly = mean longitude – apogee = $2^S\,19°41'57''\!.91 - 3^S\,9°15'21'' \approx 11^S\,10°26'36''$ The result is the same as in the next example without taking into account the Greenwich correction; see the following table.						Table VI Table VII Table IX Table XI Table XII Table XIII
Perturbations	$+23''\!.54$							
True longitude	$2^S\,20°20'16''\!.65$ [♊ $20°20'16''\!.65$ / $80°20'16''\!.65$]							

Table 13.1 Process for calculating true longitude for a given moment using ZACH's tables as given in Example 13.2. Calculation example for 11 June 1787 summarised from: Zach, 1792, Tabulae Motuum Solis, Novae et Correctae, example II, p.50. ∗ Values are interpolated.

Example 13.2 Using ZACH's tables, determine the equation of time on 11 June 1787 for the meridian through Greenwich.

We use ZACH's tables from 1792, *Tabulæ motuum solis et lunæ novæ et correctæ*; for the initial value of the mean anomaly for the year 1787, Table XXVII gives us $6^S\ 0°46'34''$. The correction of the mean anomaly for 11 June is $5^S\ 9°40'2''$ – see Table XXVIII. An additional correction can be applied for a particular time of day - see Table XIX.

For the longitude of the apogee, we find the same value and correction as in Example 13.1. We additionally consider the correction for the location of Gotha in relation to Greenwich, with a difference in meridians of $42'55''$ (Table I). The mean anomaly and the longitude of the apogee with corrections are:

mean anomaly, κ_M : $11^S\ 10°28'21.''75$
apogee, λ_Γ : $3^S\ \ 9°15'21''$

Arguments for corrections № 2 – № 5 and № 7 (Table III) and corrections (Table IV) are:

Moon	☽	372	486
Jupiter	♃	646	406
Venus	♀	994	277
Mars	♂	44	207
nutation	☊	569	422

We calculate the sum of the above arguments and ignore the thousands:

Moon	☽	858
Jupiter	♃	52
Venus	♀	271
Mars	♂	251
nutation	☊	991

From Table XXX, we interpolate the value of the equation of the centre for the calculated mean anomaly:

$$c = 2'31.''99$$

From the values of the arguments for the Moon, Jupiter, Venus, Mars and for nutation, we found the following interpolated values in Tables XXXI–XXXIV and XXXVI:

Moon	☽	$-0.''367$
Jupiter	♃	$+0.''258$
Venus	♀	$+0.''487$
Mars	♂	$-0.''022$
nutation	☊	$-0.''270$

Sum of perturbations = $+0.''356$

For the value of true longitude (see the previous example) Table XXII gives us a value of declination of $23°6'$. For the reduction of the ecliptic to the equator, Table XXXVI gives us a value of $-3'26.''05$.

$$\text{Correction} = \text{perturbations}/\cos^2\delta + \text{argument VII (nutation)} = +0.''151.$$

The equation of time is the sum of the equation of the centre + the reduction of the ecliptic to the equator + correction = $+53.''9$.

	Mean anomaly	Longitude of the apogee	№ II	№ III	№ IV	№ V	№ VII	
Initial value for the year 1787	6^{S} 0°46′34″ [♎ 0°46′34″ 180°46′34″]	3^{S} 9°14′54″ [♋ 9°14′54″ 99°14′54″]	372	646	994	44	569	Table XXVII Table III
Mean anomaly 11 June	5^{S} 9°40′2″ [♍ 9°40′2″ 159°40′2″]	+27″	486	406	277	207	422	Table XXVIII Table IV
Correction of the location of Gotha in relation to Greenwich 42′55″	+1′44″ In the example, the value is 1′44″; in the table, it is 1′45.″75			Correction of the location of Gotha in relation to Greenwich, with the difference of the meridians 42′55″				Table I
Mean anomaly	11^{S} 10°28′21.″75 [♓ 10°28′21.″75 340°28′21.″75]	Longitude of the apogee with corrections 3^{S} 9°15′21″ [♋ 9°15′21″ 99°15′21″]	858	52	271	251	991	
Equation of the centre	+2′31.″99 ∗							Table XXX
☽	−0.″367∗	= +0.″356 (perturbations)						Table XXXI
♃	+0.″258∗							Table XXXII
♀	+0.″487∗							Table XXXIII
♂	−0.″022∗							Table XXXIV
☊	−0.″270∗			nutation				Table XXXVI
True longitude	2S 20°20′16.″65 [♊ 20°20′16.″65 80°20′16.″65]	For the value of the true longitude (see previous table), the declination is: $\delta = 23°6'$						For declination, see Table XXIII
Reduction of the ecliptic to the equator	−3′26.″05							Table XXXVII
Correction	+0.″151	correction (*aequationes minores*) = perturbations/$\cos^2\delta$ + argument VII (nutation, ☊) for declination						Table XXXVI
Equation of time	+53.″9	= equation of the centre + reduction of the ecliptic to the equator + correction (*aequationes minores*) (result in seconds)						

Table 13.2 A procedure for calculating the equation of time using ZACH's tables for June 11, 1787 as given in Example 13.3. Summarised from: Zach, 1792, Tabulae Motuum Solis, Novae et Correctae, example XX, p.84. ∗ Values are interpolated.

13.3 Vienna Ephemerides

13.3.1 Maximillian Hell

MAXIMILLIAN HELL (1721–1792) studied philosophy in Vienna, including mathematics and astronomy, and received his doctorate in 1752. For several years he edited the *Vienna ephemerides*, or *Ephemerides astronomicae ad meridianum Vindobonensem*.[28] In the *Ephemerides*, HELL did not limit himself to tables, but also devoted considerable space to solving various cases in practical astronomy. He is also known for his precise calculation of the parallax of the Sun. His contribution to astronomy was much later appreciated by NEWCOMB, who refuted HELL's allegations of plagiarism concerning his observation of the transit of Venus.[29] Among HELL's pupils was FRANZ DE PAULA TRIESNECKER (1745–1817), who succeeded his teacher as head of the Vienna Observatory. TRIESNECKER is best known for his work in mathematical geography.

HELL built his ephemerides on various sources. The introduction to the 1757 edition attributes data to HALLEY and CASSINI. The 1761 edition credits the works of LACAILLE, MAYER, LALANDE and CASSINI.

The equation of time in the first editions is given in minutes, seconds and sixtieths of seconds[30] for each day of the month, in a column under the heading *Conversio temporis veri in medium* (*conversion of true time to mean time*), and the argument is the day of the month.[31] The tables indicate when the value of the equation of time is added and when it is subtracted. In later editions, the equation of time for the days of the month is found indirectly under the designation *Tempus medium meridie veri*, or "*the mean time at noon*". The argument is the day of the month or year, and the values of the equation of time are given in minutes and seconds to one decimal place.[32] The special feature of these tables is that the equation of time does not have a sign, but the values of the equation of time change by 12 hours when the sign changes, or by 24 hours in later editions. The values corresponding to the equation of time in HELL's *Ephemerides* are obtained as the difference between the true and mean solar time and have the opposite sign to the equation of time we use today.[33]

In the *Ephemerides* for 1764, *Ephemerides astronomicae anni bissexti 1764 ad meridianum Vindobonensem*, HELL published CASSINI's ephemerides of the Sun and MAYER's ephemerides of the Moon in an appendix. In the form of a separate table *Æquationis temporis prope vera, ſeu*

28 The first edition of HELL's work was the 1756 edition, and in the editions from 1793 onwards, the *Ephemerides* were edited by TRIESNECKER and BÜRG. The last edition for 1806 was published in 1805.
29 Newcomb, 1883, On Hell's Alleged Falsification of his Observations of the Transit of Venus in 1769.
30 See Chapter 3, note 58.
31 See for example Hell, 1757.
32 For the explanation see Hell, 1760, p.178.
33 The equation of time in this form is still used in contemporary French astronomical literature.

juxta vulgarem methodum compoſitæ ad annum 1750, he also published an equation of time as a function of the longitude of the Sun and values in minutes and seconds. The values in this table are taken from LACAILLE's *Ephemerides* and are apparently the same as in LACAILLE's 1758 tables.[34] The values are for 1750 and are published in two tables *Æquationis temporis prope veræ, seu juxta vulgarem methodum compositae, ad annum 1750,* the argument being the Sun's true longitude.[35]

The values of the equation of time in the *Vienna Eephemerides* are also available in the same form in the editions edited by TRIESNECKER and BÜRG after HELL's death.

13.3.2 Johann Tobias Bürg

TOBIAS BÜRG (1766–1834) is less well known because of his shadowing by ZACH, DELAMBRE and LITTROW, but his contribution to astronomy is no less important. He grew up in modest circumstances. In Vienna, he studied mathematics and astronomy with HELL and TRIESNECKER. As a student, he began to make astronomical observations. After completing his studies, he began teaching at the Lyceum in Klagenfurt in 1791, where he met JOHANN SÖLLNER, later owner of Wiesenau Castle in the Lavant Valley. He continued his career in Vienna, where he took up a post as assistant at the University Observatory. After 1802, he was astronomer-computant for two years at the Seebergen Observatory near Gotha in Thuringia, where he worked with ZACH, and in 1806 he became professor of mathematics and astronomy in Vienna. In that same year, he was awarded a prize for his lunar ephemerides by the Paris Academy and was knighted in 1808. He was one of the candidates to succeed TRIESNECKER, but LITTROW won the post. We know that he observed a solar eclipse in 1820 and determined the geographical coordinates of Klagenfurt.[36] Despite health problems that blighted his life, he spent the last years of his life creatively at Wiesenau Castle in Carinthia, where he returned in 1825. Some of the astronomical instruments that BÜRG brought with him to Wiesenau are now preserved in the Rudolfinum Museum in Klagenfurt.

Among astronomical calculators, BÜRG was one of the leading names of his time. He was also involved in geodetic surveying, but most of his work was on ephemerides. For several years, together with TRIESNECKER, he produced and published the *Vienna Ephemerides.* His most important contribution is the improved lunar ephemerides, that refined LAPLACE's perturbation theory.[37] To account for subleties in the Moon's complex motion, he added extra expressions to the perturbation calculation and took into account perturbations in the Moon's motion due to the Sun and the oblateness of the Earth. His work between 1813 and 1820 was also the basis for the lunar ephemerides in the *British Nautical Almanac.*

34 Lacaille, 1758, Tabulae Solares Ad Meridianum Parisinum, Tabula I, p.4.
35 Hell (ed.), 1764, appendix, for an explanation with the equation of time, see A3.
36 Bürg, 1823a.
37 Bürg, 1823b, 1824a, 1824b, 1825.

13.4 Joseph Johann von Littrow

JOSEPH JOHANN VON LITTROW (1781–1840) was born in Prague and, after studying in Kazan, directed the Observatory in Buda. He was one of the first astronomers to deduce the existence of the Sun's chromosphere. From 1819 to 1840, he was director of the astronomical observatory in Vienna and professor of astronomy at the University of Vienna. LITTROW wrote several astronomical works and textbooks with a high didactic value. In addition to astronomical topics, his works addressed geometry, chronometry, physics and optics, with an emphasis on the manufacture of telescopes. His popular text *Die Wunder des Himmels* places him among the most important popularisers of science, in particular astronomy, of his time. His books also encompass spherical and theoretical astronomy and gnomonics, and also discuss the equation of time in the following form:[38]

[13.11] $$E = \frac{1}{15}(\alpha_\odot - \underbrace{\bar{\lambda}}_{=\bar{\alpha}})$$

$\alpha_\odot$	right ascension
$\bar{\lambda}$	longitude of the first mean Sun circling the ecliptic
	or right ascension of the second mean Sun circling the equator, $\bar{\alpha}$

LITTROW cites the well-known relationship between right ascension and the Sun's true longitude (Appendix D, Equation D.37):

$$\tan \alpha_\odot = \tan \lambda_\odot \cos \varepsilon$$

LITTROW expanded the equation into power series, whose final result $\alpha_\odot - \lambda_\odot$ or reduction of the ecliptic to the equator, R, is a contribution to the equation of time due to the obliquity of the ecliptic:[39]

[13.12] $$R = \alpha_\odot - \lambda_\odot = -\tan^2\tfrac{\varepsilon}{2}\sin 2\lambda_\odot + \tfrac{1}{2}\tan^4\tfrac{\varepsilon}{2}\sin 4\lambda_\odot - \tfrac{1}{3}\tan^6\tfrac{\varepsilon}{2}\sin 6\lambda_\odot + \cdots$$

R	reduction of the ecliptic to the equator
$\alpha_\odot$	right ascension
$\lambda_\odot$	true longitude
ε	obliquity of the ecliptic

38 The mathematical derivation of the equation of time is given in Littrow, 1830, Vorlesungen über Astronomie, pp.65–67. The equation of time for particular days in a month is tabulated in Littrow, 1838, Gnomonik, oder Anleitung zur Verfertigung aller Arten von Sonnenuhren, pp.105–108 and as a function of the mean longitude in Littrow, 1828, Calendariographie or Anleitung alle Arten Kalender zu verfertigen, p.485, Table V.

39 For the explanation see Littrow, 1830, Vorlesung V, pp.54–67 in particular p.65. The procedure is also given in Todhunter, 1859, Plane Trigonometry, p.236, example 299, or Brünnow, 1881, Lehrbuch der Sphärischen Astronomie, pp.14–18. LITTROW's result is LAGRAGE's equation (Equation 12.18) with a negative sign. See also Chapter 15.

After applying the terms for the true and mean anomaly of the Sun and the longitude of the perihelion:[40]

$$\nu = \lambda_{\odot} - \varpi \qquad [13.13]$$

ν true anomaly of the Sun
$\lambda_{\odot}$ true longitude
ϖ longitude of perihelion

LITTROW's expression for the mean longitude as a function of the perigee and the true longitude, whereby KEPLER's equation must be solved in the process of derivation, can be written as:[41]

$$\bar{\lambda} = \lambda_{\odot} - 2e\sin(\lambda_{\odot} - \varpi) + 2\theta\left(e - \frac{1}{2}\theta\right)\sin 2(\lambda_{\odot} - \varpi) - \theta^2\left(e - \frac{2}{3}\theta\right)\sin 3(\lambda_{\odot} - \varpi) + \cdots \qquad [13.14]$$

Where:[42] $\theta = \dfrac{e}{1 + \sqrt{1 - e^2}}$

e eccentricity

Taking into account the above expressions, we obtain the final expression for the equation of time containing the contributions due to the obliquity of the ecliptic and uneven motion:

$$E = \frac{1}{15}\left(\alpha_{\odot} - \underbrace{\bar{\lambda}}_{=\bar{\alpha}}\right) = \frac{1}{15}\left(\lambda_{\odot} - \bar{\lambda} + \underbrace{\alpha_{\odot} - \lambda_{\odot}}_{\substack{R\,=\,\text{equation 13.12} \\ \text{reduction} \\ \text{of the ecliptic} \\ \text{to the equator}}}\right) \qquad [13.15]$$

$\lambda_{\odot}$ true longitude
$\alpha_{\odot}$ right ascension
$\bar{\lambda}$ longitude of the first mean Sun circling the ecliptic

Since both parts of the equation of time use different reference points in time, a correction is also required. LITTROW thus states the correction of the true longitude - the nutation in the longitude = $-16.''78 \sin \Omega$ and the correction of the right ascension of the second mean Sun - the nutation in right ascension = $-15.''39 \sin\Omega$, where Ω is the longitude of the ascending node of the Moon.

40 See also Chapter 15 or Equations 15.9 – 15.11 and 15.36. The longitude of the perihelion for individual years is published in the *Astronomical Almanac*, section C1.

41 Littrow, 1830, p.65. LITTROW has written series in terms of powers of true anomaly and then plugged in Equation 13.13. See also Chapter 15. Contrary to in modern textbooks, LITTROW uses the symbols e for obliquity of the ecliptic and ε for eccentricity in his works.

42 See Chapter 15, Equation 15.70.

Taking into account the above corrections, LITTROW's equation of time is:

$$E = \tfrac{1}{15}\big(\lambda_{\odot} - 16''.78 \sin \Omega - \underbrace{\bar{\lambda}}_{=\bar{\alpha}} + 15''.39 \sin \Omega + \alpha_{\odot} - \lambda_{\odot}\big) \qquad [13.16]$$

$$E = \tfrac{1}{15}\big(\lambda_{\odot} - \underbrace{\bar{\lambda}}_{=\bar{\alpha}} + \alpha_{\odot} - \lambda_{\odot}\big) - 0''.093 \sin \Omega$$

$\lambda_{\odot}$	true longitude
$\alpha_{\odot}$	right ascension
$\bar{\lambda}$	longitude of the first mean Sun circling the ecliptic
Ω	longitude of the ascending node of the Moon

Taking into account Equations 13.12 and 13.14, we obtain the expression for the contribution to the equation of time due to uneven motion and the obliquity of the ecliptic:[43]

$$\begin{aligned} E = \tfrac{1}{15}\Big(& 2e \sin(\lambda_{\odot} - \varpi) - 2\theta\left(e - \tfrac{1}{2}\theta\right) \sin 2(\lambda_{\odot} - \varpi) \\ & + 2\theta^2\left(e - \tfrac{2}{3}\theta\right) \sin 3(\lambda_{\odot} - \varpi) - \tan^2\tfrac{\varepsilon}{2} \sin 2\lambda_{\odot} \\ & + \tfrac{1}{2}\tan^4\tfrac{\varepsilon}{2} \sin 4\lambda_{\odot} - \tfrac{1}{3}\tan^6\tfrac{\varepsilon}{2} \sin 6\lambda_{\odot} - 0''.093 \sin \Omega\Big) \end{aligned} \qquad [13.17]$$

Taking into account the equivalence:[44]

$$\lambda_{\odot} = \bar{\lambda} + 2e \sin M + \frac{5}{4}e^2 \sin 2M + \frac{e^3}{4}\left(\frac{13}{3} \sin 3M - \sin M\right) + \cdots$$

and the data for the year 1800, the expression of the equation of time for 1800 follows, as LITTROW states in his work. For 1800 the data are:

$$e = 0.0167907 \qquad \varepsilon = 23°27'56''.7 \qquad \varpi = 279°29'0''.0$$

And, for the equation of time (compare with DELAMBRE, 1806 or Equation 13.10):[45]

$$\begin{aligned} E = {} & 79''.36 \sin \bar{\lambda} + 435''.82 \cos \bar{\lambda} - 597''.08 \sin 2\bar{\lambda} + 1''.57 \cos 2\bar{\lambda} \\ & - 3''.42 \sin 3\bar{\lambda} - 18''.80 \cos 3\bar{\lambda} + 13''.25 \sin 4\bar{\lambda} - 0''.20 \cos 4\bar{\lambda} \\ & + 0''.15 \sin 5\bar{\lambda} + 0''.88 \cos 5\bar{\lambda} - 0''.40 \sin 6\bar{\lambda} - 0''.01 \cos 6\bar{\lambda} \\ & - 0''.01 \sin 7\bar{\lambda} - 0''.04 \cos 7\bar{\lambda} - 0''.093 \sin \Omega \end{aligned} \qquad [13.18]$$

43 Littrow, 1830, p.66.
44 Littrow, 1830, p.66.
45 Littrow, 1830, p.66.

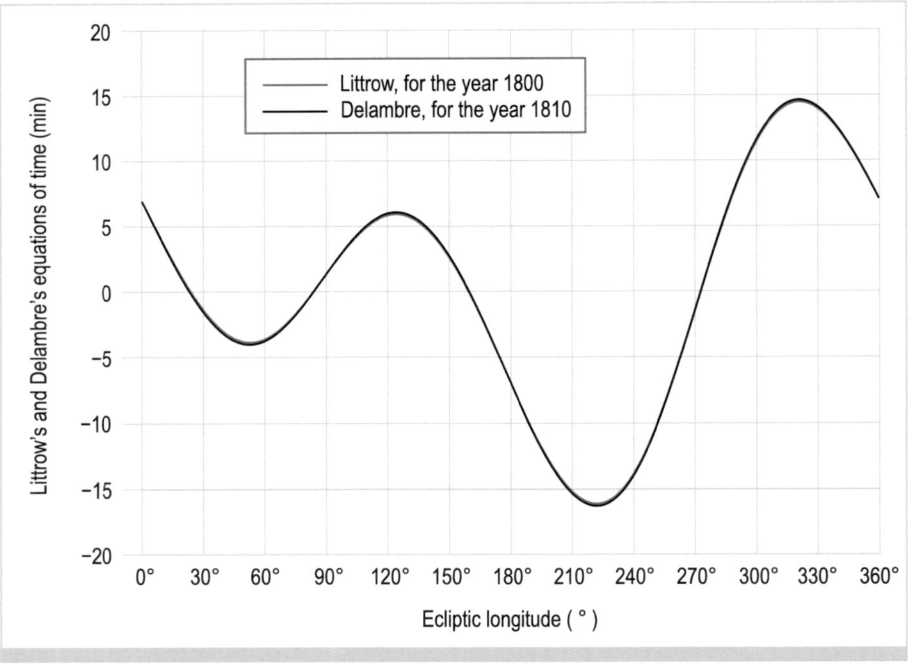

Figure 13.2 The values of LITTROW's and DELAMBRE's equations of time for the years 1800 and 1810. Calculated using Equations 13.10 and 13.18 without taking into account corrections due to nutation.

Bibliography

- ANONYMOUS. Die Declination der Sonne und Zeitgleiche (Equation of Time) für die Jahre 1868, 1869, 1870, 1871 und 1872. Hamburg: Verlag von C.D.S. Gerrits.
- BALÁZS, Lajos G., Peter BROSCHE, Hilmar W. DUERBECK and Endre ZSOLDOS (eds.). 2004. The European Scientist. Symposium on the era and work of Franz Xaver von Zach. Acta Historica Astronomiae, 24, Dick, Wolfgang R. and Jürgen Hammel (eds.). Frankfurt am Main: Harri Deutsch. ISBN 978-3-944913-15-5.
- BRADLEY, James. 1798. Astronomical Observations made at the Royal Observatory of Greenwich, from 1787 to 1789, volume III. London.
- BROSCHE, Peter. 2000. Der Astronom der Herzogin, Leben und Werk von Franz Xaver von Zach 1754-1832. Acta Historica Astronomiae, 12, Dick, Wolfgang R., Hilmar W. Duerbeck and Jürgen Hammel (eds.), 43–90. Frankfurt am Main: Harri Deutsch. ISBN 978-3-944913-06-3.
- BROSCHE, Peter. 2014. Zach-Spätlese. Acta Historica Astronomiae, 54, Dick, Wolfgang R. and Jürgen Hammel (eds.). Leipzig: Akademische Verlagsanstalt. ISBN 978-3-944913-45-2.
- BUREAU DES LONGITUDES. 1820. Connaissance des Temps, ou des mouvemens célestes, pour l'an 1823. Paris: Courcier. (Google Books, 10 December 2023).
- DELAMBRE, Jean Baptiste Joseph. 1787. Mémoire sur les élémens de l'orbite solaire, dans lequel on détermine, par de nouvelles observations, l'apogée, la longitude moyenne & la plus grande équation du soleil. In: Nouveaux Mémoires de l'Académie Royale des Sciences et Belles-Lettres de Berlin, Année 1785, 16, 191–218. Berlin: George Jacques Decker. (Google Books, 31 May 2020).
- DELAMBRE, Jean Baptiste Joseph. 1788. Second mémoire sur les élémens de l'orbite solaire. In: Nouveaux Mémoires de l'Académie Royale des Sciences et Belles-Lettres de Berlin, Année 1786, 17, 234–248. Berlin: George Jacques Decker. (Google Books, 31 May 2020).
- DELAMBRE, Jean Baptiste Joseph and Johann Tobias BÜRG. 1806. Tables astronomiques, publiées par le Bureau des Longitudes de France. Paris: Courcier. URL: https://gallica.bnf.fr/ark:/12148/bpt6k3200419j.texteImage (18 February 2024).
- DELAMBRE, Jean Baptiste Joseph. 1814. Astronomie théoretique et pratique, Book 2. Paris: Courcier. URL: https://gallica.bnf.fr/ark:/12148/bpt6k9641041m.texteImage (18 February 2024).
- DICK, Wolfgang R. and Oliver SCHWARZ. 2016. Franz Xaver von Zach und die Astronomie seiner Zeit. Acta Historica Astronomiae, 59, Dick, Wolfgang R. and Jürgen Hammel (eds.). Leipzig: Akademische Verlagsanstalt. ISBN 978-3-944913-50-6.
- FORBES, Eric. G. 1965. Dr. Bradley's Astronomical Observations. In: Quarterly Journal of the Royal Astronomical Society, 6, 321–328.
- GÖRING, Heinrich. 1862. Der Zeitmesser. Tafeln der Mittagsverbesserung für die Breitengrade 30 bis 60 und der Zeitgleichung für alle Tage. Paderborn: Verlag von Ferdinand Schöningh.
- HAMMER, E. 1893. Zeitbestimmung (Uhr-Kontrole) ohne Instrumente durch Benützung der Ergebnisse einer Landesvermessung. Stuttgart: J. B. Metzlerscher Verlag.
- HELL, Maximillian (ed.). 1757. Ephemerides astronomicae ad meridianum Vindobonensem. Wien: Trattner. URL: https://sammlungen.ulb.uni-muenster.de/hd/periodical/titleinfo/6368875?lang=en (30 September 2023).
- HELL, Maximillian (ed.). 1760. Ephemerides astronomicae ad meridianum Vindobonensem. Wien: Trattner. URL: https://www.manuscriptorium.com/apps/index.php?direct=record&pid=NKCR__-NKCR__14E0000521764FH56Q4-cs (30 September 2023).
- HELL, Maximillian (ed.). 1764. Ephemerides Astronomicae Anni Bissexti 1764 ad meridianum Vindobonensem. Wien: Trattner. (Google Books, 30 September 2023).
- LACAILLE, Nicolas Louis. 1758. Tabulae Solares Ad Meridianum Parisinum. Wien: Trattner. URL: https://www.deutsche-digitale-bibliothek.de/item/CWAAQI5JXTDK5MVM4PBRGAUNSLWZLALK (30 September 2023).
- LAPLACE, Pierre Simon. 1802. Traité de Mécanique Céleste, Part 3. Paris: J. B. M. Duprat. (Google Books, 31 May 2020).
- LITTROW, Joseph Johann. 1821. Theoretische und practische Astronomie, zweyter Theil. Wien: J. B. Wallishausser.
- LITTROW, Joseph, Johann. 1828. Calendariographie oder Anleitung alle Arten Kalender zu verfertigen. Wien: J. G. Heubner.
- LITTROW, Joseph Johann. 1830. Vorlesungen über Astronomie. Wien: J. G. Heubner.

- Littrow, Joseph Johann. 1834. Die Wunder des Himmels; oder gemeinfaßliche Darstellung des Weltsystems. Band 1. Theoretische Astronomie. Stuttgart: Carl Hoffmann.
- Littrow, Joseph Johann. 1838. Gnomonik, oder Anleitung zur Verfertigung aller Arten von Sonnenuhren. Wien: Carl Gerold.
- Littrow, Carl Ludwig. 1842. Erläuterungen zu J. J. v. Littrow's Vorlesungen über Astronomie. Wien: Carl Gerold.
- Maskelyne, Nevil. 1799. Astronomical Observations made at the Royal Observatory of Greenwich, from 1787 to 1789, volume III. London.
- Meeus, Jean. 2000. Mathematical Astronomy Morsels. Second Printing. Richmond, VA: Willmann-Bell. ISBN 0-943396-51-4.
- Newcomb, Simon. 1883. On Hell's Alleged Falsification of his Observations of the Transit of Venus in 1769. In: Monthly Notices of the Royal Astronomical Society, 43, 7, 371–381. DOI 10.1093/mnras/43.7.371a. URL: https://archive.org/details/paper-doi-10_1093_mnras_43_7_371a (18 February 2024).
- Newcomb, Simon. 1960. A compendium of spherical astronomy. Reprint of 1906 edition. New York: Dover.
- Schaub, Werner. 1950. Sphärische Astronomie. Leipzig: Geest & Portig.
- Sonndorfer, Rudolf. 1864. Theorie und Construction der Sonnen-Uhren. Wien: Wilhelm Braumüller. Reprint 1978. Osnabrück: Biblio Verlag. ISBN 3 7648 0613 3.
- Von Zach, Franz Xaver. 1792. Tabulae motuum solis novae et correctae ex theoria gravitatis et observationibus recentissimis erutae. Gotha: Carolum Guilielmum Ettinger. (Google Books. 31 May 2020).
- Von Zach, Franz Xaver. 1804. Tabulae motuum solis novae et iterum correctae ex theoria gravitatis clarissimi de la Place et ex observationibus recentissimis in specula astronomica Ernestina habitis erutae. Supplementum ad tabulas motuum solis anno 1792 editas. Gotha: Libraria Beckeriana. (Google Books. 31 May 2020).
- Von Zach, Franz Xaver. 1807. Tabulae Specialis Aberrationis et Nutationis. In: Monatliche Correspondenz zur Beförderung der Erd- und Himmelskunde, 16, 365–388. URL: https://mdz-nbn-resolving.de/urn:nbn:de:bvb:12-bsb10538607-6 (18 February 2024).
- Von Zach, Franz Xaver. 1811. Versuch einer Geschichtlichen Darstellung der Fortschritte der Sternkunde im verfloßenen Deccenio. In: Monatliche Correspondenz zur Beförderung der Erd- und Himmelskunde, 23, 3–44. URL: https://zs.thulb.uni-jena.de/receive/jportal_jparticle_00319543 (18 February 2024).
- Von Zach, Franz Xaver. 1812. Versuch die Berechnung der Zeit-Gleichung zu erleichtern. In: Monatliche Correspondenz zur Beförderung der Erd- und Himmelskunde, 25, 297–321. URL: https://zs.thulb.uni-jena.de/receive/jportal_jparticle_00321096 (18 February 2024).
- Wilson, Curtis A. 1980. Perturbations and Solar Tables from Lacaille to Delambre: The Rapprochement of Observation and Theory, Part I. In: Archive for History of Exact Sciences, 22, 1/2, 53–188. DOI 10.1007/BF00327870.
- Wilson, Curtis A. 1980. Perturbations and Solar Tables from Lacaille to Delambre: The Rapprochement of Observation and Theory, Part II. In: Archive for History of Exact Sciences, 22, 3, 189–304. DOI 10.1007/BF00348584.
- Woodhouse, Robert. 1821. A Treatise on Astronomy, Theoretical and Practical, Part 1, Volume 1. Cambridge: J. Deighton & Sons and G. & W. B. Whittaker. URL: https://archive.org/details/treatiseonastron11woodrich (18 February 2024).
- Woodhouse, Robert. 1823. A Treatise on Astronomy, Theoretical and Practical, Part 2, Volume 1. Cambridge: J. Deighton & Sons and G. & W. B. Whittaker. URL: https://archive.org/details/p2treatiseonastr01wooduoft (18 February 2024).
- Wurm, Johann Friedrich. 1817. Ueber die Zeit in welcher Beobachtungen angegeben werden. In: Zeitschrift für Astronomie und verwandte Wissenschaften, May–June, 304–307.

14

Reconstruction of ancient tables

The Equation of Time is a Correction, which added to or ſubſtracted from the apparent Time (according to its Title at the Top of the Column) gives equated or mean time, or that which ſhould be ſhewn by a good Clock or Watch.

The Nautical Almanac and Astronomical Ephemeris for the year 1767, p.150.

14.1 Values in ancient tables

14.1.1 Origin of the term equation of time

The ancient Greek astronomers did not have a term for the equation of time – the closest is *nychthemeron*, Gr. νυχθήμερον, which means the length of a combined day and night; for example, the time between two consecutive sunrises. The word was used by PTOLEMY, who distinguished between:[1]

- True or apparent solar days, Gr. ἀνώμαλον νυχθήμερον (*anomalon nychthemeron*); these are the days determined by the movement of the Sun.
- Mean solar days, Gr. ὁμαλον νυχθήμερον (*homalon nychthemeron*); these are the days as defined by clock time in the modern sense of the word.

Greek and Byzantine astronomers used the term ἡ παρὰ τὴυ ἀνισότητα τῶν νυχθημέρων διαφορά, *the difference due to the inequality of days and nights*, and the Arabic astronomers used *the correction of days and their nights*.[2] The Arabic term taʿdīl al-ayām bi laiālaihā was subsequently translated into Latin as *the equation of days with their nights*, Lat. *equatio dierum cum noctibus suis*.[3] COPERNICUS used the term *correction of natural days*, Lat. *diei naturalis differentia*;[4] the medieval term was *the equation of days*,[5] Lat. *aequatio dierum*, *aequationis dierum*, later replaced by the term *equation of natural days*, Lat. *aequatio dierum natural*, *aequationis diei naturalis*, *aequationes dierum naturalium* or Eng. *æquation of natural days*.[6] With the first appearance of clocks, the term *equation of time*, Lat. *æquatio temporis*, also emerged, and was already an established term in the professional literature of the 18th century.

1 Toomer, 1998, Ptolemy's Almagest, p.23. Evans, 1998, The History and Practice of Ancient Astronomy, pp.240–242. The combined length of day and night, the *nychthemeron*, which is today known as the day and is measured from midnight to midnight, was generally measured in antiquity as the time interval between two consecutive sunrises. The ancient sources, from which it is evident that the authors measured the length of the day from midnight to midnight, as is the case today, according to AL-BĪRŪNĪ's allegations, are the astronomical guide *Zīj-i Shāh* and *Ārya-Siddhānta*. See Kennedy, 1983, p.334, and Ansari, 1977, Āryabhata I. His Life and his Contributions, p.10.

2 Van Dalen, 1994, On Ptolemy's Table for the Equation of Time, p.100.

3 In some cases it is also called the *"conjunction of days with their nights"*, Kennedy, 1956, p.141, Chabás, Goldstein, 2003, p.185.

4 Neugebauer, Swerdlow, 1984, Mathematical Astronomy and Copernicus's Revolutionibus, Part 1, p.172.

5 Lat. *aequo*, equalise, straighten. From this comes the English word for equation. *Aequatio dierum = equation of days*. For the word formula see Birkenmajer, 1933.

6 Neugebauer, Swerdlow, 1984, p.172.

14.1.2 Values in ancient tables

The equation of time is often listed in old tables in conjunction with right ascension, and in historical sources its values are listed in several ways:

- The argument is longitude in degrees, and the equation of time is in time units (minutes and seconds). For example, PTOLEMY, AL-KHWĀRIZMĪ and KŪSHYĀR gave values in this way.
- The argument is longitude in degrees, and the values are given in (time) degrees and seconds. In this way they are listed by AL-BATTĀNĪ and in the *Toledan Tables.*
- The argument is the day of the year, and the values are given in time units. We presently give the values of the equation of time in this way, but it was an exception in the Middle Ages. We only know about three such tables from that era; those compiled by JOHN OF MURS, ABRAHAM ZACUT and an unknown author who could have been NICHOLAUS POLONIUS or ABRAHAM ZACUT.[7]

The values of the equation of time, given today in minutes and seconds, are found in the older literature in sexagesimal, and the symbols E_h and E_d (the index h = hours, d = degrees) are used to denote the equation of time. While today the values of the equation are given for individual days of the year, values in the old tables were given for the ecliptic longitudes of the Sun. The values in the old tables can be given in so-called time-degrees, taking into account that the following relationship applies for the hour angle: $^{360°}/_{24h} = 15°/h$, or even more precisely, 15.°041, or sexagesimal 15;2,28°, which helps us convert time-degrees into time units (minutes and seconds).

Regardless of the fact that in ancient times and in the Middle Ages, the equation of time was viewed primarily from the point of view of the changing length of the day – today it means the difference between solar and local time – the required mathematical tools were already at a fairly high level. Some examples of calculations of the equation of time from antiquity and the Middle Ages have been reconstructed, but knowledge of the computational procedures that led to the tables with the equation of time is still relatively modest.

7 For the tables of JOHN OF MURS from 1321, see Chabás, Goldstein, 2009, John of Murs's Tables of 1321, Journal for the History of Astronomy, Vol. 40, no. 3, the equation of time on pp.311–312, and Chabás, Goldstein, 2012, A Survey of European Astronomical Tables in the Middle Ages, pp.40–41. ABRAHAM ZACUT, *Ḥibbur* and *Tabule verificate*, see Chabas, Goldstein, 2018, New evidence on Abraham Zacut's astronomical tables. Chabás, Goldstein, 2000, Astronomy in the Iberian Peninsula: Abraham Zacut and the Transition from Manuscript to Print. Transactions of the American Philosophical Society, vol. 90, Part 2, pp.48 and 108–109. NICHOLAUS POLONIUS was a Polish astronomer who arrived in Salamanca in around 1460 and led the department of astronomy and astrology at the university there until 1464. It is likely that he brought with him the tables entitled *Tabulae resolutae*, a series of *Parisian Alfonsine Tables*. He is probably the author of *Tabule verificate*, see Chabás, Goldstein, 2014, Essays on Medieval Computational Astronomy, pp.60–61, and Chabás, 1998, Astronomy and Salamanca in the Mid-fifteenth Century: The Tabulae Resolutae, Journal for the History of Astronomy, vol. 29, no. 95, pp.167–175.

Author	Year	Jan./Feb.	May	July	Oct./Nov.
Ptolemy [a]	ca. 150 CE	0;33,23 h ♒ 18°	0;6;12 h ♉ 30°	0;16,21 h ♌ 9°	0;0,0 h ♎ 30° – ♏ 3°
Abū Ma'shar [b]	9th century	0;34,00 h ♒			0;0,0 h ♎ 21° – ♏ 4°
Al-Khwārizmī [c]	9th century	0;0,0 h ♒ 22°	0;23,28 h ♉ 30° – ♊ 2°	0;14,40 h ♌ 1° – 7°	0;34,28 h ♏ 9° – 10°
Al-Battānī [d]	9th/10th century	0;0° ♒ 18° – 19°	5;33° ♉ 28° – ♊ 5°	3;4° ♌ 1° – 9°	7;54° ♏ 8° – 10°
Al-Zarqālī [e] *Toledan Tables*	10th century	0;0,0 h ♒ 18° – 19°	0;22,12 h ♉ 27° – ♊ 3°	0;12,16 h ♌ 1° – 9°	0;31,36 h ♏ 8° – 9°
Kūshyar ibn Labbān [f]	10th/11th century	0;0,0 h ♒ 18° – 19°	0;21,56 h ♉ 23° – 26°	0;11,59 h ♌ 5° – 7°	0;31,33 h ♏ 8° – 9°
Al-Baghdādī [g]	1285	– 0;2,14 h ♒ 19° – 21°	0;20,3 h ♉ 25° – 27°	0;10,1 h ♌ 5° – 6°	0;29,22 h ♏ 8°
William of St. Cloud [h]	ca. 1292	25 January, 4 February 0;0 h	30 April, 10 May, 30 May 0;22 h	19 July 0;12 h	17 October, 27 October 0;31 h
Peter of St. Omer [i] John of Lignères [i]	1293–94 14th century	0;0° ♒ 18° – 25°	5;21° ♉ 25° – 27°	2;49° ♌ 5°	7;57° ♏ 8° – 9°
Peter of Dacia [j]	ca. 1300	0;0 h ♒ 12°	0;22 h ♉ 16°	0;12 h ♋ 28°	0;32 h ♏ 6°
John of Murs [k]	1321	0;0,0 h 24 Jan. – 12 Feb.	0;20,58 h 6 May	0;12,0 h 18 – 24 Jul.	0;32,38 h 24 Oct.
Immanuel ben Jacob Bonfils [l]	ca. 1350	0;0,0 h ♒ 15° – 20° 0;0° ♒ 15° – 19°	0;24,52 h ♉ 28° – ♊ 1° 6;13° ♉ 28° – ♊ 1°	0;15,52 h ♌ 2° – 6° 3;58° ♌ 2° – 6°	0;33,20 ♏ 3° – 8° 8;20° ♏ 3° – 8°
Levi ben Gerson [m]	14th century	0;0° ♒ 20° – 24°	5;8° ♉ 24° – 28°	2;45° ♌ 1° – 5°	8;13° ♏ 9° – 10°
Abraham Zacut? [n]	ca. 1460	0;0 h 22 Jan. – 6 Feb.	0;23 h 30 Apr.	0;12 h 12 – 26 Jul.	0;32 h 20 – 24 Oct.
Abraham Zacut [l]	15th century	0;0 h 22 Jan. – 6 Feb.	0;22 h 27 Apr. – 21 May	0;12 h 12 – 26 Jul.	0;32 h 20 – 24 Oct.

Table 14.1 Some examples of the maximum and minimum tabulated values of the equation of time from antiquity and the Middle Ages. Adapted from sources mentioned in the notes. For epochs, see Tables 14.6 and 14.7.

Notes

a *Handy Tables*, Gr. Πτολεμαίου Πρόχειοι Κανόνες. Halma, 1822, part I, pp.148–153. Mercier, Tihon, 2011, pp.89–96. Also Van Dalen, 1994, p.119.

b Pingree, Kennedy (trans., ed.), 1981, The Book of the Reasons Behind Astronomical Tables, pp.155–157.

c Suter, 1914, Die astronomischen Tafeln des Muhammed ibn Mūsā al-Khwārizmī in der Bearbeitung des Maslama ibn Ahmed al-Madjrītī und der lateinischen Übersetzung des Adelard von Bath, pp.80–81. Also, Van Dalen, 1996, Al-Khwārizmī's astronomical tables revisited: analysis of the equation of time. In: Casulleras, Samsó, (eds.), From Baghdad to Barcelona: Studies in the Islamic Exact Sciences in Honour of Prof. Juan Vernet. Universitat de Barcelona, vol. 1, pp.242–244, Table 4.

d Nallino, 1907, Al-Battānī sive Albatenii, Opus Astronomicum, II, pp.61–64. Van Dalen and Pedersen, 2008, Re-editing the tables in the Sābi' Zīj by al-Battānī, Acta Historica Leopoldina, vol. 54. Comparison with AL-KHWĀRIZMĪ's values: 7°54′ = 31 min 36 s, Suter, 1914, p.80.

e Pedersen, 2002, Toledan Tables, pp.979–980.

f The argument is mean solar motion. Van Dalen, 2021, Ptolemaic Tradition and Islamic Innovation: The Astronomical Tables of Kūshyār ibn Labbān, pp.121–122.

g Van Dalen, 1993, Ancient and Mediaeval Astronomical Tables: mathematical structure and parameter values, pp.146–149. The probable author of the values is ḤABASH AL-ḤĀSIB.

h William of St. Cloud, *Almanach planetarum*, Pedersen (ed.), 2014, p.30. Values are given for three or four days for each month.

i Pedersen, 2002, pp.984–985, BB31. Chabás, Goldstein, 2009, p.312 also Chabás, Goldstein, 2012, p.40. The real author seems to be PETER OF ST. OMER in the treatise of *Tractatus de semissis* of 1293–94. The equation is also associated with JOHN OF LIGNÈRES, who worked later. Chabás, Saby, 2023, pp.72–75. See also Pal. Lat. 1412, folio 101v, Biblioteca Apostolica Vaticana. Pietroni, 2022, p.94. Chabás, Goldstein, 2003, p.186, Bodleian Library, Oxford, MS Can. Misc. 27, f.79r. Tabulae Resolutae, a table from the Jagiellonian Library, Krakow has the same value of 7;57°, MS 1865, ff.84v–85r, Chabás, Goldstein, 2012, p.39. For the equation of time in the *Alfonsine Tables*, see also Delambre, 1819, Astronomie du Moyen Age, pp.249–250.

j Petrus de Dacia, *Tractatus instrumenti eclipsium*, Pedersen (ed.), 1978, p.49. Also Pedersen (ed.), 1983, Petri Philomenae de Dacia et Petri de S. Audomaro, volume I, p.505. Values are given only for some values of the longitude for each sign of the zodiac. Also similar is the equation of time table as used at St Albans by RICHARD OF WALLINGFORD at the beginning of the 14th century. See North, 1976, III, p.100, *Tabula equacio dierum in minuta horarum resolute.*

k Chabás, Goldstein, 2009, John of Murs's Tables of 1321, p.311. Lisbon, MS Ajuda, 52-XII-35, f.59r.

l Goldstein, Chabás, 2017, Analysis of the astronomical tables for 1340 compiled by Immanuel ben Jacob Bonfils, pp.76–78. Values in time units are from Bayerische Staatsbibliothek MS Heb. 386, f.15b in MS Heb. 343, f.185b. Values in time-degrees are from Biblioteca Medicea Laurenziana MS 2597, 45a–72a. Also Goldstein, 2017, The Introduction to Immanuel ben Jacob Bonfils' Tables for 1340, Aleph, vol. 17, no. 1, pp.167–176. Chabás, Goldstein, 2000, Astronomy in the Iberian Peninsula: Abraham Zacut and the Transition from Manuscript to Print, pp.108–109; New York, JTSA, MS Heb. 2597, f.59b, the values are in time degrees with a maximum value of 8;20°. BONFILS obviously summarised the values from the tables of ABRAHAM BAR ḤIYYA, which were based on AL-BATTĀNĪ's tables. He worked in the 14th century and very little is known about his life. He is the author of a treatise on the equation of time; Solon, 1971, p.10, and Steinschneider, 1964, p.160. He is also the author of works on solar eclipses. Solon, 1971, The Six Wings of Immanuel Bonfils and Michael Chrysokokkes, Centaurus, vol. 15, no. 1, pp.1–20. For more information on his life and work, see Steinschneider, 1964, Mathematik bei den Juden, pp.155–165, and Sarton, 1927–48, Introduction to the History of Science, III, pp.1517–1520.

m Goldstein, 1974, pp.102–104 and 167.

n ABRAHAM ZACUT, alleged author of Tabule Verificate, Salamanca. Chabás, Goldstein, 2018, New evidence on Abraham Zacut's astronomical tables. Chabás, Goldstein, 2000, Astronomy in the Iberian Peninsula: Abraham Zacut and the Transition from Manuscript to Print, pp.24, 29 and 108. Biblioteca Nacional, Madrid, MS 3385 ff.104r–113r, equation of time f.108r.

o Zacuto, 1473, Almanach Perpetuum Celestium Motuum (reprint 1915), p.42. *Tabule verificate, Salamanca*, with the alleged start date of January 1, 1464, were very likely used by ZACUT as the basis for the tables published in *Almanach perpetuum*, 1496 and *Ḥibbur*, in addition to the *Alfonsine Tables*. Chabás, Goldstein, 2000, pp.29, 56, 108.

Author	Year	Jan./Feb.	May	July	Oct./Nov.
Alfonsine Tables [a]	1483 Edition	0;0° ♒ 18° – 19°	5;33° ♉ 28° – ♊ 5°	3;4° ♌ 1° – 9°	7;54° ♏ 8° – 9°
Alfonsine Tables [b]	1492 Edition 1545 Edition	0;0° ♒ 18° – 25°	5;21° ♉ 25° – 27°	2;50° ♌ 2° – 6°	7;57° ♏ 8° – 9°
Alfonsine Tables	1492 Edition 1545 Edition	≈0;0 min ♒ 21° – 26°	20;48 min ♉ 28° – 30°	11;40 min ♌ 3° – 5°	32;52 min ♏ 7° – 8°
Ulugh Beg [c] ∗	15th century	≈0;0,0 h ♒ 21°	≈0;19,59 h ♉ 24° – 26°	≈0;9,57 h ♌ 3° – 5°	≈0;31,7 h ♏ 9° – 10°
Jamshīd al-Kāshī [d] ∗	15th century	≈0;0,0 h ♒ 20°	≈0;20,38 h ♉ 28°	≈0;10,58 h ♌ 5°	≈0;31,35 h ♏ 10°
John of Gmunden [e]	early 15th century	0;0° ♒ 18° – 25°	5;21° ♉ 25° – 27°	2;49° ♌ 5°	7;57° ♏ 8° – 9°
Joannes Holibrocus [f]	early 15th century	0;0° ♒ 14° – 21°	4;19° ♉ 23°	3;6° ♋ 30° – ♌ 5°	8;28° ♏ 9° – 10°
Georg Peuerbach [g]	1459	0;0 min ♒ 20°–23°	20;56 min ♉ 24°–30°	11;56 min ♋ 29° – ♌ 2°	32;48 min ♏ 7° – 10°
Johannes Stöffler [h]	1514	0;0 min ♒ 21° – 26°	20;48 min ♉ 28° – ♊ 1°	11;40 min ♌ 3° – 5°	32;52 min ♏ 7° – 8°
Giovanni Bianchini [i]	1523	0;0,0 h ♒ 18° – 19°	0;22,12 h ♉ 28° – ♊ 5°	0;12,16 h ♌ 1° – 9°	0;31,36 h ♏ 8° – 9°
Giovanni Bianchini [j]		♒ 20° – 23°	♉ 24° – 29°	♌ 1° – 6°	♏ 7° – 10°
Henri Baers ou Vekenstyl [k]	1528	0;0,0 h ♒ 21° – 26°	0;20,48 h ♉ 28° – 30°	0;11,44 h ♋ 29° – ♌ 2°	0;32,52 h ♏ 7° – 8°
Peter Apian [l]	ca. 1540	0;0,0 h 31 Jan.	0;20,54 h 11 May	0;12,00 h 11 – 12 Jul.	0;32,42 h 20 Oct.
Johannes Virdung [m]	1542	0;0 min ♒ 21° – 26°	20;48 min ♉ 28° – ♊ 1°	11;40 min ♌ 3° – 5°	32;52 min ♏ 7° – 8°
Rheticus, Acronius [n]	1542	0;0 min ♒ 21° – 26°	20:48min ♉ 28° – 30°	11;40 min ♌ 3° – 5°	32;52 h ♏ 7° – 8°
Pietro Pitati [o]	1544	0;0,0 h ♒ 21° – 26°	0;20,48 h ♉ 28° – 30°	0;11,40 h ♌ 3° – 5°	0;32,52 h ♏ 7° – 8°
Erasmus Reinhold [p]	1551 1585	0;0,0 h ♒ 21° – 23°	0;18,49 h ♉ 23° – 25°	0;8,20 h ♌ 3° – 5°	0;30;31 h ♏ 10° – 11°
		0;8,9 h ♒ 22°	0;10,41 h ♉ 23° – 24°	0;0,11 h ♌ 4°	0;22;22 h ♏ 10° – 11°

Table 14.2/1 Some examples of the maximum and minimum tabulated values of the equation of time from the 15th to 17th century. Adapted from sources mentioned in the notes. ∗ Calculated values. For epochs, see Tables 14.6 and 14.7.

Author	Year	Jan./Feb.	May	July	Oct./Nov.
Georg Peuerbach [q]	1553	0;0,0 h ♒ 20° – 23°	0;20,56 h ♉ 24° – 30°	0;11,52 h ♌ 1° – 6°	0;32,48 h ♏ 7° – 10°
Erasmus Oswald Schreckenfuchs [r]	1556	0;0,0 h ♒ 21° – 26°	0;20,48 h ♉ 29° – 30°	0;11,40 h ♌ 3° – 5°	0;32,51 h ♏ 7° – 8°
Michael Mästlin [s]	1580	S 8;9 min ♒ 22°	A 10;41 min ♉ 23° – 24°	A 0;11 min ♌ 4°	A 22;22 min ♏ 10° – 11°
Johannes Stadius [t]	Edition of 1585	8;9 min (negative value) ♒ 22°	10;1 min ♉ 23° – 24°	0;11 min ♌ 4°	22;22 min ♏ 10° – 11°
Giovanni Antonio Magini [u]	1607	A 8 min ♒ 15° – 29°	S 12 min ♉ 19° – ♊ 2°	S 2 min ♋ 24° – ♌ 13°	S 24 min ♎ 29° – ♏ 19°
David Origanus [v]	1609	S 7;25 min ♒ 21° – 23°	A 11;11 min ♉ 23° – 24°	A 0;39 min ♌ 4° – 6°	A 23;3 min ♏ 10° – 12°
	1609	S 8;10 min ♒ 19° – 23°	A 12;0 min ♉ 23° – 27°	A 2;0 min ♋ 30° – ♌ 6°	A 24;20 min ♏ 8° – 10°
Johannes Kepler [x]	1627	0;15 h ♒ 12° – ♓ 4°	0;4 h ♉ 16° – ♊ 4°	0;5 h ♋ 20° – ♌ 16°	0;17 h ♏ 10°
Christian Longomontanus [w]	ca. 1622	0;0,0 h ♈♋ 0°	0;9,56 h ♉♏ 15° – 17°	0;9,56 h ♒♌ 13° – 15°	0;0,0 h ♎♑ 30°
Ismael Boulliau [y]	1645	A 12;34 min ♒ 18°	S 6;55 min ♉ 20° – 21°	A 7;34 min ♌ 8° – 9°	S 13;8 min ♏ 12° – 13°
Jeremy Shakerley [z]	1653	A 12;34 min ♒ 18°	S 6;55 min ♉ 20° – 21°	A 7;34 min ♌ 8° – 9°	S 13;8 min ♏ 12° – 13°
Vincent Wing [a1]	1656	S 12;31 min ♒ 18°	A 6;50 min ♉ 21°	S 7;34 min ♌ 8° – 9°	A 13;9 min ♏ 12°

Table 14.2/2 Some examples of the maximum and minimum tabulated values of the equation of time from the 15th to 17th century. Adapted from sources mentioned in the notes. ∗ Calculated values. For epochs, see Tables 14.6 and 14.7.

Notes

a Edition from the year 1483, Chabás, 1998, pp.169–170.
b Tabulae Astronomice Alfonsi Regis, 1492, *Tabula equationis dierum com noctibus suis vetum*, values in time degrees. Table with the same minimum and maximum values is also available in Manuscript Pal. Lat. 1445, ff.222r–223v, Biblioteca Apostolica Vaticana. Pedersen, 2002, p.985. *Tabula equationis dierum com noctibus: moderna verificata ad annos Christi 1456*, values in minutes and seconds. Alfonso X, Astronomicae Tabulae, 1545, p.273.
c The values are calculated according to the procedure of Example 3.3 and are not a copy of the original tables. For more information, see Gauss, 1929, Werke 12, pp.64–68. Also Delambre, 1819, p.210.
d The values are calculated according to the procedure of Example 3.2. The argument is the true longitude of the Sun. For an explanation, see Kennedy, 1988, Two Medieval Approaches to the Equation of Time, Centaurus, vol. 31, no. 1, pp.2–3.
e Manuscript Cent. VI, 18, ff.46v–47r, Johannes de Gamundia, Tabulae, *Tabula aequationis dierum com noctibus suis*, Stadtbibliothek Nürnberg.
f Data extracted from Pietroni, 2022, p.124, Figure 49. Data source Egerton manuscript 889, ff.150v–151r.
g Peuerbach, 1514, Tabulae eclypsium magistri Georgij Peurbachij, *Tabula equationis dierum*. See also Pietroni, 2022, p.153, Figure 57.
h Johannes Stöffler, 1514, Justingensis germani Tabulae Astronomicae, *Tabula aequationis dierum cum noctibus moderna*.
i The values are from Bianchini's tables Tabulae Astronomiae, 1523. Adapted from Chabás, Goldstein, 2009, The Astronomical Tables of Giovanni Bianchini, pp.100–102. For the table from the 1526 Edition see Pietroni, 2022, pp.136–137. For another different table with Bianchini's equation of time see the manuscript Pal. Lat. 1375, f.258v, *Tabula horarum meridiei et equationis dierum ad meridianum ferarrie et bononie.*
j Manuscript Pal. Lat. 1375, f.90r, *Tabula equanoms dierum*, Biblioteca Apostolica Vaticana. See also Pietroni 2022, p.139, Figure 54.
k Tables astronomiques de Louvain, 1528. Meeus, Savoie, 1995, L'Équation du temps, p.188.
l The values are adapted from Apianus, 1540, Astronomicum Caesareum, and Ionides, 1934, Medieval Astronomical Constants, Publications of the Astronomical Society of the Pacific, 46, 274, table on pp.330–331.
m Johannes Virdung, 1542, Tabulae resolutae de supputandis siderum motibus, *Tabula aequationis dierum cum noctibus*.
n Rheticus, Acronius, 1542, Tabulae Astronomicae In Gratiam Studiosae Iuventutis Seorsim Aeditae: *I De aequatione dierum*.
o Petrus Pitatus, 1544, Almanach Novum Petri Pitati Veronensis Mathematici, p.93.
p The values are adapted from *Prutenicae tabulæ coelestium motuum*, 1551 or 1585. The work of Erasmus Reinhold contains two tables with an equation of time. One with a comment from Ptolemy and Copernicus (Reinhold, 1585, f.5), the other by Regiomontanus (Reinhold, 1585, f.6). In the table the values which are supposed to follow Regiomontanus' teachings are given. The values are always positive, while the values of the equation of time which follow the principles of Ptolemy and Copernicus are partially negative. The change in the sign is given in the table at ♈ 3° and ♑ 16°.
q Luminarium atque planetarum motuum tabulae octoginta quinque, omnium ex his quae alphonsum sequntur quàm faciles, 1553. The author of the table is Peuerbach, and the authors of the work are Bianchini, Prugnero and Peuerbach.
r Commentaria, in Nouas theoricas planetarum..., 1556, p.38.
s Mästlin, 1580, Ephemerides novæ.
t Johannes Stadius, 1585, Ephemerides Joannis Stadii, *Canon equationis dierum naturalium ex Copernici doctrina*, p.120.
u Magini, 1607, Continuatio ephemeridum coelestium motuum, *Tabella æquationis dierum naturalium Tychonica*, str. 7. Values are in whole minutes.
v David Origanus, 1609, Nouae motuum coelestium ephemerides Brandenburgicae, annorum 60, incipientes ... pp.98, 100. The first time equation is based on the *Prutenic Tables* and has similar values to those of Stadius and other proponents of the *Alfonsine Tables*, while the second time equation applies to Tycho's equation of time.
w Longomontanus' equation of time is very similar to Brahe's; it contains only one part of the equation of time – the contribution due to the obliquity of the ecliptic. Source: Astronomia Danica, 1622, Pars Altera, Book I, Chapter 3, p.59, Table *Tabula æquationis diei naturalis*. Similar to Brahe and Longomontanus, this form of the equation of time was also advocated by Ismael Boulliau, Lat. Ismael Bullialdus.
x Reichert (ed.), 2014, Rudolphine Tables, p.178.
y Boulliau, 1645, Astronomia Philolaica, *Tabula æquationis dierum ciulium* (9).
z Shakerley, 1653, Tabulæ Britanicæ, *A Compounded Table of the Æquation of Time*, microfiche, source: Bayerische Staatsbibliothek.
al Wing, 1656, Astronomia Instaurata, Table (4) and Table (5), *A Table of the Aequation of dayes*.

In addition to PTOLEMY's work, we at least basically understand the calculations of the equation of time from the Islamic period, from astronomers AL-KHWĀRIZMĪ, KŪSHYĀR, AL-KĀSHĪ and ULUGH BEG.[8] Different authors often used different arguments for the equation of time, but also common are examples of different tables from the same author, calculated for different epochs or other conditions, which make direct comparisons or mathematical reconstruction of the equations of time by various authors highly difficult if not impossible.

Example 14.1 Calculate the equation of time in degrees at the ecliptic longitude of the Sun 120°. The equation of time in the sexagesimal notation is 0;15,29, see Table 1.1 (solution: $E = 3°54'15''$):

$$0;15{,}29 = \frac{15}{60} + \frac{29}{60^2} = 0.258056$$

Taking into account that the hour angle is 15°, we get the following result:

$$15° \cdot 0.258056 = 3\overset{\circ}{.}870833 = 3°52'15''$$

In historical literature, the more precise expression 15;2,28° is sometimes used for the hour angle; see van Dalen, 1994, p.104.

The arc minutes and seconds that we get from the decimal part of the calculated result:

$$0\overset{\circ}{.}870833 \cdot 60 = 52'.25$$

$$0'.25 \cdot 60 = 15''$$

A lot of information on ancient tables is available in various commentaries from ancient authors, for example THEON's commentary on PTOLEMY's table. However, the original method of calculating values in the tables is not always apparent from the comments.[9]

The equation of time in the old sources is not indicated with a sign, but as an absolute value. However, the sources generally state whether the value of the equation of time is added to or subtracted from (solar) time. e.g. A = *add*, S = *subtract*, *Le soleil retarde* R ⊙, *Le soleil avance* A ⊙, ⊙ *before*, ⊙ *behind*, *equal time slower*, *equal time faster*, *clock or watch too slow* etc.

8 Van Dalen, 1996, al-Khwārizmī's astronomical tables revisited: analysis of the equation of time, In: Casulleras, Samso (ed.), From Baghdad to Barcelona: Studies in the Islamic Exact Sciences in Honour of Prof. Juan Vernet, vol. 1. Van Dalen, 1993, Ancient and Mediaeval Astronomical Tables: Mathematical Structure and Parameter Values. Kennedy, 1988, Two Medieval Approaches to the Equation of Time, Centaurus, vol. 31, no. 1, pp.1–8.

9 THEON's comments are available in Tihon (ed.), 1978, Le 'Petit Commentaire' de Théon d'Alexandrie aux 'Tables faciles' de Ptolémée, Tihon (ed.), 1985, Le 'Grand commentaire' de Théon d'Alexandrie aux 'Tables faciles' de Ptolémée Livre I, Tihon (ed.), 1991, Le 'Grand commentaire' de Théon d'Alexandrie aux 'Tables faciles' de Ptolémée Livre II, III, and Tihon (ed.), 1999, Le 'Grand commentaire' de Théon d'Alexandrie aux 'Tables faciles' de Ptolémée Livre IV.

Author	Year	Jan./Feb.	May	July	Oct./Nov.
CHRISTIAAN HUYGENS [a]	1665 1673	0 min 0 s 30 Jan. – 1 Feb. 0 min 0 s 9 – 11 Feb.	19 min 29 s 3 – 5 May 19 min 29 s 13 – 15 May	9 min 46 s 14 – 16 Jul. 9 min 46 s 24 – 26 Jul.	31 min 55 s 20 – 23 Oct. 31 min 55 s 30 Oct. – 2 Nov.
JOHN FLAMSTEED [b]	1672	15 min 3 s ♒ 21° – 23°	4 min 4 s ♉ 24° – 26°	5 min 41 s ♌ 3° – 4°	16 min 16 s ♏ 9° – 11°
FLAMINIO MEZZAVACCA [c]	1675	S 15 min 0 s ♒ 22° – 24°	A 4 min 13 s ♉ 24°	S 5 min 48 s ♌ 4° – 6°	A 16 min 9 s ♏ 9° – 11°
JOHN SMITH [d]	1678–1694	0 min 0 s 31 Jan. – 1 Feb.	19 min 29 s 4 – 5 May	9 min 46 s 15 – 16 Jul.	31 min 57 s 21 – 22 Oct.
Connoissance des temps [e]	1679	A 15 min 7 s 10 Feb.	T 4 min 1 s 13 – 16 May	A 5 min 54 s 25 – 28 Jul.	T 16 min 2 s 2 Nov.
THOMAS TOMPION [f]	1683	*watch too fast* 14 min 46 s 30 Jan. – 1 Feb.	*watch too slow* 4 min 14 s 2 – 5 May	*watch too fast* 5 min 56 s 15 – 17 Jul.	*watch too slow* 16 min 5 s 22 – 24 Oct.
WILLIAM MOLYNEUX [g]	1686	A 14 min 54 s 1 Feb.	S 4 min 17 s 4 May	A 5 min 49 s 15 – 17 Jul.	S 16 min 4 s 23 – 24 Oct.
PHILIPPE DE LA HIRE [h]	1687	A 10 min 2 s ♒ 20°	S 9 min 14 s ♉ 22° – 24°	A 0 min 48 s ♌ 5° – 6°	S 20 min 58 s ♏ 11° – 12°
GIOVANNI DOMENICO CASSINI [i]	1694	S 15 min 0 s ♒ 22° – 24°	A 4 min 13 s ♉ 24°	S 5 min 48 s ♌ 4° – 6°	A 16 min 9 s ♏ 9° – 11°
JOHANN PHILIPP VON WURZELBAU [j]	~1700	A 15 min 2 s ♒ 22° – 23°	S 4 min 10 s ♉ 23° – 25°	A 5 min 57 s ♌ 2° – 5°	S 16 min 23 s ♏ 9° – 11°
WILLIAM WHISTON [k]	1707 1715	A 15 min 0 s ♒ 22° – 24°	S 4 min 13 s ♉ 24°	A 5 min 48 s ♌ 4° – 6°	S 16 min 9 s ♏ 9° – 11°
JOHANN GABRIEL DOPPELMAYR [l]	1708	14 min 49 s 30 Jan. – 1 Feb.	4 min 13 s 4 May	5 min 46 s 15 – 16 Jul.	16 min 1 s 22 Oct.
JOHN SMART [m]	1702–1710	*clock or watch too fast* 14 min 49 s 30 – 31 Jan.	*clock or watch too slow* 4 min 13 s 4 May	*clock or watch too fast* 5 min 46 s 15 – 16 Jul.	*clock or watch too slow* 16 min 1 s 22 Oct.
JOHN HARRISON [n]	1717	15 min 30 Jan.	4 min 5 – 10 May	6 min 10 – 20 Jul.	16 min 15 – 25 Oct.
JOHANN LEONHARD ROST [o]	1718	A 15 min 2 s ♒ 22° – 23°	S 4 min 10 s ♉ 23° – 25°	A 5 min 57 s ♌ 2° – 5°	S 16 min 23 s ♏ 9° – 11°

Table 14.3 Some examples of the maximum and minimum tabulated values of the equation of time in the 17th and 18th centuries. Adapted from sources mentioned in the notes. See also Davis, 2003a, The Equation of Time as Shown on Sundials, Bulletin of the British Sundial Society, vol. 16, issue 4, pp.135–144.

Notes

a Huygens, 1669, Instructions Concerning the Use of Pendulum-Watches, for Finding the Longitude at Sea. Phil. Trans., 1 January 1669, vol. 4, no. 45–56, tables on pp.940–941. Horologium Oscillatorium, 1673, p.15. The values of the equation of time are the same, but offset by ten days.

b Flamsteed, 1678, De Temporio Aequatione Diatriba, pp.460-461. Kitto, 1999, John Flamsteed, Richard Towneley and the Equation of Time. Antiquarian Horology and Proceedings of the Antiquarian Horological Society, vol. 25, p.181. Kitto, 2001, Richard Towneley and the Equation of Natural Days. Bulletin of the British Sundial Society, vol. 13, issue 2, p.64. FLAMSTEED's tables with the equation of time along with other tables are available in the Cambridge archives: RGO 1/23, 224r, Mural arc stellar observations, RGO 1/46, ff.40v–53r, 59v–64r, RGO 1/47, ff.23r–25v, 31r, RGO 1/51, ff.72r–73v... In 1700, in *The Artificial Clockmaker*, a treatise of watch and clockwork, 2nd edition, WILLIAM DERHAM published tables with the equation of time, which were probably also FLAMSTEED's work.

c Flaminii de Mezavachis, 1675, Ephemerides Felsineæ Recentiores. *Tabula æquationis dierum ciuilium naturalium*, p.144. The same table is also published in the edition of 1701, on page 249.

d The values of the equation of time were published by JOHN SMITH between 1678 and 1694 in several pamphlets. His tables, which may have been summarised from HUYGENS, are a speciality; SMITH mentions, for example, the daily change of the equation of time, in mathematical terms, the first derivative of the equation of time, or the cumulative value of the changes to each day of the month, calculated from the first of the month. An example of the daily changes or increments is published in *Horological Disquisitions Concerning the Nature of Time*, published in 1678. The values are from the reprint of 1962. The cumulative changes for individual days of the month are published, for example, in Smith, 1678, Equation of time demonstrated to the meanest capacity, together with a table of equations, absolutely necessary for the true adjusting and right managing of pendulum clocks. Both humbly presented to the master, wardens, and court of assistants of the company of clock-makers: by John Smith, C. M. Philomath. The original table is kept by the Bodleian Library. With the ⨀ symbol, the tables show the local minimum or maximum values of the equation of time in the tables, or what SMITH calls days when the day (or the meaningful time between successive solar mid-days) is 24 hours long. These tables are obviously not free of errors; in *Horological Disquisitions* (reprint), they are missing values for the 31st day in May and August. If we complete this with the values of the daily increments that obviously follow from the table for the largest or the smallest local values, and we add the increments for individual days, we obtain, with some differences, the same values as the daily increments that follow from HUYGENS' table with the equation of time.

e Values are from the first edition of the *Connoissance des temps*, published in 1679, Table entry *Equation de l'horloge, avance/tarde*. Modern edition gives the values as time of the meridian passage.

f In 1683, he published a table with the equation of time in London: *A table of the equation of days, showing how much a good pendulum watch ought to be faster or slower than a true sun-dial, every day of the year*. The original is kept in the Huntington library; later, a table was published with the same title, from which the values in the table are derived. For the table from 1683, see Symonds, 1951, Thomas Tompion, his life and work, p.75, Figure 48. The 1690 table, whose authorship is also attributed to him, is also preserved: *Tabula aequationum dierum naturalium temporis intervalla Inter horas ab horologiis oscillatoriis aequabiliter moventibus, & à sciotericis solaribus accuratis indicatas, ad diem quemlibet singulum anni 1690. St. novi, exhibens: Futuris etiamannis hujus faeculi, absque sensibili differentiâ inferviens*. TOMPION's first publication of the equation of time was based on the Julian calendar, and the second on the Gregorian calendar. The most likely author of the values in TOMPION's tables is JOHN FLAMSTEED; the first table is supposed to have been given to TOMPION by ROBERT HOOKE, and the second probably originates from FLAMSTEED's *Doctrine of the Sphere*, published in 1680. For more information, see Wing, Short-title catalogue of books printed in England, Scotland, Ireland, Wales, and British America, and of English books printed in other countries, 1641–1700, and Forbes et al. (eds.), 1997, The Correspondence of John Flamsteed, The First Astronomer Royal, II, p.552.

g In 1686, MOLYNEUX published the equation of time in *Sciothericum telescopicum* in the tables called the *Tables of Æquation of Natural Days*. The explanation of the equation of time in Chapter X is based on FLAMSTEED's *De inæqualitate dierum solarium*, published in *Opera posthuma Jeremiæ Horroxcii*, of 1673, pp.443–464.

h De La Hire, 1687, Tabularum astronomicarum, section with Tables, pp.2–3, *Tabula III, æquatio temporis ad hoc tempus*. Short explanation is available in lesson 1, pp.47–50. De La Hire, 1702, Tabulæ astronomicæ Ludovici Magni, section with Tables, p.2, *Tabula III, æquatio temporis ad annum 1700. complet*. Short explanation is available in lesson 2, pp.5–7. Also available in 1727 edition.

i GIOVANNI DOMENICO CASSINI (1625–1712), Italian-French astronomer. Monsieur Cassini, 1694, His New and Exact Tables for the Eclipses of the First Satellite of Jupiter, Reduced to the Julian Stile, and Meridian of London, Phil. Trans., vol. 18, pp.237–256, 1 January 1694. *Tabula æquationis dierum*, pp.248–249. Reprint in The Philosophical Transactions and Collections to the End of the Year 1700; Abridg'd, and Dispos'd under General Heads, Volume 1, Third Edition, 1722, pp.272–273, also in Fourth Edition, 1731.

j JOHANN PHILIPP VON WURZELBAU (1651–1725), German astronomer, also known for the discovery of the analemmic curve. Before being admitted into the nobility, he was known as JOHANN PHILIPP WURZELBAUER. The values of the equation of time, *Aequationes dierum naturalium*, are from the astronomical tables, *Tabulae astronomicæ*, published around the year 1700, pp.35–36. The authors of the tables are PHILIPPE DE LA HIRE and WURZELBAU. The table was also published in *Uranies noricæ basis astronomica* of 1719, p.71, edited by WURZELBAU. For some information about his life and work see also Pilz, 1977, pp.298–303.

k WILLIAM WHISTON (1667–1752), an English astronomer, published several astronomical works. The values of the equation of time are from *Prælectiones astronomicæ publicis habitæ*, pp.331–332. The same values are also published in the 1715 English edition, *Astronomical lectures Read in the Publick Schools at Cambridge*, in the *Table of the Equation of Days derived from both the former, and depending on the Sun's Place*. Also in separate tables are the two contributions to the equation of time as a function of the Sun's mean anomaly and longitude.

l JOHANN GABRIEL DOPPELMAYR (1677–1750), German mathematician, astronomer and cartographer, provides values of the equation of time in the *Neu-vermehrte Welperische Gnomonica*, 1708, in the Appendix on pages 34–35. In the table, values for the leap year and for three consecutive years are reproduced. FLAMSTEED is said to be the author of the values - see the text on p.33.

m Smart, John, 1702, Tables of time calculated for two hundred years. vizt. the XVII. and XVIII. centuries. To which are added Mr Flamsteed's table of the equation of natural days, and other useful tables, first printing 1702, reprint 1710. *A Table of the equation of natural days showing how much a well regulated and adjusted pendulum clock or watch, goes faster or slower than a true sun dial.*

n JOHN HARRISON (1693–1776), inventor of the chronometer. The 1717 table lists the equation of time and the difference between solar and clock time at sunset for Barrow, latitude 53°18′, for each fifth day in minutes, and the 1728 table lists the values in minutes and seconds for every third day, taking into account the leap years, but it contains complete values for only five months. See Sobel, Andrewes, 1998, The Illustrated Longitude: The True Story of the Lone Genius Who Solved the Greatest Scientific Problem of His Time, pp.80 and 91.

o JOHANN LEOHARD ROST (1688–1727), also known as MELETAON, German writer and astronomer. The equation of time is published in *Astronomisches Handbuch*, 1718, table 18, pp.269–270. An equation of time with the same values is obviously also used in reprints of the astronomical manual. See, for example, the new edition, 1773, table 18, pp.316–317. ROST was a friend of WURZELBAU, from whom, according to the astronomical guide, originate the values of the equation of time. The values are in fact the same as those in the tables written by PHILIPPE DE LA HIRE and WURZELBAU. Gaab, Simons, 2008, Johann Leonhard Rost, "Romanist" und Astronom. Acta Historica Astronomiae, 36, pp.106–155. For some comments see also Pilz, 1977, pp.305–307.

Author	Year	Jan./Feb.	May	July	Oct./Nov.
John Keill[a]	1718	14 min 50 s ♒ 22 1/2°	4 min 6 s ♉ 24°	5 min 53 s ♌ 3 1/2°	16 min 11 s ♏ 10°
George Neale[b]	1733	*watch too fast, avance* 14 min 49 s 30 – 31 Jan.	*watch too slow, retarde* 4 min 12s 4 – 5 May	*watch too fast, avance* 5 min 46 s 15 – 16 Jul.	*watch too slow, retarde* 16 min 1 s 22 – 23 Oct.
James Atkinson[c]	1735	14 min 49 s	4 min 13 s	5 min 46 s	16 min 1 s
Charles Leadbetter[d]	 1737 1739 1756 1769	*clocks too fast, subtract from the equal time* 14 min 49 s 31 Jan. *watch too fast* 14 min 49 s 10 – 11 Feb.	*clocks too slow, add to the equal time* 4 min 5 s 3 – 4 May *watch too slow* 4 min 5 s 14 – 16 May	*clocks too fast, subtract from the equal time* 5 min 57 s 15 – 16 Jul. *watch too fast* 5 min 55 s 26 – 27 Jul.	*clocks too slow, add to the equal time* 16 min 13 s 22 Oct. *watch too slow* 16 min 13 s 1 – 3 Nov.
Jacques Cassini[e]	1740	A 14 min 53 s ♒ 22° – 24°	S 4 min 8 s ♉ 24°	A 5 min 50 s ♌ 4° – 5°	S 16 min 12 s ♏ 9° – 11°
Eustachio Manfredi[f]	1750	A 14 min 52 s ♒ 21° – 23°	S 4 min 5 s ♉ 24° – 25°	A 5 min 50 s ♌ 3° – 4°	S 16 min 12 s ♏ 9° – 10°
Nicolas Louis de Lacaille[g]	1755 1758	14 min 41 s 11 Feb. A 14 min 41 s ♒ 22° – 23°	4 min 2 s 15 May S 4 min 3 s ♉ 23° – 25°	5 min 57 s 27 Jul. A 5 min 57 s ♌ 2° – 5°	16 min 11 s 2 Nov. S 16 min 9 s ♏ 10° – 11°
Ferdinand Berthoud[h]	1759	*Le soleil retarde* R ⊙ 14 min 44 s 11 Feb.	*Le soleil avance* A ⊙ 4 min 4 s 14 – 15 May	*Le soleil retarde* R ⊙ 5 min 56 s 25 – 26 Jul.	*Le soleil avance* A ⊙ 16 min 9 s 31. Oct – 2. Nov.
François Bédos de Celles[i]	1760	A 14 min 40 s ♒ 22° – 23°	S 4 min 2 s ♉ 24°	A 5 min 58 s ♌ 3° – 4°	S 16 min 9 s ♏ 10° – 11°
Society of Gentlemen[j]	1763	*Equal time faster* 14 min 49 s 10 – 11 Feb.	*Equal time slower* 4 min 5 s 14 – 16 May	*Equal time faster* 5 min 55 s 26 – 27 Jul.	*Equal time slower* 16 min 13 s 1 – 3 Nov.
Roger Long[k]	1764	⊙ *behind* 14 min 43 s 9 – 11 Feb.	⊙ *before* 4 min 4 s 14 – 16 May	⊙ *behind* 5 min 55 s 25 – 28 Jul.	⊙ *before* 16 min 9 s 1 – 3 Nov.
Joseph Jérôme Lefrançais de Lalande[l] *	1764	≈14 min 46 s 323°	≈4 min 6 s 53°	≈6 min 3 s 123°	≈16 min 16 s 220°

Table 14.4 Some examples of the maximum and minimum tabulated values of the equation of time in the 18th century. Adapted from sources mentioned in the notes. * Calculated values.

Notes

a The given values of the equation of time are summarised from the KEILL's book *Introductio ad veram astronomiam*, p.413. The book also appeared in English; see, for example, *An Introduction to the True Astronomy or Astronomical lectures Read in the Astronomical School of the University of Oxford*, Second Edition, 1730, pp.321–322.

b GEORGE NEALE was a clockmaker, the tables with the equation of time were part of the equipment of his clocks. The tables had the same or very similar values as the tables of J. SMART. Turner, 2002, Sundials in History. In: Sundials at Greenwich, A Catalogue of Sundials, Nocturnals and Horary Quadrants in the National Maritime Museum, Greenwich, p.18. Turner, 2002, Eighteenth century scientific instruments and their makers. In: The Cambridge History of Science, IV, Eighteenth Century Science. Porter (ed.).

c JAMES ATKINSON SENIOR published the tables in a supplement to his book *Epitome of the Art of Navigation*. Davis, 2003a, p.138.

d CHARLES LEADBETTER, teacher of mathematics, navigation and astronomy from London, published an explanation of the equation of time, probably based on MOLYNEUX's work, in the book *Uranoscopia or, the Contemplation of the Heavens*, published in 1735. The first table with values of the equation of time was published in 1737, in *Mechanick Dialling or the New Art of Shadows*, and two years later was reprinted in the book *The Young Mathematician's Companion, A table of equation of time, for regulating of the pendulum clocks with the Sun*, pp.290–291. The table was published in the reprint of *Mechanick Dialling or the New Art of Shadows*, 1756 and 1769, Table II, pp.72–73, but this time for the Gregorian calendar. The same values were also published in the *Dictionary of Arts and Sciences*, Society of Gentlemen, 1763.

e JACQUES CASSINI, 1740, Tables Astronomiques du Soleil, De La Lune, Des Planetes, Des Etoiles Fixes, Et Des Satelites De Jupiter Et De Saturne, Avec l'explication & l'usage des ces memes tables, Table II, pp.8–9.

f EUSTACHIO MANFREDI, 1750, Eustachii Manfredii Introductio In Ephemerides Cum Opportunis Tabulis Ad Usum Bononiensis Scientiarum Instituti.

g NICOLAS LOUIS DE LACAILLE, 1755, Ephimerides des Mouvements Celestes. LACAILLE lists values in sixtieths of a second. The table lists the values in seconds per day with the local maximum or minimum values of the equation of time. The values for 1758 are from the tables *Tabulae solares ad meridianum parisinum, Tabula I*, p.4. For a brief commentary on LACAILLE's equation of time see Melander, 1777, Anmerkungen über die Zeitgleichung Von Herrn Daniel Melander, Prof. der Astronomie von Upsal. The comment is from a letter from Melander to Bernoulli, 7 March 1777. Astronomisches Jahrbuch oder Ephemeriden für das Jahr 1780, pp.20–22.

h FERDINAND BERTHOUD (1727–1807), L'art de conduire et de régler les pendules et les montres, 1759, pp.67–78. In German translation (1791), a local maximum of 5 min 56 s is given for 25 July. In 1775, BERTHOUD also published extensive tables with the equation of time in *Les longitudes par la mesure du temps, ou méthode pour déterminer les longitudes en mer; avec le secours des horloges marines, Équation du temps, Table I – Table IV*; each of the four tables has values for every fourth year.

i FRANÇOIS BÉDOS DE CELLES (1709–1779), La gnomonique pratique ou l'art de tracer des cardans solaires, 1760, *Table de l'Equation du temps, calculée pour chaque dégré de l'Ecliptique*, pp.380–381. The table contains values for each degree of longitude. Pages 252–253 in the *Table de la décision du Soleil, & de l'équation du temps convenable aux dégrés de l'Ecliptique pris de trois en trois,* hold the value of the equation of time in seconds for each third degree of the individual signs of the zodiac.

j The second edition of *A New and Complete Dictionary of Arts and Sciences* of 1763 also contains a table with the equation of time whose values are the same as in CHARLES LEADBETTER's table of 1756. For the table with the equation of time, see Vol. II, p.1119.

k ROGER LONG (1680–1770), English astronomer. For the table with the equation of time see *Astronomy in Five Books*, Book 3, A table of the equation of time; showing how much the sun is ahead of or behind the clock every day, p.508. The values are summarised in the French tables published in *Connoissance des temps* of 1752.

l LALANDE's maximum or minimum values are an approximate estimate based on both the contributions of the equation of time from the book *Astronomie* of 1764, tables XIII and XIV. The values of the equation of time are calculated by taking into account the value of the apogee 98°6′, and the intermediate values are calculated using linear interpolation.

Notes

a DOMINIQUE FRANÇOIS RIVARD (1697–1778), French mathematician and philosopher. The values of the equation of time are from *La gnomonique, ou L'art de faire des cadrans*, 1742, pp.305–306, or from the edition of 1767, pp.292–293. The values are given in the tables in steps of three degrees of ecliptic longitude.

b *The Nautical Almanac and Astronomical Ephemeris for the year 1767*, London, 1766, tables 2, 14, 26, 38, 50 ...

c TYCHO WING (1696–1750), grandnephew of VINCENT WING, taught mathematics, geodesy and navigation. *Collection of English Almanacs for the Years 1702–1835, Table of the equation of natural days, exactly calculated for the year 1770.*

d Values for the years 1716 and 1771. *Tabulæ motuum solis et lunæ novæ et correctæ*. Tables II and III.

e Values for 1771. *Sammlung Astronomischer Taffeln*, 1776. Band 1, on pages 286–287, contains tables with the equation of time for the years 1716 and 1771 for the true longitude of the Sun.

f Vogel, 1774, Practischer Unterricht von Taschenuhren, sowohl für die Verfertiger, als auch für die Liebhaber derselben, *Aequationstabelle der gleichen und scheinbaren Zeit*, pp.341–352. Values are given for days of the year. In German clockmaker's literature, the term *Aequationstabelle* was often used for tables with the equation of time.

Author	Year	Jan./Feb.	May	July	Oct./Nov.
DOMINIQUE FRANÇOIS RIVARD [a]	1742 1767	*Equation* A 14 min 53 s ♒ 24°	*Equation* S 4 min 8 s ♉ 24°	*Equation* A 5 min 49 s ♌ 3° and 6°	*Equation* S 16 min 12 s ♏ 9°
NEVIL MASKELYNE [b]	1767	A 14 min 42 s 10 – 12 Feb.	S 4 min 2 s 14 – 16 May	A 6 min 2 s 26 – 28 Jul.	S 16 min 14 s 2 – 3 Nov.
TYCHO WING [c]	1770	A 14 min 46 s 10 – 12 Feb.	S 4 min 2 s 14 – 16 May	A 5 min 57 s 25 – 28 Jul.	S 16 min 14 s 1 – 3 Nov.
TOBIAS MAYER [d]	1716 1771	+14 min 48 s ♒ 22° – 23° +14 min 42 s ♒ 21° – 23°	−4 min 8 s ♉ 23° – 25° −4 min 2 s ♉ 23° – 25°	+5 min 55 s ♌ 3° – 4° +6 min 2 s ♌ 3° – 5°	−16 min 10 s ♏ 10° – 11° −16 min 14 s ♏ 10° – 11°
Sammlung Astronomischer Taffeln, 1776 [e]	1771	+14 min 42 s ♒ 21° – 23°	−4 min 2 s ♉ 23° – 25°	+6 min 3 s ♌ 3°	−16 min 14 s ♏ 10° – 11°
CHRISTIAN F. VOGEL [f]	1774	14 min 44 s 11 Feb.	4 min 4 s 14 – 15 May	5 min 56 s 25 Jul.	16 min 9 s 31 Oct. – 2 Nov.
GEORG VEGA [g]	1797	+14 min 40 s 11 Feb.	−4 min 0 s 14 – 16 May	+6 min 7 s 27 Jul.	−16 min 15 s 1 – 2 Nov.
JEAN-BAPTISTE JOSEPH DELAMBRE [h]	1806	+14 min 38.5 s ♒ 21° or 321°	−3 min 59.0 s ♉ 23° or 53°	+6 min 4.8 s ♌ 4° or 124°	−16 min 14.7 s ♏ 12° or 222°
ALEXANDER EWING, JOHN GUMMERE [i]	1812	+14 min 42 s ♒ 22°	−4 min 1 s ♉ 23° – 25°	+6 min 1 s ♌ 3° – 4°	−16 min 13 s ♏ 9° – 11°
JOHN GUMMERE [j]	1822	+14 min 36 s ♒ 18° – 19°	−3 min 57 s ♉ 20° – 22°	+6 min 7 s ♌ 1° – 3°	−16 min 16 s ♏ 10°
JOSEPH JOHANN VON LITTROW [k]	1829	+14 min 36 s 320° 10 Feb.	−3 min 54 s 50° – 55° 11 – 16 May	+6 min 6 s 125° 25 – 30 Jul.	−16 min 12 s 220° – 225° 2 – 7 Nov.

Table 14.5 Some examples of the maximum and minimum tabulated values of the equation of time in the 18th and 19th centuries. Adapted from sources mentioned in the notes.

g Vega, 1797, Logarithmisch – trigonometrische nebst andern zum Gebrauch der Mathematik eingerichteten Tafeln und Formeln, Band II, 2nd Edition, p.197.

h Delambre, 1806, Tables Astronomiques, publiées par le Bureau des Longitudes de France, first edition, table VIII, also Delambre, 1814, Astronomie Theoretique et Pratique, table III, pp.210–213.

i Ewing, revised by Gummere, 1812, Practical Astronomy: Containing a Description of the Solar System; the Doctrine of the Sphere; the Principal Problems in Astronomy illustrated with many Examples, table XI, *Of the equation of time fitted to each degree of the eclipticK.*

j Gummere, 1823, An Elementary Treatise on Astronomy, table XIX.

k The values of the mean longitude of the Sun are summarised from Littrow, 1829, Calendariographie, p.485, table V, and the calendar dates are summarised from the book Littrow, 1837, Die Wunder des Himmels - Gemeinverständliche Darstellung des astronomischen Weltbildes. Also published in Jüttner, 1842, Anweisung zum Gebrauche der Himmelskugel, p.24. Some of the values of the equation of time in his books vary slightly in the various tables.

14.1.3 Periods and epochs

Since different calendars were used in antiquity, different periods with corresponding epochs were used to facilitate chronological tracking. Calendar periods were usually associated with coronation or similar state events. In astronomy, epochs allowed for the constant change in astronomical parameters: for reliable calculation of the position of celestial bodies in the future, astronomers need to know where those bodies were at a specific moment.

Author	Epoch constant	λ_{Υ}	e	ε
PTOLEMY	+4;23° [a] −3;18° or −3;33° [b]	♊ 5;30° 65;30°	2;30 $e/R = 1/24$	23;51,20°
ĀRYABHAṬA I [c]		78°		24°
VARĀHAMIHIRA [d]		♊ 20° 80°	2;20	23;40° 24°
BRAHMAGUPTA [e]		80°		24°
AL-KHWĀRIZMĪ [f]	4;30°	82;39°	2;30	23;51°
AL-BATTĀNĪ [g]	4;6,30°	82;15°	2;4,45	23;35°
KŪSHYĀR IBN LABBĀN [h]	4;4°	♊ 24° 84°	2;4,45	23;35°
AL-BAGHDĀDĪ [i]	3;33°	82;39°	2;4,35,30	23;35°
JAMSHĪD AL-KĀSHI [j]		90°	2;6,9	23;30°
ULUGH BEG [k]	55114‴	93;0,18°	0.0237037	23;30,16.8°
LEVI BEN GERSON [l]	4;2° ♒ 20°	♋ 3°	2;14 $e/R = 2;8,0$	23;33°

Table 14.6 Values for calculating the equation of time used by some medieval astronomers. The sources are listed in the notes. For the epochs see Table 14.7.

Notes

a In the *Almagest*, the epoch is 1 Thoth, Nabonassar 1. Almagest III.7, Toomer, 1998, p.168. Values of the right ascension and longitude at the epoch are $\alpha_0 = 335;8°$ and $\lambda_{M0} = 330;45°$. Neugebauer, 1975, pp.63–64, 67. For the calculation see Example 1.1.

b In the *Handy Tables*, the epoch is PHILIP's era. Tihon, Mercier, 2011, 1b, p.52. Apparently, the value of 66° is used for longitude of the apogee instead of 65;30°. For more information see Tihon, Mercier, 2011, p.96 and van Dalen, 1994, p.114. For the calculation of right ascension and longitude at the epoch and the epoch constant see Example 1.3.

c Shukla, 1976, Āryabhaṭa, Āryabhaṭīya, pp.17 and 20.

d March 22, 505, Tuesday, 18:00 at sunset, beginning of year 428 of the Śaka era. Neugebauer, Pingree, 1971, The Pañcasiddhāntikā of Varāhamihira, Part 2, pp.8, 24, 69. The VARĀHAMIHIRA lists the values 23;20°, 23;40° and 24° for the obliquity of the ecliptic. For the eccentricity and apogee, see also Part 2, pp.59, 70, III, 20–22, p.33, and XVI, 12–14, p.102, Table 22.

e Khaṇḍakhādyaka, pp.48, 50, Rao, 2000, pp.27–28. Beginning of year 587 of the Śaka era.

f Van Dalen, 1996, Al-Khwārizmī's Astronomical Tables Revisited, pp.245–246. From the available sources, it is not possible to reliably determine whether the longitude of the apogee was at 82;39°, or perhaps at the rounded value of 82;40°. AL-MAJRĪṬĪ uses the value of 77°55′ in his comments on the longitude of the apogee λ_Γ. Epoch is the Hijra era, longitude of the Sun for the epoch ♋ 24°. Al-Khwārizmī, Neugebauer, 1962, p.65. AL-KHWĀRIZMĪ's epoch is also on July 14, 622 at noon, Chabás, Goldstein, 2003, pp.144 and 155.

g Pedersen, 2002, The Toledan Tables, p.970, a tabulated equation of time from the *Toledan Tables* equal to AL-BATTĀNĪ's values, pp.972–975. AL-BATTĀNĪ is also supposed to have used the value of 82;17°, see Chabás, Goldstein, 2003, p.188, or Nallino, 1903, pp.44, 214, 246. The epoch of the Seleucid era, April 3, 311 BCE, Mesopotamia (October 1, 311 BCE, Syria), the longitude of the Sun for the epoch ♓ 4°. Al-Khwārizmī, Neugebauer, 1962, pp.64–65.

h Kennedy, 1988, pp.2–3 or van Dalen, 1993, Ancient and Mediaeval Astronomical Tables: mathematical structure and parameter values, p.151.

i Van Dalen, 1993, p.151.

j AL-KĀSHI supposedly used the same epoch as AL-ṬŪSĪ, Kennedy, 1988, p.5. Additionally, the following apply: $\lambda_0 = 320°$, $\alpha_0 = 322;25,17°$ and $\lambda_{M0} = 318;27,42,37°$. See also Kennedy, 1960, p.170.

k Zach, 1799, Allgemeine Geographische Ephemeriden, 3. Band, pp.179–183, $\lambda_\Gamma = 92;36,37°$ see Gauss, 1929, Werke 12, pp.64–68.

l Goldstein, 1974, pp.93–94 and 102–103.

Era	Epoch	Julian day number / Julian date
Hebrew era [a]	Molad tishri, October 6, 3761 BCE (−3760), 5 hrs 204 ch after sunset Tekufat nisan, April 1, 3760 BCE (−3759), 6:00 pm	~347 997.5 348 174.25
Kaliyuga [b]	February 17/18, 3102 BCE (−3101), midnight	588 465.5
Era of NABONASSAR [c]	February 26, 747 BCE (−746), midday	1 448 638
Era of PHILIP [d]	November 12, 324 BCE (−323)	1 603 398
Seleucid era [e] (era of ALEXANDER)	October 1, 312 BCE (−311)	1 607 739
Seleucid era [f]	April 3, 311 BCE (−310)	1 607 923
Era of AUGUSTUS [g]	August 31, 30 BCE (−29)	1 710 708
Era Christi [h]	December 25, 1 BCE (?) March 25, 1 CE	1 721 417 1 721 507
Śaka era [i]	March 15, 78	1 749 621
VARĀHAMIHIRA [j]	March 22, 505, at sunset	~1 905 590.25
Hijra [k]	July 15, 622, midday	1 948 439
YAZDEGERD [l]	June 16, 632	1 952 063
BRAHMAGUPTA [m]	March 22/23, 665, midnight	1 964 030.5
Era of ALFONSO [n]	January 1, 1252 June 1, 1252	2 178 351 2 178 503
ULUGH BEG [o]	July 5, 1437	2 246 108

Table 14.7 Some of the epochs most commonly used in historical sources. Adapted from Ginzel, 1906, Handbuch der mathematischen und technischen Chronologie and from sources mentioned in the notes. For a summary, see Neugebauer, 1929, Astronomische Chronologie, Part 1, pp.43–45 or Neugebauer, 1937, pp.153–158.

Notes

a Mercier, 2014, Astronomical Tables of Abraham Bar Hiyya, p.155. *Molad* (Eng. birth) is the time of conjunction of the mean Sun and Moon. *Tekufat* (Eng. cycle) is the final position of the mean Sun in a quadrant or 0°, 90°, 180°, 270°. First after *molad* is *tekufat nisan*. The time unit chelek (plural chalakim, here abbreviated ch) is $^1/_{1080}$ hour. The next *molad* or *molad nisan* occurs at 3 hrs 642 ch after midnight of 1/2 April 3760 BCE.

b Shukla, 1960, Bhāskara I and his works, Part II, Mahā-Bhāskarīya, p.3.

c The NABONASSAR era was used by PTOLEMY in the *Almagest*, where it is also stated that HIPPARCHUS already knew or used this epoch. Toomer, 1998, Almagest IV.11, H340, p.212. See also Swerdlow, Neugebauer, 1984, pp.183–188.

d The era of PHILIP coincides with the beginning of the 425th year of the NABONASSAR era. Beginning on Thoth 1, 425. Ginzel, 1909, p.147. Tihon, Mercier, 2011, 1b, p.52. The Era of PHILIP is named after ALEXANDER's older half-brother PHILIP ARRHIDAEUS, or PHILIP III OF MACEDON.

e Also the Era of ALEXANDER. Ginzel, 1906, pp.136–138, 263. Tihon, Mercier, 2011, 1b, pp.72–74.

f Tihon, Mercier, 2011, p.73.

g Thoth 1, 30 BCE. PTOLEMY used this epoch in the *Canobic Inscription*, Lat. *Inscriptio canobi*, for more information see Chapter 1, note 54. For additional comments on the Era of AUGUSTUS see also Neugebauer, 1936, Hilfstafeln zur technischen Chronologie, Astronomische Nachrichten, vol. 261, no.10, p.155.

h Declercq, 2000, Anno Domini, The Origins of the Christian Era, pp.143–147.

i Ginzel, 1906, p.391.

j Tuesday, March 22, 505 at sunset (18:00). Neugebauer, Pingree, 1970/71, Part I, p.8, Part II, p.35. Beginning of year 428 of the Śaka era.

k Ginzel, 1906, p.258. July 15, 622 is the epoch in the astronomical sense, July 16, 622 is the epoch in the historical sense, $JDN = 1{,}948{,}440$.

l Persian era, the beginning of the rule of the last SASSANID king YAZDEGERD III. Ginzel, 1906, pp.298–300.

m Rao, 2000, p.28. Beginning of year 587 of the Śaka era.

n In *Castilian Alfonsine Tables* January 1, 1252. Chabás, Goldstein, 2003, p.144.

o Akhmedov, 2000, Astronomy, Astrology, Observatories and Calendars, p.204.

14.2 Mathematical reconstruction

14.2.1 Parameters relevant to reconstructing tables

For precise analysis or mathematical reconstruction of tables, the following parameters must be known:

- An independent variable, either the true or mean longitude of the Sun.
- The obliquity of the ecliptic ε, longitude of the apogee λ_Γ, eccentricity e, the conversion factor between degrees and time units, which is, as a rule, rounded to a value of $15° \equiv 1$ h; the exact value is 15;2,28.
- The accuracy of the available tables with the value of the equation of time and the mathematical procedures available, for example interpolation, in its original calculation.

Due to the symmetry of the right ascension α and solar equation c, the following are valid:[10]

[14.1] $$\alpha(180° - \lambda_\odot) = 180° - \alpha(\lambda_\odot) \qquad \alpha(180° + \lambda_\odot) = 180° + \alpha(\lambda_\odot)$$

[14.2] $$c(\underbrace{\lambda_\Gamma + \kappa}_{\lambda_\odot}) = -c(\lambda_\Gamma - \kappa) \qquad c(\lambda_\Gamma + 180° + \kappa) = -c(\lambda_\Gamma + \kappa)$$

Similarly for the equation of time E (see Equations 1.2 and 1.11):

[14.3] $$E(\lambda_\odot) = \tfrac{1}{15}(\lambda_\odot - \alpha(\lambda_\odot) + c(\lambda_\odot - \lambda_\Gamma) + k)$$

[14.4] $$E(180° + \lambda_\odot) = \tfrac{1}{15}(180° + \lambda_\odot - \alpha(180° + \lambda_\odot) + c(180° + \lambda_\odot - \lambda_\Gamma) + k)$$

$\lambda_\odot$ ecliptic longitude
λ_Γ longitude of the apogee
$\alpha(\lambda_\odot)$ right ascension as function of ecliptic longitude
k epoch constant

If we sum up both equations, we can express the right ascension. If two values of the equations of time for arguments 180° apart are subtracted (Equations 14.3 and 14.4), we can express the solar equation.[11] The epoch constant k can be reconstructed from the tables of the equation of time, where:[12]

[14.5] $$E(\lambda_\odot) + E(180° - \lambda_\odot) + E(180° + \lambda_\odot) + E(360° - \lambda_\odot) = \tfrac{4}{15}k$$

10 The procedure below is summarised by Van Dalen, 1996, al-Khwārizmī's astronomical tables revisited: analysis of the equation of time in Casulleras and Samsó (eds.), From Baghdad to Barcelona: Studies in the Islamic Exact Sciences in Honour of Prof. Juan Vernet. Universitat de Barcelona, vol. 1, pp.220–224 also Van Dalen, 1993, pp.104–105.

11 See van Dalen, 1996, p.221, Equation 7 and p.222, Equation 8.

12 For a detailed explanation and mathematical derivation see Van Dalen, 1996, pp.222–223, also Van Dalen, 1993, pp.104–105.

If n-values are tabulated for the equation of time and n is divisible by 4, then the following expression applies:[13]

$$k = \frac{15}{n}\sum_{i=1}^{n} E\left(\frac{i \cdot 360°}{n}\right) \qquad [14.6]$$

Reconstructions generated in this way are considered accurate to the same precision as the table under consideration.

Example 14.2 Determine the approximation of the epoch used to calculate the equation of time from PTOLEMY's *Handy Tables* (solution: $k = 3.°565$ or $k = 3;33°$).

From the values in Table 1.1, we choose, for example, $\lambda_{\odot} = 20°$, and we get the following values for the equation of time: $E(\lambda_{\odot}) = 14\,\text{min}\ 41\,\text{s}$, $E(180° - \lambda_{\odot}) = 11\,\text{min}\ 5\,\text{s}$, $E(180° + \lambda_{\odot}) = 1\,\text{min}\ 10\,\text{s}$ and $E(360° - \lambda_{\odot}) = 30\,\text{min}\ 6\,\text{s}$. We convert these into hours: $E(\lambda_{\odot}) = 0.2447$, $E(180° - \lambda_{\odot}) = 0.1847$, $E(180° + \lambda_{\odot}) = 0.01944$, $E(360° - \lambda_{\odot}) = 0.5017$.

From Equation 14.5:

$$\begin{aligned}\frac{4}{15}k &= E(\lambda_{\odot}) + E(180° - \lambda_{\odot}) + E(180° + \lambda_{\odot}) + E(360° - \lambda_{\odot}) \\ &= 0.2447 + 0.1847 + 0.01944 + 0.5017 \\ k &= 3.°565 \approx 3°33'\end{aligned}$$

A very similar result is also obtained using Equation 14.6, where we also use Table 1.1:

$$\begin{aligned}k &= \frac{15}{n}\sum_{i=1}^{n} E\left(\frac{i \cdot 360°}{n}\right) = \frac{15}{36}\left(E\left(\frac{1 \cdot 360°}{36}\right) + E\left(\frac{2 \cdot 360°}{36}\right) + \cdots + E\left(\frac{36 \cdot 360°}{36}\right)\right) \\ &= \frac{15}{36}(E(10°) + E(20°) + \cdots + E(360°)) = 3.°563\end{aligned}$$

Example 14.3 Determine AL-BATTĀNĪ's equation of the centre (solar equation) for the equation of time in the *Toledan Tables* for $\lambda_{\odot} = 10° - \lambda_{\Gamma}$ or 287;45°. AL-BATTĀNĪ's apogee value or λ_{Γ} is 82;15°, see Table 14.6 (solution: $c \approx -1°52'$, see Nallino, 1907, p.80).

From the values in Table 5.1, the values of the equation of time are 3;3° for $\lambda_{\odot} = 10°$ and 6;49° for $\lambda_{\odot} = 190°$. In hours $E(\lambda_{\odot}) = 0.2033$ and $E(180° + \lambda_{\odot}) = 0.4544$. From Equations 14.3 and 14.4:

$$\begin{aligned}c(\lambda_{\odot} - \lambda_{\Gamma}) &= \tfrac{15}{2}\big(E(\lambda_{\odot}) - E(180° + \lambda_{\odot})\big) = \tfrac{15}{2}(E(10°) - E(190°)) \\ &= \tfrac{15}{2}(0.2033 - 0.4544) = -1.°88 = -1°52'\end{aligned}$$

The sign corresponds to Equation 1.5, see Chapter 1. Other values of the equation of the centre are calculated in the same way, and the intermediate values can be interpolated.

13 Van Dalen, 1996, p.224, Equation 11.

14.2.2 Approximation of discrete points using the method of least squares

When approximating a set of points or finding a function that most closely approximates a set of points, a useful approach is the least squares method, with which we can find a continuous function, for example a polynomial, which best corresponds to a given sequence of points. Developed and first used by GAUSS, the method is suitable for analysing various old tables since it allows approximation with a polynomial whose degree is less than the number of samples.[14]

A sequence of m points is given:

[14.7] $$(x_1, y_1), (x_2, y_2), \dots, (x_m, y_m) \qquad \text{or} \qquad (x_i, y_i),\ i = 1, 2, \dots, m$$

which we want to approximate with the n-th order polynomial, where $n < m$:

[14.8] $$P_n(x) = \sum_{j=0}^{n} a_j x^j \qquad \text{or} \qquad P_n(x) = a_0 + a_1 x + a_2 x^2 + a_3 x^3 + \cdots + a_n x^n$$

A given curve or polynomial must pass through at least some of the tabulated points, which condition is expressed as a system of equations. This can generally be written as $\boldsymbol{Aa} = \boldsymbol{y}$, where the matrix $\boldsymbol{A}$ is:

$$\boldsymbol{A} = \begin{bmatrix} 1 & x_1 & x_1^2 & \cdots & x_1^n \\ 1 & x_2 & x_2^2 & \cdots & x_2^n \\ \vdots & \vdots & \vdots & \ddots & \vdots \\ 1 & x_m & x_m^2 & \cdots & x_m^n \end{bmatrix} \qquad \boldsymbol{a} = \begin{bmatrix} a_0 \\ a_1 \\ a_2 \\ \vdots \\ a_n \end{bmatrix} \qquad \boldsymbol{y} = \begin{bmatrix} y_1 \\ y_2 \\ \vdots \\ y_m \end{bmatrix}$$

or:

[14.9] $$\begin{aligned} a_0 + a_1 x_1 + a_2 x_1^2 + \cdots + a_n x_1^n &= y_1 \\ &\vdots \\ a_0 + a_1 x_m + a_2 x_m^2 + \cdots + a_n x_m^n &= y_m \end{aligned}$$

The least squares method gives a result that, with respect to the given points, generates an approximating function with the smallest deviation or error. The approximation's accuracy depends on the degree of the polynomial in use, and is assessed against measurements or observations that are also exposed to errors. The overall error is calculated as the sum of the squares of the individual discrepancies:

$$E = e_1^2 + e_2^2 + e_3^2 + \cdots + e_m^2$$

14 Hamming, 2016, Numerical Methods for Scientists and Engineers, pp.459–469. Hamming, 2004, Methods of Mathematics, pp.782–786.

We are looking for coefficients of the polynomial that minimise the sum of the squares:

$$E(a_0, a_1, \dots, a_n) = \sum_{i=1}^{m} (y_i - a_0 - a_1 x_i - a_2 x_i^2 - \dots - a_n x_i^n)^2$$

For the ***n***-th degree polynomial:

$$E(a) = \sum_{i=1}^{m} (y_i - P_n(x_i))^2 = \sum_{i=1}^{m} \sum_{j=1}^{n} \left(y_i - a_j x_i^j\right)^2$$

The sum of the squares is minimised by calculating partial derivatives for each ***a***:

$$\frac{\partial E}{\partial a_0} = -2 \sum_{i=1}^{m} (y_i - a_0 - a_1 x_i - \dots - a_n x_i^n)$$

$$\frac{\partial E}{\partial a_1} = -2 \sum_{i=1}^{m} x_i (y_i - a_0 - a_1 x_i - \dots - a_n x_i^n)$$

$$\vdots$$

$$\frac{\partial E}{\partial a_n} = -2 \sum_{i=1}^{m} x_i^n (y_i - a_0 - a_1 x_i - \dots - a_n x_i^n)$$

The general expression for coefficient a_k is, therefore:

$$\frac{\partial E}{\partial a_k} = -2 \sum_{j=0}^{n} x_i^k \left(y_i - a_j x_i^j\right) \qquad k = 0, 1, \dots, n$$

If each partial derivative is equated with zero and divided by −2, we obtain a system of normal equations:

$$\sum_{i=1}^{m} x_i^k y_i = \sum_{i=1}^{m} \sum_{j=1}^{n} a_j x_i^{j+k} \qquad k = 0, 1, \dots, n$$

For the system of normal equations, the following law applies:

$$A^{\mathrm{T}} A a = A^{\mathrm{T}} y$$

$$A^{\mathrm{T}}A = \begin{bmatrix} 1 & 1 & \cdots & 1 \\ x_1 & x_2 & \cdots & x_m \\ x_1^2 & x_2^2 & \cdots & x_m^2 \\ \vdots & \vdots & \ddots & \vdots \\ x_1^n & x_2^n & \cdots & x_m^n \end{bmatrix} \begin{bmatrix} 1 & x_1 & x_1^2 & \cdots & x_1^n \\ 1 & x_2 & x_2^2 & \cdots & x_2^n \\ \vdots & \vdots & \vdots & \ddots & \vdots \\ 1 & x_m & x_m^2 & \cdots & x_m^n \end{bmatrix}$$

$$= \begin{bmatrix} m & \sum_{i=1}^{m} x_i^1 & \sum_{i=1}^{m} x_i^2 & \cdots & \sum_{i=1}^{m} x_i^n \\ \sum_{i=1}^{m} x_i^1 & \sum_{i=1}^{m} x_i^2 & \sum_{i=1}^{m} x_i^3 & \cdots & \sum_{i=1}^{m} x_i^{n+1} \\ \sum_{i=1}^{m} x_i^2 & \sum_{i=1}^{m} x_i^3 & \sum_{i=1}^{m} x_i^4 & \cdots & \sum_{i=1}^{m} x_i^{n+2} \\ \vdots & \vdots & \vdots & \ddots & \vdots \\ \sum_{i=1}^{m} x_i^n & \sum_{i=1}^{m} x_i^{n+1} & \sum_{i=1}^{m} x_i^{n+2} & \cdots & \sum_{i=1}^{m} x_i^{2n} \end{bmatrix}$$

The coefficients a_i are then calculated from the following:

[14.10] $$\boldsymbol{a} = (\boldsymbol{A}^{\mathrm{T}}\boldsymbol{A})^{-1}\boldsymbol{A}^{\mathrm{T}}\boldsymbol{y}$$

Example 14.4 Approximate AL-KHWĀRIZMĪ's equation of time using the least squares method with a 10th-degree polynomial and calculate the error between the tabulated and the approximated values. The source of the data is Suter, 1914, Tab. 67 and 68, pp.181–182, including the sample for 0° longitude, which has the same value as for 360°. The coefficients of the approximating polynomial $a_0 \ldots a_{10}$ are determined using the least squares method. The matrix calculation (Equation 14.10) can be carried out using a spreadsheet programme; examples of the array functions for matrix manipulation are MMULT, TRANSPOSE and MINVERSE.

The coefficients of the 5th- and 10th-degree approximating polynomial (see Equation 14.8), calculated by the least squares method and shown in the figure, are (with some rounding):

$a_0 = 3.76825532$	$a_1 = 65.86930503$	$a_2 = -77.17815361$
$a_3 = 35.24507188$	$a_4 = -6.655434982$	$a_5 = 0.436149089$

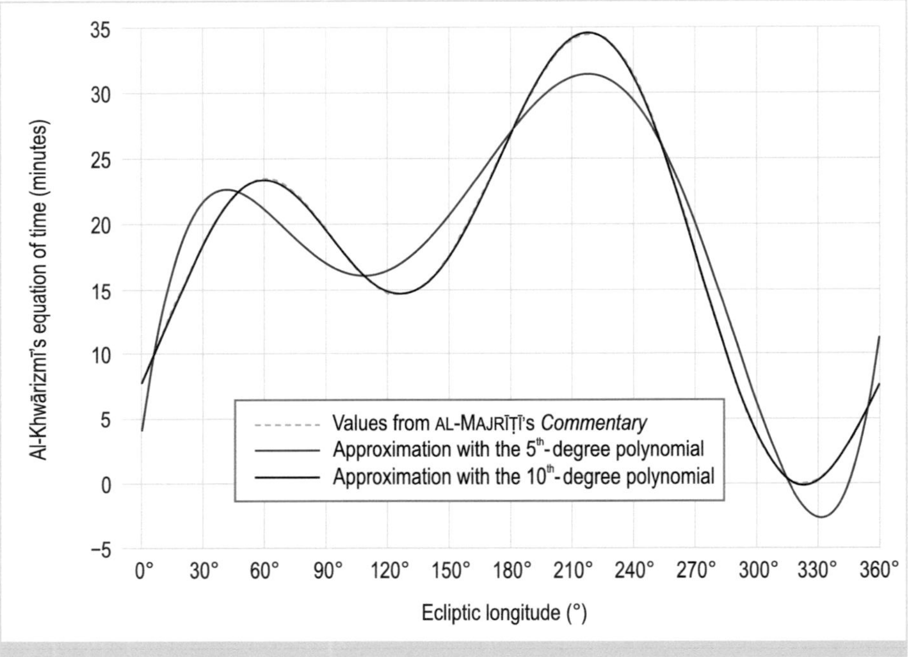

Figure 14.1 Approximation of discrete values of the equation of time from AL-MAJRĪṬĪ's *Commentary* (Suter, 1914) with the polynomials of the 5th and 10th degree.

Coefficients of the 10th-degree polynomial:

$a_0 = 7.732834671$	$a_1 = 20.06682234$	$a_2 = 2.642342019$
$a_3 = 7.664634243$	$a_4 = -33.63562685$	$a_5 = 26.73047942$
$a_6 = -9.464038245$	$a_7 = 1.748453965$	$a_8 = -0.170284609$
$a_9 = 0.007796907$	$a_{10} = -0.00010921.$	

Examples of approximations for the 5th- and 10th-degree polynomials, and the deviations of the approximated values from the values in the original tables are shown in Figures 14.1 and 14.2 for some historical examples.[15]

15 For an explanation of the reason for the major error at the end of the intervals and how to minimise it, see Hamming, 2016, p.469.

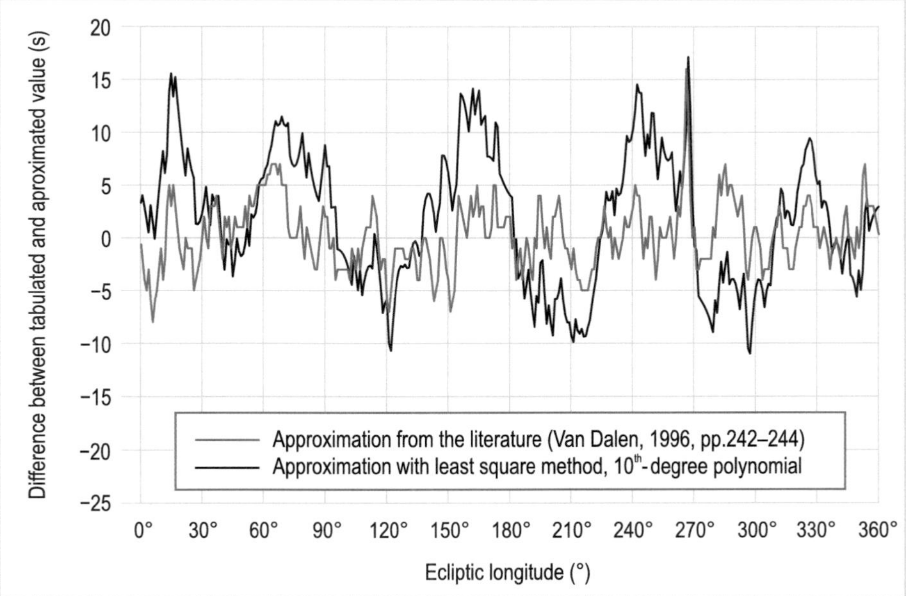

Figure 14.2 Approximation erors for the equation of time from AL-MAJRĪṬĪ's *Commentary* (Suter, 1914) with a 10th-degree polynomial approximation and an example of approximation from the literature (Van Dalen, 1996, pp.242–244).

	PTOLEMY	AL-KHWĀRIZMĪ	KŪSHYĀR	ULUGH BEG
a_0	22.5217	7.7323	9.7970	7.2091
a_1	−22.61467	20.0668	20.8732	21.2267
a_2	−4.3906	2.6423	19.1235	−7.2006
a_3	1.4213	7.6646	−50.8799	16.3240
a_4	23.8686	−33.6356	26.3365	−37.5523
a_5	−21.3397	26.7305	−7.3828	28.1076
a_6	7.6910	−9.4640	5.0316	−9.8940
a_7	−1.3821	1.7485	−2.8940	1.8482
a_8	0.1231	−0.1703	0.7833	−0.1846
a_9	−0.0043	0.0078	−0.0983	0.0089
a_{10}	−0.000002	−0.000109	0.0047	−0.000144

Table 14.8 Coefficients $a_0 \dots a_{10}$ of 10th-degree approximating polynomials calculated using the least squares method for the equations of time of PTOLEMY, AL-KHWĀRIZMĪ, KŪSHYĀR and ULUGH BEG. Calculated values are rounded. The values of the functions that are approximated by these polynomials are listed in Tables 1.1, 3.1, 3.2 and 3.4. See also van Dalen, 1993, p.136.

14.3 Series expansions for a circular eccentric orbit

14.3.1 Equation of time as a function of mean anomaly

In the case of a simple eccentric circular orbit (Figure 1.3), the equation of time can also be written in the form of a series.[16] Series expansions are useful in many cases, and the circular eccentric orbit was in use from HIPPARCHUS to COPERNICUS or up to ERASMUS REINHOLD's *Prutenic Tables*. From the right spherical triangle (Figure 14.3) between the vernal equinox, Sun and the Sun's projection onto celestial equator, we can derive the following basic equations:[17]

[14.11] $$\tan \alpha_{\odot} = \tan \lambda_{\odot} \cos \varepsilon$$

[14.12] $$\sin \delta = \sin \lambda_{\odot} \sin \varepsilon$$

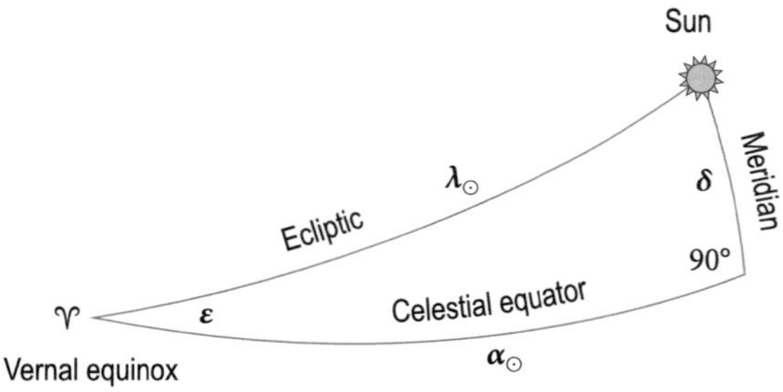

Figure 14.3 The spherical triangle composed of the vernal equinox, the Sun and the Sun's projection on the celestial equator, from which we derive the equations connecting the ecliptic longitude and right ascension or declination.

The equations are obtained by means of rules that apply to spherical triangles, whereby we can use the law of cosines or NAPIER's rules.[18] NAPIER's rules for a right spherical triangle (see Appendix D, Section D.1 or Figure 14.4) are:

- The sine of the middle part is the same as the product of the tangents of the adjacent parts.
- The sine of the middle part is the same as the product of the cosines of the opposite parts.

16 The process below is summarised from Smart, 1999, Textbook on Spherical Astronomy, pp.117–120 and North, 1976, Richard of Wallingford: An Edition of His Writings with Introduction, English Translation and Commentary. Volume 3, Appendix 30, pp.201–205. See also Smart, 1999, Textbook on Spherical Astronomy, pp.146–150.

17 See also Appendix D, section D.6.1.

18 For the derivations of the rules for spherical triangles, see Kells, Kern, Bland, 1940, Plane and Spherical Trigonometry, in particular pp.271–280. Ayres, 1954, Schaum's Outline of Theory and Problems of Plane and Spherical Trigonometry, pp.155–166. Selby (ed.), 1973, CRC Standard Mathematical Tables, 21st Edition, pp.238–241. Dörrie, 1950, Ebene in Sphärische Trigonometrie, pp.310–329. Lietzmann, 1949, Elementare Kugelgeometrie mit numerischen und konstruktiven Methoden, pp.135–151.

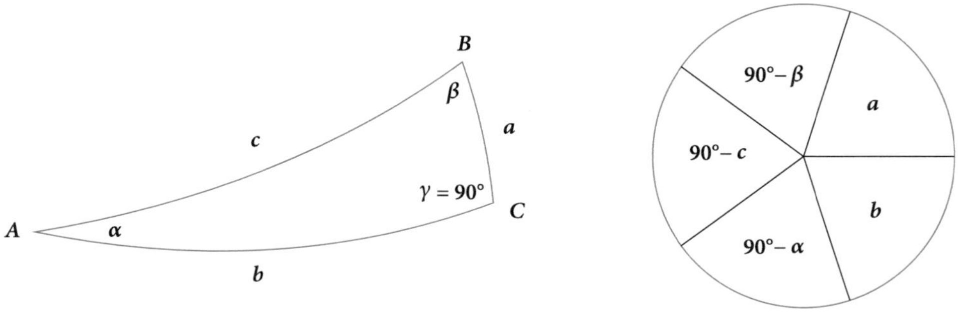

Figure 14.4 NAPIER's rules for right spherical triangles.

Using NAPIER's rules (Figure 14.4), we obtain:

$$\sin(90° - \varepsilon) = \tan(90° - \lambda_{\odot}) \tan \alpha_{\odot}$$

and taking into account the equivalences

$$\sin(90° - \varepsilon) = \cos \varepsilon \qquad \text{and} \qquad \tan(90° - \lambda_{\odot}) = \cot \lambda_{\odot}$$

we find the following:

$$\cos \varepsilon = \cot \lambda_{\odot} \tan \alpha_{\odot} \qquad \text{or} \qquad \tan \alpha_{\odot} = \tan \lambda_{\odot} \cos \varepsilon$$

which corresponds to Equation 14.11 connecting the ecliptic longitude and the right ascension. In a similar way, we obtain the expression that connects the declination and the ecliptic longitude.

Applying NAPIER's rules, we obtain: $\sin \delta_{\odot} = \cos(90° - \lambda_{\odot}) \cos(90° - \varepsilon)$

and taking into account the equivalences

$$\cos(90° - \lambda_{\odot}) = \sin \lambda_{\odot} \qquad \text{and} \qquad \cos(90° - \varepsilon) = \sin \varepsilon$$

the following applies (Equation 14.12): $\sin \delta_{\odot} = \sin \lambda_{\odot} \sin \varepsilon$

Using EULER's formulas:

[14.13] $$e^{ix} = \cos x + i \sin x \qquad \text{and} \qquad e^{-ix} = \cos x - i \sin x$$

i imaginary unit, $i^2 = -1$ or $i = \pm\sqrt{-1}$

from which it follows that:

[14.14] $$\sin x = \frac{e^{ix} - e^{-ix}}{2i} \quad \text{and} \quad \cos x = \frac{e^{ix} + e^{-ix}}{2}$$

or:

[14.15] $$\tan x = \frac{\sin x}{\cos x} = -i\frac{e^{ix} - e^{-ix}}{e^{ix} + e^{-ix}}$$

we can express the contribution to the equation of time due to the obliquity of the ecliptic in the form of a series.[19]

The $\cos x$ function is written in the form $\cos x = \frac{1 - t^2}{1 + t^2}$, where $t = \tan\frac{\varepsilon}{2}$.

If we substitute $\tan\alpha_\odot$ and $\tan\lambda_\odot$ in Equation 14.11 by expression 14.15, we obtain:

$$-i\frac{e^{2i\alpha_\odot} - 1}{e^{2i\alpha_\odot} + 1} = -i\frac{e^{2i\lambda_\odot} - 1}{e^{2i\lambda_\odot} + 1}\cos\varepsilon$$

or:

$$-i\frac{e^{2i\alpha_\odot} - 1}{e^{2i\alpha_\odot} + 1} = -i\frac{e^{2i\lambda_\odot} - 1}{e^{2i\lambda_\odot} + 1} \cdot \frac{1 - t^2}{1 + t^2}$$

Rearranging, we obtain:

[14.16] $$e^{2i\alpha_\odot} = \frac{1 + t^2 e^{-2i\lambda_\odot}}{1 + t^2 e^{2i\lambda_\odot}} e^{2i\lambda_\odot}$$

Taking natural logarithms of both sides of the equation yields:

$$\ln e^{2i\alpha_\odot} = \ln\left(1 + t^2 e^{-2i\lambda_\odot}\right) + \ln e^{2i\lambda_\odot} - \ln\left(1 + t^2 e^{2i\lambda_\odot}\right)$$

Note that the general expression $\ln(1+x)$ can be written in the form of a series:[20]

$$\ln(1 + x) = x - \frac{1}{2}x^2 + \frac{1}{3}x^3 - \frac{1}{5}x^5 + \cdots$$

19 Summarised from North, 1976, Volume III, p.201, Appendix 30. For series expansion see Flotow, 1911, Einleitung in die Astronomie, pp.15–20.

20 The series for ln(1+x) was discovered by NEWTON, however, NICOLAUS MERCATOR (c.1620–1687) was the first to publish it in *Logarithmotechnia* in 1668. See Ferraro, 2008, The Rise and Development of the Theory of Series up to the Early 1820s, pp.19, 56. For his contribution to clock making see Chapter 8, note 46. For his mathematical works see Hofmann, 1950.

Substituting $x = t^2 e^{-2i\lambda_\odot}$ or $x = t^2 e^{2i\lambda_\odot}$, we obtain:

$$2i\alpha_\odot = t^2 e^{-2i\lambda_\odot} - \frac{1}{2} t^4 e^{-4i\lambda_\odot} + \frac{1}{3} t^6 e^{-6i\lambda_\odot} - \cdots$$
$$+ 2i\lambda_\odot - \left(t^2 e^{-2i\lambda_\odot} - \frac{1}{2} t^4 e^{-4i\lambda_\odot} + \frac{1}{3} t^6 e^{-6i\lambda_\odot} - \cdots\right)$$

Rearranging, we get:

$$2i\alpha_\odot = -t^2\left(e^{2i\lambda_\odot} - e^{-2i\lambda_\odot}\right) + \frac{t^4}{2}\left(e^{4i\lambda_\odot} - e^{-4i\lambda_\odot}\right) - \frac{t^6}{3}\left(e^{6i\lambda_\odot} - e^{-6i\lambda_\odot}\right) + \cdots$$

and by applying EULER's formulas (Equation 14.14):

$$2i\alpha_\odot = 2i\lambda_\odot - 2it^2 \sin 2\lambda_\odot + 2i\frac{t^4}{2} \sin 4\lambda_\odot - 2i\frac{t^6}{3} \sin 6\lambda_\odot + \cdots$$

After rearranging the above equation we get the difference $\lambda_\odot - \alpha_\odot$, which is part of the equation of time due to the obliquity of the ecliptic.[21]

[14.17] $$\lambda_\odot - \alpha_\odot = t^2 \sin 2\lambda_\odot - \frac{t^4}{2} \sin 4\lambda_\odot + \frac{t^6}{3} \sin 6\lambda_\odot - \cdots$$

For the expression $\lambda_\odot - \alpha_\odot$, older sources also use the term reduction of ecliptic to the equator, and the values of the reduction of ecliptic to the equator are often also stated in the form of tables.

In a similar way, we also express the equation of the centre as a function of the mean anomaly. The equation of the centre is defined by Equation 1.7:

$$\tan c = \frac{e \sin \kappa_M}{R + e \cos \kappa_M}$$

Taking into account EULER's formulas (Equation 14.14) or the equivalence $\tan c = -i\frac{e^{2ic} - 1}{e^{2ic} + 1}$ (Equation 14.15), we obtain:

$$-i\frac{e^{2ic} - 1}{e^{2ic} + 1} = \frac{e\left(\frac{e^{i\kappa_M} - e^{-i\kappa_M}}{2i}\right)}{60 + e\left(\frac{e^{i\kappa_M} + e^{-i\kappa_M}}{2}\right)}$$

After rearranging, we get $e^{2ic}\left(1 + \frac{e}{60} e^{-i\kappa_M}\right) = 1 + \frac{e}{60} e^{i\kappa_M}$

21 This is LAGRANGE's expression or Equation 12.18.

Taking natural logarithms of both sides yields:

$$\ln \mathrm{e}^{2ic} + \ln\left(1 + \frac{e}{60}\mathrm{e}^{-i\kappa_M}\right) = \ln\left(1 + \frac{e}{60}\mathrm{e}^{i\kappa_M}\right)$$

Again we apply the series expansion for ln(1+x) and letting $x = \frac{e}{60}\mathrm{e}^{-i\kappa_M}$ or $x = \frac{e}{60}\mathrm{e}^{i\kappa_M}$:

$$2ic + \left(\frac{e}{60}\mathrm{e}^{-i\kappa_M} - \frac{1}{2}\cdot\frac{e^2}{60^2}\mathrm{e}^{-2i\kappa_M} + \frac{1}{3}\cdot\frac{e^3}{60^3}\mathrm{e}^{-3i\kappa_M} - \cdots\right)$$

$$= \frac{e}{60}\mathrm{e}^{i\kappa_M} - \frac{1}{2}\cdot\frac{e^2}{60^2}\mathrm{e}^{2i\kappa_M} + \frac{1}{3}\cdot\frac{e^3}{60^3}\mathrm{e}^{3i\kappa_M} - \cdots$$

or:

$$2ic = \frac{e}{60}\left(e^{i\kappa_M} - e^{-i\kappa_M}\right) - \frac{1}{2}\cdot\frac{e^2}{60^2}\left(e^{2i\kappa_M} - e^{-2i\kappa_M}\right) + \frac{1}{3}\cdot\frac{e^3}{60^3}\left(e^{3i\kappa_M} - e^{-3i\kappa_M}\right) - \cdots$$

Euler's formula (Equation 14.14) gives us:

$$2ic = \frac{e}{60}2i\sin\kappa_M - \frac{1}{2}\cdot\frac{e^2}{60^2}2i\sin 2\kappa_M + \frac{1}{3}\cdot\frac{e^3}{60^3}\sin 3\kappa_M + \cdots$$

We then obtain the final result for the equation of the centre as a function of the mean anomaly:[22]

$$c = \frac{e}{60}\sin\kappa_M - \frac{1}{2}\cdot\frac{e^2}{60^2}\sin 2\kappa_M + \frac{1}{3}\cdot\frac{e^3}{60^3}\sin 3\kappa_M + \cdots \quad [14.18]$$

The equation of time as a function of the mean anomaly is obtained by first replacing $\lambda_\odot$ with $\lambda_M - c$ in Equation 14.17 ($\lambda_\odot - \alpha_\odot$), from which we obtain:

$$\lambda_\odot - \alpha_\odot = t^2\sin 2(\lambda_M - c) - \frac{t^4}{2}\sin 4(\lambda_M - c) + \frac{t^6}{3}\sin 6(\lambda_M - c) - \cdots$$

By applying the trigonometric identity $\sin(x \pm y) = \sin x\cos y \pm \cos x\sin y$ where $x = 2\lambda_M$, $4\lambda_M \ldots$ or $y = c$ and by considering that for $x \ll 1$, $\sin x \approx x$ and $\cos x \approx 1$ apply, we get:[23]

$$\lambda_\odot - \alpha_\odot = t^2\left(\sin 2\lambda_M \underbrace{\cos 2c}_{\approx 1} - \cos 2\lambda_M \underbrace{\sin 2c}_{\approx 2c}\right) - \frac{t^4}{2}\left(\sin 4\lambda_M \underbrace{\cos 4c}_{\approx 1} - \cos 4\lambda_M \underbrace{\sin 4c}_{\approx 4c}\right)$$

$$+ \frac{t^6}{3}(\sin 6\lambda_M\cos 6c - \cos 6\lambda_M\sin 6c)$$

22 North, 1976, p.202.

23 For an explanation see Smart, 1999, pp.148–149. The element t^2 has a magnitude of $\frac{1}{25}$ or 0.04 and $e \approx \frac{1}{60}$. From this follow the simplifications in the equation. See also North, 1976, p.202.

The result for the contribution to the equation of time due to the obliquity of the ecliptic, as a function of mean longitude, is:

[14.19] $$\lambda_{\odot} - \alpha_{\odot} = t^2 \sin 2\lambda_M - \frac{2et^2}{60} \sin \kappa_M \cos 2\lambda_M - \frac{t^4}{2} \sin 4\lambda_M + \cdots$$

If both parts of the equation of time ($\lambda_{\odot} - \alpha_{\odot}$) and c or the reduction of the ecliptic to the equator (Equation 14.19) and the equation of the centre (Equation 14.18) are joined, and λ_M is written as $\lambda_\Gamma - \kappa_M$ (Equation 1.5), we get the equation of time $E = \lambda_{\odot} - \alpha_{\odot} + c$ as a function of the mean anomaly:[24]

[14.20] $$E = \frac{e}{60} \sin \kappa_M - \frac{e^2}{2 \cdot 60^2} \sin 2\kappa_M + t^2 \sin 2(\lambda_\Gamma + \kappa_M)$$

$$- 2t^2 \frac{e}{60} \cos 2(\lambda_\Gamma + \kappa_M) \sin \kappa_M - \frac{t^4}{2} \sin 4(\lambda_\Gamma + \kappa_M) + \cdots$$

14.3.2 Equation of time as a function of mean longitude

The equation of time can also be written as a function of the mean longitude. In Equation 14.20, we replace κ_M with $\lambda_M - \lambda_\Gamma$ (Equation 1.5) to obtain:[25]

[14.21] $$E = \frac{e}{60} \sin(\lambda_M - \lambda_\Gamma) - \frac{e^2}{2 \cdot 60^2} \sin 2(\lambda_M - \lambda_\Gamma) + t^2 \sin 2\lambda_M$$

$$- 2t^2 \frac{e}{60} \cos 2\lambda_M \sin(\lambda_M - \lambda_\Gamma) - \frac{t^4}{2} \sin 4\lambda_M + \cdots$$

We can expand individual elements of the above equation:

$$\frac{e}{60} \sin(\lambda_M - \lambda_\Gamma) = \frac{e}{60} (\sin \lambda_M \cos \lambda_\Gamma - \cos \lambda_M \sin \lambda_\Gamma)$$

$$-\frac{e^2}{2 \cdot 60^2} \sin 2(\lambda_M - \lambda_\Gamma) = -\frac{e^2}{2 \cdot 60^2} (\sin 2\lambda_M \cos 2\lambda_\Gamma - \cos 2\lambda_M \sin 2\lambda_\Gamma)$$

$$-2t^2 \frac{e}{60} \cos 2\lambda_M \sin(\lambda_M - \lambda_\Gamma) = -2t^2 \frac{e}{60} \cos 2\lambda_M (\sin \lambda_M \cos \lambda_\Gamma - \cos \lambda_M \sin \lambda_\Gamma)$$

24 North, 1976, p.202. For the sign of the equation of the centre or c as a function of the mean anomaly, see Chapter 1, Equations 1.5 or 1.6 or Pedersen, 2011, pp.138–141.

25 Summarised from North, 1976, Volume III, Appendix 30, p.203. See also Smart, 1999, p.149.

Taking into account the trigonometric identities:

$$\cos 2\lambda_M = \cos^2 \lambda_M - \sin^2 \lambda_M \qquad \text{or} \qquad \cos 2\lambda_M = 1 - 2\sin^2 \lambda_M$$

$$\cos^3 \lambda_M = \frac{1}{4}(\cos 3\lambda_M + 3\cos \lambda_M)$$

$$\sin^3 \lambda_M = \frac{1}{4}(3\sin \lambda_M - \sin 3\lambda_M)$$

for the fourth element, we thus obtain:

$$-2t^2 \frac{e}{60} \cos 2\lambda_M \sin(\lambda_M - \lambda_\Gamma)$$
$$= -t^2 \frac{e}{60}(-\sin \lambda_M \cos \lambda_\Gamma + \sin 3\lambda_M \cos \lambda_\Gamma - \cos 3\lambda_M \sin \lambda_\Gamma - \cos \lambda_M \sin \lambda_\Gamma)$$

Merging all expanded parts gives the equation of time as a function of mean longitude:[26]

[14.22]
$$E = \frac{e}{60}\left(1 + t^2\right) \cos \lambda_\Gamma \sin \lambda_M - \frac{e}{60}\left(1 - t^2\right) \sin \lambda_\Gamma \sin \lambda_M$$
$$- \left(t^2 - \frac{e^2}{2 \cdot 60^2} \cos 2\lambda_\Gamma\right) \sin 2\lambda_M + \frac{e^2}{2 \cdot 60^2} \sin 2\lambda_\Gamma \cos 2\lambda_M$$
$$- \frac{t^2 e}{60} \cos \lambda_\Gamma \sin 3\lambda_M + \frac{t^2 e}{60} \sin \lambda_\Gamma \cos 3\lambda_M - \frac{t^4}{2} \sin 4\lambda_M$$

26 See North, 1976, p.202. The correct coefficient for $\sin 4\lambda_M$ is $-t^4/2$, see North, 1976, p.203, coefficient h''.

14.3.3 Equation of time as a function of true longitude

The equation of time can also be written as a function of the true longitude. For the equation of the centre, the following applies (Equation 1.3):

$$\sin c = \frac{e}{60}\sin(\lambda_\odot - \lambda_\Gamma) = \frac{e}{60}(\sin\lambda_\odot \cos\lambda_\Gamma - \cos\lambda_\odot \sin\lambda_\Gamma)$$

By combining the equation of the centre, and aproximating $\sin c \approx c$, and Equation 14.17, we obtain the equation of time as a function of the true longitude:

$$E = t^2 \sin 2\lambda_\odot - \frac{t^4}{2}\sin 4\lambda_\odot + \frac{t^6}{3}\sin 6\lambda_\odot + \cdots$$
$$+ \frac{e}{60}(\sin\lambda_\odot \cos\lambda_\Gamma - \cos\lambda_\odot \sin\lambda_\Gamma)$$

We now rearrange to get the final result:[27]

[14.23] $$E = \frac{e}{60}\cos\lambda_\Gamma \sin\lambda_\odot - \frac{e}{60}\sin\lambda_\Gamma \cos\lambda_\odot + t^2 \sin 2\lambda_\odot - \frac{t^4}{2}\sin 4\lambda_\odot + \frac{t^6}{3}\sin 6\lambda_\odot - \cdots$$

The equation of time, either as a function of the mean anomaly, mean longitude or true longitude (Equations 14.20, 14.22 and 14.23), can generally be written as:[28]

[14.24] $$E = a_0 \sin x + a_1 \cos x + a_2 \sin 2x + a_3 \cos 2x + a_4 \sin 3x$$
$$+ a_5 \cos 3x + a_6 \sin 4x + a_7 \cos 4x + \cdots$$

The angles are in radians. The expressions for coefficients a_0 ... a_7 for all three cases follow from the procedure above and are given in Table 14.9.[29]

The calculated coefficients for PTOLEMY's, AL-KHWĀRIZMĪ's and AL-BATTĀNĪ's values are shown in Table 14.10.[30]

27 Summarised from North, 1976, Volume III, Appendix 30, p.205.

28 North, 1976, p.202.

29 North, 1976, pp.202–203. It seems that the a_0 and a_6 coefficients for $E=f(\kappa_M)$ or $E=f(\lambda_M)$ in North's table are typo errors; see *a'* p.202 or *h''* p.203.

30 North, 1976, p.205. In the first line of the table on page 205, the values for *D* and *L* should be $40.'01$ instead of $4.'01$ and $+1.'68$ instead of $-1.'681$. The value of parameter *A* is obviously also a typo error. The recalculated value is about $109'$ instead of $118'$. For AL-BATTĀNĪ's values and comments see North, 1976, p.205 or Table 14.6 note g.

	$f(\kappa_M)$	$f(\lambda_M)$	$f(\lambda_\odot)$
a_0 $\sin x$	$\frac{e}{60}(1+\tan^2\frac{\varepsilon}{2}\cos 2\lambda_\Gamma)$ *	$\frac{e}{60}(1+\tan^2\frac{\varepsilon}{2})\cos\lambda_\Gamma$	$\frac{e}{60}\cos\lambda_\Gamma$
a_1 $\cos x$	$\frac{e}{60}\tan^2\frac{\varepsilon}{2}\sin 2\lambda_\Gamma$	$-\frac{e}{60}(1-\tan^2\frac{\varepsilon}{2})\sin\lambda_\Gamma$	$-\frac{e}{60}(1-\tan^2\frac{\varepsilon}{2})\sin\lambda_\Gamma$
a_2 $\sin 2x$	$-\frac{e^2}{2\cdot 60^2}+\tan^2\frac{\varepsilon}{2}\cos 2\lambda_\Gamma$	$\tan^2\frac{\varepsilon}{2}-\frac{e^2}{2\cdot 60^2}\cos 2\lambda_\Gamma$	$\tan^2\frac{\varepsilon}{2}$
a_3 $\cos 2x$	$\tan^2\frac{\varepsilon}{2}\sin 2\lambda_\Gamma$	$\frac{e^2}{2\cdot 60^2}\sin 2\lambda_\Gamma$	
a_4 $\sin 3x$	$-\frac{e}{60}\tan^2\frac{\varepsilon}{2}\cos 2\lambda_\Gamma$	$-\frac{e}{60}\tan^2\frac{\varepsilon}{2}\cos\lambda_\Gamma$	
a_5 $\cos 3x$	$-\frac{e}{60}\tan^2\frac{\varepsilon}{2}\sin 2\lambda_\Gamma$	$\frac{e}{60}\tan^2\frac{\varepsilon}{2}\sin\lambda_\Gamma$	
a_6 $\sin 4x$	$-\frac{1}{2}\tan^4\frac{\varepsilon}{2}\cos 4\lambda_\Gamma$	$-\frac{1}{4}\tan^4\frac{\varepsilon}{2}$ *	$-\frac{1}{2}\tan^4\frac{\varepsilon}{2}$
a_7 $\cos 4x$	$-\frac{1}{2}\tan^4\frac{\varepsilon}{2}\sin 4\lambda_\Gamma$		

Table 14.9 Coefficients $a_0 \dots a_7$ in Equation 14.24 as functions of mean anomaly $f(\kappa_M)$, mean longitude $f(\lambda_M)$ and true longitude $f(\lambda_\odot)$. The parameters in this table are from North, 1976, Volume III, Appendix 30, pp.202–204 or derived based on procedures in the text. Values are in radians. $\ast$ It seems that North's values include two small typographic errors at parameter a_0 for the equation of time as a function of the mean anomaly $f(\kappa_M)$ and at parameter a_6 for the equation of time as a function of the mean longitude $E = f(\lambda_M)$. Corrected values are presented in this table.

	a_0	a_1	a_2	a_3	a_4	a_5	a_6	a_7
			PTOLEMY $\lambda_\Gamma = 65;30°$, $e = 2;30°$, $\varepsilon = 23;51,20°$					
$E(\lambda_M)$	139.05	4.82	−103.62	115.77	4.19	−4.82	0.48	3.39
$E(\kappa_M)$	62.05	−124.53	155.36	2.25	−2.65	5.82	−1.71	
$E(\lambda_\odot)$	59.40	−130.34	153.40				−3.42	
			AL-KHWĀRIZMĪ $\lambda_\Gamma = 82;39°$, $e = 2;30°$, $\varepsilon = 23;51°$					
$E(\lambda_M)$	137.06	1.62	−151.29	38.91	6.18	−1.62	−2.98	1.68
$E(\kappa_M)$	19.14	−135.73	156.21	0.76	−0.82	6.34	−1.71	
$E(\lambda_\odot)$	18.32	−142.06	153.33				−3.42	
			AL-BATTĀNĪ $\lambda_\Gamma = 82;14°$, $e = 2;4,45°$, $\varepsilon = 23;35°$					
$E(\lambda_M)$	114.13	1.39	−146.41	40.12	5.00	−1.39	−2.80	1.68
$E(\kappa_M)$	16.80	−112.89	151.81	0.55	−0.70	5.14	−1.63	
$E(\lambda_\odot)$	16.10	−118.03	149.82				−3.26	

Table 14.10 Calculated coefficients a_0 ... a_7 for the equation of time as a function of mean longitude, mean anomaly and true longitude, Equation 14.24. The parameters in this table are derived or calculated based on procedures given in North, 1976, Volume III, Appendix 30, pp.202–204. AL-BATTĀNĪ's values are calculated by using different parameters than in North, 1976. It seems that North has used 1;59 as value the for eccentricity.

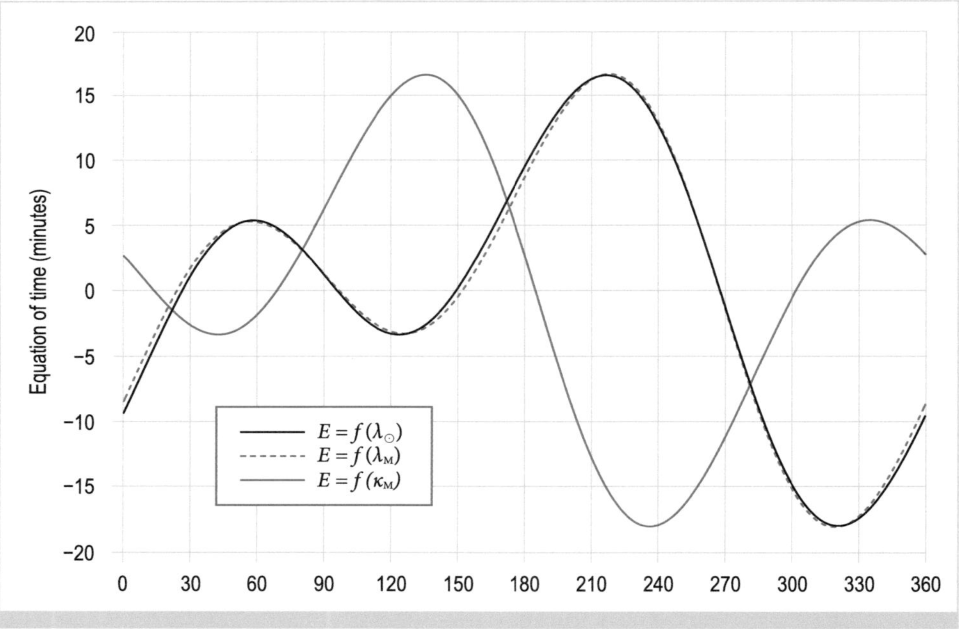

Figure 14.5 Example of AL-KHWĀRIZMĪ's equation of time as a function of the true longitude, mean longitude or mean anomaly. Calculation based on Table 14.10 and Equation 14.24.

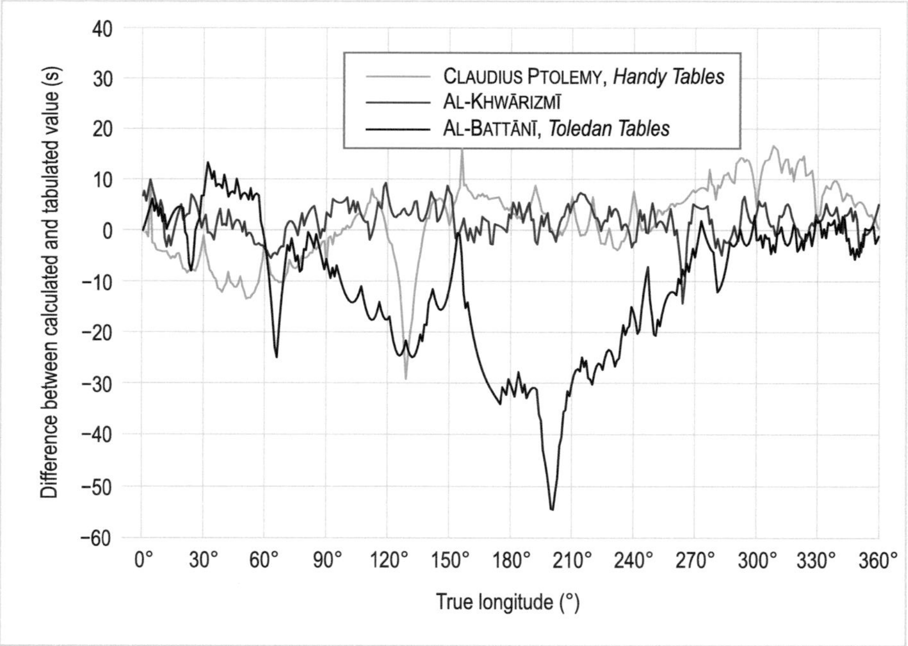

Figure 14.6 The difference between the tabulated and calculated values of the equation of time for the true longitude of the Sun and for the tables of PTOLEMY, AL-BATTĀNĪ and AL-KHWĀRIZMĪ for the coefficients given in Table 14.10. Sources of data for tables: Halma, 1822, pp.148–155, Suter, 1914, pp.181–182 and Pedersen, 2002, pp.972–975.

Bibliography

- Akhmedov, A. 2000. Astronomy, Astrology, Observatories and Calendars, History of Civilizations of Central Asia, Bosworth, C. E. and M. S. Asimov (eds.), Vol. IV, Part 2. Paris: United Nations Educational, Scientific and Cultural Organization. ISBN 978-92-3-103654-5. URL: https://unesdoc.unesco.org/images/0012/001204/120455e.pdf (24 February 2024).
- Alfonso X. 1492. Tabulae Astronomice Alfonsi Regis. (Google Books, 9 November 2022).
- Alfonso X. 1545. Astronomicae Tabulae. (Google Books, 9 November 2022).
- Ansari, S. M. Razaullah. 1977. Āryabhaṭa I, His Life and his Contributions. In: Bulletin of the Astronomical Society of India, 5, 10–18.
- Apianus, Peter. 1540. Astronomicum Caesareum. Ingolstadt: Arnt von Aych. URL: https://archive.org/details/astronomicumcsar00apia/page/n4/mode/2up (24 February 2024).
- Apianus, Peter and S. A. Ionides. 1936. Caesars' Astronomy (Astronomicum Caesareum). In: Osiris, 1, 356–389. DOI 10.1086/368431.
- Atkinson, James. 1753. Epitome of the Art of Navigation. London: W. & J. Mount, T. Page. (Google Books, 31 May 2020).
- Ayres, Frank. 1954. Schaum's Outline of Theory and Problems of Plane and Spherical Trigonometry. New York: Schaum Publishing.
- Baers ou Vekenstyl, Henri. 1976. Les Tables Astronomiques de Louvain de 1528, Emanuell Poullet and Alfred de Smet (eds.). Bruxelles: Edition Cultur et Civilisation.
- Bédos de Celles, François. 1760. La gnomonique pratique ou l'art de tracer des cardans solaires. Paris: Briasson, Despilly, Hardy. DOI 10.3931/e-rara-13798.
- Berthoud, Ferdinand. 1759. L'art de conduire et de régler les pendules et les montres. Paris: Michel Lambert. DOI 10.3931/e-rara-13665.
- Berthoud, Ferdinand. 1775. Les longitudes par la mesure du temps, ou méthode pour déterminer les longitudes en mer; avec le secours des horloges marines. Paris: J. B. G. Musier. (Google Books, 7 April 2020).
- Berthoud, Ferdinand. 1791. Anweisung zur Kentniss, Gebrauch und guten Haltung der Hand und Taschenuhren. Meissen: Karl Friedrich Willhelm Erbstein. (Google Books, 30 August 2021).
- Bianchini, Giovanni. Manuscript Pal. Lat. 1375. Biblioteca Apostolica Vaticana [online source]. URL: https://digi.vatlib.it/view/MSS_Pal.lat.1375 (30 October 2022).
- Birkenmajer, Aleksander. 1933. Formula. In: Isis, 19, 2, 364–378. DOI 10.1086/346752. URL: https://www.jstor.org/stable/225106.
- Blanchino, Ioanne, Nicolao Prvgnero and Georgio Pevrbachio. 1553. Luminarium atque planetarum motuum tabulae octoginta quinque, omnium ex his quae Alphonsum sequuntur quam faciles. (Google Books, 5 April 2020).
- Boulliau, Ismael. 1645. Astronomia Philolaica. Paris: Piget. (Google Books, 6 August 2023).
- Cassini, Giovanni Domenico. 1694. Monsieur Cassini New and Exact Tables for the Eclipses of the First Satellite of Jupiter, Reduced to the Julian Stile, and Meridian of London. In: Philosophical Transactions, 18, 237–256. DOI 10.1098/rstl.1694.0048.
- Cassini, Giovanni Domenico. 1705. The Philosophical Transactions and Collections to the End of the Year 1700; Abridg'd, and Dispos'd under General Heads, Volume I. London: John Lowthorp. URL: https://archive.org/details/s3id13683230/page/272/mode/2up (24 February 2024).
- Cassini, Jacques. 1740. Tables Astronomiques du Soleil, De La Lune, Des Planètes, Des Étoiles Fixes, Et Des Satelites De Jupiter Et De Saturne, Avec l'explication & l'usage des ces mêmes tables. Paris: Imprimerie Royale. URL: https://mdz-nbn-resolving.de/urn:nbn:de:bvb:12-bsb10049331-8 (24 February 2024).
- Chabás, José. 1998. Astronomy in Salamanca in the Mid-fifteenth Century: The Tabulae resolutae. In: Journal for the History of Astronomy, 29, 95, 167–175. DOI 10.1177/002182869802900209.
- Chabás, José and Bernard R. Goldstein. 2000. Astronomy in the Iberian Peninsula: Abraham Zacut and the Transition from Manuscript to Print. In: Transactions of the American Philosophical Society, 90, Part 2. American Philosophical Society.
- Chabás, José and Bernard R. Goldstein. 2003. The Alfonsine Tables of Toledo. Dordrecht: Kluwer Academic Publisher. ISBN 1-4020-1572-0.
- Chabás, José and Bernard R. Goldstein. 2009. John of Murs's Tables of 1321. In: Journal for the History of Astronomy, 40, 3, 297–320. DOI 10.1177/002182860904000303.

- Chabás, José and Bernard R. Goldstein. 2009. The Astronomical Tables of Giovanni Bianchini. Leiden: Brill Academic Pub. ISBN 9789004176157.
- Chabás, José and Bernard R. Goldstein. 2012. A Survey of European Astronomical Tables in the Late Middle Ages. Leiden: Brill Academic Pub. ISBN 978-9004230583.
- Chabás, José and Bernard R. Goldstein, R. 2014. Essays on Medieval Computational Astronomy. Leiden: Brill. ISBN 9789004281745.
- Chabás, José and Bernard R. Goldstein. 2018. New evidence on Abraham Zacut's astronomical tables. In: Archive for History of Exact Sciences, 72, 1, 21–62. DOI 10.1007/s00407-017-0199-2.
- Chabás, José and Marie-Madeleine Saby. 2023. The Tables of 1322 by John of Lignères. An Edition with Commentary. Turnhout: Brepols. DOI 10.1484/M.ALFA-EB.5.133667. e-ISBN 978-2-503-59610-5.
- Collection of English Almanacs for the Years 1702–1835. 1761. London: A. Wilde. URL: https://archive.org/details/collectionengli03unkngoog/mode/2up (24 February 2024).
- Connoissance des Temps. 1679. Paris: Jean Baptiste Coignard. URL: https://gallica.bnf.fr/ark:/12148/bpt6k6506709d (24 February 2024).
- Davis, John. 2003a. The Equation of Time as shown on Sundials. In: Bulletin of the British Sundial Society, 16, 4, 135–144.
- Davis, John. 2003b. More on the Equation of Time on Sundials. In: Bulletin of the British Sundial Society, 17, 2, 66–74.
- De la Hire, Philippe. 1702. Tabulæ astronomicæ Ludovici Magni. Paris: Jean Boudot. (Google Books, 16 June 2023).
- De la Hire, Philippe and Johann Phillip von Wurzelbau. About 1700. Tabulae Astronomicae. URL: https://reader.digitale-sammlungen.de/resolve/display/bsb10796771.html (24 February 2024).
- De Lacaille, Nicolas-Louis. 1755. Ephémérides des Mouvemens Célestes: pour le méridien de Paris. Paris: J. J. Etienne Collombat. (Google Books, 6 April 2020).
- De Lacaille, Nicolas-Louis. 1758. Tabulae Solares. Paris: H. L. Guerin & L. F. Delatour. URL: https://www.digitale-sammlungen.de/de/view/bsb10049176 (24 February 2024).
- De Lalande, Joseph Jérôme Le Français. 1764. Astronomie, Book 1. Paris: Desaint & Saillant, Libraires, rue S. Jean-de-Beauvais. URL: https://www.digitale-sammlungen.de/de/view/bsb11347059 (24 February 2024).
- De Mezavachis, Flaminii. 1675. Ephemerides Felsineæ Recentiores. Bologna. URL: https://archive.org/details/bub_gb_iXPs8r8iV2UC (24 February 2024).
- De Mezavachis, Flaminii. 1701. Ephemerides Felsineæ Recentiores. Bologna: Constantini Pisarij. (Google Books, 31 May 2020).
- Declercq, Georges. 2000. Anno Domini. The Origins of the Christian Era. Turnhout: Brepols. ISBN 2-503-51050-7.
- Delambre, Jean-Baptiste Joseph. 1806. Tables Astronomiques, publiées par le Bureau des Longitudes de France. Paris: Courcier. (Google Books, 31 May 2020).
- Delambre, Jean-Baptiste. 1819. Histoire de l'Astronomie du Moyen Age. Reprint 2006. Paris: Éditions Jacques Gabay. ISBN 2-87647-251-1.
- Delambre, Jean-Baptiste. 1821. Historie de l'Astronomie Moderne, volume I. Reprint 2006. Paris: Éditions Jacques Gabay. ISBN 2-87647-252-X.
- Derham, William. 1714. The Artificial Clock-maker, a Treatise of Watch and Clock-worth. 3rd Edition. London: James Knapton. (Google Books, 31 May 2020).
- Dörrie, Heinrich. 1950. Ebene und Sphärische Trigonometrie. München: R. Oldenbourg
- Ewing, Alexander. 1812. Practical Astronomy: Containing a Description of the Solar System; the Doctrine of the Sphere; the Principal Problems in Astronomy illustrated with many Examples. Revised by Gummere, John. Burlington, NJ: David Allinson & Co. (Google Books, 11 September 2021).
- Ferraro, Giovanni. 2008. The Rise and Development of the Theory of Series up to the Early 1820s. New York: Springer. ISBN 978-0-387-73467-5. DOI 10.1007/978-0-387-73468-2.
- Flamsteed, John. 1678. De Temporis Aequatione Diatriba. In: Jerremiah Horrox, Opera Posthuma, 441–464. London. URL: https://www.e-rara.ch/zut/content/pageview/854825 (24 February 2024).
- Forbes, Eric Gray, Lesley Murdin and Frances Wilmoth (eds.). 1997. The Correspondence of John Flamsteed, the First Astronomer Royal, Volume 2. Boca Raton, FL: CRC Press. ISBN 978-0750303910.
- Gaab, Hans and Olaf Simons. 2008. Johann Leonhard Rost, "Romanist" und Astronom. In: Acta Historica Astronomiae 36, Beiträge zur Astronomiegeschichte, Band 9, 106–155. Frankfurt am Main: Harri Deutsch. ISBN 978-3-8171-1831-1.

- Gauss, Carl Friedrich. 1929. Werke 12. Leipzig: Teubner.
- Ginzel, Friedrich Karl. 1906. Handbuch der mathematischen und technischen Chronologie, Band 1. Leipzig: J. C. Hinrichs'sche Buchhandlung. Reprint ALO Arbeitsgemeinschaft. ISBN 3-226-00478-6.
- Goldstein, Bernard R. 1974. The Astronomical Tables of Levi Ben Gerson. (= Translation, 45). New Haven: The Connecticut Academy of Arts and Sciences. ISBN 0-208-01425-X.
- Goldstein, Bernard R. 2017. The Introduction to Immanuel ben Jacob Bonfils' Tables for 1340. In: Aleph, 17, 1, 167–176. DOI 10.2979/aleph.17.1.0167.
- Goldstein, Bernard R. and José Chabás. 2017. Analysis of the Astronomical Tables for 1340 compiled by Immanuel ben Jacob Bonfils. In: Archive for History of Exact Sciences, 71, 71–108. DOI 10.1007/s00407-016-0181-4.
- Gummere, John. 1822. An Elementary Treatise on Astronomy. Philadelphia: Kimber & Sharpless. (Google Books, 11 September 2021).
- Halma, Nicolas. 1822. Commentaire de Théon d'Alexandrie, sur le Livre III de l'Almageste de Ptolemée; Tables Manuelles des Mouvements des Astres, traduites pour la première fois du grec en français, sur les manuscrits de la bibliothèque du Roi, par M. L'Abbé Halma. Reprint 1990. Paris: Librarie A. Blanchard. ISBN 2-85367-169-0.
- Hamming, Richard Wesley. 2016. Numerical Methods for Scientists and Engineers, 2nd Revised Edition. New York: Dover Publications. ISBN 978-0486652412.
- Hamming, Richard Wesley. 2004. Methods of Mathematics. New York: Dover Publications. ISBN 978-0486439457.
- Hofmann, Jos. E. 1950. Nicolaus Mercator (Kaufmann). Abhandlungen der Mathematisch-Naturwissenschaftlichen Klasse, Nr. 3. Akademie der Wissenschaften und der Literatur, Mainz. Wiesbaden: Franz Steiner Verlag.
- Huygens, Christiaan. 1669. Instructions Concerning the Use of Pendulum-Watches, for Finding the Longitude at Sea. In: Philosophical Transactions, 1 January 1669, 4, 45–56, 937–976. DOI 10.1098/rstl.1669.0013.
- Huygens, Christiaan. 1673. Horologium Oscillatorium. Paris: F. Muguet. (Google Books, 31 May 2020).
- Johannes de Gamundia. Tabula aequationis dierum com noctibus suis. In: Manuscript Cent. VI, 18, ff.46v–47r. Stadtbibliothek Nürnberg.
- Jüttner, Joseph. 1842. Anweisung zum Gebrauche der Himmelskugel. Olmütz: Alois Skarnitzl.
- Keill, John. 1718. Introductio ad Veram Astronomiam, seu Lectiones Astronomicae Habitæ in Schola Astronomica Acadamiæ Oxoniensis. London: Hen. Clements. URL: https://reader.digitale-sammlungen.de//resolve/display/bsb10060987.html (24 February 2024).
- Keill, John. 1730. An Introduction to the True Astronomy or Astronomical Lectures Read in the Astronomical School of the University of Oxford. 2nd Edition. London: Bernard Lintot. (Google Books, 6 April 2020).
- Kells, Lyman M., Willis F. Kern and James R. Bland. 1940. Plane and Spherical Trigonometry. New York, London: McGraw-Hill.
- Kennedy, Edward S. 1956. A Survey of Islamic Astronomical Tables. Transactions of the American Philosophical Society. New Series, 46, 2. Philadelphia: The American Philosophical Society.
- Kennedy, Edward S. 1958. The Sasanian Astronomical Handbook Zīj-I Shāh the Astrological Doctrine of "Transit". In: Journal of the American Oriental Society, 78, 4, 246–262. DOI 10.2307/595788. Reprinted in Kennedy, 1983, Studies in the Islamic Exact Sciences. Beirut: American University of Beirut.
- Kennedy, Edward S. 1960. The Planetary Equatorium of Jamshīd Ghiyāth al-Dīn al-Kāshī. Princeton Oriental Studies, Volume 18. Princeton, NJ: Princeton University Press.
- Kennedy, Edward S. 1988. Two Medieval Approaches to the Equation of Time. In: Centaurus, 31, 1, 1–8. DOI 10.1111/j.1600-0498.1988.tb00713.x.
- Kitto, Tony. 1999. John Flamsteed, Richard Towneley and the Equation of Time. In: Antiquarian Horology and Proceedings of the Antiquarian Horological Society, 25, 180–184.
- Kitto, Tony. 2001. Richard Towneley and the Equation of Natural Days. In: Bulletin of the British Sundial Society, 13, 2, 60–65.
- Leadbetter, Charles. 1735. Uranoscopia or, the contemplation of the heavens. London: J. Wilcox. URL: https://archive.org/details/BUSA297_083/page/n5 (24 February 2024).
- Leadbetter, Charles. 1739. The Young Mathematician's Companion. London: J. Hodges. URL: https://archive.org/details/bim_eighteenth-century_the-young-mathematician_leadbetter-charles_1739 (25 August 2024).

- Leadbetter, Charles. 1769. Mechanick Dialling or the New Art of Shadows. London: G. Pearch. URL: https://archive.org/details/mechanickdiallin00lead/page/n2/mode/2up (24 February 2024).
- Lietzmann, Walter. 1949. Elementare Kugelgeometrie mit numerischen und kontruktiven Methoden. Göttingen: Vandenhoeck & Ruprecht.
- Littrow, Joseph Johann. 1828. Calendariographie. Wien: Heubner.
- Long, Roger. 1764. Astronomy in Five Books. 3rd Book. Cambridge. (Google Books, 7 April 2020).
- Longomontani, Christiaanii S. 1640. Astronomia Danica. Amsterdam: Ioh. & Cornelivm Blaev. URL: https://reader.digitale-sammlungen.de/en/fs1/object/display/bsb10049403_00001.html (24 February 2024).
- Magini, Giovanni Antonio. 1607. Continuatio ephemeridum coelestium motuum. Venice: Damian Zenari. (Google Books, 17 June 2023).
- Manfredi, Eustachio. 1750. Eustachii Manfredii Introductio In Ephemerides Cum Opportunis Tabulis Ad Usum Bononiensis Scientiarum Instituti. DOI 10.3931/e-rara-52480.
- Maskelyne, Nevil. 1767. The Nautical Almanac and Astronomical Ephemeris for the Year 1767. London: Printed by W. Richardson and S. Clark. URL: https://archive.org/details/the-nautical-almanac-and-astronomical-ep (24 September 2024).
- Mästlin, Michael. 1580. Ephemerides novæ. Tübingen: Georg Gruppenbach. (Google Books, 17 June 2023).
- Mayer, Tobias. 1770. Tabulæ Motuum Solis et Lunæ Novæ et Correctæ. Londini: Gulielmi et Johannis Richardson. DOI 10.3931/e-rara-467.
- Meeus, Jean and Denis Savoie. 1995. L'Équation du temps. In: l'Astronomie fondamentale, 109, 188–193.
- Melander, Daniel. 1777. Anmerkungen über die Zeitgleichung Von Herrn Daniel Melander, Prof. der Astronomie von Upsal. In: Astronomisches Jahrbuch oder Ephemeriden für das Jahr 1780, 20–22. (Google Books, 2 June 2023).
- Mercier, Raymond. 2014. Astronomical Tables of Abraham Bar Hiyya. In: Time, Astronomy, and Calendars in the Jewish Tradition, Stern, Sacha and Charles Burnett (eds.), 155–207. Lleiden: Brill. DOI 10.1163/9789004259669_009. e-ISBN 9789004259669.
- Millás, J. M. 1959. La obra Séfer Ḥešbón mahlekot ha-kokabim de R. Abraham Bar Ḥiyya ha-Bargeloni. Barcelona: Consejo Superior de Investigationes Científicas.
- Molyneux, William. 1686. Sciothericum Telescopicum. Dublin: Andrew Crook and Samuel Helsham. (Google Books, 25 February 2020).
- Nallino, Carlo Alfonso. 1903. Al-Battānī sive Albatenii, Opus Astronomicum. Pubblicazioni del reale osservatorio di Brera in Milano, XL, I. Reprint 1997. Frankfurt: Institut fur Geschichte der Arabisch-Islamischen Wissenschaften. ISBN 3-8298-4011-X.
- Nallino, Carlo Alfonso. 1907. Al-Battānī sive Albatenii, Opus Astronomicum. Pubblicazioni del reale osservatorio di Brera in Milano, XL, II. Reprint 1997. Frankfurt: Institut fur Geschichte der Arabisch-Islamischen Wissenschaften. ISBN 3-8298-4011-X.
- Nallino, Carlo Alfonso. 1907. Al-Battānī sive Albatenii, Opus Astronomicum. Pubblicazioni del reale osservatorio di Brera in Milano, XL, Part II. Reprint 1969. Frankfurt: Minerva.
- Neugebauer, Otto. 1962. The astronomical tables of al-Khwārizmī, Translation with Commentaries on the Latin Version, Suter, Heinrich (ed.), Supplemented by Corpus Christi College MS 283 by O. Neugebauer. København: Ejnar Munksgaard.
- Neugebauer, Otto and David Pingree. 1970–1971. The Pañcasiddhāntikā of Varāhamihira, I-II. København: Munksgaard.
- Neugebauer, Otto and Noel M. Swerdlow. 1984. Mathematical Astronomy in Copernicus's de Revolutionibus, Part I-II. New York: Springer. ISBN 3-540-90939-7.
- Neugebauer, Paul Viktor. 1929. Astronomische Chronologie, Band 1. Berlin: Walter de Gruyter & Co.
- Neugebauer, Paul Viktor. 1936. Hifstafeln zur technischen Chronologie, Teil I. In: Astronomische Nachrichten, 261, 10, 153–194. DOI 10.1002/asna.19362611002.
- North, John D. 1976. Richard of Wallingford, An edition of his writings with introductions, English translation and commentary, Volume 3, Illustrations, Tables, Appendixes, Glossaries, Bibliography and Indexes. Oxford: Clarendon Press. ISBN 0 19 858139 4.
- Origanus, David. 1609. Nouae motuum coelestium ephemerides Brandenburgicae. Frankfurt/Oder: Eichorn. (Google Books, 6 August 2023).
- Pedersen, Fritz Saaby (ed.). 1978. Petrus de Dacia: Tractatus Instrumenti Eclipsium. In: Cahiers de l'Institut du Moyen-Âge Grec et Latin, 25, 1–102. URL: https://cimagl.saxo.ku.dk/download/25/25pedersen1-102.pdf (30 October 2022).

- PEDERSEN, Fritz Saaby. 1983. Petri Philomenae de Dacia et Petri de S. Audomaro opera quadrivialia, Part I. Copenhagen: Societas Linguae & Litterarum Danicarum. ISBN 87-12-15896-8.
- PEDERSEN, Fritz S. 2002. The Toledan Tables: A review of the manuscripts and the textual versions with an edition. Copenhagen: The Royal Danish Academy of Sciences and Letters, C.A. Reitzeils Forlag. ISBN 87-7876-267-7.
- PEDERSEN, Fritz Saaby (ed.). 2014. William of Saint-Cloud: Almanach Planetarum, An edition of the canons, a few samples from the tables, and a foray into the numbers. Copenhagen: Centre for the Aristotelian Tradition, Saxo Institute. URL: https://cimagl.saxo.ku.dk/download/83/83Pedersen1-135.pdf (30 October 2022).
- PEDERSEN, Olaf. 2011. A Survey of the Almagest. New York, Dordrecht, Heidelberg, London: Springer. ISBN 978-0-378-84825-9. DOI 10.1007/978-0-387-84826-6.
- PEUERBACH, Georg. 1514. Tabulae eclypsium magistri Georgij Peurbachij. (Google Books, 6 August 2023).
- PIETRONI, Silvia. 2022. A Survey of the Evolution of the Equation of Time During the Centuries. ISSN 2038-3657. URL: https://www.researchgate.net/publication/361618308_A_survey_on_the_evolution_of_the_equation_of_time_during_the_centuries (24 February 2024).
- PILZ, Kurt. 1977. 600 Jahre Astronomie in Nürnberg. Nürnberg: Hans Carl. ISBN 9783418004471.
- PINGREE, David. 1985. The Astronomical Works of Gregory Chioniades, The Zīj al-'Alā ī. Part I: Text, Translation and Commentary. Amsterdam: Gieben. ISBN 9070265656.
- PINGREE, David and Edward KENNEDY (trans., eds.). 1981. The Book of the Reasons Behind Astronomical Tables. New York: Scholar's Facsimiles & Reprints Delmar. ISBN 0-8201-1298-4.
- PITATI, Pietro. 1544. Almanach Novvm Petri Pitati Veronensis Mathematici. Tübingen: Mor. URL: https://archive.org/details/bub_gb_aot7Xgcf_-4C/mode/2up (24 February 2024).
- POULLE, Emmanuel and Antoine DE SMET (eds.). 1976. Les Tables Astronomiques De Louvain De 1528. Introduction, Traduction Et Commentaire. Bruxelles: Culture et Civilisation.
- REICHERT, Jürgen (ed.). 2014. Die Rudolphinischen Tafeln. Würzburg: Königshausen & Neumann. ISBN 978-3-8260-5352-8.
- REINHOLD, Erasmus. 1551. Prutenicae Tabulæ Coelestium Motuum. Tübingen. (Google Books, 5 April 2020).
- REINHOLD, Erasmus. 1585. Prutenicae Tabulæ Coelestium Motuum. Wittenberg: Matthaeus Velack. URL: https://archive.org/details/ARes34403/page/n6/mode/2up (24 February 2024).
- RHETICUS and ACRONIUS. 1542. Tabulae Astronomicae In Gratiam Studiosae Iuventutis Seorsim Aeditae: I De aequatione dierum.
- RICCIOLI, Giovanni Battista. 1665. Almagestum Novum. (Google Books, 25 February 2020).
- RIVARD, Dominique François. 1742. La gnomonique, ou L'art de faire des cadrans. Paris: Jean Desaint & Charles Saillant. URL: https://gallica.bnf.fr/ark:/12148/bpt6k855724k.image (24 February 2024).
- RIVARD, Dominique François. 1767. La gnomonique, ou L'art de faire des cadrans. Paris: Jean Desaint. URL: https://numelyo.bm-lyon.fr/f_view/BML:BML_00GOO0100137001101382583 (24 February 2024).
- ROST, Johann Leonhard. 1718. Astronomisches Handbuch. Nürnberg: Peter Conrad Monath. URL: https://www.digitale-sammlungen.de/en/view/bsb11109612 (24 February 2024).
- ROST, Johann Leonhard. 1773. Astronomisches Handbuch, Neue Auflage, Volume III. Nürnberg: George Peter Monath. URL: https://diglib.hab.de/drucke/ne-271-3/start.htm (24 February 2024).
- Sammlung Astronomischer Taffeln: Unter Aufsicht der Königlich-Preussischen Akademie der Wissenschaften. 1776. Band 1. Berlin: Georg Jacob Decker. DOI 10.3931/e-rara-2166.
- SARTON, George. 1948. Introduction to the History of Science, III, part 2. Washington DC: Carnegie Institution.
- SCHAUB, Werner. 1950. Vorlesungen über Sphärische Astronomie. Leipzig: Geest & Portig.
- SCHRECKENFUCHS, Erasmus Oswald. 1556. Eras. Oswaldi Schreckenfuchsii Commentaria, in Nouas theoricas planetarum Georgii Purbachii, quas etiam breuibus tabulis pro eliciendis tùm ueris motibus omnium planetarum, item tabulis coniunctionum & oppositionum ac eclipsium luminarium ad summum illustrauit, lucemq maximam ijs adiecit. Basileae: Henrichvm Petri. URL: https://archive.org/details/bub_gb_yXN3L-9owtUC/mode/2up (24 February 2024).
- SELBY, Samuel. 1973. CRC Standard Mathematical Tables, 21st Edition. Cleveland, OH: The Chemical Rubber Co. ISBN 0-87819-621-8.
- SHAKERLEY, Jeremy. 1653. Tabulæ Britanicæ: The British Tables. London: R. and W. Leybourn. (Google Books, 6 August 2023).

- Shukla, Kripa Shankar and K. V. Sarma. 1976. Āryabhaṭīya of Āryabhaṭa. New Delhi: Indian National Science Academy.
- Smart, John. 1702. Tables of time calculated for two hundred years. vizt. the XVII. and XVIII. centuries. To which are added Mr. Flamsteed's table of the equation of natural days, and other useful tables. London. URL: https://archive.org/details/b32995519/mode/2up (24 February 2024).
- Smart, William Marshall. 1931. Textbook on Spherical Astronomy. Reprint 1999. Cambridge: Cambridge University Press. ISBN 0 521 29180 1.
- Smith, John. 1678. Equation of time demonstrated to the meanest capacity together with a table of equations, absolutely necessary for the true adjusting and right managing of pendulum clocks. Both humbly presented to the master, wardens, and court of assistants of the company of clock-makers: by John Smith, C.M. Philomath.
- Smith, John. 1694. Horological Disquisitions Concerning the Nature of Time. London: printed for Richard Cumberland. Reprint 1962. Reading: The Thames Facsimile Company.
- Sobel, Dava and William J. Andrewes. 1998. The Illustrated Longitude: The True Story of the Lone Genius Who Solved the Greatest Scientific Problem of His Time. London: Fourth Estate. ISBN 1-85702-714-0.
- Society of Gentlemen. 1763. A new and complete dictionary of arts and sciences. 2nd Edition. London: W. Owen. (Google Books, 7 April 2020).
- Solon, Peter. 1971. The Six Wings of Immanuel Bonfils and Michael Chrysokokkes. In: Centaurus, 15, 1, 1–20. DOI 10.1111/j.1600-0498.1971.tb00147.x.
- Stadius, Johannes. 1585. Ephemerides Joannis Stadii. Lyon: Beraud, Michaelem. (Google Books, 6 August 2023).
- Stigler, Stephen M. 1981. Gauss and the Invention of Least Squares. In: The Annals of Statistics, 9, 3, 465–474. DOI 10.1214/aos/1176345451.
- Stöffler, Johannes. 1514. Justingensis germani Tabulae Astronomicae. URL: https://www.digitale-sammlungen.de/en/view/bsb00001937 (24 February 2024).
- Tihon, Anne and Raymond Mercier. 2011. Ptolemaiou Procheiroi Kanones. Ptolemy's Handy Tables. Volume 1a, 1b: Tables A1-A2. Transcription and Commentary. Leuven: Peeters Publishers. ISBN 978-90-429-2436-9.
- Turner, G. L. E. 2002. Eighteenth century scientific instruments and their makers. In: The Cambridge History of Science, IV, Eighteenth Century Science. Porter, Roy (ed.). Cambridge: Cambridge University Press. ISBN 978-0521572439.
- Van Dalen, Benno. 1989. A Statistical Method for Recovering Unknown Parameters from Medieval Astronomical Tables. In: Centaurus, 32, 2, 85–145. DOI 10.1111/j.1600-0498.1989.tb00842.x.
- Van Dalen, Benno. 1993. Ancient and Mediaeval Astronomical Tables: Mathematical Structure and Parameter Values. Doctoral dissertation. Utrecht: University of Utrecht. ISBN 90-393-0167-0.
- Van Dalen, Benno. 1994. On Ptolemy's Table for the Equation of Time. In: Centaurus, 37, 97–153. DOI 10.1111/j.1600-0498.1994.tb00005.x.
- Van Dalen, Benno. 1996. Al-Khwārizmī's astronomical tables revisited: analysis of the equation of time. In: Casulleras, J. and J. Samsó (eds.). From Baghdad to Barcelona: Studies in the Islamic Exact Sciences in Honour of Prof. Juan Vernet. Volume 1. Barcelona: Universitat de Barcelona.
- Van Dalen, Benno. 2014. Islamic Astronomical Tables: Mathematical Analysis and Historical Investigation (Variorum Collected Studies). London: Routledge. ISBN 978-1472422385.
- Van Dalen, Benno and Fritz S. Pedersen. 2008. Re-Editing the Tables in the Sâbi' Zîj by al-Battânî (ca. AD 900). In: Mathematics Celestial and Terrestrial Festschrift für Menso Folkerts zum 65. Geburtstag. Dauben, Joseph W., Stefan Kirschner, Paul Kunitzsch, Andreas Kühne and Richard P. Lorch (eds.). Halle a/d Saale: Deutsche Akademie der Naturforscher Leopoldina. ISBN 978-3-8047-2482-2.
- Virdung, Johannes. 1542. Tabulae resolutae de supputandis siderum motibus. URL: https://www.deutsche-digitale-bibliothek.de/item/3K6FMKWKMVBA4HG43FCC4Z6AA56D3SW4 (24 February 2024).
- Vogel, Christian F. 1774. Practischer Unterricht von Taschenuhren, sowohl für die Verfertiger, als auch für die Liebhaber derselben, Aequationstabelle der gleichen und scheinbaren Zeit. Reprint 1977. München: Callwey. ISBN 9783766703972.
- Von Flotow, Albrecht. 1911. Einleitung in die Astronomie. Leipzig: G. J. Göschen'sche Verlagshandlung.
- Von Wurzelbau, Johann Phillip (ed.). 1719. Uranies Noricæ Basis Astronomica. Nürnberg: Johannis Ernesti Adelbulneri. (Google Books, 31 May 2020).

- VON ZACH, Franz Xaver. 1799. Aus mehreren Briefen des Dr. Burckhardts. In: Allgemeine Geographische Ephemeriden, 3. Band, February, 179–183. Weimar: Industrie Comptoirs. (Google Books, 7 April 2020).
- WHISTON, William. 1707. Prælectiones Astronomicæ Publicis Habitæ. London: Benjamin Tooke. (Google Books, 11 September 2021).
- WHISTON, William. 1715. Astronomical Lectures Read in the Publick Schools at Cambridge. London: R. Senex, W. Taylor. (Google Books, 11 September 2021).
- WING, Vincent. 1656. Astronomia Instaurata. London: R. and W. Leybourn. URL: https://archive.org/details/bim_early-english-books-1641-1700_astronomia-instaurata-_wing-vincent_1656/mode/2up (25 August 2024).
- WING, Tycho. 1768. Almanack for the Year of Our Lord, 1768. London: Printed by T. Parker. In: Collection of English Almanacs for the years 1702–1835. URL: https://archive.org/details/collectionengli01unkngoog/mode/2up (24 September 2024).
- ZACUTO, Abraham. 1493. Almanach Perpetuum Celestium Motuum. Reprint 1915. München: J. B. Obernetter.

This page is intentionally left blank.

Errata in the second edition

Errata in the second edition

The most important corrections for the second edition are listed below. Correct entries are:

p.31	Figure 0.1	... Current value between the winter solstice and the spring equinox is 88.94 days ... and between autumn equinox and winter solstice 89.86 days ...
p.34	Note 53	... The value 4500 was not chosen by Theon at random; it is the square of the distance *EB*, see Figure 0.5, and it is necessary in calculation of the chord of an arc of 36°.

p.44 Comment: Apollonius of Rhodes and Apollonius of Alexandria seem to be the same person.

p.52 ... lunar theory and the 8 and 19-year lunisolar ...

p.86 Based on different translations in the literature, the Indian word for intercalary months is not consistently used in the second chapter. Two different expressions appear in the literature: *Adhimāsas* and *adhikamāsas*.

p.99 note 58 $21\,659' = 21\,600' + 59'$, $59'$ = mean diurnal motion of the Sun. Vaṭeśvara also used the value $21\,659'$.

p.122 **[3.13]** $E = (\lambda_{\odot} - \alpha_{\odot}) - c + k$

p.131 Ibn Ṣā'id should be Ṣā'id al-Andalusi

p.135 Example 4.1 p = 19°.546

p.160 He estimated Greenwich's latitude as 51°28′34″, ...

p.171 Let arc $\widehat{♈s}$ or $\widehat{♎s}$ be equal to $\widehat{♈S}$ or $\widehat{♎S}$ on the ecliptic.

p.176 Table 6.1 Correct symbols under equation of time are:
☽
♀
♃
☊

p.179 ... so value of the angle $\widehat{PTQ}$ cannot be greater than ...

p.180 … that he observed from the Luxembourg palace in 1769.

p.198 A sign for partial derivative is missing in some equations. Correct reading is:

$$\frac{\partial^2 f(y)}{\partial x^2} = \frac{\partial^2 F_1(y)}{\partial x \partial z} = \frac{\partial}{\partial z}\frac{\partial F_1(y)}{\partial x} = \frac{\partial}{\partial z}\phi(y)\frac{\partial F_1(y)}{\partial z}$$

$$\frac{\partial^2 f(y)}{\partial x^2} = \frac{\partial}{\partial z}\left((\phi(y))^2\frac{\partial f(y)}{\partial z}\right) \quad [7.14]$$

$$\left(\frac{\partial^2 f(y)}{\partial x^2}\right)_0 = \frac{\partial}{\partial z}\left((\phi(z))^2\frac{\partial f(z)}{\partial z}\right)$$

$$(\phi(y))^2\frac{\partial f(y)}{\partial z} = \frac{\partial F_2(y)}{\partial z}$$

p.238	Table 8.5	Nevil Maskelyne	S 4 min 2 s 14 – 16 May
		Tycho Wing	S 4 min 2 s 14 – 16 May

p.241 VARĀHAMIHIRA March 22, 505, at sunset ~1 905 590.25

p.252 Taking into account EULER's formulas (Equation 14.14) or the equivalence

$\tan c = -\mathrm{i}\dfrac{\mathrm{e}^{2ic}-1}{\mathrm{e}^{2ic}+1}$ (Equation 14.15), we obtain: $-\mathrm{i}\dfrac{\mathrm{e}^{2ic}-1}{\mathrm{e}^{2ic}+1} = \dfrac{e\left(\dfrac{\mathrm{e}^{\mathrm{i}\kappa_M} - \mathrm{e}^{-\mathrm{i}\kappa_M}}{2\mathrm{i}}\right)}{60 + e\left(\dfrac{\mathrm{e}^{\mathrm{i}\kappa_M} + \mathrm{e}^{-\mathrm{i}\kappa_M}}{2}\right)}$

p.255

$$E = \frac{e}{60}\left(1+t^2\right)\cos\lambda_\Gamma \sin\lambda_M - \frac{e}{60}\left(1-t^2\right)\sin\lambda_\Gamma \sin\lambda_M$$
$$- \left(t^2 - \frac{e^2}{2\cdot 60^2}\cos 2\lambda_\Gamma\right)\sin 2\lambda_M + \frac{e^2}{2\cdot 60^2}\sin 2\lambda_\Gamma \cos 2\lambda_M$$
$$- \frac{t^2 e}{60}\cos\lambda_\Gamma \sin 3\lambda_M + \frac{t^2 e}{60}\sin\lambda_\Gamma \cos 3\lambda_M - \frac{t^4}{2}\sin 4\lambda_M \quad [8.22]$$

p.256 … AL-BATTĀNĪ's values $\varepsilon = 23°35'$, $c_{\max} = 1°59'$ and longitude of the apogee $\lambda_\Gamma = 82°14'$ …

Symbols for the equation of time in the first column are exchanged and, additionally, some values in the Table 8.9 were not calculated correctly. The corrected entries should be read as:

	a_0	a_1	a_2	a_3	a_4	a_5	a_6	a_7
				PTOLEMY ... $e = 2;30°$ ...				
$E(\kappa_M)$	139.05						0.48	
$E(\lambda_M)$							−1.71	
				AL-KHWĀRIZMĪ ... $e = 2;30°$ ...				
$E(\kappa_M)$								
$E(\lambda_M)$							−1.71	

Index

Index of names

C

D

H

I

J

M

N

O

P

Q

R

S

W

X

Y

Z

Index of subjects

B

C

F

G

M

N

S

T